8
Springer Series on Fluorescence

Methods and Applications

Series Editor: O.S. Wolfbeis

For further volumes:
http://www.springer.com/series/4243

Springer Series on Fluorescence
Series Editor: O.S.Wolfbeis

Recently Published and Forthcoming Volumes

Advanced Fluorescence Reporters in Chemistry and Biology I

Fundamentals and Molecular Design

Volume Editor: Alexander P. Demchenko

With contributions by

P.R. Callis · P.-T. Chou · R.J. Clarke · M. Dakanali · I. Demachy ·
A.P. Demchenko · T. Gonçalves · M.A. Haidekker · D.J. Hagan ·
C.-C. Hsieh · M.-L. Ho · H. Hu · A.D. Kachkovski · E. Kim · B. Levy ·
D. Lichlyter · F. Merola · A. Mustafic · M. Nipper · L.A. Padilha ·
S.B. Park · H. Pasquier · L.D. Patsenker · O.V. Przhonska · M. Sameiro ·
E.W. Van Stryland · A.L. Tatarets · E.A. Theodorakis ·
E.A. Terpetschnig · V.I. Tomin · S. Webster

 Springer

Volume Editor
Prof. Dr. Alexander P. Demchenko
Palladin Institute of Biochemistry
National Academy of Sciences of Ukraine
Kyiv 01601
Ukraine
alexdem@ukr.net

ISSN 1617-1306   e-ISSN 1865-1313
ISBN 978-3-642-04700-8   e-ISBN 978-3-642-04702-2
DOI 10.1007/978-3-642-04702-2
Springer Heidelberg Dordrecht London New York

Library of Congress Control Number: 2010934374

Cover design: WMXDesign GmbH, Heidelberg, Germany

Printed on acid-free paper

Springer is part of Springer Science+Business Media (www.springer.com)

Series Editor

Prof. Dr. Otto S.Wolfbeis

Institute of Analytical Chemistry
Chemo- and Biosensors
University of Regensburg
93040 Regensburg
Germany
otto.wolfbeis@chemie.uni-regensburg.de

Aims and Scope

Fluorescence spectroscopy, fluorescence imaging and fluorescent probes are indispensible tools in numerous fields of modern medicine and science, including molecular biology, biophysics, biochemistry, clinical diagnosis and analytical and environmental chemistry. Applications stretch from spectroscopy and sensor technology to microscopy and imaging, to single molecule detection, to the development of novel fluorescent probes, and to proteomics and genomics. The *Springer Series on Fluorescence* aims at publishing state-of-the-art articles that can serve as invaluable tools for both practitioners and researchers being active in this highly interdisciplinary field. The carefully edited collection of papers in each volume will give continuous inspiration for new research and will point to exciting new trends.

Preface

Fluorescence reporter is the key element of any sensing or imaging technology. Its optimal choice and implementation is very important for increasing the sensitivity, precision, multiplexing power, and also the spectral, temporal, and spatial resolution in different methods of research and practical analysis. Therefore, design of fluorescence reporters with advanced properties is one of the most important problems. In this volume, top experts in this field provide advanced knowledge on the design and properties of fluorescent dyes. Organic dyes were the first fluorescent materials used for analytical purposes, and we observe that they retain their leading positions against strong competition of new materials – conjugated polymers, semiconductor nanocrystals, and metal chelating complexes. Recently, molecular and cellular biology got a valuable tool of organic fluorophores synthesized by cell machinery and incorporated into green fluorescent protein and its analogs.

Demands of various fluorescence techniques operating in spectral, anisotropy, and time domains require focused design of fluorescence reporters well adapted to these techniques. Near-IR spectral range becomes more and more attractive for various applications, and new dyes emitting in this range are strongly requested. Two-photonic fluorescence has become one of the major tools in bioimaging, and fluorescence reporters well adapted to this technique are in urgent need. These problems cannot be solved without the knowledge of fundamental principles of dye design and of physical phenomena behind their fluorescence response. Therefore, this book describes the progress in understanding these phenomena and demonstrates the pathways for improving the response to polarity, viscosity, and electric field in dye environment that can be efficiently used in sensing and imaging. Prospective pathways of synthesis of new dyes, including creation of their combinatorial libraries, and of their incorporation into molecular and supramolecular sensor elements are highlighted in this book.

Demonstrating the progress in an interdisciplinary field of research and development, this book is primarily addressed to specialists with different background – physicists, organic and analytical chemists, and photochemists – to those who develop and apply new fluorescence reporters. It will also be useful to specialists in bioanalysis and biomedical diagnostics – the areas where these techniques are most extensively used.

Kyiv, Ukraine   Alexander P. Demchenko
June 2010

Contents

Part III   Organic Dyes with Response Function

Part IV   Fluorophores of Visible Fluorescent Proteins

Part I
General Aspects

Comparative Analysis of Fluorescence Reporter Signals Based on Intensity, Anisotropy, Time-Resolution, and Wavelength-Ratiometry

Alexander P. Demchenko

Abstract The response signal of an immense number of fluorescence reporters with a broad variety of structures and properties can be realized through the observation in changes of a very limited number of fluorescence parameters. They are the variations in intensity, anisotropy (or polarization), lifetime, and the spectral changes that allow wavelength-ratiometric detection. Here, these detection methods are overviewed, and specific demands addressed to fluorescence emitters for optimization of their response are discussed.

Keywords Anisotropy · Intensity sensing · Time-resolved fluorimetry · Wavelength ratiometry

Contents

A.P. Demchenko
Palladin Institute of Biochemistry, National Academy of Sciences of Ukraine, Kyiv 01601, Ukraine
e-mail: alexdem@ukr.net

A.P. Demchenko (ed.), *Advanced Fluorescence Reporters in Chemistry and Biology I:*
Fundamentals and Molecular Design, Springer Ser Fluoresc (2010) 8: 3–24,
DOI 10.1007/978-3-642-04702-2_1, © Springer-Verlag Berlin Heidelberg 2010

1   Why Fluorescence?

Fluorescence is the basic reporting technique in many chemical sensors and bio-sensors with a broad range of applications in clinical diagnostics, monitoring the environment, agriculture, and in various industrial technologies. Being an efficient method of transforming the act of target binding into readable signal already on molecular level, it puts virtually no limit to target chemical nature and size. The range of its applications extends to imaging the living cells and tissues with the possibility of recording the target spatial distribution. In all these applications, fluorescence competes successfully with other detection methods that are based on electrochemical response or on the change in mass, heat, or refractive index on target binding [1]. There are many reasons for such great popularity:

- Fluorescence techniques are the *most sensitive*. With proper dye selection and proper experimental conditions, the absolute sensitivity may reach the limit of single molecules. This feature is especially needed if the target exists in trace amounts. High sensitivity may allow avoiding time-consuming and costly target-enrichment steps.
- They offer very high *spatial resolution* on the level of hundreds of nanometers, which is achieved by light microscopy. Moreover, with recent developments on overcoming the light diffraction limit, it has reached molecular scale. This allows obtaining detailed cellular images and operating with dense multianalyte sensor arrays.
- Their distinguishing feature is the *high speed* of response. This response develops on the scale of fluorescence lifetime of photophysical or photochemical events that provide the response and can be as short as 10^{-8}–10^{-10} s. Because of that, the fluorescence reporting is never time-limiting, so that this limit comes from other factors, such as the rate of target – sensor mutual diffusion and the establishment of dynamic equilibrium between bound and unbound target.
- They allow sensing *at a distance* from analyzed object. Because the fluorescence reporter and the detecting instrument are connected via emission of light, the sensing may occur in an essentially noninvasive manner and allow formation of images.
- The greatest advantage of fluorescence technique that derives from these features is its *versatility*. Fluorescence sensing can be provided in solid, liquid, and gas media, and at all kinds of interfaces between these phases. It can trace rare events with high spatial and temporal resolution. Fluorescence detection can be equally well-suited for remote industrial control and for sensing different targets within the living cells.

To our benefit, fluorescence is a well-observed phenomenon characteristic for many materials. This allows providing broad selection of fluorescence reporters that have to be chosen according to different criteria: high molar absorbance and fluorescence quantum yield, convenient wavelengths of excitation and emission, high chemical stability, and photostability. They are well-described in other chapters of this book and in other books of these series. As we will see subsequently, they should be adapted to particular technique of target detection and to particular method of observation of fluorescence response, which needs establishing additional criteria for their selection.

In this regard, it has to be stressed that fluorescence reporters have to be divided into two broad categories according to two major trends of technologies in which they are used. This division is necessary because some criteria for the choice of optimal reporters are quite the opposite.

One category is the reporters serving as *labels* and *tags*. Their only response should be based on their presence in particular medium or at particular site. Ideally, the response should be *directly proportional* to reporter concentration and *independent* of any factors that influence fluorescence parameters (quenching or enhancing of emission, wavelength shifting). Such emitting dyes or nanoparticles are extensively applied in imaging techniques based on their affinity to particular components of a complex system (e.g., living cell) and also in sensing different soluble targets that uses separation of bound and unbound labeled components. The most advantageous in these applications are the dyes that are nonfluorescent in a free state but become strongly fluorescent on their binding; this allows avoiding separation of labeled compounds and free reporters. The common observation in the application of labels and tags is the detection of fluorescence intensity, so that high spectral resolution may not be needed.

The second category is the reporters serving as *probes* or that involved in *molecular sensors*. As probes, they should respond to the changes of their molecular environment, and as essential parts of the sensors, they should be coupled to recognition units and respond to target binding by the change of parameters of their fluorescence. Ideally, this response should be *independent* of their concentration, and the valuable information should be derived from the concentration-independent *change* of their fluorescence parameters. Therefore, the reporters should be selected with the properties that provide these changes in the broadest dynamic range.

Accordingly, we have to consider two types of sensitivity in fluorescence reporting. One is the absolute sensitivity, which is the ability to detect fluorescence signal with the necessary level of precision. The other, which should be applied to probes and sensors, is the sensitivity in detecting the difference between the probes interacting differently with their environment or between the sensors with bound and unbound target. This type of sensitivity is determined by dynamic range of variation of the recorded fluorescence parameters. Developing such reporters is a much harder task, and it deserves a more detailed discussion.

Several parameters of fluorescence emission can be used as outputs in fluorescence sensing and imaging. Fluorescence intensity F can be measured at given wavelengths of excitation and emission (usually, band maxima). Its dependence on emission wavelength, $F(\lambda^{em})$ gives the fluorescence *emission spectrum*. If this intensity is measured over the excitation wavelength, one can get the fluorescence *excitation spectrum* $F(\lambda^{ex})$. Emission *anisotropy*, r (or similar parameter, polarization, P) is a function of the fluorescence intensities obtained at two different polarizations, vertical and horizontal. Finally, emission can be characterized by the *fluorescence lifetime* τ_F, which is the reverse function of the rate of emissive depopulation of the excited state. All these parameters can be determined as a function of excitation and emission wavelengths. They can be used for reporting on sensor-target interactions and a variety of possibilities exist for their employment in sensor constructs. The major concern here is obtaining reproducible analytical information free from interferences and background signals.

2 Sensing Based on Emission Intensity

Emission intensity measurements with low spectral resolution are frequently used in all types of techniques that involve fluorescence labeling and also in different sensing and imaging technologies that use fluorescence quenching as the reporter signal. Fluorescence reporters in the form of molecules or nanoparticles are either covalently conjugated to molecules of interest or used as stains to detect quantitatively the target compounds by noncovalent attachment. In cellular research, they can penetrate spontaneously into the cell and label genetically prepared protein-binding sites.

The change from light to dark (or the reverse) in fluorescence signal is easily observed and recorded as the change of fluorescence intensity at a single wavelength so that high spectral resolution is commonly not needed. For providing the coupling of sensing event with a change in fluorescence intensity from very high values to zero or almost zero values a variety of quenching effects can be used. The quenching may occur due to *conformational flexibility* in reporter molecule [2], intramolecular *photoinduced electron transfer* (PET) between its electron-donor and electron-acceptor fragments [3], on interaction with other chromophores [4], or with heavy [5] and transition metal [6] ions. Formation–disruption of hydrogen bonds with solvent molecules and different solvent-dependent changes of dye geometry can be observed in many organic dyes. Dramatic quenching in water (and to lesser extent in some alcohols) may occur due to formation by these molecules the traps for solvated electrons [7]. In addition, the solvent can influence the dye energetics, particularly the inversion of $n*$ (non-fluorescent) and $\pi*$ (fluorescent) energy levels [8]. Thus, the researcher has a lot of choice for constructing a sensor with response based on the principle of intensity sensing [9, 10].

Connection between the reversible target binding and the change in fluorescence intensity can be easily established based on the mass action law. In the simplest case of binding with stoichiometry 1:1, the target analyte concentration $[A]$ can be obtained from the measured fluorescence intensity F as:

$$[A] = K_d \left(\frac{F - F_{\min}}{F_{\max} - F} \right) \tag{1}$$

Here $F_{\min}$ is the fluorescence intensity without binding and $F_{\max}$ is the intensity when the sensor molecules are totally occupied. K_d is the dissociation constant. The differences in intensities in the numerator and denominator allow compensating for the background signal, and the obtained ratio can be calibrated in target concentration. But since F, $F_{\min}$ and $F_{\max}$ are expressed in relative units, they have to be determined in the same test and in exactly the same experimental conditions. This requires proper calibration, which is difficult and often not possible.

Calibration in fluorescence sensing means the operation, as a result of which at every sensing element (molecule, nanoparticle, etc.) or at every site of the image the fluorescence signal becomes independent of any other factor except the concentration of bound target. It is needed because the fluorescence intensity is commonly measured in relative units that have no absolute meaning if not compared with some standard measurement, and therefore, the problem of calibration in intensity sensing is very important [11]. Thus, the recorded changes of intensity always vary from instrument to instrument, and the proper reference even for compensating these instrumental effects is difficult to apply. Additional problems appear on obtaining information from cellular images and sensor arrays where the distribution in reporter concentration within the image or between different array spots cannot be easily measured. Moreover, their number can decrease due to chemical degradation and photobleaching. Therefore, internal calibration and photostability become a great concern in these applications. These difficulties justify strong efforts of the researchers to develop fluorescence dyes and sensing methods that allow excluding or compensating these factors. Those are the "intrinsically referenced" fluorescence detection methods [12, 13] that will be considered below.

3 Variation of Emission Anisotropy

Like other methods of fluorescence sensing, the anisotropy sensing is based on the existence of two states of the sensor, so that the switching between them depends on the concentration of bound target. Anisotropy sensing allows providing direct response to target binding that is independent of reporter concentration. This is because the measured anisotropy (or polarization) does not depend on absolute fluorescence intensity.

The measurement of *steady-state anisotropy* r is simple and needs two polarizers, one in excitation and the other in emission beams. When the sample is excited by vertically polarized light (indexed as $_V$) and the intensity of emission is measured at vertical (F_{VV}) and horizontal (F_{VH}) polarizations, then one can obtain r from the following relation:

$$r = \frac{F_{VV} - G \times F_{VH}}{F_{VV} + G \times 2F_{VH}} = \frac{1 - G \times (F_{VH}/F_{VV})}{1 + G \times 2(F_{VH}/F_{VV})}, \tag{2}$$

where G is an instrumental factor. Anisotropy has substituted *polarization P*, which was also used for characterizing polarized emission, and their relation is $r = 2P/(3 - P)$.

Equation (2) shows why r is in fact a *ratiometric parameter*: this is because the variations of intensity influence proportionally the F_{VV} and F_{VH} values. Therefore, the anisotropy allows obtaining self-referencing information on sensing event from a single reporter dye. This information is independent on reporter concentration.

Anisotropy describes the rotational dynamics of reporter molecules or of any sensor segments to which the reporter is rigidly fixed. In the simplest case when both the rotation and the fluorescence decay can be represented by single-exponential functions, the range of variation of anisotropy (r) is determined by variation of the ratio of fluorescence lifetime (τ_F) and *rotational correlation time* (φ) describing the dye rotation:

$$r = \frac{r_0}{1 + \tau_F/\varphi} \tag{3}$$

Here r_0 is the limiting anisotropy obtained in the absence of rotational motion. The dynamic range of anisotropy sensing is determined by the difference of this parameter observed for free sensor, which is commonly the rapidly rotating unit and the sensor-target complex that exhibits a strongly decreased rate of rotation.

As follows from (3), the variation of anisotropy can be observed if φ and τ_F are of comparable magnitude, and on target binding, there is the variation of *rotational mobility* of fluorophore (the change of φ) or the variation of its *emission lifetime* τ_F. At given τ_F, the rate of molecular motions determines the change of r, so that in the limit of slow molecular motions ($\varphi \gg \tau_F$) r approaches r_0, and in the limit of fast molecular motions ($\varphi \ll \tau_F$) r is close to 0. This determines dynamic range of the assay, which will decrease if φ and τ_F change in the same direction. Thus, there are three possibilities for using the fluorescence anisotropy in sensing:

- When anisotropy increases with the increase of molecular mass of rotating unit. For instance, the sensor segment rotates rapidly and massive target binding decreases this rate. The target binding can also displace small competitor to solution with increase of its rotation rate.
- When anisotropy increases due to increase of local viscosity producing higher friction on rotating unit. This can happen, for instance, in micelles or lipid

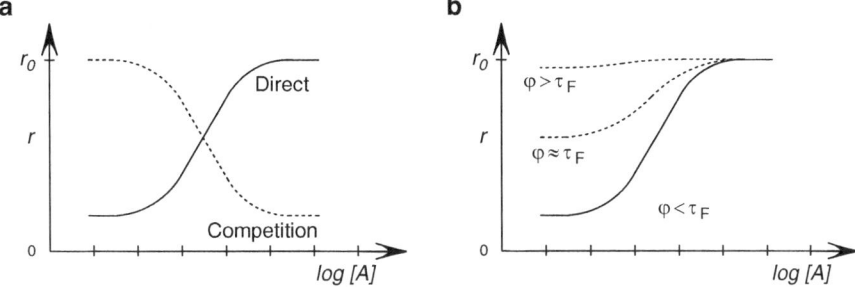

Fig. 1 Dependence of response of anisotropy sensor on analyte concentration in direct and competition assays (**a**) and this dependence for direct assay at different correlations between φ and τ_F (**b**)

vesicles that change their dynamics and order on target binding, and incorporated dye senses that.

• When anisotropy increases due to fluorescence lifetime decrease being coupled to any effect of dynamic quenching.

The differences between two (free and with the bound target) sensor states are detected when they possess different values of anisotropy, r_f of free and r_b of bound state (Fig. 1a). Their fractional contributions depend also on the relative intensities of correspondent forms. Since the additivity law is valid only for the intensities, the parameters derived in anisotropy sensing appear to be weighted by fractional intensities of these forms, F_f and F_b:

$$r = F_f r_f + F_b r_b \qquad (4)$$

This means that if the intensity of one of the forms is zero (static quenching), such anisotropy sensor is useless since it will show anisotropy of only one of the forms. The account of fractional intensity factor $R = F_b/F_f$ (the ratio of intensities of bound and free forms) leads to a more complicated function for the fraction of bound target, f:

$$f = \frac{r - r_f}{(r - r_f) + R(r_b - r)} \qquad (5)$$

Advantages and disadvantages of sensing technologies based on the measurement of anisotropy were discussed many times [14], and we will address only the questions related to the choice of optimal reporters. The limiting r_0 value 0.4 is theoretically achieved only for fluorophores with collinear absorption and emission transition dipole moments, and this limits the dynamic range of response. But the most important is fitting τ_F to the range of variations of φ (Fig. 1b). The fact is that with typical dyes possessing τ_F of several nanoseconds, the sensors can detect the binding of only small labeled molecules, or labeled receptors should be very flexible without targets. In the case of sensing of high molecular weight targets, τ_F should be

10–100 ns or longer [15]. It should satisfy the best sensing conditions, which correspond to $\varphi < \tau_F$ before the target binding and $\varphi > \tau_F$ after the binding. The possibility to achieve this range with large molecular rotating units is offered only by long-lifetime luminophors and only by those of them, which possess high r_0 values.

The weak point of anisotropy sensing is its great sensitivity to light-scattering effects. This occurs because the scattered light is always 100% polarized, and its contribution can be a problem if there is a spectral overlap between scattered and fluorescent light. For avoiding the light-scattering artifacts, the dyes with large Stokes shifts should be preferably used together with sufficient spectral resolution.

4 Time-Resolved and Time-Gated Detection

Fluorescence decays as a function of time, and the derived lifetimes can be used in fluorescence reporting. In an ideal case, the decay is exponential and it can be described by initial amplitude α and lifetime τ_F for each of the two, free (with index[F]) and bound (with index[B]), forms. If both of these forms are present in emission, we observe the result of additive contributions of two decays:

$$F(t) = \alpha_F \exp\left(-t/\tau_F^F\right) + \alpha_B \exp\left(-t/\tau_F^B\right) \tag{6}$$

To be detected, the presence of target should provide significant change of τ_F recorded within the time resolution of the method. Application of lifetime detection in sensing is based on several principles:

- Modulation of τ_F by dynamic quencher. Here, the effect of quenching competes with the emission in time and is determined by the diffusion of a quencher in the medium and its collisions with the excited dye. In this case, the relative change of intensity, F_0/F, is strictly proportional to correspondent change of fluorescence lifetime, τ_0/τ_F, where F_0 and τ_0 correspond to conditions without quencher [16]. Successfully this approach was applied only to oxygen sensing using the long-lifetime luminescence emitters [17]. In this case, the decrease of τ_F occurs gradually with oxygen concentration (Fig. 2a).
- The switch between discrete emitter forms with fixed but different lifetimes corresponding to free (F) and bound (B) forms of the sensor. Belonging to the same dye, these two forms can be excited at the same wavelength. When excited, they emit light independently, and the observed nonexponential decay can be deconvolved into two different individual decays with lifetimes τ_F^F and τ_F^B (Fig. 2b). The ratio of preexponential factors α_F and α_B will determine the target concentration [18]:

$$\frac{\alpha_B}{\alpha_F} = \frac{\varepsilon_B}{\varepsilon_F} \frac{\Phi_B \tau_F^F}{\Phi_F \tau_F^B} \frac{[LR]}{[L]} \tag{7}$$

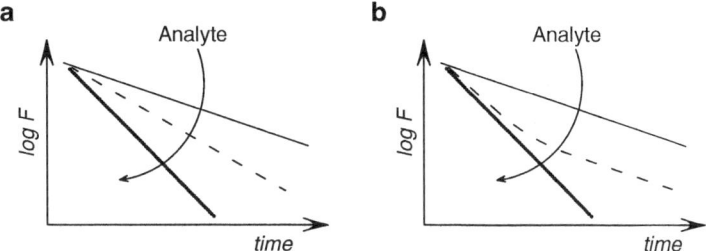

Fig. 2 The changes in fluorescence decay kinetics on binding the analyte. (**a**) The analyte is the dynamic quencher. The decay becomes shorter gradually as a function of its concentration. (**b**) The analyte binding changes the lifetime. Superposition of decay kinetics of bound and unbound forms is observed

It can be seen that the ratio of concentrations of free and occupied receptors is determined not only by α_F and α_B values but also by correspondent lifetimes τ_F^F and τ_F^B and the products of molar absorbances ε_F or ε_B and quantum yields Φ_F or Φ_B.

- Using the long-lifetime emission as a reference in intensity sensing by short-lifetime dye. This approach known as dual luminophore referencing (DLR) will be considered in the next section.

The lifetime detection techniques are self-referenced in a sense that fluorescence decay is one of the characteristics of the emitter and of its environment and does not depend upon its concentration. Moreover, the results are not sensitive to optical parameters of the instrument, so that the attenuation of the signal in the optical path does not distort it. The light scattering produces also much lesser problems, since the scattered light decays on a very fast time scale and does not interfere with fluorescence decay observed at longer times.

Summarizing, we stress that the anisotropy and the fluorescence decay functions change in a complex way as a function of target concentration. Species that fluoresce more intensely contribute disproportionably stronger to the measured parameters. Simultaneous measurements of steady-state intensities allow accounting this effect.

5 Wavelength Ratiometry with Two Emitters

Simultaneous application of two emitting reporters allows providing the self-referenced reporter signal based on simple intensity measurements, without application of anisotropy or lifetime sensing that impose stringent requirements on fluorescence reporters. Usually, the two dyes are excited at a single wavelength with the absence or in the presence of interaction between them.

5.1 Intensity Sensing with the Reference

In intensity sensing, the most efficient and commonly used method of "intrinsic referencing" is the introduction of a *reference dye* into a sensor molecule (or into support layer, the same nanoparticle, etc.) so that it can be excited together with the reporter dye and provide the reference signal [1]. The reference dye should conform to stringent requirements:

- It should absorb at the same wavelength as the reporting dye. The less common is the use of two channels of excitation since this requires more sophisticated instrumentation.
- For recording the intensity ratio at two emission wavelengths, it should possess strongly different emission spectrum but a comparable intensity to that of reporter band.
- In contrast to that of reporting dye, the reference emission should be completely insensitive to the presence of target.
- Direct interactions between the reference and reporter dyes leading to PET or FRET in this approach should be avoided.

If the reference dye is properly selected, then it can provide an additional independent channel of information and two peaks in fluorescence spectrum can be observed – one from the reporter with a maximum at λ_1 and the other from the reference with a maximum at λ_2 (Fig. 3). Their intensity ratio can be calibrated in concentration of the bound target. Thus, if we divide both the numerator and denominator of (1) by $F_{ref}(\lambda_2)$, the intensity of the reference measured in the same conditions but at different wavelength (λ_2) from that of reporter, we can obtain target concentration from the following equation that contains only the intensity ratios $R = F(\lambda_1)/F_{ref}(\lambda_2)$, $R_{min} = F_{min}(\lambda_1)/F_{ref}(\lambda_2)$, and $R_{max} = F_{max}(\lambda_1)/F_{ref}(\lambda_2)$:

$$[A] = K_d \left(\frac{R - R_{min}}{R_{max} - R} \right) \tag{8}$$

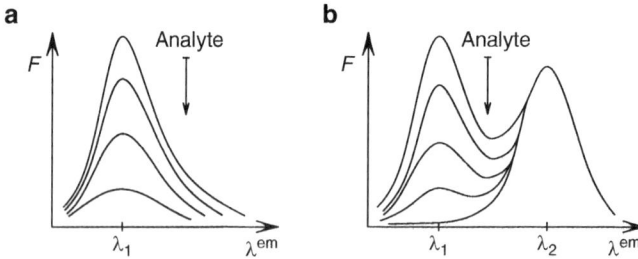

Fig. 3 Intensity sensing (**a**) and this sensing with the reference dye (**b**). The fluorescence intensity with the band maximum at λ_1 decreases as a function of analyte concentration. The reference dye allows providing the ratio of two intensities detected at wavelengths λ_1 and λ_2

Separate detection of these two signals, one from the reporter dye and the other from the reference, can be provided based not only on the difference of their fluorescence band positions but also on the difference in anisotropy [15] or lifetime [15, 19]. The change of these parameters with the variation of intensity of reporter dye is based on the fact that the measured anisotropy or lifetime is a sum of intensity-weighted anisotropies or lifetimes of contributing species. This type of referencing can be used even if the reporter and the reference dyes possess strongly overlapping fluorescence spectra. The intensity calibration in the lifetime domain has an advantage in the studies in highly light scattering media.

An interesting development in this respect is the dual luminophore referencing (DLR) in phase-modulation detection technique [19]. Phosphorescent lumino-phore with long lifetime serves as the reference producing strong and stable phase shift that can be measured using inexpensive device using LED light source. Reporter dye excited simultaneously with the reference can exhibit short lifetime, but its quenching/dequenching generates the change in phase shift of modulated emission. In this way, the phase angle reflects directly the intensity change of the reporter and consequently the concentration of the target. Here, the two-dye ratiometry combines the advantages of time-resolved detection with simplicity of instrumentation using single filter-detector arrangement and operating at low modulation frequencies. This method was extended recently for detecting two analytes [20].

Summarizing, we outline what is achieved with the introduction of reference dye. The two dyes, responsive and nonresponsive to target binding, can be excited and their fluorescence emission detected simultaneously, which compensates the variability and instability of instrumental factors. In principle, the results should be reproducible on the instruments with a different optical arrangement, light source intensity, slit widths, etc. The two-band ratiometric signal can be calibrated in target concentration. This calibration, in some range of target concentrations, will be insensitive to the concentration of sensor (and reporter dye) molecules.

5.2   Formation of Excimers

When molecule absorbs light, it can make a complex with the ground-state mole-cule like itself. These excited dimeric complexes are called the excimers. Excimer emission spectrum is very different from that of monomer; it is usually broad, shifted to longer wavelengths, and it does not contain vibrational structure. The double labeling is needed for this technique, which is facilitated by the fact that the dyes are of the same structure. Meantime, a researcher is limited in their selection. Usually pyrene derivatives are used because of unique property of this fluorophore to form stable excimers with fluorescence spectra and lifetimes that are very different from that of monomers. The structured band of monomer is observed at about 400 nm, whereas that of excimer located at 485 nm is broad, structureless, and long-wavelength shifted. Long lifetimes ($\sim$300 ns for monomer and $\sim$40 ns for excimer) allow easy rejection of background emission and application of lifetime sensing [21].

There are many possibilities to use these complex formations in fluorescence sensing. If the excimer is not formed, we observe emission of the monomer only, and upon its formation there appears characteristic emission of the excimer. We just need to make a sensor, in which its free and target-bound forms differ in the ability of reporter dye to form excimers and the fluorescence spectra will report on the sensing event. Since we will observe transition between two spectroscopic forms, the analyte binding will result in increase in intensity of one of the forms and decrease of the other form with the observation of isoemissive point [22].

Meantime, we have to keep in mind that monomer and excimer are independent emitters possessing different lifetimes and that nonspecific influence of quenchers may be different for these two forms. For instance, dissolved oxygen may quench the long-lifetime emission of monomer but not of the excimer.

5.3 Förster Resonance Energy Transfer

Two or more dye molecules or light absorbing particles with similar excited-state energies can exchange their energies due to long-range dipole–dipole resonance interaction between them. One molecule, the *donor*, can absorb light and the other, the *acceptor*, can accept this energy with or without emission. This phenomenon known as Förster resonance energy transfer (FRET) has found many applications in sensing [23, 24]. The FRET sensing usually needs labeling with two dyes serving as donor and acceptor. Only in rare, lucky cases, intrinsic fluorescent group of sensor or target molecules can be used as one of the partners in FRET sensing.

FRET to nonfluorescent acceptor provides a single-channel response in intensity with all disadvantages that were described above. Meantime, there are two merits in this approach. One is over traditional intensity sensing: the quenching can occur at a long distance, which allows exploring conformational changes in large sensor molecules, such as proteins [25] or DNA hairpins [26]. The other is over the FRET techniques using fluorescent acceptor: a direct excitation of the acceptor is not observed in emission.

FRET to fluorescent acceptor is obviously more popular because of its two-channel self-calibrating nature. Sensing may result in switching between two fluorescent states, so that in one of them a predominant emission of the donor can be observed and in the other – of the acceptor. This type of FRET can be extended to time domain with the benefit of using simple instrumentation with the long-lifetime donors [27].

FRET can take place if the emission spectrum of the donor overlaps with the absorption spectrum of the acceptor and they are located at separation distances within 1–10 nm from each other. The efficiency of energy transfer E can be defined as the number of quanta transferred from the donor to the acceptor divided by all the quanta absorbed by the donor. According to this definition, $E = 1 - F_{DA}/F_D$, where F_{DA} and F_D are the donor intensities in the presence and absence of the acceptor. Both have to be normalized to the same donor concentration. If the time-resolved measurements are used, then the knowledge of donor concentration is not required, and $E = 1 - <\tau_{DA}>/<\tau_D>$, where $<\tau_{DA}>$ and $<\tau_D>$ are the average lifetimes in the presence and absence of the acceptor [28].

The energy transfer efficiency exhibits a very steep dependence on the distance separating two fluorophores, R:

$$E = R_0^6 / (R_0^6 + R^6), \qquad (9)$$

here, R_0 is the parameter that corresponds to a distance with 50% transfer efficiency (the Förster radius). Such steep dependence on the nanometer scale allows diversity of possibilities in sensor development. We list several of them:

- FRET sensing based on heterotransfer (the transfer between different molecules or nanoparticles) with reporting to the change of donor–acceptor distance. Since this distance is comparable with the dimensions of many biological macromolecules and of their complexes, many possibilities can be realized for coupling the response with the changes in sensor geometry. The most popular approaches use conformational change in double labeled sensor [29], enzymatic splitting of covalent bond between two labeled units [30] and competitive substitution of labeled competitor in a complex with labeled sensor [31].
- Exploration of collective effects in multiple transfers that appear when the donor and acceptor are the same molecules and display the so-called homotransfer. In this case, the presence of only one molecular quencher can quench fluorescence of the whole ensemble of emitters coupled by homotransfer [32]. The other possibility of using homo-FRET is the detection of intermolecular interactions by the decrease of anisotropy [33].
- FRET modulation by photobleaching. Photobleaching can specifically destroy the acceptor giving rise to fluorescence of the donor. This approach is useful in some sensing technologies and especially in cellular imaging where it is important to compare two signals or images, with and without FRET, with the same composition and configuration in the system [34].
- FRET sensing based on protic equilibrium in the acceptor that changes its absorption spectrum and thus modulates the overlap integral [35]. There are many fluorescent pH indicators that display pH-dependent absorption spectra in the visible with their different positions depending on ionization state. Thus, the change in pH can be translated into the change of FRET efficiency.
- Photochromic FRET using as acceptors the photochromic compounds such as spiropyrans [36]. They have the ability to undergo a reversible transformation between two different structural forms in response to illumination at appropriate wavelengths. These forms may have different absorption (and in some cases,

fluorescence) spectra. Thus, they offer a possibility of reversible switching of FRET effect between "on" and "off" states without any chemical intervention, just by light.

Realization of all these possibilities is traditionally performed with organic dyes [28]. There are many variants in choosing the dye donor–acceptor pair in which two correspondent bands are well separated on the wavelength scale or produce different lifetimes. Meantime, we observe increasing popularity of lanthanide chelates [37] and Quantum Dots [38, 39] as FRET donors, which is mainly because of their increased brightness and longer emission lifetimes [40]. If the acceptor is excited not directly but by the energy transferred from the donor, its lifetime increases to that of the donor [41]. This allows providing many improvements in sensing technologies especially in view that organic dyes are much more "responsive" but are behind these emitters in lifetime and brightness.

Concluding the section on wavelength ratiometry with two emitters, we stress that they provide the two-channel informative signal in sensing, in which these channels are independent or, as in the case of FRET, partially dependent. In the latter case, quenching of fluorescence of the donor quenches also the acceptor emission but the quenching of the acceptor emission does not influence the emission of the donor. Independence of quenching effects may cause a nonspecific and nonaccountable effect on ratiometric reporter signal [42]. It should be also remembered that the reporter molecules can exhibit different degradation and photobleaching as a function of time. These effects may provide the time-dependent but target-independent changes of the measured intensity ratios. In addition, because the sensitivity to quenching (by temperature, ions, etc.) can be different for reporter and reference dyes and they emit independently, every effect of fluorescence quenching unrelated to target binding will interfere with the measured result. This can make the sensor nonreproducible in terms of obtaining precise quantitative data even in serial measurements with the same instrument.

6 Wavelength Ratiometry with Single Emitter

In sensor technologies, the use of a single emitter is more attractive than of two emitters. This is not because of avoiding the necessity of double labeling alone. Chemical degradation and photobleaching producing nonfluorescent products from the reporter dye in this case will not distort its wavelength-ratiometric signal. Meantime, the reporter dyes should conform to stringent requirements: they should possess spectrally recognizable ground-state and/or excited-state forms and the switching between these forms should occur on target binding. Ground-state inter-actions resulting in differences in excitation energies generate the differences in excitation spectra (Fig. 4a). The excited-state reactions offer additional possibilities

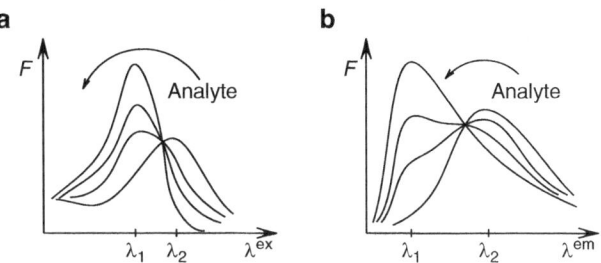

Fig. 4 The changes in excitation (**a**) and emission (**b**) spectra on analyte binding when this binding generates new ground-state or excited-state forms. λ_1 and λ_2 are the positions of the band maxima of the analyte-bound and analyte-free forms

for observing new bands in fluorescence emission spectra belonging to reactant and reaction product forms (Fig. 4b).

In contrast to intensity sensing with the reference, where the reference provides the signal of constant intensity, the two forms in a single reporter molecule interconvert reporting to target binding. We then observe interplay of intensities at two selected wavelengths, λ_1 and λ_2, with their change in converse manner and the generation of isobestic and isoemissive points. If such a point is chosen as the reference, then (8) can be used. In a more general case, when λ_2 is a different wavelength, (e.g., it is the maximum of the second band), the result has to be corrected to include the factor that accounts for this intensity redistribution, which is the ratio of intensities of free and bound forms at wavelength λ_2:

$$[A] = K_d \left(\frac{R - R_{min}}{R_{max} - R} \right) \left(\frac{F_F(\lambda_2)}{F_B(\lambda_2)} \right) \qquad (10)$$

The change in noncovalent intermolecular interactions with the environment changes the energies of electronic transitions resulting in the shifts of electronic absorption and emission bands. If these interactions are stronger in the ground state, then with their increase, the difference in energy between the states increases resulting in the shift of spectra to shorter wavelengths. On the opposite, if the interaction energy is stronger in the excited state, then on increase of this interaction, the spectra shift to longer wavelengths.

6.1 Transitions Between Ground-State Forms

Spectacular differences in absorption/excitation spectra are often observed for the dyes that exist in protonation–deprotonation equilibria. Their straightforward application is for pH sensing and also for designing the reporters, in which the shifting of such equilibrium by external proton donor or acceptor group is involved in sensing event.

Next in importance is the response based on the shifts between H-bond free and H-bonded forms. Formation of intermolecular H-bonds leads to spectral changes in the same direction as the protonation but of smaller magnitude. H-bonding requires steric arrangement between donor and acceptor groups. In certain cases, it is coupled with conformational changes stabilizing one of the conformers. Spectral shifts can be also observed with the formation of ground-state intramolecular charge transfer (ICT) state that originates from polarization of π-electrons and can be stimulated by increased polarity of the medium. The intermolecular H-bonding can be involved also in this case: being formed at an acceptor site of ICT compounds, it causes red shifts in the absorption and emission bands, whereas the interaction at a donor site produces the shift in opposite direction.

One of the applications of these ground-state effects is the sensing of local electric fields with highly polarizable electrochromic dyes [43, 44]. The stilbene-like dyes exhibiting ICT are the popular sensors for Ca^{2+} ions that exhibit interaction of chelated Ca^{2+} ion with the electron-donor nitrogen atom [45]. There are many reports on the construction of chemical sensors for other, beside calcium, ions based on ICT mechanism [8] but those that exhibit ratiometric response are still rare cases. Promising are the systems that use the switching between two ground-state tautomers of the dye [46]. Commonly in all these cases, the wavelength-ratiometric signal is recorded at two excitation wavelengths with the detection of fluorescence at single wavelength (Fig. 4a).

6.2   Transitions Between Excited-State Forms

Being richer in energy than the ground states, the excited states allow broader range of electronic transformations resulting in shifts of fluorescence spectra and in the appearance of new bands that allows ratiometric detection of intensities (Fig. 4b). Unfortunately, many of these reactions result in quenching with the loss of benefits of wavelength-ratiometric recording. Therefore, efforts in sensor design should be directed at achieving the highest brightness of both initially excited and reaction product forms.

Proton dissociation in the excited states commonly occurs much easier than in the ground states, and the great difference in proton dissociation constants by several orders of magnitude is characteristic for 'photoacids' [47]. These dyes exist as neutral molecules and their excited-state deprotonation with the rate faster than the emission results in new red-shifted bands in emission spectra [48]. Such properties can be explored in the same manner as the ground-state deprotonation with the shift of observed spectral effect to more acidic pH values.

Excited-state intramolecular proton transfer (ESIPT) exhibits different regularities [49, 50]. Commonly, this is a very fast and practically irreversible reaction proceeding along the H-bonds preexisting in the ground state. Therefore, only the reaction product band is seen in fluorescence spectra. Such cases are not interesting for designing the fluorescence reporters. The more attractive dual emission is observed in two cases. (a) When the initially excited state becomes the ICT state stabilized to be of similar energy as the ESIPT product state, then the ESIPT reaction becomes reversible and an equilibrium between two forms can be established on a timescale faster than the emission. This is the case of designed 3-hydroxyflavone derivatives [51]. (b) When the ESIPT reaction exhibits slow kinetics on the timescale of emission. This can be due to intermolecular H-bonding perturbations as observed for parent 3-hydroxyflavone [52] and 3-hydroxyquinolones [53]. In the case of rapidly established equilibrium between two forms, the internally calibrated signal is resistant to any uncontrolled quenching effect produced by collisional quenchers or the temperature. This is because their lifetimes are equal and they are quenched proportionally with the retention of the same intensity ratio [54–56]. Since the ESIPT reaction can provide dramatic shifts in fluorescence spectra (by 100 nm and more), finding new systems exhibiting dual emission based on ESIPT is a great concern.

Usually, the π-electronic system in highly fluorescent organic dyes becomes in the excited state a stronger dipole that interacts stronger with polar environment resulting in long-wavelength shifts [57, 58]. This effect can be enhanced by generating the ICT states by introducing into the π-electronic system the chemical substitutions donating and withdrawing electronic density [50]. Such dyes known as polarity sensors can be used for ratiometric reporting if the difference in polarity of reporter environment can be induced by target binding. If water, protic solvents, or H-bond forming groups of atoms are involved in interactions with reporter dye, then the H-bond formation with its acceptor group (that is usually carbonyl) results in the spectral shifts in the same direction as the increase of polarity [59]. Combination of these effects may result in dramatic spectral shifts that allow ratiometric reporting. The ICT states can be further stabilized with the formation of TICT (twisted intramolecular charge transfer) states that form distinct fluorescence bands [60]. Important point is the finding of such reporters, in which these states are strongly emissive.

6.3   Multiparametric Reporters Combining the Transitions Between Ground-State and Excited-State Forms

From basic photophysics, we may derive that in order to obtain the effects of switching in excitation spectra, we need to operate with two or more ground-state forms of the reporter dyes and to couple the sensing event with the transitions between them. In contrast, for obtaining two or more bands in fluorescence spectra, one ground-state form can be enough (and often preferable), but there should be excited-state reactions generating new species, so that both the reactant and the product in this reaction should emit fluorescence at different wavelengths. The ground-state and excited-state transformations can be governed by different types of molecular forces. Therefore, by proper reporter design, there is a possibility for

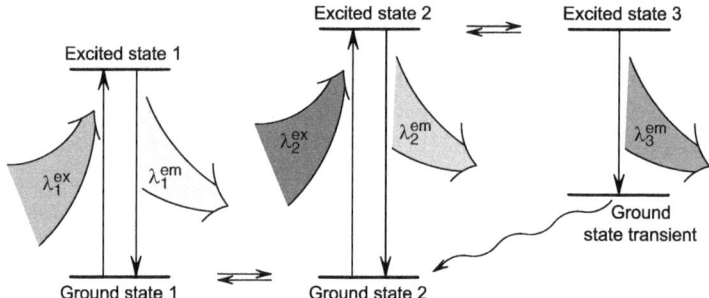

Fig. 5 General scheme of ground-state and excited-state transformations and emissions in the case of a reversible excited-state reaction involving one of the two ground-state species (from Ref. [62], modified)

not only combining these effects but also for obtaining separate information on these interactions. This idea is illustrated in Fig. 5. It was realized with 3-hydroxy-flavone dyes exhibiting ground-state equilibrium between the species with and without intermolecular H-bonds [51] and also an excited-state equilibrium between ICT and ESIPT states indicating polarity of the environment [61]. When these dyes are applied to test the unknown properties of their environment, one can observe three partially overlapped emission bands, and their excitation-wavelength-dependent deconvolution allows obtaining independently the polarity and the hydrogen bonding potential [62]. Following this approach and with proper selection of reporter dye, one can design the reporter responding differently to two different properties of their environment and to construct sensors based on this response.

7   Concluding Remarks

From this short survey, one can derive that many possibilities for technology design can be realized based on proper selection of reporters within the limited number of fluorescence detection methods. Each of these methods offers its own advantage in sensing but puts its special demands on photophysical and spectroscopic properties of reporters. Intensity sensing is the simplest technique that is least demanding regarding the dye properties. But it features many disadvantages due to the need for internal calibration of response signal. Such self-referenced signal can be provided by second dye partner serving as the reference or participating in excimer formation and FRET.

The calibration may not be needed in anisotropy and lifetime sensing. In lifetime sensing, the single-channel response allows obtaining the signal that does not need calibration. In anisotropy sensing, the two (vertical and horizontal) polarizations provide the necessary two channels, and in FRET to fluorescent acceptor, these two channels are selected as the intensities at two wavelengths.

Being the most convenient way of providing the self-referenced signal, the two-band wavelength-ratiometric recording can be realized not only by the application of two dyes but also with a single dye exhibiting ground-state or excited-state reaction leading to wavelength-shifting and generation of new bands. In two-band ratiometric sensing because the signal comes from a single type of the dye and the forms emitting at two wavelengths may have the same lifetimes, the internally calibrated signal has the advantage to be resistant to any uncontrolled quenching effect.

Every one of these techniques needs proper selection of reporter dyes. Many requests therefore should be addressed to synthetic chemists and photochemists.

References

1. Demchenko AP (2009) Introduction to fluorescence sensing. Springer, Amsterdam
2. McFarland SA, Finney NS (2001) Fluorescent chemosensors based on conformational restriction of a biaryl fluorophore. J Am Chem Soc 123:1260–1261
3. de Silva AP, Fox DB, Moody TS, Weir SM (2001) The development of molecular fluorescent switches. Trends Biotechnol 19:29–34
4. Marme N, Knemeyer JP, Sauer M, Wolfrum J (2003) Inter- and intramolecular fluorescence quenching of organic dyes by tryptophan. Bioconjug Chem 14:1133–1139
5. Lebold TP, Yeow EK, Steer RP (2004) Fluorescence quenching of the S1 and S2 states of zinc meso-tetrakis(4-sulfonatophenyl)porphyrin by halide ions. Photochem Photobiol Sci 3: 160–166
6. Chen YG, Zhao D, He ZK, Ai XP (2007) Fluorescence quenching of water-soluble conjugated polymer by metal cations and its application in sensor. Spectrochim Acta A Mol Biomol Spectrosc 66:448–452
7. Ellison EH, Moodley D, Hime J (2006) Fluorescence study of arene probe microenvironment in the intraparticle void volume of zeolites interfaced with bathing polar solvents. J Phys Chem B 110:4772–4781
8. de Silva AP, Gunaratne HQN, Gunnaugsson T, Huxley AJM, McRoy CP, Rademacher JT, Rice TE (1997) Signaling recognition events with fluorescent sensors and switches. Chem Rev 97:1515–1566
9. Demchenko AP (2005) Optimization of fluorescence response in the design of molecular biosensors. Anal Biochem 343:1–22
10. Descalzo AB, Zhu S, Fischer T, Rurack K (2010) Optimization of the coupling of target recognition and signal generation. In: Demchenko AP (ed) Advanced Fluorescence Reporters in Chemistry and Biology II. Springer Ser Fluoresc 9:41–105
11. Vogt RFJ, Marti GE, Zenger V (2008) Quantitative fluorescence calibration: a tool for assessing the quality of data obtained by fluorescence measurements. In: Resch-Genger U (ed) Standardization and quality assurance in fluorescence measurements I: Springer Ser Fluoresc 5:3–31
12. Demchenko AP (2005) The problem of self-calibration of fluorescence signal in microscale sensor systems. Lab Chip 5:1210–1223
13. Schaferling M, Duerkop A (2008) Intrinsically referenced fluorimetric sensing and detection schemes: methods, advantages and applications. In: Resch-Genger U (ed) Standardization and quality assurance in fluorescence measurements I: Springer Ser Fluoresc 5:373–414
14. Jameson DM, Croney JC (2003) Fluorescence polarization: past, present and future. Comb Chem High Throughput Screen 6:167–173

15. Guo XQ, Castellano FN, Li L, Lakowicz JR (1998) Use of a long lifetime Re(I) complex in fluorescence polarization immunoassays of high-molecular weight analytes. Anal Chem 70:632–637
16. Lakowicz JR (2007) Principles of fluorescence spectroscopy, 3rd edn. Springer, New York
17. Maliwal BP, Gryczynski Z, Lakowicz JR (2001) Long-wavelength long-lifetime lumino-phores. Anal Chem 73:4277–4285
18. Lakowicz JR (1999) Principles of fluorescence spectroscopy. Kluwer Academic, New York
19. Liebsch G, Klimant I, Krause C, Wolfbeis OS (2001) Fluorescent imaging of pH with optical sensors using time domain dual lifetime referencing. Anal Chem 73:4354–4363
20. Borisov SM, Neurauter G, Schroeder C, Klimant I, Wolfbeis OS (2006) Modified dual lifetime referencing method for simultaneous optical determination and sensing of two analytes. Appl Spectrosc 60:1167–1173
21. Yang CJ, Jockusch S, Vicens M, Turro NJ, Tan W (2005) Light-switching excimer probes for rapid protein monitoring in complex biological fluids. Proc Natl Acad Sci USA 102: 17278–17283
22. Yang RH, Chan WH, Lee AWM, Xia PF, Zhang HK, Li KA (2003) A ratiometric fluorescent sensor for Ag-1 with high selectivity and sensitivity. J Am Chem Soc 125:2884–2885
23. Clegg RM (1996) Fluorescence resonance energy transfer. In: Wang XF, Herman B (eds) Fluorescence imaging spectroscopy and microscopy. John Wiley, New York, pp 179–252
24. Selvin PR (2000) The renaissance of fluorescence resonance energy transfer. Nat Struct Biol 7:730–734
25. Tahtaoui C, Guillier F, Klotz P, Galzi JL, Hibert M, Ilien B (2005) On the use of nonfluo-rescent dye labeled ligands in FRET-based receptor binding studies. J Med Chem 48: 7847–7859
26. Tyagi S, Kramer FR (1996) Molecular beacons: probes that fluoresce upon hybridization. Nat Biotechnol 14:303–308
27. Hildebrandt N, Charbonniere LJ, Lohmannsroben HG (2007) Time-resolved analysis of a highly sensitive forster resonance energy transfer immunoassay using terbium complexes as donors and quantum dots as acceptors. J Biomed Biotechnol 2007:79169
28. Wu PG, Brand L (1994) Resonance energy-transfer - methods and applications. Anal Bio-chem 218:1–13
29. Petitjean A, Lehn JM (2007) Conformational switching of the pyridine-pyrimidine-pyridine scaffold for ion-controlled FRET. Inorganica Chim Acta 360:849–856
30. Gershkovich AA, Kholodovych VV (1996) Fluorogenic substrates for proteases based on intramolecular fluorescence energy transfer (IFETS). J Biochem Biophys Methods 33: 135–162
31. Xu H, Wu HP, Huang F, Song SP, Li WX, Cao Y, Fan CH (2005) Magnetically assisted DNA assays: high selectivity using conjugated polymers for amplified fluorescent transduction. Nucleic Acids Res 33:e83
32. Johansson MK, Cook RM (2003) Intramolecular dimers: a new design strategy for fluores-cence-quenched probes. Chemistry 9:3466–3471
33. Tramier M, Coppey-Moisan M (2008) Fluorescence anisotropy imaging microscopy for homo-FRET in living cells. Methods Cell Biol 85:395–414
34. Jares-Erijman EA, Jovin TM (2003) FRET imaging. Nat Biotechnol 21:1387–1395
35. Takakusa H, Kikuchi K, Urano Y, Kojima H, Nagano T (2003) A novel design method of ratiometric fluorescent probes based on fluorescence resonance energy transfer switching by spectral overlap integral. Chemistry 9:1479–1485
36. Giordano L, Jovin TM, Irie M, Jares-Erijman EA (2002) Diheteroarylethenes as thermally stable photoswitchable acceptors in photochromic fluorescence resonance energy transfer (pcFRET). J Am Chem Soc 124:7481–7489
37. Selvin PR (2002) Principles and biophysical applications of lanthanide-based probes. Annu Rev Biophys Biomol Struct 31:275–302

38. Algar WR, Krull UJ (2008) Quantum dots as donors in fluorescence resonance energy transfer for the bioanalysis of nucleic acids, proteins, and other biological molecules. Anal Bioanal Chem 391:1609–1618
39. Medintz IL, Mattoussi H (2009) Quantum dot-based resonance energy transfer and its growing application in biology. Phys Chem Chem Phys 11:17–45
40. Resch-Genger U, Grabolle M, Nitschke R, Nann T (2010) Nanocrystals and nanoparticles vs. molecular fluorescent labels as reporters for bioanalysis and the life sciences. A critical comparison. In: Demchenko AP (ed) Advanced Fluorescence Reporters in Chemistry and Biology II. Springer Ser Fluoresc 9:3–40
41. Charbonniere LJ, Hildebrandt N, Ziessel RF, Lohmannsroben HG (2006) Lanthanides to quantum dots resonance energy transfer in time-resolved fluoro-immunoassays and luminescence microscopy. J Am Chem Soc 128:12800–12809
42. Demchenko AP (2005) The future of fluorescence sensor arrays. Trends Biotechnol 23: 456–460
43. Clarke RJ, Zouni A, Holzwarth JF (1995) Voltage sensitivity of the fluorescent probe RH421 in a model membrane system. Biophys J 68:1406–1415
44. Callis PR (2010) Electrochromism and solvatochromism in fluorescence response of organic dyes. A nanoscopic view. In: Demchenko AP (ed) Advanced Fluorescence Reporters in Chemistry and Biology I. Springer Ser Fluoresc 8:309–330
45. Grynkiewicz G, Poenie M, Tsien RY (1985) A new generation of Ca^{2+} indicators with greatly improved fluorescence properties. J Biol Chem 260:3440–3450
46. Chang CJ, Javorski J, Nolan EM, Shaeng M, Lippard SJ (2004) A tautomeric zinc sensor for ratiometric fluorescence imaging: application to nitric oxide-release of intracellular zinc. Proc Natl Acad Sci USA 101:1129–1134
47. Arnaut LG, Formosinho SJ (1993) Excited-state proton-transfer reactions. 1. Fundamentals and intermolecular reactions. J Photochem Photobiol A Chem 75:1–20
48. Davenport LD, Knutson JR, Brand L (1986) Excited-state proton transfer of equilenin and dihydro equilenin: inreractions with bilayer vesicles. Biochemistry 25:1186–1195
49. Formosinho SJ, Arnaut LG (1993) Excited-state proton-transfer reactions. 2. Intramolecular reactions. J Photochem Photobiol A Chem 75:21–48
50. Hsieh C-C, Ho M-L, Chou P-T (2010) Organic dyes with excited-state transforma-tions (electron, charge and proton transfers). In: Demchenko AP (ed) Advanced Fluorescence Reporters in Chemistry and Biology I. Springer Ser Fluoresc 8:225–266
51. Shynkar VV, Klymchenko AS, Piemont E, Demchenko AP, Mely Y (2004) Dynamics of intermolecular hydrogen bonds in the excited states of 4′-dialkylamino-3-hydroxyflavones. On the pathway to an ideal fluorescent hydrogen bonding sensor. J Phys Chem A 108: 8151–8159
52. Strandjord AJG, Barbara PF (1985) Proton-transfer kinetics of 3-Hydroxyflavone – solvent effects. J Phys Chem 89:2355–2361
53. Yushchenko DA, Shvadchak VV, Bilokin MD, Klymchenko AS, Duportail G, Mely Y, Pivovarenko VG (2006) Modulation of dual fluorescence in a 3-hydroxyquinolone dye by perturbation of its intramolecular proton transfer with solvent polarity and basicity. Photochem Photobiol Sci 5:1038–1044
54. Tomin VI, Oncul S, Smolarczyk G, Demchenko AP (2007) Dynamic quenching as a simple test for the mechanism of excited-state reaction. Chem Phys 342:126–134
55. Altschuh D, Oncul S, Demchenko AP (2006) Fluorescence sensing of intermolecular interactions and development of direct molecular biosensors. J Mol Recognit 19:459–477
56. Oncul S, Demchenko AP (2006) The effects of thermal quenching on the excited-state intramolecular proton transfer reaction in 3-hydroxyflavones. Spectrochim Acta A Mol Biomol Spectrosc 65:179–183
57. Valeur B (2002) Molecular fluorescence. Wiley VCH, Weinheim
58. Tomin VI (2010) Physical principles behind spectroscopic response of organic fluorophores to intermolecular interactions. In: Demchenko AP (ed) Advanced Fluorescence Reporters in Chemistry and Biology I. Springer Ser Fluoresc 8:189–224

59. Vazquez ME, Blanco JB, Imperiali B (2005) Photophysics and biological applications of the environment-sensitive fluorophore 6-N, N-Dimethylamino-2, 3-naphthalimide. J Am Chem Soc 127:1300–1306
60. Grabowski ZR, Rotkiewicz K, Rettig W (2003) Structural changes accompanying intramolecular electron transfer: Focus on twisted intramolecular charge-transfer states and structures. Chem Rev 103:3899–4031
61. Klymchenko AS, Demchenko AP (2003) Multiparametric probing of intermolecular interactions with fluorescent dye exhibiting excited state intramolecular proton transfer. Phys Chem Chem Phys 5:461–468
62. Caarls W, Celej MS, Demchenko AP, Jovin TM (2009) Characterization of coupled ground state and excited state equilibria by fluorescence spectral deconvolution. J Fluorescence 20:181–190

Part II
Design of Organic Dyes

Optimized UV/Visible Fluorescent Markers

M. Sameiro T. Gonçalves

Abstract Fluorescent molecules have been widely used as biomolecular labels, enzyme substrates, environmental indicators, and cellular stains and thus constitute indispensable tools in chemistry, physics, biology, and medicinal sciences. The large variation in the photophysics of the available fluorophores connected with chemical alterations give fluorescent probe techniques an almost unlimited scope for the detection of specific molecules and the investigation of intermolecular interactions on a molecular scale.

This chapter focuses on recent developments in the design and applications of fluorescent organic markers, such as coumarins, benzoxadiazoles, acridones, acridines, polyaromatics (naphthalene, anthracene, and pyrene), fluorescein, and rhodamine derivatives, which display maximum fluorescence emission in the UV/visible region and have been applied in the labeling of relevant biomolecules, namely DNA, RNA, proteins, peptides, and amino acids, among others.

Keywords Benzoxadiazoles · Coumarins · Fluorescein · Fluorescent probes · Rhodamine

Contents

M.S.T. Gonçalves
Centro de Química, Universidade do Minho, Campus de Gualtar, 4710-057 Braga, Portugal
e-mail: msameiro@quimica.uminho.pt

A.P. Demchenko (ed.), *Advanced Fluorescence Reporters in Chemistry and Biology I:*   27
Fundamentals and Molecular Design, Springer Ser Fluoresc (2010) 8: 27–64,
DOI 10.1007/978-3-642-04702-2_2, © Springer-Verlag Berlin Heidelberg 2010

1 Introduction

Over the last years, fluorescent molecules have been widely used as biomolecular labels, enzyme substrates, environmental indicators, and cellular stains, and
thus constitute indispensable tools in chemistry, physics, biology, and medicinal
sciences [1–10]. Owing to their high sensitivity, the detection of single fluorescent molecules and investigation of the interaction of these molecules with their
local environment, the visualization of a biochemical or biological process, have
all become routinely possible through the use of appropriate instrumentation,
near-field microscopy, or confocal techniques [11]. In addition, the large variation in the photophysics of the available fluorophores connected with chemical
modifications give fluorescent probe techniques an almost unlimited scope for
the detection of specific molecules and the investigation of intermolecular interactions on a molecular scale.

This chapter focuses on recent developments in the design and applications of
fluorescent organic markers, such as coumarins, benzoxadiazoles, acridones, acridines, polyaromatics (naphthalene, anthracene, pyrene), fluoresceins, and rhodamines, which display maximum fluorescence emission in the UV/visible and have
been applied in the labeling of relevant biomolecules, namely DNA, RNA, proteins,
peptides, and amino acids, among others.

2 Coumarin Markers

2-Oxo-2*H*-benzopyrans, trivially designated as coumarins, represent one of the most
widespread and interesting class of heteroaromatic reagents for fluorescent labeling.
Organic probes built on the coumarin scaffold have been reported in the derivatization
of amino acids [12–14], peptides [15], nucleic acids [16, 17], as well as in studies with
proteins, namely enzymes [18–22]. Examples include 7-amino-4-methylcoumarin
(AMC) **1** derivatives, such as 7-amino-4-methylcoumarin-3-acetic acid (AMCA) **2**,
having a carboxylic acid as reactive group for derivatization and wavelength of
maximum excitation (λ_{ex}) at 350 nm, which was a widely used UV-excitable probe
for the fluorescent labeling of proteins [23]. AMCA **2** and its more recent analogue,
Alexa Fluor 350, displayed an intense blue fluorescence with a narrow emission peak
between 440 and 460 nm and showed excellent photostability (AMCA **2** is over three
times more photostable when compared with fluorescein).

AMC
1

AMCA
2

AMCA **1** is still being used. For example, Han and co-workers recently reported a fluorescence-based procedure designated as the "AMCA switch method," in which the S-nitrosylated cysteines are converted into AMCA fluorophore-labeled cysteines [24]. AMCA-HPDP **3** was used in the labeling step. The labeled proteins were then analyzed by nonreducing SDS-PAGE, and the S-nitrosylated molecules could be readily detected as brilliant blue bands under UV light. Furthermore, when combined with liquid chromatography-tandem mass spectrometry (LC-MS/MS), the S-nitrosocysteines can be identified with the recognizable AMC tag in the MS spectra.

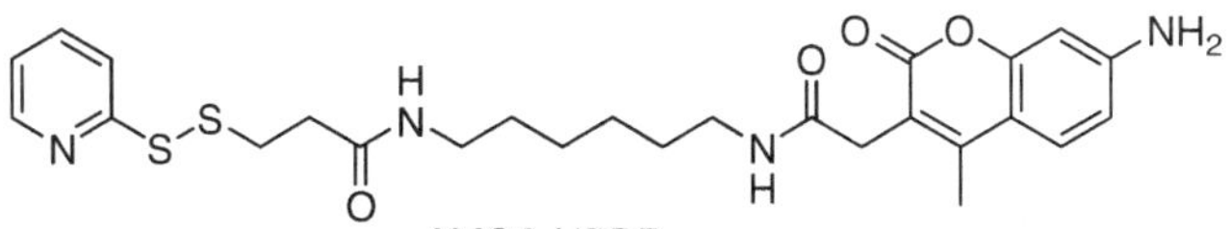

AMCA-HPDP
3

Violet-excitable derivatives of 7-hydroxycoumarin were developed and widely used in the preparation of fluorescent protein conjugates and enzyme substrates [25, 26]. Several 7-hydroxycoumarins conjugated to enzyme substrates were used for assays of phosphatases, β-galactosidases, and β-lactamases [27]. The 3-carboxy-7-hydroxycoumarin **4** has an excitation maximum at 386 nm, and the pK_a of its phenolic hydroxyl group is at about 7.5. In the physiological pH range of 6–8, the dye molecules are not fully deprotonated and, consequently, do not display their maximum fluorescence intensity. Studies of the influence of fluorination in the photochemical properties of 7-hydroxycoumarins demonstrated that fluorinated derivatives, namely 3-carboxy-6,8-difluoro-7-hydroxycoumarin (Pacific Blue **5**, wavelength of maximum absorption (λ_{abs})/wavelength of maximum emission (λ_{em}) 401/452 nm), have higher fluorescent quantum yields [28, 29] and red-shifted in excitation wavelengths $(\lambda_{ex} \sim 400$ nm).

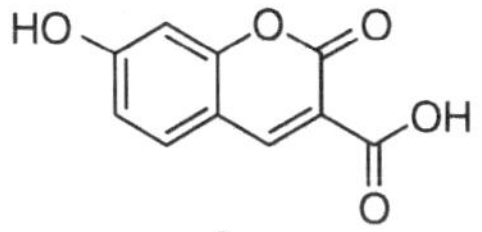

4 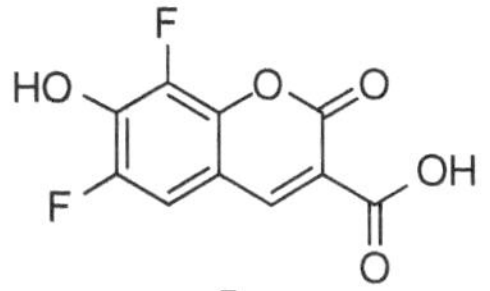   **5**

More recently, with the aim of searching for new violet-excitable dyes with improved photophysical and photochemical properties, three mono- and bis-halogenated hydroxycoumarins **6a–c** were synthesized, conjugated with antibodies, and cell analysis was screened using flow cytometry [30]. The monochlorinated hydroxycoumarin (V450) **6a** $(\lambda_{abs}/\lambda_{em}$ 404/448 nm, after reacting with 1 M glycine at pH 9.6 to stabilize its absorption maxima) was found to have a high fluorescence quantum yield $(\Phi_F \sim 0.98)$, and human leucocyte-specific monoclonal antibodies (CD3, CD4 and CD45) conjugated with this dye displayed reliable performance in flow cytometry assays.

The results reported showed that V450 **6a** is as fluorescent as, or more fluorescent than, the existing fluorophores with similar spectral characteristics (e.g., Pacific Blue **5**), whereas two of the three antibody clones tested demonstrated that a 20–30% gain in signal could be obtained by using V450 **6a**.

In addition, comparisons of photostability with conjugates made from existing dyes revealed good results for V450 **6a**. V450–antibody conjugates are also appropriate for use in multicolor immunophenotyping panels. Furthermore, this fluorophore proved to be compatible with protocols employing both BD FACS Lysing Solution and BD PharmLyse, and multicolour reagent mixtures containing V450–antibody conjugates were found to be functional and stable.

a R = H (V450)
b R = F
c R = Cl

6

The use of "click" chemistry is a promising strategy as a postsynthetic ligation for nucleic acids in order to circumvent the time-consuming synthesis of phosphoramidites as DNA building blocks [31, 32]. This is particularly relevant for several fluorophores that are unstable under the acidic, oxidative, or basic conditions of automated DNA phosphoramidite chemistry and DNA workup.

Coumarins are an example of brightly emitting organic fluorophores that are unstable under the typically strong basic conditions usually used during DNA cleavage and deprotection. Thus, the incorporation of these labels into oligonucleotides through the conventional phosphoramidite chemistry is difficult or unpraticable. Berndl and co-workers [33] reported the use of postsynthetic "click" chemistry in the modification of presynthesised alkynylated oligonucleotides **7** (bearing the alkyne group at the 2′-position of uridine) with the fluorescent azide **8** to prepare the corresponding modified duplexes **DNA1Y-DNA3Y**.

The UV/vis spectra of the modified single strands and duplexes showed an absorption maximum in the range between 515 and 534 nm. Duplexes bearing guanine as the counterbase (**DNA1G**) or as the base adjacent to the coumarin modification site (**DNA3Y**) showed the most red-shifted absorption, particularly significant in **DNA3G** (534 nm). The steady-state fluorescence spectra of the coumarin-modified duplexes displayed maxima in the range 606–637 nm. All modified duplexes exhibited a significant Stokes' shift of approximately 100 nm. The duplexes **DNA1Y** showed quantum yields in the range between 0.30 and 0.35, while Φ_F of the duplexes with adjacent G–C base pairs (**DNA3Y**) were lower (0.20–0.27).

Overall, the significant Stokes' shift of ~100 nm and the good quantum yields make the coumarin dye a powerful fluorescent probe for nucleic acids assays or cell biology. The postsynthetic "click" chemistry makes this fluorophore readily accessible for fluorescent labeling of nucleic acids.

7

8

5′ G-C-A-G-T-C-T-T-X- T-T-C-A-C-T-G-A 3′
3′ C-G-T-C-A-G-A-A-Y-A-A-G-T-G-A-C-T 5′

for Y = A X = X^1: **DNA1A**
X = T: **DNA2A**

5′ G-C-A-G-T-C-T-G-X-G-T-C-A-C-T-G-A 3′
3′ C-G-T-C-A-G-A-C-Y-C-A-G-T-G-A-C-T 5′

for Y = A X = X^1: **DNA3A**
X = T: **DNA4A**

X^1 =

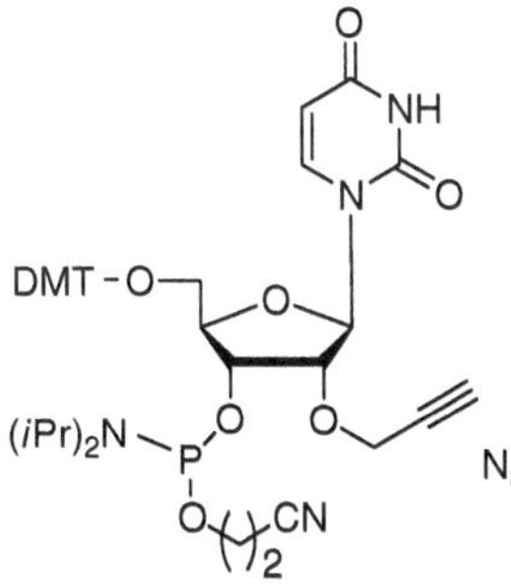

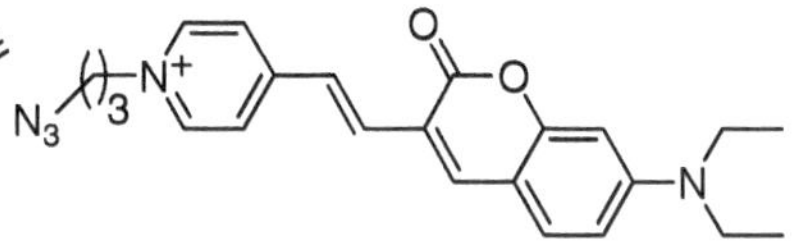

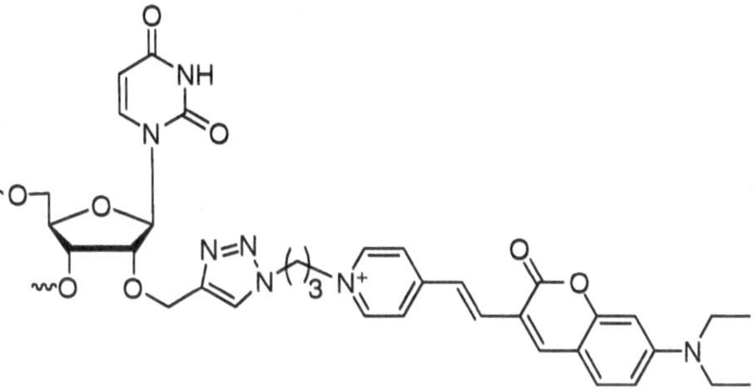

Recently, novel polymethine carbonyl-dyes based on coumarin moiety and their boron difluoride complexes **9a–d** and **10a–d** [34–36] were evaluated as fluorescent dyes for the detection of native proteins using bovine serum albumin (BSA) as a model protein, and as probes for the nonspecific detection of proteins using a BSA/sodium dodecyl sulfate (SDS) mixture [37]. Optical properties of these compounds in the absence and presence of BSA, as well as in SDS and BSA/SDS mixture, were measured in Tris–HCl buffer (pH 8.0) (Table 1).

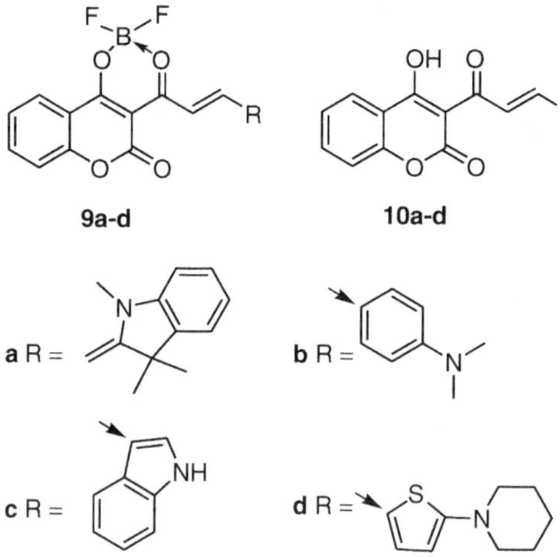

9a-d   10a-d

a R =   b R =

c R =   d R =

In the presence of BSA, coumarins **9a**, **10a**, **9b**, and **10c** showed a considerable shift of the bands in excitation and/or emission spectra, or with the appearance of new ones, when compared to the dyes spectra in buffer (λ_{ex} 413–585 nm, λ_{em} 462–712 nm, **9a–d** and **10a–d**). For the remaining coumarins, corresponding maxima positions were close to those observed in buffer.

In the excitation/emission spectra of coumarins, two bands were observed, with the exception of dyes **9a**, **10b**, and **10d**, which possess a single band. Fluorescence excitation maxima of studied dyes were found to be between 402

Table 1 Fluorescence properties of dyes **9a–d** and **10a–d** (5×10^{-6} M) solutions in buffer (0.05%), BSA (0.2 mg/mL), and BSA–SDS mixture (0.05% and 0.2 mg/mL, respectively)

Dye	In buffer		In SDS presence		In BSA presence		In BSA/SDS presence	
	λ_{ex} (nm)	λ_{em} (nm)	λ_{ex} (nm)	λ_{em} (nm)	λ_{ex} (nm)	λ_{em} (nm)	λ_{ex} (nm)	λ_{em} (nm)
9a	585	712	574	586	576	590	572	588
10a	453	525	–	–	462	520	–	–
	496	588	–	–	–	–	–	–
	–	–	540	564	523	555	540	562
9b	413	462	400	458	432	524	405	467
	–	–	–	–	574	606	540	680
10b	418	552	435	551	429	533	–	–
	520	560	520	588	–	–	511	578
9c	–	–	–	–	402	480	–	–
	517	539	515	550	518	552	471	537
10c	404	493	–	–	407	481	408	510
	470	522	472	505	464	514	471	516
9d	470	556	–	–	470	552	–	–
	564	610	560	607	574	614	603	611
10d	464	548	523	591	471	552	–	–
	547	586	576	596	–	–	558	589

and 576 nm and emission maxima were located in the range 480–614 nm. Stokes shifts values were between 14 and 104 nm. All coumarins increased their fluorescence intensity in the presence of BSA but displayed a low to moderate intensity level of fluorescence. The highest intensity increase and the brightest emission in the BSA complex were observed for the unsubstituted dyes **10d** (about 63 times) and **10a**, respectively.

Studies of fluorescence properties of the dye pair (i.e., boron difluoride complexes dye **9a–d** and the nonsubstituted one **10a–d**) in the BSA/SDS mixture revealed that coumarins **9b** and **10c** showed two excitation and emission bands, while other heterocycles showed single bands in the corresponding spectra. For all studied compounds, excitation and emission maxima occurred in the range 405–603 nm and 467–680 nm, respectively.

In the emission spectra of coumarins **9a**, **10a**, **10c**, and **9d** in BSA/SDS mixture, the most pronounced bands revealed similar wavelengths to those in the presence of BSA. Observed Stokes' shifts values for dyes **9a–d** and **10a–d** occurred between 8 and 102 nm. The fluorescence increase of the dye upon BSA/SDS addition varied from 2.5 times (**9b**) to 330 times (**10d**), while the brightest complex resulted from dye **10a**, its quantum yield being about 0.27. Despite the observed emission enhancement value, the fluorescence intensity of other coumarin dyes in the presence of BSA/SDS mixture was only low to moderate.

From the results obtained, it was found that compound **10a** showed very high fluorescence intensity in the presence of the BSA and BSA/SDS mixture (Φ_F 0.27) together with a noticeable emission enhancement. The presence of dimethyl indolenyl increased the affinity of the dyes to both native and denatured proteins. The authors proposed compound **10a** for further studies as fluorescent probes for protein detection.

Functionalized benzocoumarins **11a–c** were reported as fluorogenic reagents for several L-amino acids, namely phenylalanine, glycine, alanine, and valine. The resulting fluorescent ester conjugates displayed λ_{abs} values in the range 345–360 nm and λ_{em} values in the range 411–478 nm with high Stokes' shifts (66–131 nm) and Φ_F from 0.13 to 0.70 [38].

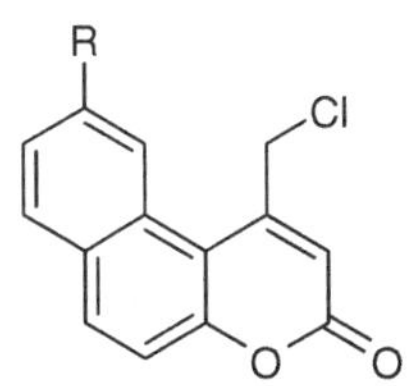

a R = H Obb-Cl
b R = OH Obh-Cl
c R = OMe Obm-Cl

11

Recently, chloromethylated benzocoumarin **11c**, hydroxylmethylated benzocoumarin **12**, and chloromethylated coumarin **13** were used in the efficient preparation of several fluorescent ester conjugates of N-benzyloxycarbonyl–neurotransmitter amino acids, such as β-alanine, tyrosine, 3,4-dihydroxyphenylalanine (DOPA), glutamic acid, and γ-aminobutyric acid (GABA) [39, 40].

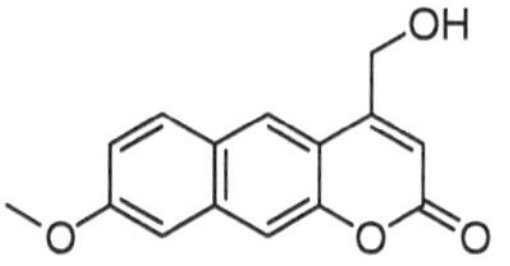

The resultant bioconjugates displayed absorption and emission maxima in the range 320–349 nm and 393–503 nm, respectively, with high Stokes' shifts (73–158 nm), and Φ_F were moderate to excellent (0.14–0.76).

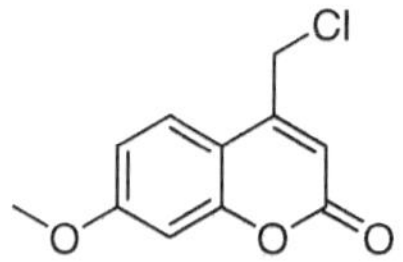

Considering photophysical properties, all labels are appropriate fluorogenic reagents for the derivatization of nonfluorescent molecules, such as the neurotransmitter amino acids used in the study, benzocoumarin **11c** probably being the most interesting fluorophore owing to the obtained results in the fluorescent GABA conjugates.

3 Benzoxadiazole, Acridone, and Acridine Markers

Nitrogen heterocycles are especially interesting, since they constitute an important class of natural and nonnatural products, many of which exhibit useful optical properties; these have been recently synthesized and evaluated as probes in bioassays purposes [41–44]. Despite the diversity of these compounds, benzoxadiazole, acridone, and acridine fluorophores were chosen as a focus in biolabeling applications.

Heterocyclic fluorophores based on the benzoxadiazole nucleous, namely 4-nitrobenz-2-oxa-1,3-diazole (NBD) **14** derivatives/analogs, have been widely used as derivatization reagents for analysis purposes. Examples include the amino- or thiol reactive 4-fluoro-7-nitrobenz-2-oxa-1,3-diazole (NBD-F) **15** and 4-chloro-7-nitrobenz-2-oxa-1,3-diazole (NBD-Cl) **16** [45–50] and the thiol-reactive *N*-((2-(iodoacetoxy)ethyl)-*N*-methyl)amino-7-nitrobenz-2-oxa-1,3-diazole (IANBD ester) **17** [51] and 7-chlorobenz-2-oxa-1,3-diazole-4-sulfonate (SBD-Cl) **18** [52]. NBD-F and NBD-Cl derivatives can be excited at about 470 nm by using the relatively inexpensive and reliable argon ion lasers or newer diode pumped solid state (DPSS) lasers. NBD-F has been used as a labeling tag in various capillary electrophoresis (CE) experiments for amino acids [53–57] including the monitorization of in vivo dynamics of amino acids neurotransmitters [58].

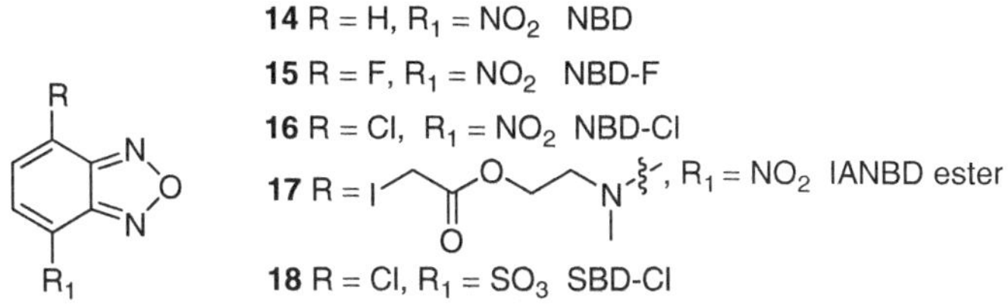

14 R = H, R$_1$ = NO$_2$ NBD
15 R = F, R$_1$ = NO$_2$ NBD-F
16 R = Cl, R$_1$ = NO$_2$ NBD-Cl
17 R = [structure], R$_1$ = NO$_2$ IANBD ester
18 R = Cl, R$_1$ = SO$_3$ SBD-Cl

Recently, Guminski and co-workers [59] reported the synthesis of a series of conjugated spermine (natural polyamide, which is involved in cellular metabolism) derivatives with benzoxadiazole (NBD-Cl), phenylxanthene, and bodipy fluorophores. As it is well known, internalization of polyamines from extracellular sources can compensate the low availability of such substances within the cell. This cellular uptake process involves the highly regulated polyamine transport system (PTS), which is hyperactive in most cancer cells. By using these probes, fluorescence detected within the cells would be expected to be dependent upon the efficiency of the PTS activity. As a result, the positively marked cancer cells would be expected to be more sensitive to polyamine conjugated antitumor compounds. The fluorescent probes described were comparatively evaluated to identify the activity of the PTS. Owing to the results obtained, N_1-methylspermine-NBD conjugate **19** was identified and selected as the superior tool for identifying cancer cells with high PTS activity being recently used in a clinical context.

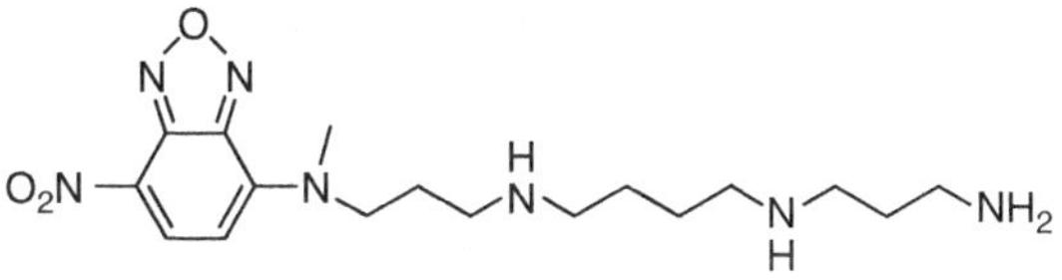

19

N-Acetyl-L-aspartyl-L-glutamate (NAAG) is a dipeptide neurotransmitter, abundantly present in the mammalian brain, which acts as an endogenous agonist in the type3 metabotropic glutamate receptor, cleaved enzymatically to *N*-acetyl-L-aspartate (NAA) and glutamate (Glu) by glutamate carboxypeptidase (GCP) II or III [60]. Fukushima and co-workers [61] have reported a sensitive determination method of NAA by using high-performance liquid chromatography (HPLC), with precolumn fluorescent derivatization using 4-*N*,*N*-dimethylaminosulfonyl-7-*N*-(2-aminoethyl)amino-2,1,3-benzoxadiazole (DBD-ED) **20** [62].

NAAG and NAA are of extreme importance in neuronal functions and their simultaneous determination can provide useful information on the etiology of neurological diseases. Arai and collaborators [63] extended the method previously reported in the simultaneous determination of NAAG and NAA in the rat brain homogenate sample. Detection wavelength was set at 562 nm with 438 nm of excitation for both DBD-NAAG **21** and DBD-NAA **22**. The detection limits of NAAG and NAA were about 12 and 34 fmol on the column, respectively (signal to noise ratio 3).

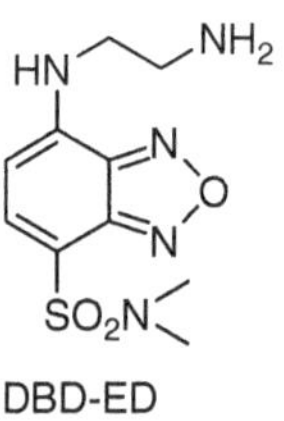

DBD-ED
20

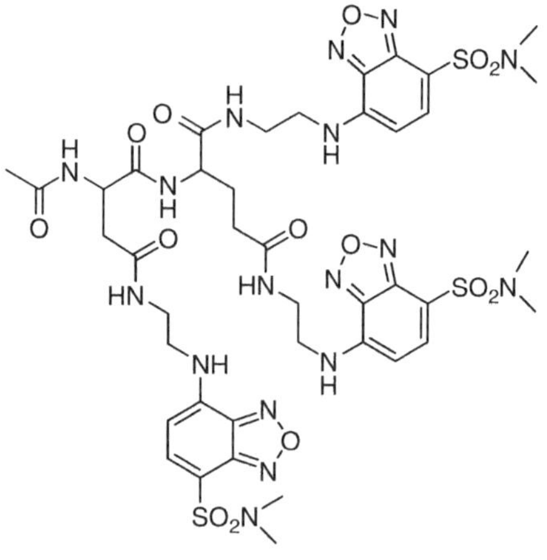

DBD-NAAG
21

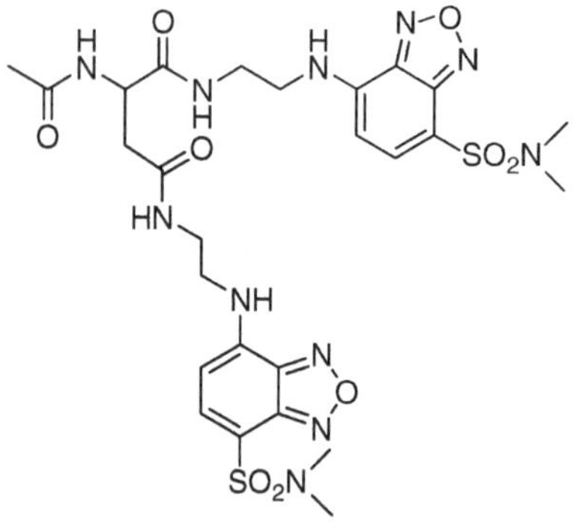

DBD-NAA
22

Acridone and its derivatives are planar tricyclic heteroaromatic molecules with no charge, showing intense fluorescence and stability against photodegradation, oxidation, and heat [64, 65]. Several acridone derivatives have been synthesized and used in the fluorescent labeling of amino acids [66], peptides [67], antibodies [68], and DNA [69, 70]. Shoji and co-workers [69] showed that the triphosphate of a thymine analog bearing the acridone moiety can be incorporated into DNA during the polymerase chain reaction, enzymatically forming multiacridone-labeled DNA. Although there are some sequence limitations, it was also reported that acridone-tagged DNA [66] can be used as a base-discriminating fluorescent label for single-nucleotide polymorphism.

Following previous work, Hagiwara and collaborators [71] recently prepared 5′-terminal acridone-labeled DNAs, using the succinimidyl ester **24** of the acridone acetic acid **23** reported before [69], and evaluated their use as donors for a fluorescence resonance energy transfer (FRET) system in combination with 3′-dabcyl-tagged DNA as an acceptor, which can detect the target DNA by emission-quenching caused by FRET. Acridone **23** showed maximum absorption at 388 and 407 nm, and high fluorescence with maximum emission at 422 and 448 nm (λ_{ex} 388 nm) and Φ_F 0.89 in water. For the corresponding acridone-tagged DNAs, two peaks were also found in the fluorescence spectra at about 425 and 450 nm in response to excitation at 388 nm (in aqueous saline solution at pH 7.6).

Considerable amounts of quenching of the acridone emissions by guanine in the DNA occurred when guanine was close to acridone, which can be applied as a quencher-free probe (no additional quencher is required) for the detection of a special sequence of DNA. The DNA bearing acridone at the C5 position of inner thymidine could distinguish the opposite T–T base mismatch, while enhancement of discrimination ability is needed for the practical use of single nucleotide polymorphism (SNP) typing.

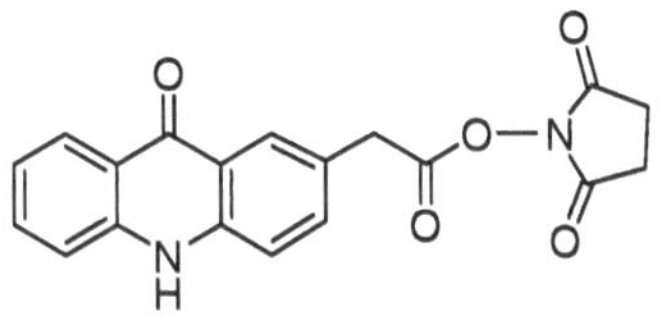

24 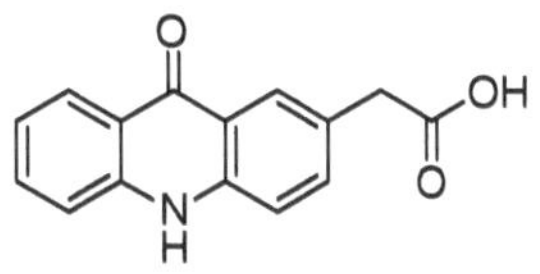   **23**

Although biotin-fluorophores have been described in the detection of biomolecules, namely proteins, peptides, or DNA, most of these loose part of their fluorescence when binding to the tetrameric proteins avidin and streptavidin.

Agiamarnioti and co-workers [72] synthesized a novel biotinylated fluorophore, 10-(2-biotinyloxyethyl)-9-acridone **25** with favorable properties for bioanalytical applications. In aqueous solutions, it displayed high fluorescence (Φ_F 0.45, λ_{ex} 395 nm and λ_{em} 425 nm) and retains about 60% or 25% of its fluorescence after binding to avidin or streptavidin, respectively.

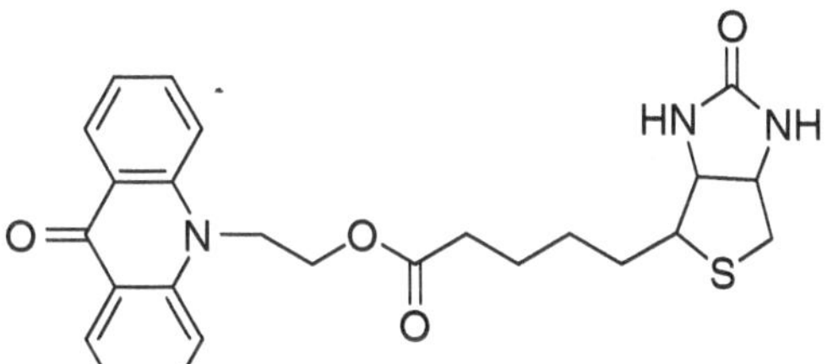

25

Acridine and its derivatives are also fused nitrogen heterocycles similar to acridones, which display a high fluorescence quantum yield and possess the ability to intercalate tightly, though reversively, to the DNA helical structure [73], with large binding constants [74]. As a result, acridine dyes are recognized in the field of the development of probes for nucleic acid structure and conformational determination [75–77].

Recently, Wu and co-workers [78, 79] reported the synthesis of dyes **26** [78] and **27** [79] based on the acridine skeleton and the investigation of their interaction with nucleic acids. These two dyes displayed intense fluorescence (Φ_F 0.28 **26** and 0.26 **27**, in buffer solution, pH 7.2) and water solubility. It was found that the fluorescence intensity of dyes **26** and **27** were quenched in the presence of DNA. A method for DNA determination based on the quenching fluorescence of these dyes (λ_{ex} 260 nm, λ_{em} 451 nm, **26**; λ_{ex} 258 nm, λ_{em} 451 nm, **27**) was established. Under optimal conditions, the linear range was $0.05–2.0$ µgmL^{-1} **26** and $0.1–4.0$ µgmL^{-1} **27** of both fish semen (fs DNA) and calf thymus DNA (ct DNA). The corresponding determination limits are 9.1 ngmL^{-1} **26** or 4.6 ngmL^{-1} **27** for fs DNA and 8.7 ngmL^{-1} **26** or 5.1 ngmL^{-1} **27** for ct-DNA. The results suggested that the interaction mode between dye **26** and DNA was intercalative binding (with a large binding constant), while in the case of compound **27** groove binding occurred.

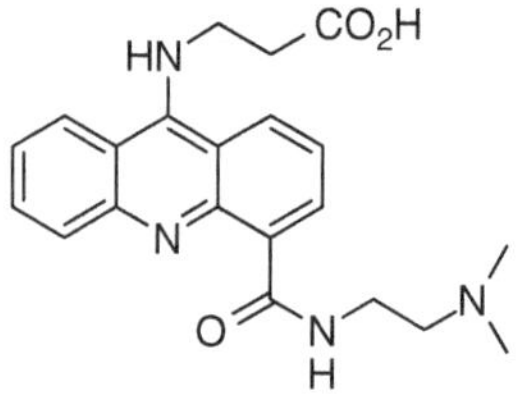

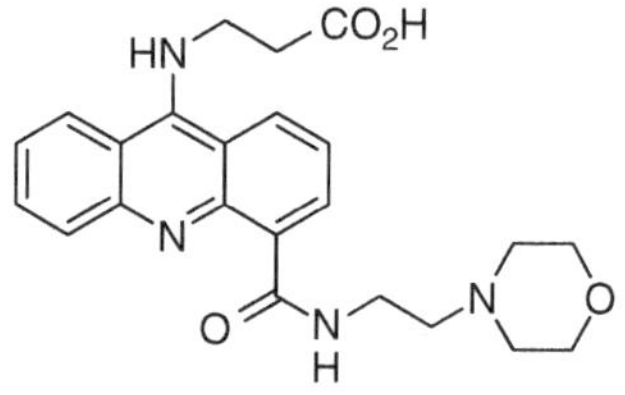

4 Polyaromatic Markers

Polycyclic aromatic compounds, namely naphthalene, anthracene and pyrene derivatives are widely used as fluorescent probes in relevant biomolecules.

The amine-reactive 5-(dimethylamino)naphthalene-1-sulfonyl (dansyl) chloride **28** [80] and related fluorophores [81, 82], as well as the 5-((2 aminoethyl)amino) naphthalene-1-sulfonic acid (EDANS) **29**, are included in the naphthalene fluorophore family. Derivatives of the latter, such as compound **30**, exhibit a λ_{max}/λ_{em} 336/520 nm, molar absorptivity (ε) of 6.1 × 10^3 M^{-1} cm^{-1}, and a fluorescent quantum yield of 0.27 in water [83]. The use of EDANS is particularly interesting in FRET experiments [84, 85]. Furthermore, 4-amino-3,6-disulfonylnaphthalimides (e.g., Lucifer yellow **31**), associated to a longer absorption (λ_{max} 428 nm) [86] are suitable polar tracers [87].

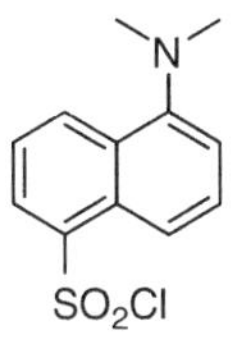

dansyl-chloride
28

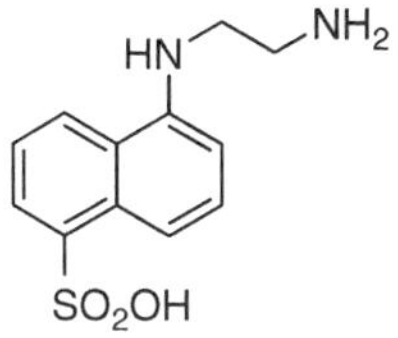

EDANS
29

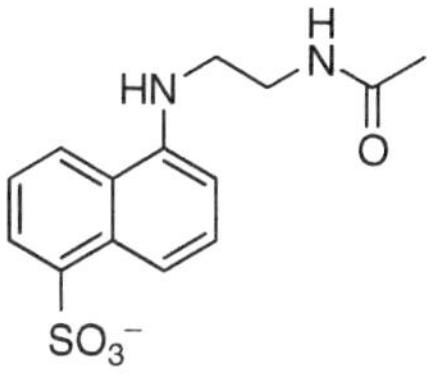

30

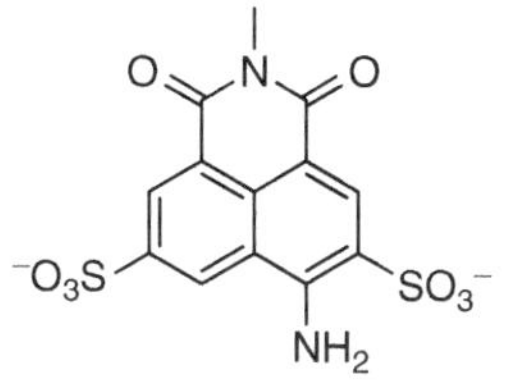

Lucifer yellow
31

New ratiometric fluorescent cysteine probes based on naphthalene–thiourea–thiadiazole (NTTA) **32** fluorescent organic nanoparticles (FONs) were reported by Li and co-workers [88]. NTTA FONs show good water solubility, high sensing selectivity, and a bathochromic shift (74 nm) of fluorescence emission upon binding cysteine in aqueous media was detected. Under optimum conditions, linear relationships were found to exist between the relative fluorescence intensity ratio I_{430}/I_{356} and the logarithmic concentration of cysteine in the range of 0–100 μM; the detection limit was found to be 1.5×10^{-9} M, while the fluorescence intensity of NTTA FONs to other amino acids was insignificant.

This is the first Cys fluorescent sensor derived from FONs, in which the fluorescence enhancing property is in conjunction with a remarkable red-shifted fluorescence emission. Despite the potential sources of error when considering complicated clinical samples, the authors believe that this probe can be applied to study the effects of Cys in a biological system.

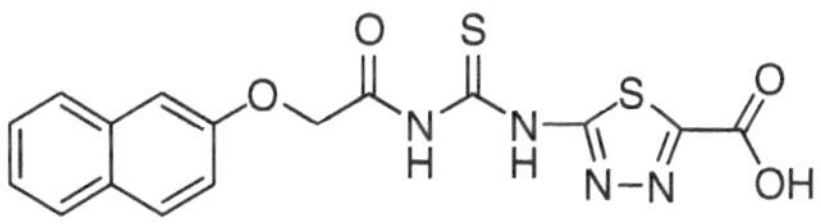

NTTA
32

Also recently, Liao and collaborators [89] proposed a homogeneous noncompetitive assay of a protein in biological samples based on FRET by using its tryptophan residues as intrinsic donors and its specific fluorescent ligand as the FRET acceptor, which was defined as an analytical FRET probe. To evaluate this method, a naphthylamine derivative, namely *N*-biotinyl-*N'*-(1-naphthyl)-ethylenediamine (BNEDA) **33** was used as an analytical FRET probe for the homogeneous noncompetitive assay of streptavidin.

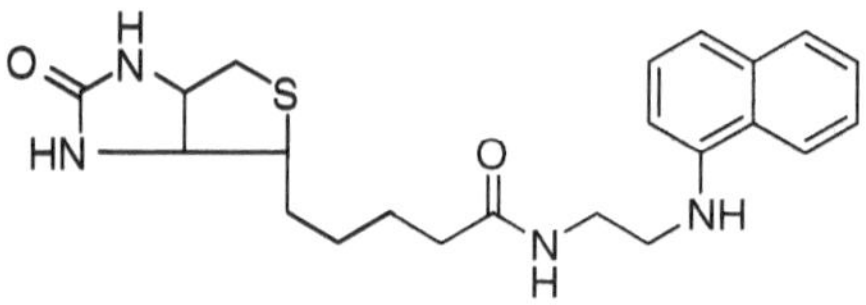

BNEDA
33

Through excitation at 280 nm, free BNEDA produced negligible fluorescence at 430 nm. However, the bound BNEDA produced a much higher stable fluorescence at 430 nm after 2 min of a binding reaction. The competitive binding between BNEDA and biotin gave the dissociation constant of (16 ± 3) fM for BNEDA. Through excitation at 280 nm, fluorescence at 430 nm of reaction mixtures containing 32.0 nM BNEDA responded linearly to streptavidin subunit concentrations ranging from 0.40 to 30.0 nM with the desirable resistance to common interferences in biological samples. As a result, by using tryptophan residue(s) as intrinsic donor (s) in a protein of interest and its fluorescent ligand as the corresponding FRET acceptor, this homogeneous noncompetitive assay of the protein in biological samples was effective and advantageous.

A novel and simple method based on site-directed fluorescence labeling using the BADAN label, allowing for the examination of protein–lipid interactions in great detail, was described by Koehorst and co-workers [90]. This methodology was applied to an embedded membrane, the M13 major coat protein. In a high-throughput approach, a total of 40 site-specific cysteine mutants were prepared and labeled with BADAN **34**. These mutants covered 80% of the total number of amino acid residues in the primary sequence of the 50-residues long protein. BADAN-labeled M13 coat protein mutants were reconstituted into phospholipid bilayers. The steady-state fluorescence spectra were analyzed using a three-component spectral model that enabled the separation of Stokes' shift contributions from water and internal label dynamics, and protein topology. Analysis of these data revealed the embedment and topology of the labeled protein in the membrane bilayer under various conditions of a headgroup charge and lipid chain-length, as well as key characteristics of the membrane such as hydration level and local polarity, provided by the local dielectric constant.

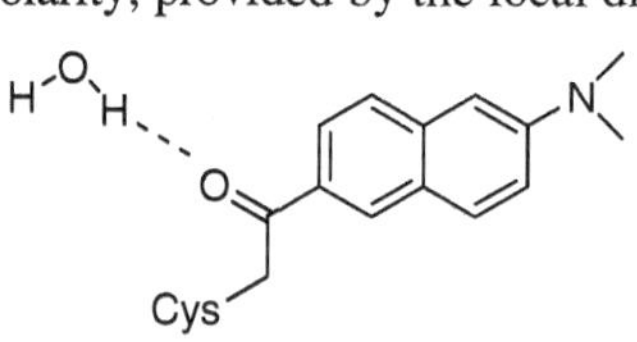

34

The dansyl derivative 9-azidononyl-5-(dimethylamino)naphthalene-1-sulfonate **35** was used by Yi and collaborators [91] as an azido-fluorescent label in a tandem method of sulfonium alkylation and "click" chemistry for the modification of biomolecules. Fluorescent labeling of a protein was successfully carried out after simple incubation of BSA with sulfonium salt **36** followed by azido-containing fluorophore **35**, at room temperature.

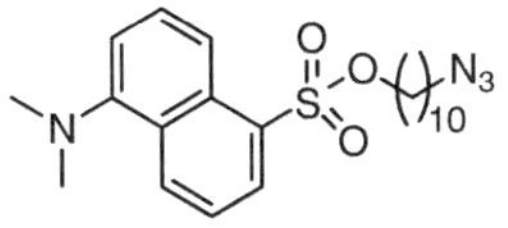

35 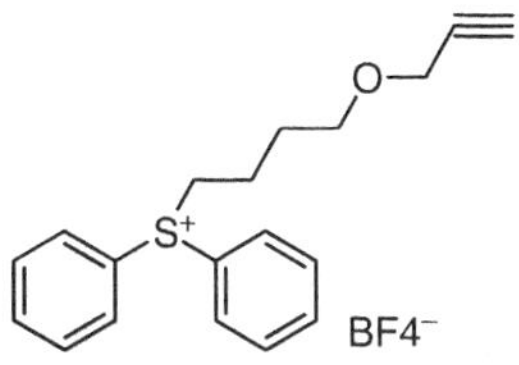   **36**

The photoinduced electron transfer (PET) process of anthracenes has been widely applied in ion sensing [92], singlet oxygen detection [93], screening catalysts [94], molecular logic gate construction [95], and cell labeling by recognition of carbohydrates [96].

Xie and co-workers [97] investigated the possibility of designing a new type of fluorogenic reaction based on the PET process of anthracenes. Owing to the high electron density of the α-nitrogen of azido group [98, 99], it was envisioned that the introduction of the azido group close to the anthryl core via a nonconjugated linker would lead to a favorable electron transfer from the azido to the excited anthryl core and induce the quenching of fluorescence. After the Cu (I)-catalyzed azide-alkyne cycloaddition (CuAAC), the lone pair electrons on nitrogen will become part of the aromatic system and become a much weaker electron donor. Therefore, the fluorescence of the anthryl core will be remediated resulting in a fluorogenic phenomenon. Thus, the authors reported the synthesis of five novel azido derivatized anthracenes **37–41** and used the CuAAC reaction to activate their fluorescence. This method was applied to label multiple alkyne organic molecules and large alkyne functionalized biological macromolecules, such as the cowpea virus (CPMV) and the tobacco mosaic virus (TMV).

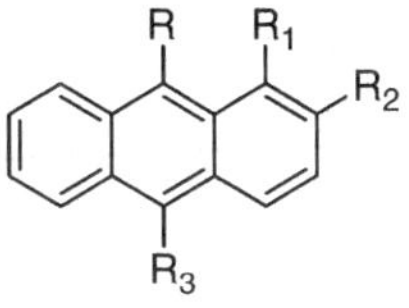

37 R = R$_1$ = R$_2$ = H, R$_3$ = CH$_2$N$_3$
38 R = CH$_2$OH, R$_1$ = R$_2$ = H, R$_3$ = CH$_2$N$_3$
39 R = CN, R$_1$ = R$_2$ = H, R$_3$ = CH$_2$N$_3$
40 R = H, R$_1$ = CH$_2$N$_3$, R$_2$ = R$_3$ = H
41 R = R$_1$ = H, R$_2$ = N$_3$, R$_3$ = H

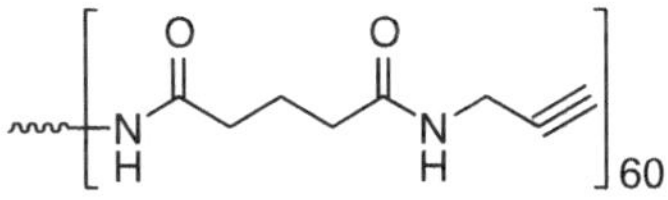

CPMV- Alkyne
42

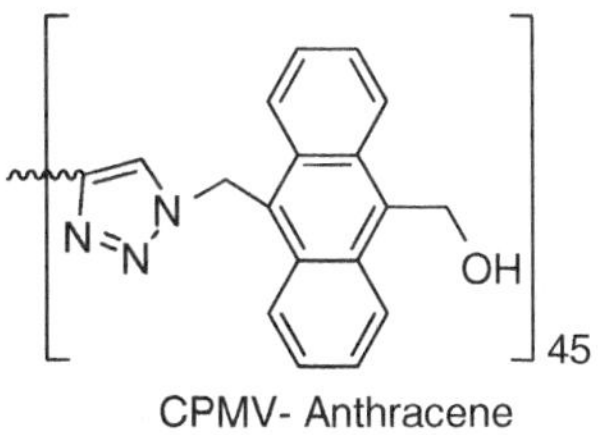

CPMV- Anthracene
43

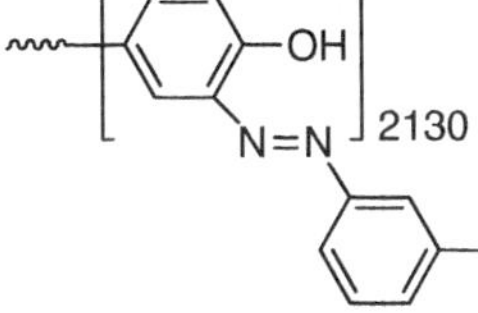

TMV- Alkyne
44

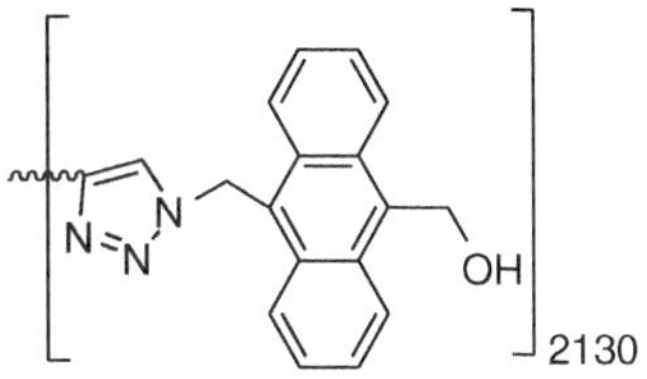

TMV- Anthracene
45

The synthesized CPMV-alkyne **42** was subjected to the CuAAC reaction with **38**. Due to the strong fluorescence of the cycloaddition product **43** as low as 0.5 nM, it could be detected without the interference of starting materials. TMV was initially subjected to an electrophilic substitution reaction at the *ortho*-position of the phenol ring of tyrosine-139 residues with diazonium salts to insert the alkyne functionality, giving derivative **44** [100]. The sequential CuAAC reaction was achieved with greatest efficiency yielding compound **45**, and it was found that the TMV remained intact and stable throughout the reaction.

Pyrene fluorophores are also used as probes. Derivatives of pyrene show λ_{max}/λ_{em} 340/376 nm, ε 4.3 × 10^4 M^{-1} cm^{-1}, and Φ_F 0.75 [87, 101], and due to its environmental sensitivity, this fluorophore can be used to report on RNA folding [102]. Pyrene also displays a long-lived excited state ($\tau > 100$ ns), which allows for an excited pyrene molecule to associate with a pyrene in the ground state. The resulting eximer exhibits a red-shift in fluorescence intensity (λ_{em} ∼490 nm). This characteristic can be used to study important biomolecular processes, such as protein conformation [103].

DNA and RNA quantification, SNP typing, hybridization, and structural alteration have been widely carried out by modified oligonucleotides possessing pyrene derivatives [104–113]. As is known, pyrene-1-carboxaldehyde fluorescence is considerably dependent on solvent polarity [114], being strong in methanol but insignificant in nonpolar solvents [115]. Owing to this property, Tanaka and collaborators developed a pyrenecarboxamide-tethered modified DNA base, ^{Py}U **46**, and applied it to SNP discrimination in DNA [116–120].

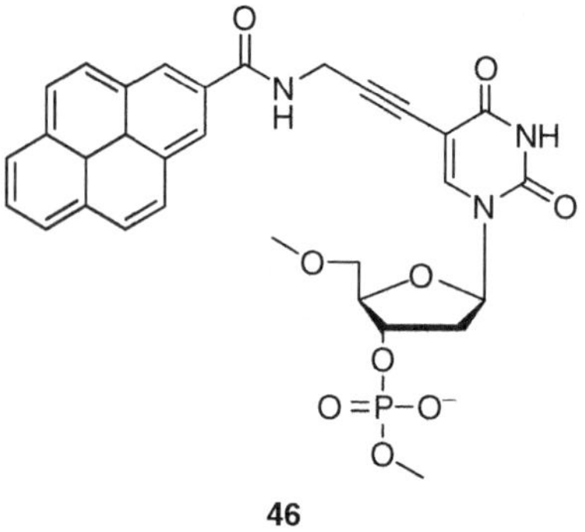

46

Fluorescence probes possessing the ^{Py}U base **46** selectively emit fluorescence only when the complementary base is adenine. In this case, the chromophore of ^{Py}U is extruded to the outside of the duplex because of Watson–Crick base pair formation, and exposed to a highly polar aqueous phase. On the contrary, the duplex containing a ^{Py}U/N (N = G, C and T) mismatched base pair shows a structure in which the glycosyl bond of uridine is rotated to the syn conformation. In this conformation, the fluorophore is located at a hydrophobic site of the duplex. The control of base-specific fluorescence emission is based on the polarity change in the microenvironment where the fluorophore locates are dependent on the ^{Py}U/A base-pair formation.

The fluorescence of the ^{Py}U-containing probe ((pODN 5'-d(GGGCGGGTTTT TTTTTCCC ^{Py}U CCCTTTTTTTTTT)-3')) was quenched in the absence of poly

(A) tracts, whereas conformation of the probe changed in the presence of an RNA poly(A)tract (alteration on the micropolarity around a pyrenecarboxamide chromophore caused by hybridization with the targeted RNA) and a much higher fluorescence was observed.

Fluorescence measurement using this probe does not require a fluorescence quencher or washing process to suppress the fluorescence emission from nonbinding probes and nonspecific binding probes, which would be advantageous for the detection of mRNAs with poly(A) tracts in cells.

8-(Pyren-1-yl)-2′-deoxyguanosine (Py-G) 47 [121, 122] was incorporated synthetically as a modified DNA base and optical probe into oligonucleotides. A variety of Py-G-modified DNA duplexes have been studied by methods of optical spectroscopy.

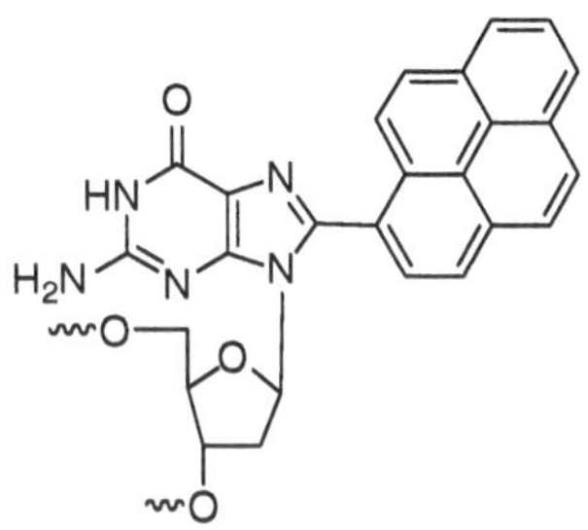

Py-G
47

The DNA duplex hybridization can be observed by both fluorescence and absorption spectroscopy since the Py-G group 47 exhibits altered properties in single strands versus double strands for both spectroscopy methods. In maximum absorption, a bathochromic shift of 5–20 nm was observed from single-stranded to duplex DNA (λ_{max} 350–370 nm; in buffer 5 mM Na_2HPO_4, pH 7.0). The fluorescence (λ_{em} ~450 nm, λ_{ex} 360 nm, in the buffer solution mentioned before) enhancement upon DNA hybridization can be improved significantly by the presence of 7-deazaguanin as an additional modification and charge acceptor three bases away from the Py-G 47 modification site. Furthermore, Py-G 47 can be applied in DNA as a photoinducable donor for charge transfer processes when indol is present as an artificial DNA base and charge acceptor. Correctly base-paired duplexes can be discriminated from mismatched ones by the comparison of their fluorescence quenching.

Overall, it can be envisioned that the Py-G group 47 represents an important label for the time-resolved studies of DNA dynamics and stacking interaction [123] and could be applied especially for assays in which conformational changes or base-flipping processes are essential in observation, such as in the investigation of DNA–protein complexes with DNA repair proteins.

5 Fluorescein Markers

The xanthene dye fluorescein 48 ($\lambda_{max}/\lambda_{em}$ 490/512 nm, ε 9.3 $\times$ 10^{-4} M^{-1} cm^{-1}, Φ_F 0.95, in water) [124] is one of the most widely used fluorophores in modern biochemistry and biological and medicinal research. This is due to the facility of obtaining it at a low price, its high absorptivity, and excellent fluorescence quantum yield. Furthermore, fluorescein has an excitation maximum that exactly matches the 488 nm argon-ion laser. However, because of the transformation between quinone and spirolactone forms [125, 126], fluorescein presents some drawbacks, such as photobleaching and pH-dependence, which limits its sensitivity in several situations.

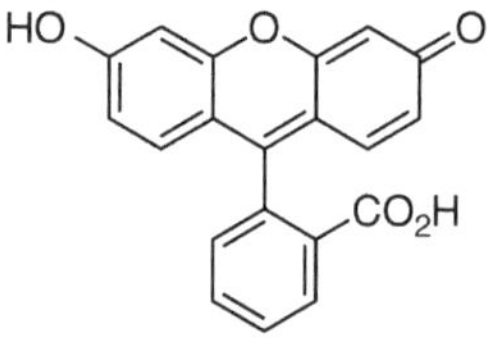

48

Among the commercially available fluorescein derivatives that have been widely used are fluorescein isothiocyanate (FITC) **49**, carboxyfluorescein (FAM) succini-midyl ester **50**, and fluorescein dichlorotriazine (DTAF) **51**. FITCs are the most commonly used fluorescein derivatives. They have been used to react with sulfhy-dryl [127], targeting reduced cysteine chains and especially amino groups in peptides or proteins [128].

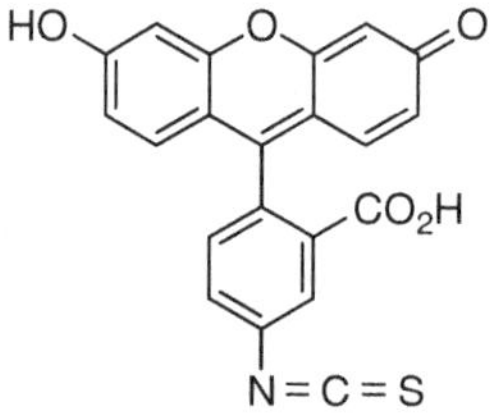

FITC
49

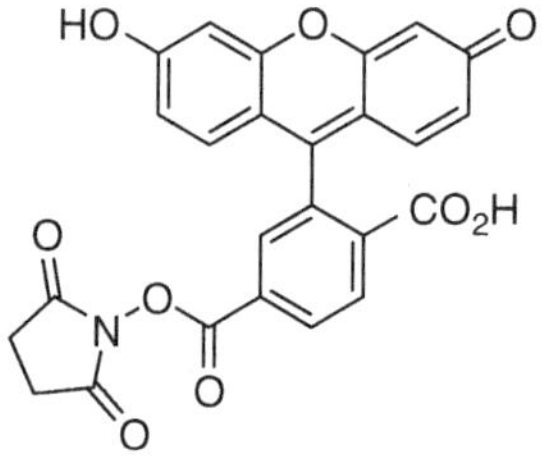

FAM-derivative
50

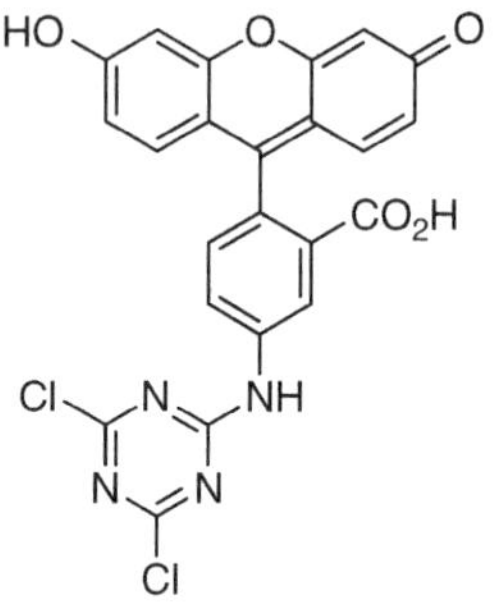

5-DTAF
51

For numerous applications, it is convenient to introduce the fluorescent label during chemical synthesis [129, 130]. FITC may react during the course of the solid phase peptide synthesis, either with a lysine or an ornithine side chain after the selective unmasking of the protecting group [131] or with a primary amino group at the N-terminal side of the growing peptide. In the latter case, an alkyl spacer such as aminohexanoic acid (Ahx) is introduced between the last amino acid and the thiourea linkage generated through the reaction of isothiocyanate and amine [132].

This is traditionally justified since it removes the bulky fluorescent dye from the bioactive sequence. Nonacidic cleavage conditions in the release of targeted peptide from the resin is also a strategy when FTIC is used in solid phase peptide synthesis.

Oligonucleotide conjugates of FITC are frequently employed as hybridization probes [133]. Peptide conjugates of FITC and other fluorescent isothiocyanates are susceptible to Edman degradation (method where the amino-terminal residue is labeled and cleaved from the peptide without cleaving the peptide bonds between other amino acid residues), making them useful in high-sensitivity amino acid sequencing [134]. FITC-labeled amino acids and peptides have been separated by capillary electrophoresis, with a detection limit of fewer than 1,000 molecules [135, 136]. FITC has also been used to detect proteins in gels [137, 138] and on nitrocellulose membranes [139, 140]; FITC is also a selective inhibitor of several membrane ATPases [141, 142].

FAM **50** reacts faster than FITC **49** and produces conjugates with a higher resistance against hydrolysis [143]. It has been used in bioconjugation approaches, including DNA detection [144]. DTAF **51** is highly reactive with proteins [143, 145]. Unlike other reactive fluoresceins, it reacts both with amino groups as well as thiol groups and even directly with hydroxyl groups, such as polysaccharides and other alcohols in an aqueous solution at pH above 9 [146, 147].

Different substituents on the carboxy-functionalized fluorescein can be introduced to produce marked alterations in the absorbance and fluorescence emission wavelengths, as well as in other physical properties. The selective substitution of chlorine for aromatic hydrogen has been found to increase fluorescence efficiency and to narrow emission and absorbance maxima when compared with fluorescein **48**, which is useful in multicolor imaging.

Tian and co-workers [148] have reported the synthesis of two chlorinated fluoresceins probes: 4,7,2′,7′-tetrachloro-6-(5-carboxypentyl) fluorescein and 4,7,4′,5′-tetrachloro-6-(5-carboxypentyl) fluorescein for labeling proteins. More recently, the same research group has described the synthesis of two novel chlorinated fluoresceins, namely 2′,4′,5′,7′-tetrachloro-6-(5-carboxypentyl)-4,7-dichloro fluorescein succinimidyl ester **52** and 2′,4′,5′,7-tetrachloro-6-(3-carboxypropyl)-4,7-dichlorofluorescein succinimidyl ester **53** [149].

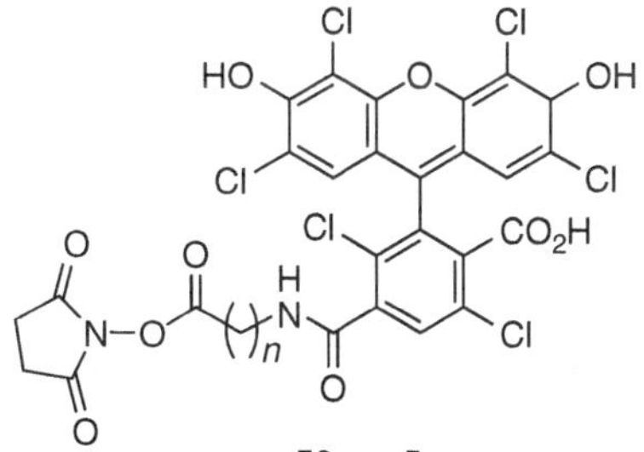

52 $n = 5$
53 $n = 3$

The novel chlorinated fluorescein succinimidyl esters **52** and **53** are considerably stable if properly stored. They exhibit intermediate reactivity toward amines, with high selectivity toward aliphatic amines. Their reaction rate with aromatic amines, alcohols, phenols, and histidine is relatively low.

These compounds have the same absorption (535 nm) and emission (550 nm) maxima. It was observed that the presence of a spacer of 6-aminohexanoic acid or 4-aminobutanoic acid significantly reduces the fluorescence quenching effect of chlorinated fluoresceins on protein, even at relatively high degrees of labeling. Compound **52** was used in the labeling of U2OS cells, and the results revealed that it has strong fluorescence and biocompatibility, thus constituting a useful substitute for fluorescein **48** in fluorescent imaging applications.

Kamoto and collaborators [150] designed and synthesized a new "turn-on" fluorescent probe **54**, which is excitable by visible light (λ_{ex} 490 nm), based on a fluorescein derivative **55**. It is composed of a metal–nitrilotriacetic (NTA) complex as the hexahistidine tag recognition site, fluorescein as the fluorophore, and a linker.

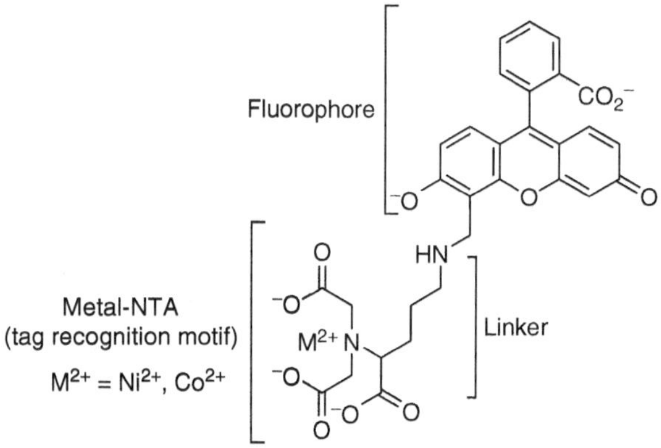

54

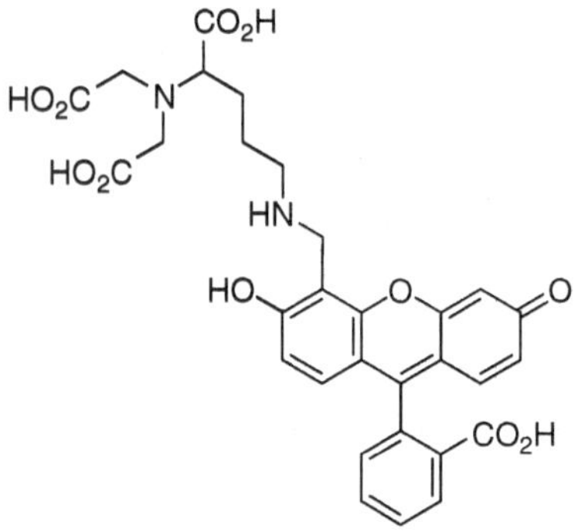

55

The novel fluorescent probe binds selectively to the tag of a hexahistidine-tagged protein, with a substantial enhancement of its fluorescence. This probe showed intrinsic fluorescence, but the addition of a hexahistidine-tagged peptide in neutral aqueous buffer resulted in an up to five, six-fold increase in the fluorescence quantum yield. The new probe is useful for the labeling of the hexahistidine tag on a protein surface, as well as on the hexahistidine peptide. Based on the results, probe **54** would be suitable for detecting tagged proteins in biological applications, such as cell imaging.

The investigation of dynamic movement and interaction of proteins within living cells in real time is important for a better understanding of cellular mechanisms and functions in molecular detail. Genetically encoded fluorescent protein(s) (FPs) have been described for this purpose [151]. In order to circumvent some of the drawbacks associated with the use of FPs [152], fluorescent biarsenical dyes (e.g., FlAsH **56**, F2FlAsH **57**, and F4FlAsH **58**) have been described [153–156].

Complex	λ_{abs}, nm (ε, M^{-1} cm^{-1})	λ_{em} (nm)	τ (ns)
FlAsH-P12 **59**	511 (52,000)	527	4.88
F2FlAsH-P12 **60**	500 (65,500)	522	4.78
F4FlAsH-P12 **61**	528 (35,100)	544	5.18

In this approach, a genetically encodable motif of four cysteines in the sequence Cys–Cys–Xaa–Xaa–Cys–Cys (where Xaa is any amino acid except cysteine) attached itself with high affinity to a fluorescent dye possessing two arsenic moieties. The 4′,5′-bis(1,3,2-dithioasolan-2-yl) fluorescein (FlAsH, **56**) [153] had two As(III) substituents that paired with the four cysteine thiol groups of the motif. As a result, the dye fluorescence intensity increased. An optimized tetracysteine (TC) sequence (–Cys–Cys–Pro–Gly–Cys–Cys–) gave high-affinity binding with a dissociation constant of 10 pM [157]. Quantitative labeling of the target occurred due to the specificity of the genetically encoded tetracysteine binding site as well as the high affinity of the binding itself [153, 157]. The wavelength of excitation and emission maxima of the FlAsH-TC complex were 508 and 528 nm, respectively.

Spagnuolo and co-workers [158] investigated the use of biarsenical dyes, namely the fluoro-substituted F2FlAsH **57** and F4FlAsH **58** in combination with visible fluorescence proteins (VFPs), such as the FRET [159] donor–acceptor (DA) pairs. F2FlAsH **57** exhibits higher absorbance, a larger Stokes' shift, a higher fluorescence quantum yield, higher photostability, and reduced pH dependence when compared to FlAsH **56**. FlAsH-EDT$_2$ **56**, F2FlAsH-EDT$_2$ **57**, and F4FlAsH-EDT$_2$ **58** were almost nonfluorescent, but on the formation of a complex between F2FlAsH **57** and a 12-mer peptidic sequence (P12) as a model target [160], a remarkable increase in fluorescence was observed, the emission peak occurring at 522 nm. The absorption maximum of F2FlAsHP12 **60** shifts 11 nm to the blue compared to FlAsH-P12 **59** whereas the maximum of F4FlAsH-P12 **61** shifted 17 nm to the red. The Stokes' shift for F2FlAsH-P12 **60** was 22 nm, 6 nm greater than that of FlAsH complex **59**. The fluorescence intensity of the peptide adduct ($\lambda_{ex}/\lambda_{em}$ 490/522 nm) was four times brighter than that of the complex with the parent FlAsH probe **59**. This enhancement was attributed to a larger extinction coefficient at 490 nm ($2\times$) and a greater emission quantum yield ($2\times$). The radiative lifetime of F2FlAsH-P12 **60** (4.78 ns) was similar to that of the corresponding FlAsH complex **59** (4.88 ns). The emission peak of

F4FlAsH-P12 **61** at 544 nm expanded the spectral range of the biarsenical dyes. Furthermore, the fluorescence lifetime increased to a value of 5.2 ns. Owing to the results of this study, the authors suggested that the two compounds (**57** and **58**) would form an excellent FRET pair with a large critical distance. In addition, they envisage that the observed characteristics should facilitate improved structural and dynamic studies of proteins in living cells.

Due to the fact that biarsenical-TC complex is stable under the denaturing conditions typically used for gel electrophoresis of proteins and has a molecular weight of less than 2 KDa, when bonded to the biarsenical dye [157], Kottegoda and collaborators [161] studied the biarsenical dyes, as fluorescent probes for in vitro, cellular peptide, and proteins studies using capillary electrophoresis.

Chattopadhaya and co-workers [162] recently reported another approach used to avoid some of the drawbacks associated with the use of FPs. The authors described a small molecule-based procedure that makes use of the unique reactivity between the cysteine residue at the N-terminus of a target protein and cell-permeable, thioester-based small molecule probes resulting in site-specific, covalent tagging of proteins.

This procedure takes advantage of intein-mediated protein splicing for the in vivo generation of the target protein bearing an N-terminal cysteine residue. Subsequent site-specific labeling of the protein occurs via the well-known native chemical ligation (NCL) reaction between this cysteine and a membrane-permeant thioester-containing small molecule probe resulting in the formation of a covalent protein-probe adduct.

Among the cell-permeable, thioester-containing small molecules are the fluorescein derivative (FL) **62**, carboxynaphthofluorescein (CF) **63** and the "caged" form of FL wherein the two acetates were replaced by photo-labile 2-nitrobenzyl groups (C2FL) **64**. Selective "uncaging" of C2FL by photolysis makes this probe useful for bioimaging techniques where spatiotemporal activation of fluorescence is required.

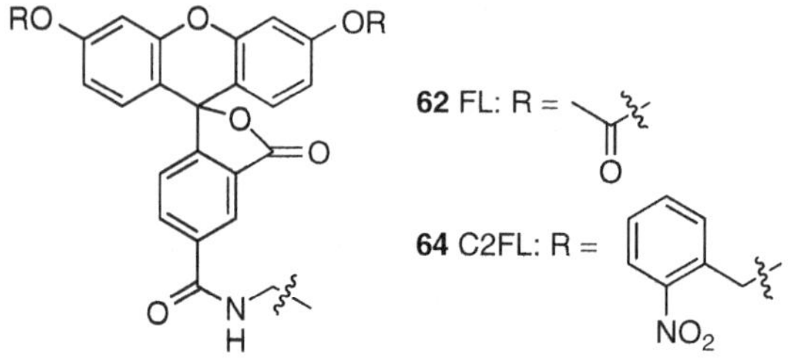

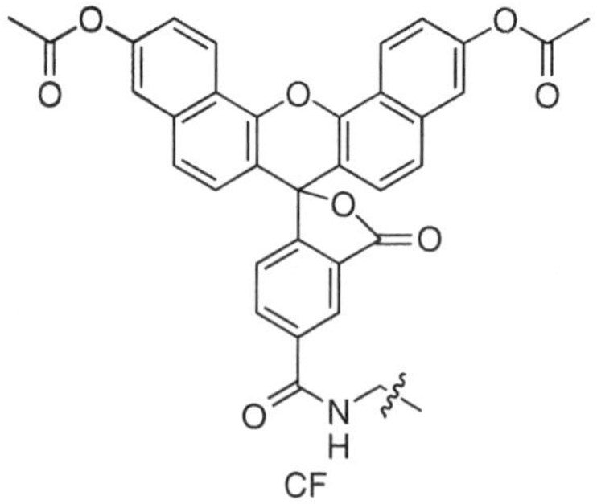

The strategy described has been demonstrated by the in vivo labeling of proteins in both bacterial and mammalian systems thereby making it potentially useful for future bioimaging and proteomics applications [163].

6 Rhodamine Markers

Rhodamine dyes, which belong to the xanthene family, along with fluorescein dyes, are important fluorophores due to their favorable photochemical and photophysical properties [164, 165]. They presented long wavelength absorption and emission maxima (450–700 nm), high molar absorptivities, and high quantum yields [87]. When compared to fluorescein isologs, rhodamines are relatively resistant to photobleaching, and their emission is pH-independent over a range from 4 to 10 [166]. As a result, they have a broad application, including in biotechnology for fluorescent labeling or single molecule detection and in medicine for imaging living cells or live animals in preclinical research [167–169], among others, as was recently revised [170].

In the last years, a large variety of rhodamine (Rh) dyes have been commercialized, including Rhodamine 110 **65**, Rhodamine 123 **66**, Rhodamine B **67**, Rhodamine 6G **68**, Rhodamine 19 **69**, Rhodamine 116 **70**, Rhodamine 101 **71**, as well as several of their corresponding reactive derivatives (succinmidyl esters, maleimides, and isothiocyanates) suitable for the covalent linkage to biomolecules [87].

65-70

Rh 101
71

Compound	R	R_1	R_2	R_3	R_4	R_5	R_6	X^-
Rh 110 **65**	H	H	H	H	H	H	H	Cl^-
Rh 123 **66**	H	H	H	H	H	H	Me	Cl^-
Rh B **67**	Et	Et	Et	Et	H	H	H	Cl^-
Rh 6G **68**	Et	H	Et	H	Me	Me	Me	Cl^-
Rh 19 **69**	Et	H	Et	H	Me	Me	H	ClO_4^-
Rh 116 **70**	Me	H	Me	H	H	H	H	ClO_4^-

The high cost of these compounds, mainly those of the reactive derivatives pose restrictions, when a considerable quantity is needed. However, the less expensive commercially available rhodamines have been used either as such or as precursors in the preparation of other derivatives for various assays.

Among the most used Rhodamines as fluorescent probes in biological studies are Rhodamine 110 **65** [171], Rhodamine 123 **66** [172], Rhodamine B **67** [173] and Rhodamine 6G **68** [174], as well as Rhodamine 800 **72** [175, 176] and Texas Red **73** [177, 178]. Recent examples of the use of Rhodamine 110 **65** and Rhodamine B **67** and their derivatives will be mentioned.

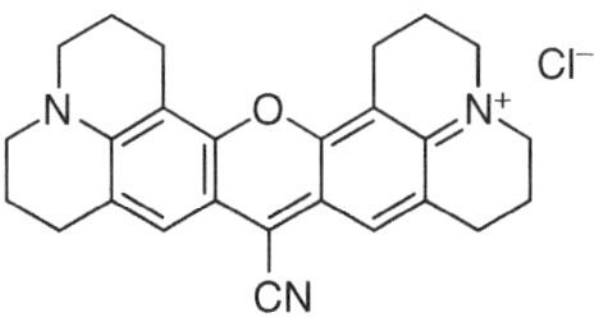

72

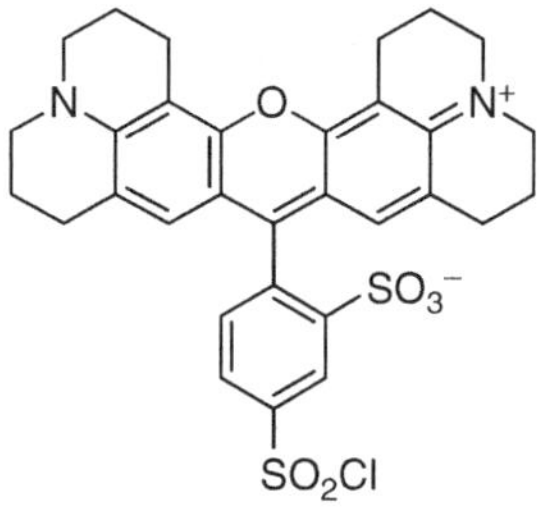

73

Yatzeck and co-workers [179] described the synthesis and evaluation of a derivative of Rhodamine 110 **74** as a probe for cytochrome P450 activity [180]. This probe is the first to use a "trimethyl lock" [181, 182] that is triggered by the cleavage of an ether bond. The trimethyl lock is an *o*-hydroxycinnamic acid derivative in which severe crowding of three methyl groups induces rapid lactonization to form a hydro-coumarin [183]. In this strategy, the phenolic oxygen of the *o*-hydroxycinnamic acid is modified to create a functional group that is a substrate for a designated enzyme, and the carboxyl group is condensed with the amino group of a dye. The unmasking of the phenolic oxygen leads to rapid lactonization with the concomitant release of the dye. An important attribute of this strategy is that the fluorescence/absorbance of the dye is completely masked by amidic resonance and the resulting lactonization within the rhodamine moiety [181]. The morpholino–urea derivative of rhodamine 110 **75**, precursor of probe **74**, was bright ($\varepsilon \times \Phi$ 2.38 $\times$ 10^4 M^{-1} cm^{-1}) but has no measurable fluorescence after N-acylation.

Owing to the fact that ethyl ethers are especially effective substrates for CYP1A1 [184], the probe possesses an ethyl group on the phenolic oxygen of the trimethyl lock. In vitro, fluorescence was manifested by CYP1A1 isozyme with K_{cat}/K_M 8.8 $\times$ 10^3 $M^{-1}s^{-1}$ and K_M 0.09 μM. In cellulo, the probe revealed the induction of cytochrome P450 activity by the carcinogen 2,3,7,8-tetrachlorodi-benzo-*p*-dioxin (TCDD), and its repression by the chemoprotectant resveratrol.

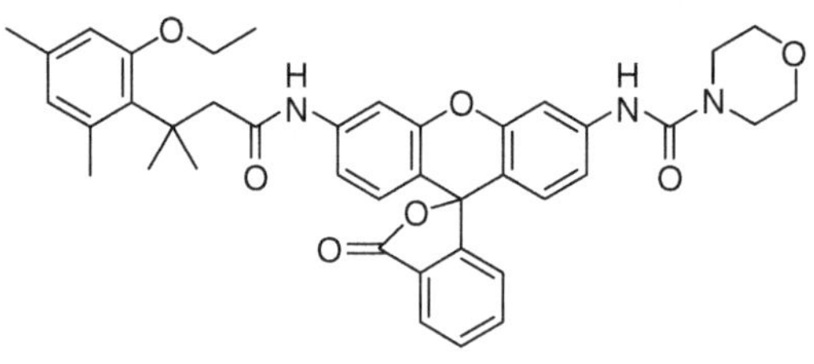

74

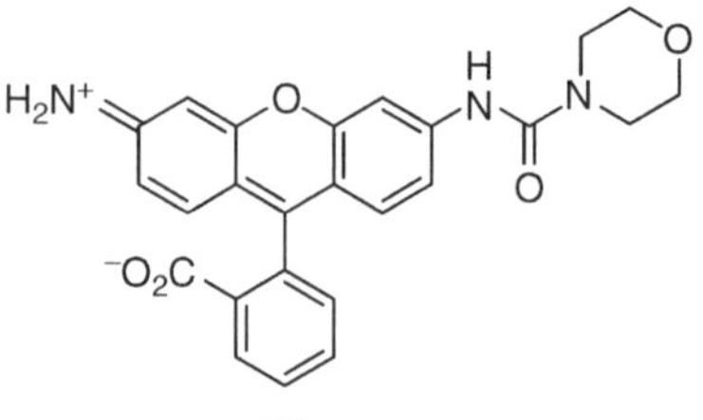

75

Recently, a series of fluorogenic rhodamine peptide substrates were synthesized to develop a flow cytometry assay (FACS) to monitor the proteolytic activity of cathepsin C in live cells [185]. Cathepsin C (Cat C), also known as depeptidyl peptidase I (DPPI), is a widely expressed lysosomal cysteine protease of the papain-fold. Rhodamine-based peptide substrates have been reported to measure the cellular activity of proteases of the papain-fold [186]. From the sixteen symmetric dipeptidic and nonsymmetric monopeptidic (e.g., **76**) substrates tested by Li and collaborators [185], (NH$_2$-aminobutyric-homophenylalanine)$_2$-rhodamine ((H$_2$N-Abu-Hph)$_2$-Rd) **77** showed the best reactivity and selectivity profile in the FACS assay using the B721 human B-lymphoblastoid cell line. The resulting FACS assay was validated through correlation of the IC$_{50}$ values with a competitive radiolabeling assay against a series of small molecule inhibitors of cathepsin C.

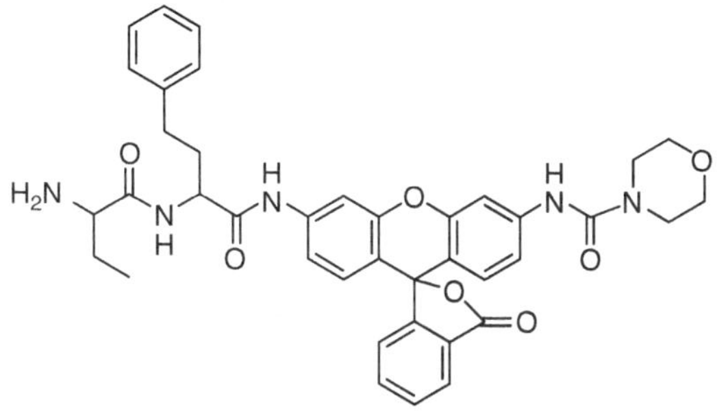

76

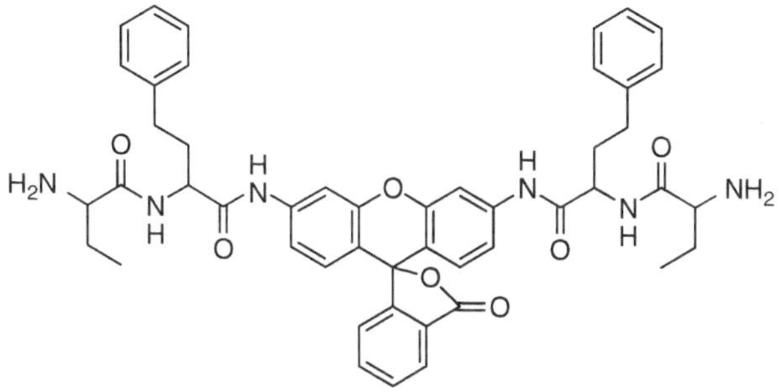

77

Changes in the levels of cellular thiol molecules, such as glutathione, homocysteine, and cysteine are related to oxidative stress connected with toxic agents and disease [187, 188]. Considering that the detection of intracellular thiols is important for the investigation of cellular function, Shibata and co-workers [189] have described a new fluorescent probe derived from Rhodamine 110 (Rh) **65** and 2,4-dinitrobenzenesulfonyl group **78**. The nucleophilic attack of the thiol group on the sulfonamide group results in the cleavage of the sulfone-amide bond, and subsequently the Rh in its open lactone form emits a fluorescence signal.

In contrast with the high yield of Rh **65** (0.645), probe **78** and monosulfonamide Rh **79** exhibit very low fluorescence quantum yields (Φ_F 0.0007 **78** and 0.003 **79**). This result suggested that the observed fluorescence signal arrives from Rh **65** after the cleavage of the bis-sulfonamide (DBN) group in the probe when sensing biological thiol.

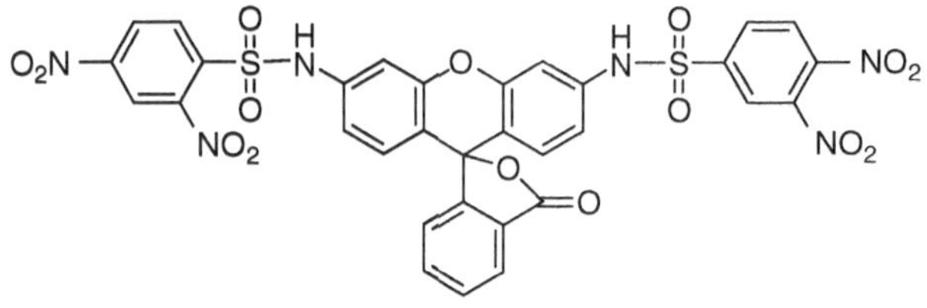

78

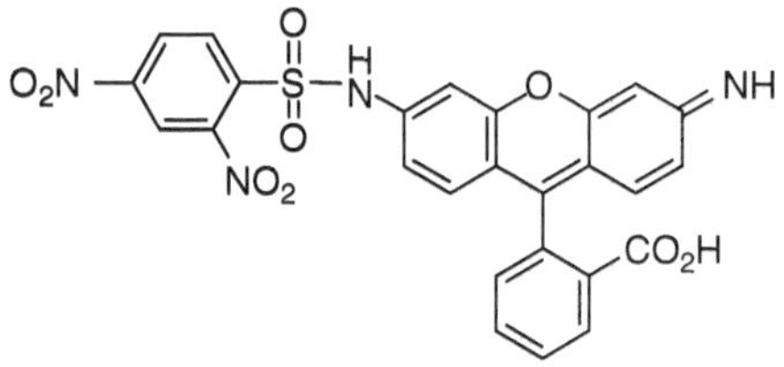

79

The selective detection of thiol species, such as glutathione or cysteine, was carried out in biological conditions (Tris–HCl buffer, 50 mM, pH 7.4). When the probe **78** solution was incubated without Cys, the maximum absorption was at 516 nm, which derived from the DNB group. By adding of Cys to the probe solution, the λ_{max} registered a hypsochromic shift to 498 nm, which results from the cleavage of the DNB group. Furthermore, no significant fluorescence with excitation at 490 nm was observed for the probe without Cys. However, the addition of Cys (10 mM) to the solution resulted in a strong emission around 520 nm, and the emission was enhanced about 5,800-fold.

It was found that the concentration limit for the detection of Cys was 100 μM with 100 nM of the probe, where the signal to background rations reached 12 times. In addition, the probe was effectively applied to the imaging of thiol species in living human cells (HeLa cells).

Rhodamine B **67** is frequently used in the quantitative determination of DNA or RNA and fluorescent labeling for DNA [190–192]. This dye was assembled onto the surface of a quartz substrate by electrostatic interaction between the fluorescence reagent RB and γ-aminopropyltriethoxysilane (APES), and the Quartz/APES/RB film was constructed (Fig. 1) [193].

Fluorescence studies and the binding interaction of Quartz/APES/RB with single- and double-stranded oligonucleotides (ssDNA and dsDNA) in Tris-HCl buffer solution of pH 7.4 were carried out. Quartz/APES/RB exhibited emission at 576 nm, whereas Quartz/APES without BB where nonfluorescent, suggesting that RB successfully assembled on the surface of quartz wafers. By comparison with λ_{em} of 5×10^{-5} M RB solution, which was 588 nm, a hypsochromic shift was found. Considering the fluorescence microscopical image of Quartz/APES/RB, it

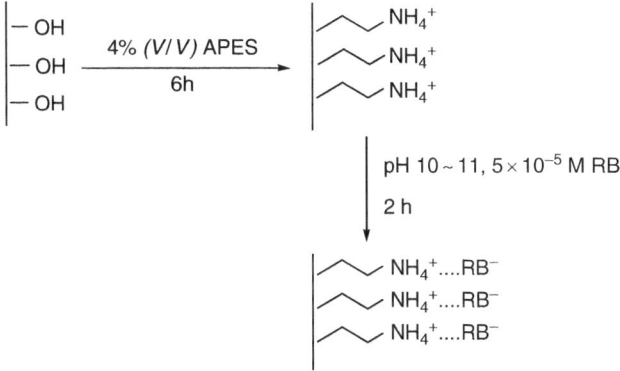

Fig. 1 Establishment mechanism of Quartz/APES/RB

was perceptible that RB is symmetrically distributed with great yellow-green fluorescence on the surface of silanized quartz wafer. The results indicated that RB on this film was present mainly as a monomer since the fluorescent color is the same as that of monomer RB in solution.

Overall, Quartz/APES/RB allowed for extremely high sensitive fluorescence recognition for ssDNA and dsDNA with the detection limit of 2.4 ngL^{-1} and 0.85 ngL^{-1}, respectively.

The proteome undergoes complex changes in response to disease, drug treatment, and normal cellular signalling processes. Characterization of such changes requires methods for time-resolved protein identification and imaging. Two-color labeling of protein populations in cells can provide new insight into global processes that rely on spatial and temporal control of protein synthesis [194], such as bacterial infection [195], cancer [196], or secretion [197]. Similarly, reactive amino acid analogs can be used to track spatial and temporal changes in protein synthesis [194, 198]. In the previous work, the methionine (Met) analogs azidohomoalanine (Aha) and homopropargylglycine (Hpg) were used both to identify [199] and to visualize [200] temporally defined subsets of the proteome.

Recently, Beatty and Tirrell [201] relied on the simultaneous or sequential addition of two reactive Met analogs, Aha and Hpg, to enable the fluorescent tagging of two protein populations within cells. The first demonstration of two-dye labeling of metabolically tagged cells was described in 2007 by Chang and co-workers [202], who used flow cytometry to show that cells treated with two reactive sugars could be labeled with distinct fluorophores.

Three types of reactive spectrally distinct fluorophores, namely lissamine rhodamine (LR) **80**, 7-dimethylaminocoumarin (DMAC) **81**, and bodipy-630 (BDPY) **82** dyes, prepared by coupling 3-azidopropylamine or propargylamine to commercially available amine-reactive dyes were evaluated for the use in selective dye-labeling of newly synthesised proteins in Rat-1 fibroblasts.

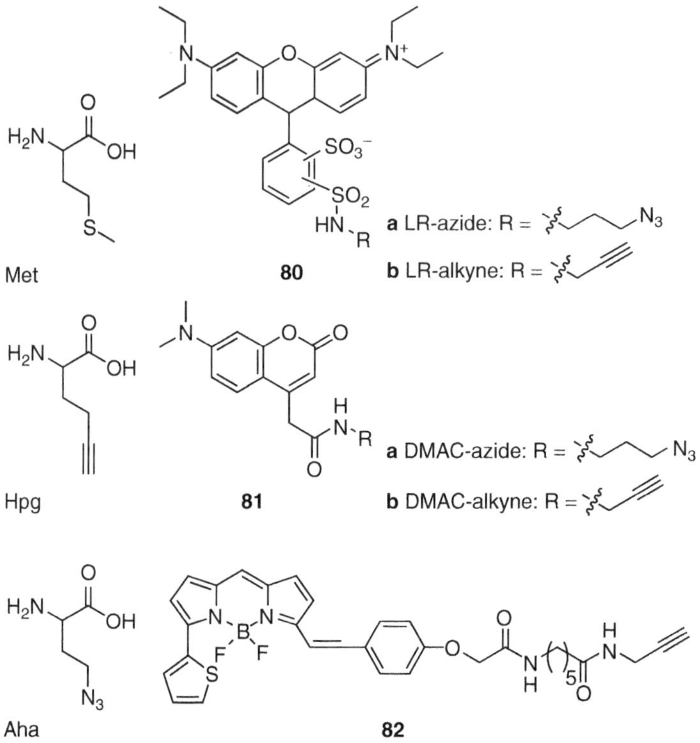

Met   80   **a** LR-azide: R = ⟶N₃
              b LR-alkyne: R = ⟶

Hpg   81   **a** DMAC-azide: R = ⟶N₃
              b DMAC-alkyne: R = ⟶

Aha   82

The combined use of two reactive fluorophores (e.g., LR **80** and DMAC **81**) to dye-label proteins displaying Aha or Hpg enables the two-color fluorescence imaging of cells. The LR **80** and DMAC **81**, but not BDPY **82**, fluorophores enable selective, efficient labeling of subsets of the proteome; cells labeled with Aha and Hpg displayed fluorescence emission three- to seven-fold more intense than that of control cells treated with Met. It also studied the simultaneous and sequential pulse-labeling of cells with Aha and Hpg. After pulse-labeling, cells were treated with reactive LR **80** and DMAC **81** dyes, and labeled cells were imaged by fluorescence microscopy and analyzed by flow cytometry. The results of these studies demonstrate that amino acid labeling can be used to achieve selective two-color imaging of temporally defined protein populations in mammalian cells.

7 Concluding Remarks

As was mentioned, there is an enormous variety of fluorophores with fluorescence emission in the UV/visible region associated with a wide range of applications related to life sciences. These compounds possess a diversity of heteroaromatic or aromatic structures, being usually polycyclic molecules with different photophysical properties, which can be changed by structural chemical modifications.

The design of fluorescent markers depends on the specific required application, and the behavior of these compounds is also dependent on the target molecule and its environment. The reviewed material includes examples of various types of biomolecule labeling, namely proteins, peptides, amino acids, or oligonucleotides with coumarin, benzoxadiazole, acridone, acridine, naphthalene, anthracene, pyrene, fluorescein, and rhodamine derivatives. However, several other classes of compounds, namely those which exhibit fluorescence in the near-infrared of the electromagnetic spectrum, are also extremely important in biological applications. They will be reviewed in subsequent chapters of this book. Despite the high number of fluorophores commercially available and reported until now, the need to circumvent some limitations of the existing compounds and the interdisciplinarity of the biomolecular fluorescent labeling field continues to be a challenge for scientists around the world.

References

1. Giepmans BNG, Adams SR, Ellisman MH, Tsien RY (2006) The fluorescent toolbox for assessing protein location and function. Science 312:217–224
2. Sadaghiani AM, Verhelst SHL, Bogyo M (2007) Tagging and detection strategies for activity-based proteomics. Curr Opin Chem Biol 11:20–28
3. Johnsson N, Johnsson K (2007) Chemical tools for biomolecular imaging. ACS Chem Biol 2:31–38
4. VanEngelenburg SB, Palmer AE (2008) Fluorescent biosensors of protein function. Curr Opin Chem Biol 12:60–65
5. Yano Y, Matsuzaki K (2009) Tag-probe labeling methods for live-cell imaging of membrane proteins. Biochim Biophys Acta-Biomembranes 1788:2124–2131
6. Rasmussen JAM, Hermetter A (2008) Chemical synthesis of fluorescent glycero- and sphingolipids. Prog Lipid Res 47:436–460
7. Monostori P, Wittmann G, Karg E, Túri S (2009) Determination of glutathione and glutathione disulfide in biological samples: an in-dept review. J Chromatogr B 877:3331–3346
8. Leopoldo M, Lacivita E, Berardi F, Perrone R (2009) Developments in fluorescent probes for receptor research. Drug Discov Today 14:706–712
9. Fei X, Gu Y (2009) Progress in modifications and applications of fluorescent dye probes. Prog Nat Sci 19:1–7
10. Gonçalves MST (2009) Fluorescent labeling of biomolecules with organic probes. Chem Rev 109:190–212
11. Vallee RAL, Tomczak N, Kuipers L et al (2003) Single molecule lifetime fluctuations reveal segmental dynamics in polymers. Phys Rev Lett 91:038301–038304
12. Berthelot T, Lain G, Latxague L, Déleris G (2004) Synthesis of novel fluorogenic L-Fmoc lysine derivatives as potential tools for imaging cells. J Fluoresc 14:671–675
13. Soares AMS, Costa SPG, Gonçalves MST (2010) 2-Oxo-2H-benzo[h]benzopyran as a new light sensitive protecting group for neurotransmitter amino acids. Amino Acids. doi:10.1007/s00726-009-0383-z
14. Fonseca ASC, Gonçalves MST, Costa SPG (2007) Photocleavage studies of fluorescent amino acid conjugates bearing different types of linkages. Tetrahedron 63:1353–1359

15. Ohuchi Y, Katayama Y, Maeda M (2001) Fluorescence-based sensing of protein kinase A activity using the dual fluorescent-labeled peptide. Anal Sci 17:i1465–i1467
16. Kosiova I, Janicova A, Kois P (2006) Synthesis of coumarin or ferrocene labeled nucleosides via staudinger ligation. Beilstein J Org Chem 2:2–23
17. Webb MR, Corrie JET (2001) Fluorescent coumarin-labeled nucleotides to measure ADP release from actomyosin. Biophys J 81:1562–1569
18. Shutes A, Der CJ (2005) Real-time in vitro measurements of GTP hydrolysis. Methods 37:183–189
19. Babiak P, Reymond JL (2005) A high-throughput, low volume enzyme assay on solid support. Anal Chem 77:373–377
20. Salisbury CM, Maly DJ, Ellman JA (2002) Peptide microarrays for the determination of protease substrate specificity. J Am Chem Soc 124:14868–14870
21. Wegener D, Wirsching F, Riester D, Schwienhorst A (2003) A fluorogenic histone deacetylase assay well suited for high-throughput activity screening. Chem Biol 10:61–68
22. Lavis LD, Chao TY, Raines RT (2006) Latent blue and red fluorophores based on the trimethyl lock. Chembiochem 7:1151–1154
23. Lavis LD, Raines RT (2008) Bright ideas for chemical biology. ACS Chem Biol 3:142–155
24. Han P, Zhou X, Huang B et al (2008) On-gel fluorescent visualization and the site identification of S-nitrosylated proteins. Anal Biochem 377:150–155
25. Anderson MT, Baumgarth N, Haugland RP et al (1998) Pairs of violet light excited fluorochromes for flow cytometry analysis. Cytometry 33:435–444
26. Telford W, Kapoor V, Jackson J et al (2006) Violet laser diodes in flow cytometry: an update. Cytometry A 69:1153–1160
27. Zlokarnik G, Negulescu PA, Knapp TE et al (1998) Quantitation of transcription and clonal selection of single living cells with β-lactamase as reporter. Science 279:84–88
28. Sun W-C, Gee KR, Haugland RP (1998) Synthesis of novel fluorinated coumarins: excellent UV light-excitable fluorescent dyes. Bioorg Med Chem Lett 8:3107–3110
29. Gee KR, Haugland RP, Sun W-C (1998) Derivatives of 6,8-difluoro-7-hydroxycoumarin. US Patent 5,830,912
30. Abrams B, Diwu Z, Guryev O et al (2009) 3-Carboxy-6-chloro-7-hydroxycoumarin: a highly fluorescent, water-soluble violet-excitable dye for cell analysis. Anal Biochem 386:262–269
31. Weisbrod SH, Marx A (2008) Novel strategies for the sitespecific labelling of nucleic acids. Chem Commun 5675–5685
32. Gramlich PME, Wirges CT, Manetto A, Carell T (2008) Postsynthetic DNA modification through the copper-catalyzed azide-alkyne cycloaddition reaction. Angew Chem Int Ed 47:8350–8358
33. Berndl S, Herzig N, Kele P et al (2009) Comparison of a nucleosidic vs non-nucleosidic postsynthetic "click" modification of DNA with base-labile fluorescent probes. Bioconjug Chem 20:558–564
34. Traven VF, Chibisova TA, Manaev AV (2003) Polymethine dyes derived from boron complexes of acetylhydroxycoumarins. Dyes Pigm 58:41–46
35. Manaev AV, Chisova TA, Traven VF (2006) Boron chelates in the synthesis of α, β-unsaturated ketones of the coumarin series. Russ Chem Bull 55:2226–2232
36. Traven VF, Manaev AV, Voevodina IV, Okhrimenko IN (2008) Synthesis and structure of new 3-pyrazolinylcoumarins and 3-pyrazolinyl-2-quinolones. Russ Chem Bull 57:1508–1515
37. Kovalska VB, Volkova KD, Manaev AV et al (2010) 2-Quinolone and coumarin polymethines for the detection of proteins using fluorescence. Dyes Pigm 84:159–164
38. Piloto AM, Rovira D, Costa SPG, Gonçalves MST (2006) Oxobenzo[f]benzopyrans as new fluorescent photolabile protecting groups for the carboxylic function. Tetrahedron 62:11955–11962
39. Fernandes MJG, Gonçalves MST, Costa SPG (2008) Comparative study of polyaromatic and polyheteroaromatic fluorescent photocleavable protecting groups. Tetrahedron 64:3032–3038

40. Fernandes MJG, Gonçalves MST, Costa SPG (2008) Neurotransmitter amino acid – oxo-benzo[f]benzopyran conjugates: synthesis and photorelease studies. Tetrahedron 64: 11175–11179

41. Wang M, Gao M, Miller KD et al (2009) Simple synthesis of carbon-11 labeled styryl dyes as new potential PET RNA-specific, living cell imaging probes. Eur J Med Chem 44: 2300–2306

42. Wen-Chen Z, Ling-Jun L, Xian-En Z et al (2008) Application of 2-(11*H*-benzo[*a*]carbazol-11-yl) ethyl carbonochloridate as a precolumn derivatization reagent of amino acid by high performance liquid chromatography with fluorescence detection. Chin J Anal Chem 36:1071–1076

43. Sugimoto T, Itagaki K, Irie K (2008) Design and physicochemical properties of new fluorescent ligands of protein kinase C isozymes focused on CH/π interaction. Bioorg Med Chem 16:650–657

44. Wang S, Wang X, Shi W et al (2008) Detection of local polarity and conformational changes at the active site of rabbit muscle creatine kinase with a new arginine-specific fluorescent probe. Biochim Biophys Acta 1784:415–422

45. Hiratsuka T (1986) Involvement of the 50-kDa peptide of myosin heads in the ATPase activity revealed by fluorescent modification with 4-fluoro-7-nitrobenzo-2-oxa-1,3-diazole. J Biol Chem 261:7294–7299

46. Watanabe Y, Imai K (1981) High-performance liquid chromatography and sensitive detection of amino acids derivatized with 7-fluoro-4-nitrobenzo-2-oxa-1,3-diazole. Anal Biochem 116:471–472

47. Luo J, Fukuda E, Takase H et al (2009) Identification of the lysine residue responsible for coenzyme A binding in the heterodimeric 2-oxoacid:ferredoxin oxidoreductase from *Sulfolobus tokodaii*, a thermoacidophilic archaeon, using 4-fluoro-7-nitrobenzofurazan as an affinity label. Biochim Biophys Acta 1794:335–340

48. Nemati M, Oveisi MR, Abdollahi H, Sabzevari O (2004) Differentiation of bovine and porcine gelatins using principal component analysis. J Pharm Biomed Anal 34:485–492

49. Solínová V, Kasicka V, Koval D et al (2004) Analysis of synthetic derivatives of peptide hormones by capillary zone electrophoresis and micellar electrokinetic chromatography with ultraviolet-absorption and laser-induced fluorescence detection. J Chromatogr B 808:75–82

50. Ghosh PB, Whitehouse MW (1968) 7-Chloro-4-nitrobenzo-2-oxa-1,3-diazole: A new fluorogenic reagent for amino acids and other amines. Biochem J 108:155–156

51. Simard JR, Getlik M, Grütter C et al (2009) Development of a fluorescent-tagged kinase assay system for the detection and characterization of allosteric kinase inhibitors. J Am Chem Soc 131:13286–13296

52. Andrews JL, Ghosh P, Ternai B, Whitehouse MW (1982) Ammonium 4-chloro-7-sulfobenzofurazan: A new fluorigenic [*sic*]thiol-specific reagent. Arch Biochem Biophys 214: 386–396

53. Hu S, Li PCH (2000) Micellar electrokinetic capillary chromatographic separation and fluorescent detection of amino acids derivatized with 4-fluoro-7-nitro-2,1,3-benzoxadiazole. J Chromatogr A 876:183–191

54. O'Brien KB, Esguerra M, Miller RF, Bowser MT (2004) Monitoring neurotransmitter release from isolated retinas using online microdialysis-capillary electrophoresis. Anal Chem 76:5069–5074

55. Zhang H, Le Potier I, Smadja C et al (2006) Fluorescent detection of peptides and amino acids for capillary electrophoresis via on-line derivatization with 4-fluoro-7-nitro-2,1,3-benzoxadiazole. Anal Bioanal Chem 386:1387–1394

56. Tseng H-M, Li Y, Barrett DA (2007) Profiling of amine metabolites in human biofluids by micellar electrokinetic chromatography with laser-induced fluorescence detection. Anal Bioanal Chem 388:433–439

57. Lapos JA, Ewing AG (2000) Injection of fluorescently labeled analytes into microfabricated chips using optically gated electrophoresis. Anal Chem 72:4598–4602

58. Klinker CC, Bowser MT (2007) 4-Fluoro-7-nitro-2, 1, 3-benzoxadiazole as a fluorogenic labeling reagent for the in vivo analysis of amino acid neurotransmitters using online microdialysis-capillary electrophoresis. Anal Chem 79:8747–8754

59. Guminski Y, Grousseaud M, Cugnasse S et al (2009) Synthesis of conjugated spermine derivatives with 7-nitrobenzoxadiazole (NBD), rhodamine and bodipy as new fluorescent probes for the polyamine transport system. Bioorg Med Chem Lett 19:2474–2477

60. Baslow MH (2000) Functions of N-acetyl-L-aspartate and N-acetyl-L-aspartylglutamate in the vertebrate brain: role in glial cell-specific signaling. J Neurochem 75:453–459

61. Fukushima T, Arai K, Tomiya M et al (2008) Fluorescence determination of N-acetylaspartic acid in the rat cerebrum homogenate using high-performance liquid chromatography with pre-column fluorescence derivatization. Biomed Chromatogr 22:100–105

62. Prados P, Fukushima T, Santa T et al (1997) 4-N,N-Dimethylaminosulfonyl-7-N-(2-aminoethyl)amino-benzofurazan as a new precolumn fluorescence derivatization reagent for carboxylic acids (fatty acids and drugs containing a carboxyl moiety) in liquid chromatography. Anal Chim Acta 344:227–232

63. Arai K, Fukushima T, Tomiya M et al (2008) Simultaneous determination of N-acetylaspartylglutamate and N-acetylaspartate in rat brain homogenate using high-performance liquid chromatography with pre-column fluorescence derivatization. J Chromatogr B 875:358–362

64. Siegmund M, Bendig J (1980) The solvent dependence of the electronic spectra and the change of the properties of N-substituted acridones at electronic excitation (In German). Z Naturforsch 35a:1076–1086

65. Chen J, Zhang J, Huang L et al (2008) Hybridization biosensor using 2-nitroacridone as electrochemical indicator for detection of short DNA species of Chronic Myelogenous Leukemia. Biosens Bioelectron 24:349–355

66. Szymanska A, Wegner K, Lankiewitz L (2003) Synthesis of N-[(tert-Butoxy)carbonyl]-3-(9, 10-dihydro-9-oxoacridin-2-yl)-L-alanine, a new fluorescent amino acid derivative. Helv Chim Acta 86:3326–3331

67. Faller T, Hutton K, Okafo G, et al (1997) A novel acridone derivative for the fluorescence tagging and mass spectrometric sequencing of peptides. Chem Commun 1529–1530

68. Bahr N, Tierney E, Reymond JL (1997) Highly Photoresistant chemosensors using acridone as fluorescent label. Tetrahedron Lett 38:1489–1492

69. Shoji A, Hasegawa T, Kuwahara M et al (2007) Chemico-enzymatic synthesis of a new fluorescent-labeled DNA by PCR with a thymidine nucleotide analogue bearing an acridone derivative. Bioorg Med Chem Lett 17:776–779

70. Saito Y, Hanawa K, Kawasaki N et al (2006) Acridone-labeled base-discriminating fluorescence (BDF) nucleoside: synthesis and their photophysical properties. Chem Lett 35:1182–1183

71. Hagiwara Y, Hasegawa T, Shoji A et al (2008) Acridone-tagged DNA as a new probe for DNA detection by fluorescence resonance energy transfer and for mismatch DNA recognition. Bioorg Med Chem 16:7013–7020

72. Agiamarnioti K, Triantis T, Papadopoulos K, Scorilas A (2006) 10-(2-Biotinyloxyethyl)-9-acridone a novel fluorescent label for (strept)avidin–biotin based assays. J Photoch Photobio A 181:126–131

73. Di GC, De MM, Chiron J, Delmas F (2005) Synthesis and antileishmanial activities of 4, 5-di-substituted acridines as compared to their 4-mono-substituted homologues. Bioorg Med Chem 13:5560–5568

74. Graves DE, Velea LM (2000) Intercalative binding of small molecules to nucleic acids. Curr Org Chem 4:915–929

75. Liu RT, Yang JH, Sun CX et al (2003) Study of the interaction of nucleic acids with acridine orange-CTMAB and determination of nucleic acids at nanogram levels based on the enhancement of resonance light scattering. Chem Phys Lett 376:108–115

76. Liu SP, Chen S, Liu ZF et al (2005) Resonance Rayleigh scattering spectra of interaction of sodium carboxymethylcellulose with cationic acridine dyes and their analytical applications. Anal Chim Acta 535:169–175

77. Wang M, Yang JH, Wu X, Huang F (2000) Study of the interaction of nucleic acids with acridine red and CTMAB by a resonance light scattering technique and determination of nucleic acids at nanogram levels. Anal Chim Acta 422:151–158

78. Wu M, Wu W, Gao X et al (2008) Synthesis of a novel fluorescent probe based on acridine skeleton used for sensitive determination of DNA. Talanta 75:995–1001

79. Wu M, Wu W, Lian X et al (2008) Synthesis of a novel fluorescent probe and investigation on its interaction with nucleic acid and analytical application. Spectrochim Acta A 71:1333–1340

80. Weber G (1952) Polarization of the fluorescence of macromolecules II. Fluorescent conjugates of ovalbumin and bovine serum albumin. Biochem J 51:155–167

81. Daniel E, Weber G (1966) Cooperative effects in binding by bovine serum albumin. I. The binding of 1-anilino-8-naphthalenesulfonate. Fluorimetric titrations. Biochemistry 5: 1893–1900

82. Weber G, Farris FJ (1979) Synthesis and spectral properties of a hydrophobic fluorescent probe: 6-propionyl-2-(dimethylamino)naphthalene. Biochemistry 18:3075–3078

83. Hudson EN, Weber G (1973) Synthesis and characterization of two fluorescent sulfhydryl reagents. Biochemistry 12:4154–4161

84. Maggiora LL, Smith CW, Zhang ZY (1992) A general method for the preparation of internally quenched fluorogenic protease substrates using solid-phase peptide synthesis. J Med Chem 35:3727–3730

85. Tyagi S, Kramer FR (1996) Molecular beacons: probes that fluoresce upon hybridization. Nat Biotechnol 14:303–308

86. Stewart WW (1981) Synthesis of 3, 6-disulfonated 4-aminonaphthalimides. J Am Chem Soc 103:7615–7620

87. Haugland RP, Spence MTZ, Johnson ID, Basey A (2005) The handbook: a guide to fluorescent probes and labeling technologies, 10th edn. Molecular Probes, Eugene, OR

88. Li H, Xu J, Yan H (2009) Ratiometric fluorescent determination of cysteine based on organic nanoparticles of naphthalene–thiourea–thiadiazole-linked molecule. Sensor Actuat B-Chem 139:483–487

89. Liao F, Xie Y, Yang X et al (2009) Homogeneous noncompetitive assay of protein via Förster-resonance-energy-transfer with tryptophan residue(s) as intrinsic donor(s) and fluorescent ligand as acceptor. Biosens Bioelectron 25:112–117

90. Koehorst RBM, Spruijt RB, Hemminga MA (2008) Site-directed fluorescence labeling of a membrane protein with BADAN: probing protein topology and local environment. Biophys J 94:3945–3955

91. Yi L, Shi J, Gao S et al (2009) Sulfonium alkylation followed by "click" chemistry for facile surface modification of proteins and tobacco mosaic virus. Tetrahedron Lett 50:759–762

92. Chang KC, Su I, Senthilvelan A, Chung W (2007) Triazole-modified calix[4]crown as a novel fluorescent on−off switchable chemosensor. Org Lett 9:3363–3366

93. Li X, Zhang G, Ma H et al (2004) 4, 5-Dimethylthio-4'-[2-(9-anthryloxy)ethylthio]tetrathiafulvalene, a highly selective and sensitive chemiluminescence probe for singlet oxygen. J Am Chem Soc 126:11543–11548

94. Harris RF, Nation AJ, Copeland GT, Miller SJ (2000) A polymeric and fluorescent gel for combinatorial screening of catalysts. J Am Chem Soc 122:11270–11271

95. Magri DC, Brown GJ, McClean GD, de Silva AP (2006) Communicating chemical congregation: a molecular and logic gate with three chemical inputs as a "lab-on-a-molecule" prototype. J Am Chem Soc 128:4950–4951

96. Yang W, Fan H, Gao X et al (2004) The first fluorescent diboronic acid sensor specific for hepatocellular carcinoma cells expressing sialyl Lewis X. Chem Biol 11:439–448

97. Xie F, Sivakumar K, Zeng Q et al (2008) A fluorogenic "click" reaction of azidoanthracene derivatives. Tetrahedron 64:2906–2914

98. Sivakumar K, Xie F, Cash BM et al (2004) A fluorogenic 1, 3-dipolar cycloaddition reaction of 3-azidocoumarins and acetylenes. Org Lett 6:4603–4606

99. Sawa M, Hsu TL, Itoh T et al (2006) Glycoproteomic probes for fluorescent imaging of fucosylated glycans in vivo. Proc Natl Acad Sci USA 103:12371–12376

100. Schlick TL, Ding Z, Kovacs EW, Francis MB (2005) Dual-surface modification of the tobacco mosaic virus. J Am Chem Soc 127:3718–3723

101. Karpovich DS, Blanchard GJ (1995) Relating the polarity dependent fluorescence response of pyrene to vibronic coupling. Achieving a fundamental understanding of the py polarity scale. J Phys Chem 99:3951–3958

102. Smalley MK, Silverman SK (2006) Fluorescence of covalently attached pyrene as a general RNA folding probe. Nucleic Acids Res 34:152–166

103. Sahoo D, Narayanaswami V, Kay CM, Ryan RO (2000) Pyrene excimer fluorescence: a spatially sensitive probe to monitor lipid-induced helical rearrangement of apolipophorin III. Biochemistry 39:6594–6601

104. Okamoto A, Ichiba T, Saito I (2004) Pyrene-labeled oligodeoxynucleotide probe for detecting base insertion by excimer fluorescence emission. J Am Chem Soc 126:8364–8365

105. Hwang GT, Seo YJ, Kim BH (2004) A Highly discriminating quencher-free molecular beacon for probing DNA. J Am Chem Soc 126:6528–6529

106. Fujimoto K, Shimizu H, Inouye M (2004) Unambiguous detection of target DNAs by excimer–monomer switching molecular beacons. J Org Chem 69:3271–3275

107. Hrdlicka PJ, Babu BR, Sorensen MD, Wengel J (2004) Interstrand communication between 2-N-(pyren-1-yl)methyl-2-amino-LNA monomers in nucleic acid duplexes: directional control and signalling of full complementarity. Chem Commun:1478–1479

108. Langenegger SM, Haner R (2004) Excimer formation by interstrand stacked pyrenes. Chem Commun:2792–2793

109. Christensen UB, Pedersen EB (2002) Intercalating nucleic acids containing insertions of 1-O-(1-pyrenylmethyl)glycerol: stabilisation of dsDNA and discrimination of DNA over RNA. Nucleic Acids Res 30:4918–4925

110. Kostenko E, Dobrikov M, Phshnyi D, Petyuk V (2001) 5′-Bis-pyrenylated oligonucleotides displaying excimer fluorescence provide sensitive probes of RNA sequence and structure. Nucleic Acids Res 29:3611–3620

111. Masuko M, Ohtani H, Ebatal K, Shimadzu A (1998) Optimization of excimer-forming two-probe nucleic acid hybridization method with pyrene as a fluorophore. Nucleic Acids Res 26:5409–5416

112. Paris PL, Langenhan JM, Kool ET (1998) Probing DNA sequences in solution with a monomer-excimer fluorescence color change. Nucleic Acids Res 26:3789–3793

113. Lewis FD, Zhang Y, Letsinger RL (1997) Bispyrenyl excimer fluorescence: a sensitive oligonucleotide probe. J Am Chem Soc 119:5451–5452

114. Kalyanasundaram K, Thomas JK (1997) Solvent-dependent fluorescence of pyrene-3-carboxaldehyde and its applications in the estimation of polarity at micelle-water interfaces. J Phys Chem 81:2176–2180

115. de Silva AP, Gunaratne HQN, Gunnlaugsson T et al (1997) Signaling recognition events with fluorescent sensors and switches. Chem Rev 97:1515–1566

116. Tanaka K, Okamoto A (2008) Design of a pyrene-containing fluorescence probe for labelling of RNA poly(A) tracts. Bioorg Med Chem 16:400–404

117. Okamoto A, Kanatani K, Saito I (2004) Pyrene-labeled base-discriminating fluorescent DNA probes for homogeneous SNP typing. J Am Chem Soc 126:4820–4827

118. Okamoto A, Saito Y, Saito I (2005) Design of base-discriminating fluorescent nucleosides. J Photochem Photobiol C-Photochem Rev 6:108–122

119. Okamoto A, Tainaka K, Ochi Y et al (2006) Simple SNP typing assay using a base-discriminating fluorescent probe. Mol BioSyst 2:122–127

120. Okamoto A (2005) Synthesis of highly functional nucleic acids and their application to DNA technol. Bull Chem Soc Jpn 78:2083–2097
121. Valis L, Mayer-Enthart E, Wagenknecht H-A (2006) 8-(Pyren-1-yl)-2'-deoxyguanosine as an optical probe for DNA hybridization. Bioorg Med Chem Lett 16:3184–3187
122. Wanninger-Weiß C, Valis L, Wagenknecht H-A (2008) Pyrene-modified guanosine as fluorescent probe for DNA modulated by charge transfer. Bioorg Med Chem 16:100–106
123. Trifonov A, Raytchev M, Buchvarov I et al (2005) Ultrafast energy transfer and structural dynamics in DNA. J Phys Chem B 109:19490–19495
124. Lakowicz JR (2006) Principles of fluorescence spectroscopy, 3rd edn. Springer, New York
125. Song A-M, Zhang J-H, Zhang M-H et al (2000) Spectral properties and structure of fluorescein and its alkyl derivatives in micelles. Colloids Surf A 167:253–262
126. Orndorff WR, Hemmer AJ (1927) Fluorescein and some of its derivatives. J Am Chem Soc 49:1272–1280
127. Miki M, Dosremedios CG (1988) Fluorescence quenching studies of fluorescein attached to Lys-61 or Cys-374 in actin: effects of polymerization, myosin subfragment-1 binding, and tropomyosin-troponin binding. J Biochem Tokyo 104:232–235
128. Boturyn D, Coll J-L, Garanger E et al (2004) Template assembled cyclopeptides as multi-meric system for integrin targeting and endocytosis. J Am Chem Soc 126:5730–5739
129. Carrigan CN, Imperiali B (2005) The engineering of membrane-permeable peptides. Anal Biochem 341:290–298
130. Song A, Wang X, Zhang J et al (2004) Synthesis of hydrophilic and flexible linkers for peptide derivatization in solid phase. Bioorg Med Chem Lett 14:161–165
131. Park SI, Renil M, Vikstrom B et al (2002) The use of one-bead one-compound combinatorial library method to identify peptide ligands for α4β1 integrin receptor in non-Hodgkin's lymphoma. Lett Pept Sci 8:171–178
132. Zhai D, Jin C, C-w S et al (2008) Gambogic acid is an antagonist of antiapoptotic Bcl-2 family proteins. Mol Cancer Ther 7:1639–1646
133. Dirks RW, van Gijlswijk RP, Tullis RH et al (1990) Simultaneous detection of different mRNA sequences coding for neuropeptide hormones by double in situ hybridization using FITC- and biotin-labeled oligonucleotides. J Histochem Cytochem 38:467–473
134. Muramoto K, Nokihara K, Ueda A, Kamiya H (1994) Gas-phase microsequencing of peptides and proteins with a fluorescent Edman-type reagent, fluorescein isothiocyanate. Biosci Biotechnol Biochem 58:300–304
135. Wu S, Dovichi NJ (1989) High-sensitivity fluorescence detector for fluorescein isothiocya-nate derivatives of amino acids separated by capillary zone electrophoresis. J Chromatogr 480:141–155
136. Cheng YF, Dovichi NJ (1988) Subattomole amino acid analysis by capillary zone electro-phoresis and laser-induced fluorescence. Science 242:562–564
137. Vera JC, Rivas CI, Cortés PA et al (1988) Purification, amino terminal analysis, and peptide mapping of proteins after in situ postelectrophoretic fluorescent labelling. Anal Biochem 174:38–45
138. Yang WC, Schmerr MJ, Jackma R et al (2005) Capillary electrophoresis-based noncompeti-tive immunoassay for the prion protein using fluorescein-labeled protein A as a fluorescent probe. Anal Chem 77:4489–4494
139. Szewczyk B, Summers DF (1987) Fluorescent staining of proteins transferred to nitrocellu-lose allowing for subsequent probing with antisera. Anal Biochem 164:303–306
140. Houston B, Peddie D (1989) A method for detecting proteins immobilized on nitrocellulose membranes by in situ derivatization with fluorescein isothiocyanate. Anal Biochem 177:263–267
141. Farley RA, Tran CM, Carilli CT et al (1984) The amino acid sequence of a fluorescein-labeled peptide from the active site of (Na, K)-ATPase. J Biol Chem 259:9532–9535
142. Jackson RJ, Mendlein J, Sachs G (1983) Interaction of fluorescein isothiocyanate with the $(H^+ + K^+)$-ATPase. Biochim Biophys Acta 731:9–15

143. Blakeslee D (1977) Immunofluorescence using dichlorotriazinylaminofluorescein (DTAF). II. Preparation, purity and stability of the compound. J Immunol Methods 17:361–364
144. Dose C, Seitz O (2008) Single nucleotide specific detection of DNA by native chemical ligation of fluorescence labeled PNA-probes. Bioorg Med Chem 16:65–77
145. Blakeslee D, Baines MG (1976) Immunofluorescence using dichlorotriazinylaminofluorescein (DTAF). I. Preparation and fractionation of labelled IgG. J Immunol Methods 13:305–320
146. de Belder AN (1975) Wik KO (1975) Preparation and properties of fluorescein-labelled hyaluronate. Carbohydr Res 44:251–257
147. Prigent-Richard S, Cansell M, Vassy J et al (1998) Fluorescent and radiolabeling of polysaccharides: binding and internalization experiments on vascular cells. J Biomed Mater Res 40:275–281
148. Tian M, Wu XL, Zhang B et al (2008) Synthesis of chlorinated fluoresceins for labeling proteins. Bioorg Med Chem Lett 18:1977–1979
149. Wu XL, Tian M, He HZ et al (2009) Synthesis and biological applications of two novel fluorescent proteins-labeling probes. Bioorg Med Chem Lett 19:2957–2959
150. Kamoto M, Umezawa N, Kato N, Higuchi T (2009) Turn-on fluorescent probe with visible light excitation for labelling of hexahistidine tagged protein. Bioorg Med Chem Lett 19:2285–2288
151. Tsien RY (1998) The green fluorescent protein. Annu Rev Biochem 67:509–544
152. Marks KM, Nolan GP (2006) Chemical labeling strategies for cell biology. Nat Methods 3:591–596
153. Griffin BA, Adams SR, Tsien RY (1998) Specific covalent labeling of recombinant protein molecules inside live cells. Science 281:269–272
154. Liu B, Archer CT, Burdine L et al (2007) Label transfer chemistry for the characterization of protein–protein interactions. J Am Chem Soc 129:12348–12349
155. Zhang X-Y, Bishop AC (2007) Site-specific incorporation of allosteric-inhibition sites in a protein tyrosine phosphatase. J Am Chem Soc 129:3812–3813
156. Hearps AC, Pryor MJ, Kuusisto HV et al (2007) The biarsenical dye Lumio™ exhibits a reduced ability to specifically detect tetracysteine-containing proteins within live cells. J Fluoresc 17:593–597
157. Adams SR, Campbell RE, Gross LA et al (2002) New Biarsenical ligands and tetracysteine motifs for protein labeling in vitro and in vivo: synthesis and biological applications. J Am Chem Soc 124:6063–6076
158. Spagnuolo CC, Vermeij RJ, Jares-Erijman EA (2006) Improved Photostable FRET-competent biarsenical–tetracysteine probes based on fluorinated fluoresceins. J Am Chem Soc 128:12040–12041
159. Hoffmann C, Gaietta G, Bunemann M et al (2005) A FlAsH-based FRET approach to determine G protein–coupled receptor activation in living cells. Nat Methods 2:171–176
160. Martin BR, Giepmans BNG, Adams SR, Tsien RY (2005) Mammalian cell–based optimization of the biarsenical-binding tetracysteine motif for improved fluorescence and affinity. Nat Biotechnol 23:1308–1314
161. Kottegoda S, Aoto PC, Sims CE, Allbritton NL (2008) Biarsenical–tetracysteine motif as a fluorescent tag for detection in capillary electrophoresis. Anal Chem 80:5358–5366
162. Chattopadhaya S, Srinivasan R, Yeo DSY et al (2009) Site-specific covalent labeling of proteins inside live cells using small molecule probes. Bioorg Med Chem 17:981–989
163. Chen GYJ, Uttamchandani M, Lue RYP et al (2003) Array-based technologies and their applications in proteomics. Curr Top Med Chem 3:705–724
164. He J, Ritalahti KM, Yang KL et al (2003) Detoxification of vinyl chloride to ethene coupled to growth of an anaerobic bacterium. Nature 424:62–65
165. Georgi A, Mottola-Hartshorn C, Warner A et al (1990) Detection of individual fluorescently labeled reovirions in living cells. Proc Natl Acad Sci USA 87:6579–6583

166. Valeur B (2002) Molecular fluorescence: principles and applications. Wiley-VCH, Weinheim, Germany
167. Johnson LV, Walsh ML, Chen LB (1980) Localization of mitochondria in living cells with rhodamine 123. Proc Natl Acad Sci USA 77:990–994
168. Belmont LD, Hyman AA, Sawin KE, Mitchison TJ (1990) Real-time visualization of cell cycle-dependent changes in microtubule dynamics in cytoplasmic extracts. Cell 62:579–589
169. Marnett LJ, Uddin MJ, Crews BC (2007) Methods and composition for diagnostic and therapeutic targeting of COX-2. US Patent WO/2007149546
170. Beija M, Afonso CAM, Martinho JMG (2009) Synthesis and applications of Rhodamine derivatives as fluorescent probes. Chem Soc Rev 38:2410–2433
171. Guzikowski AP, Naleway JJ, Shipp CT, Schutte RC (2000) Synthesis of a macrocyclic rhodamine 110 enzyme substrate as an intracellular probe for caspase 3 activity. Tetrahedron Lett 41:4733–4735
172. Wang Y, Hao D, Stein WD, Yang L (2006) A kinetic study of Rhodamine 123 pumping by P-glycoprotein. Biochim Biophys Acta 1758:1671–1676
173. Meng Q, Yu M, Zhang H et al (2007) Synthesis and application of N-hydroxysuccinimidyl rhodamine B ester as an amine-reactive fluorescent probe. Dyes Pigm 73:254–260
174. Mandalá M, Serck-Hanssen G, Martino G, Helle KB (1999) The fluorescent cationic dye rhodamine 6G as a probe for membrane potential in bovine aortic endothelial cells. Anal Biochem 274:1–6
175. Walsh RJ, Reinot T, Hayes JM et al (2002) Nonphotochemical hole burning spectroscopy of a mitochondrial selective rhodamine dye molecule in normal and cancerous ovarian surface epithelial cells. J Lumin 98:115–121
176. Abugo OO, Nair R, Lakowicz JR (2000) Fluorescence properties of rhodamine 800 in whole blood and plasma. Anal Biochem 279:142–150
177. Wippersteg V, Ribeiro F, Liedtke S et al (2003) The uptake of Texas Red-BSA in the excretory system of schistosomes and its colocalisation with ER60 promoter-induced GFP in transiently transformed adult males. Int J Parasitol 33:1139–1143
178. Horneffer V, Forsmann A, Strupat K et al (2001) Localization of analyte molecules in MALDI preparations by confocal laser scanning microscopy. Anal Chem 73:1016–1022
179. Yatzeck MM, Lavis LD, Chao T-Y et al (2008) A highly sensitive fluorogenic probe for cytochrome P450 activity in live cells. Bioorg Med Chem Lett 18:5864–5866
180. Gungerich FP (2008) Cytochrome P450 and chemical toxicology. Chem Res Toxicol 21:70–83
181. Levine MN, Lavis LD, Raines RT (2008) Trimethyl lock: a stable chromogenic substrate for esterases. Molecules 13:204–211
182. Mangold SL, Carpenter RT, Kiessling LL (2008) Synthesis of fluorogenic polymers for visualizing cellular internalization. Org Lett 10:2997–3000
183. Milstein S, Cohen LA (1972) Stereopopulation control. I. Rate enhancement in the lactonizations of o-hydroxyhydrocinnamic acids. J Am Chem Soc 94:9158–9165
184. Liu J, Ericksen SS, Besspiata D et al (2003) Characterization of substrate binding to cytochrome P450 1A1 using molecular modeling and kinetic analyses: case of residue 382. Drug Metab Dispos 31:412–420
185. Li J, Petrassi HM, Tumanut C et al (2009) Substrate optimization for monitoring cathepsin C activity in live cells. Bioorg Med Chem 17:1064–1070
186. Liu J, Bhalgat M, Zhang C et al (1999) Fluorescent molecular probes V: a sensitive caspase-3 substrate for fluorometric assays. Bioorg Med Chem Letts 9:3231–3236
187. Hong R, Han G, Fernández JM et al (2006) Glutathione-mediated drug release using monolayer protected nanoparticle carriers. J Am Chem Soc 128:1078–1079
188. Wang W, Rusin O, Xu X et al (2005) Detection of homocysteine and cysteine. J Am Chem Soc 127:15949–15958
189. Shibata A, Furukawa K, Abe H et al (2008) Rhodamine-based fluorogenic probe for imaging biological thiol. Bioorg Med Chem Lett 18:2246–2249

190. Ma Y, Zhou M, Jin X et al (2004) Flow-injection chemiluminescence assay for ultra-trace determination of DNA using rhodamine B–Ce(IV)-DNA ternary system in sulfuric acid media. Anal Chim Acta 501:25–30
191. Holeman LA, Robinson SL, Szostak JW, Wilson C (1998) Isolation and characterization of fluorophore-binding RNA aptamers. Fold Des 3:423–431
192. Fang C, Agarwal A, Devi Buddharaju K et al (2008) DNA detection using nanostructured SERS substrates with Rhodamine B as Raman label. Biosens Bioelectron 24:216–221
193. Ying Z, Xiang-Ying S, Bin L (2009) Fluorescent recognition for single- and double-stranded oligonucleotides based on rhodamine B-modified self-assembled bilayers. Chin J Anal Chem 37:665–670
194. Mathews MB, Sonenberg N, Hershey JWB (2007) Translational control in biology and medicine. Cold Spring Harbor Laboratory Press, Cold Spring Harbor, New York
195. Angelichio MJ, Camilli A (2002) In vivo expression technology. Infect Immun 70:6518–6523
196. Hay N, Sonenberg N (2004) Upstream and downstream of Mtor. Genes Dev 18:1926–1945
197. Lippincott-Schwartz J, Roberts TH, Hirschberg K (2000) Secretory protein trafficking and organelle dynamics in living cells. Annu Rev Cell Dev Biol 16:557–589
198. Budisa N (2006) Engineering the genetic code: expanding the amino acid repertoire for the design of novel proteins. Wiley-VCH, New York
199. Dieterich DC, Link AJ, Graumann J et al (2006) Selective identification of newly synthesized proteins in mammalian cells using bioorthogonal noncanonical amino acid tagging (BONCAT). Proc Natl Acad Sci USA 103:9482–9487
200. Beatty KE, Liu JC, Xie F et al (2006) Fluorescence visualization of newly synthesized proteins in mammalian cells. Angew Chem Int Ed 45:7364–7367
201. Beatty KE, Tirrell DA (2008) Two-color labeling of temporally defined protein populations in mammalian cells. Bioorg Med Chem Lett 18:5995–5999
202. Chang PV, Prescher JA, Hangauer MJ, Bertozzi CR (2007) Imaging cell surface glycans with bioorthogonal chemical reporters. J Am Chem Soc 129:8400–8401

Long-Wavelength Probes and Labels Based on Cyanines and Squaraines

Leonid D. Patsenker, Anatoliy L. Tatarets and Ewald A. Terpetschnig

Abstract In this review, we make an attempt to compare the characteristics and applications of red and near infrared cyanine and squaraine dyes used for biological research, biomedical assays, and high-throughput screening. While the favorable photophysical properties of cyanine dyes makes them predestined as covalent labels, the environmentally sensitive squaraine dyes are utilizable as both florescent probes and labels. Reducing the aggregation tendencies of these dyes in aqueous media seems to be one of the most promising ways to improve their brightness, fluorescence lifetimes, and photostability. Indolenine-based squaraines including ring-substituted squaraines exhibit great potential for the design of bright and sensitive fluorescent probes and labels with increased photostability.

Keywords Cyanines · Labels · Probes · Squaraines

Contents

L.D. Patsenker (✉)
State Scientific Institution "Institute for Single Crystals", National Academy of Sciences of Ukraine, 60 Lenin Ave., Kharkiv 61001, Ukraine
SETA BioMedicals, LLC, 2014 Silver Ct East, Urbana, IL 61801, USA
e-mail: patsenker@isc.kharkov.com

A.L. Tatarets
State Scientific Institution "Institute for Single Crystals", National Academy of Sciences of Ukraine, 60 Lenin Ave., Kharkiv 61001, Ukraine

E.A. Terpetschnig
SETA BioMedicals, LLC, 2014 Silver Ct East, Urbana, IL 61801, USA

A.P. Demchenko (ed.), *Advanced Fluorescence Reporters in Chemistry and Biology I:*   65
Fundamentals and Molecular Design, Springer Ser Fluoresc (2010) 8: 65–104,
DOI 10.1007/978-3-642-04702-2_3, © Springer-Verlag Berlin Heidelberg 2010

1   Introduction

The favorable spectral and photophysical properties of cyanine and squaraine dyes including *long-wavelength absorption and emission* maxima, *adequate quantum yields* (Φ_F) and, in particular, *high molar absorption coefficients* (ε_M) made them desirable chromophoric systems for the design of fluorescent probes and reactive labels for biomedical applications. Other advantages of dyes absorbing in the red and near-infrared (NIR) such as the reduction in unwanted background fluorescence and scattered light signals and compatibility with simple and robust excitation light sources (diode lasers and LEDs) are well known [1]. In addition the interferences from Raman scattering are usually negligible in the long wavelength region. Almost all long-wavelength fluorophores have relatively small Stokes shifts, and in some cases, only about 10 nm. Long-wavelength organic fluorophores are expected to have shorter fluorescence lifetimes as compared to those emitting in the UV/visible. The lifetime properties are in particular important when it comes to measuring fluorescence polarization as the molecular weight of the antigen that can be measured in an assay is directly related to the lifetime of the fluorescently labeled tracer molecule [2]. Again, new developments in this field are showing promising results also in this direction and are discussed below.

There are several important features that determine the performance of organic dyes as fluorescent probes and labels:

- *Brightness* defined as the quantum yield multiplied by the molar absorption coefficient,
- Sufficient *chemical stability* under assay conditions, and
- Adequate *photostability*.

Fluorescence lifetime-based applications require probes and labels with *environment-sensitive lifetimes*, while immunoassays or hybridization-based analysis require fluorescent tracers preferably labeled with a single, *mono-reactive* fluorescent label.

The number of new NIR fluorophores that can be used in biological systems has grown substantially in recent years as a consequence of extensive research efforts to improve the properties of available dyes. A brief overview of the various types of long-wavelength (above 600 nm) fluorophores including phycobiliproteins, **BODIPY**, and **Alexa Fluor** dyes (*Life Technologies*), **Cy** dyes (*GE Healthcare*),

Atto dyes (*Atto-Tec*, *Sigma-Aldrich*), **DY** dyes (*Dyomics GmbH*), fluorescent labels offered by *Li-Cor Biosciences*, and **Puretime** dyes (*Assaymetrics*), and their applications in biomedical research is provided in a recent review [3]. A comprehensive review [4] discusses the literature on the synthesis, spectral properties, aggregation, *cis-trans*-photoisomerization, behavior with surfactants and cyclodextrins, and applications of cyanine dyes as fluorescent sensors and in photodynamic therapy up to the year 2000. Labeling of biomolecules with visible and NIR-emitting dyes particularly cyanines and squaraines was reviewed in [5]. Various classes of dyes including the bioconjugation of dyes are very well reviewed in the book by Hermanson [6]. Other recent reviews describing the synthesis, properties, and numerous applications of cyanines and squaraines are [7, 8].

2 Cyanines: Fluorescent Labels for Low- and High-Molecular-Weight Analytes

The history of cyanine dyes goes back already more than 150 years with the first reported synthesis of a blue solid by Williams in 1856 [9]. Cyanines have found widespread use as sensitizers for color photography, as fluorescent probes and labels for biomolecules such as proteins and nucleic acids, and environmentally sensitive probes for reporting on the local properties such as viscosity or polarity [4]. The absorption and emission wavelengths of cyanines can be tuned to span the entire visible and the near infrared (NIR) by changing the number of methine groups [10], and they are soluble in a variety of different solvent systems and allow introduction of flexible chemistry that enables *covalent* attachment to other molecules [11]. *Indolenine*-based cyanines (indocyanines) are more attractive for the design of fluorescent probes and labels due to their *higher photostability* compared to cyanines containing benzoxazole, benzothiazole, and benzoselenazole end-groups [7].

In the early nineties, Alan Waggoner and his group at Carnegie Mellon University developed the first *reactive* cyanine dyes for biological applications [11]. These dyes were based on *water-soluble* structures that enabled *covalent* labeling of proteins and later also oligonucleotides. The design was simple and it was found that sulfo-groups introduced directly into the indolenine backbone would show best performance in regards to aggregation tendencies and brightness and later became the **Cy** series of dyes that was commercially available from *Amersham* and now *GE*. This series of dyes is represented by *trimethines* (**Cy3**, **Cy3.5**), *pentamethines* (**Cy5**, **Cy5.5**), and *heptamethines* (**Cy7**). These dyes absorb/emit at 550/570 nm (**Cy3**), 649/670 nm (**Cy5**), 675/694 nm (**Cy5.5**), and 747/776 nm (**Cy7**) and have high molar absorptivities (ε_M) 150,000, 250,000, 250,000, and 200,000 M^{-1}cm^{-1}, respectively, in aqueous media (Fig. 1). Cy labels contain two sulfo groups at *positions 5* and one reactive group on an alkyl spacer at *position 1* of indolenine moiety. One of the obvious *shortcomings* of the **Cy** dye series is their *higher aggregation* tendency

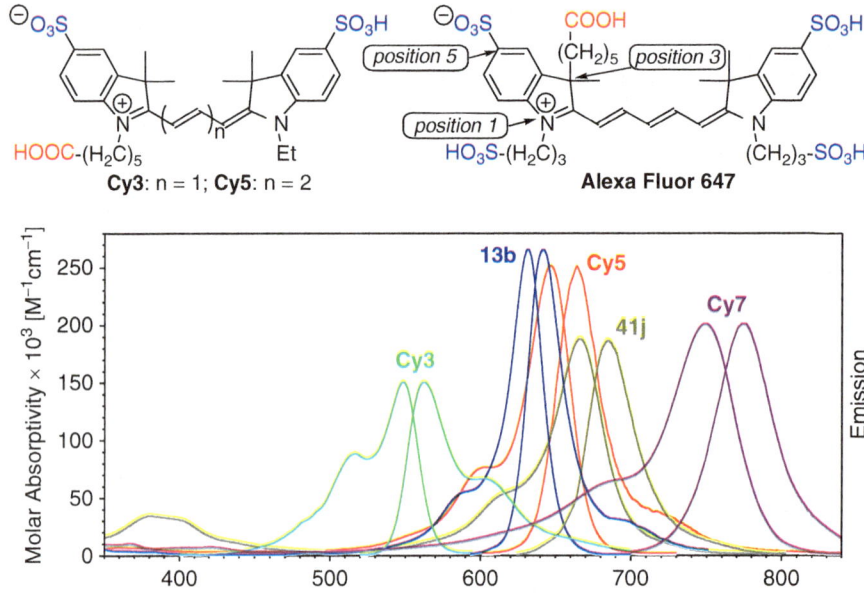

Fig. 1 Absorption and emission spectra of **Cy3**, **Cy5**, **Cy7**, oxo-squaraine **13b** (part 3 of this chapter) and dicyanomethylene squaraine **41j** (part 4 of this chapter) in water (pH 7.4)

upon binding of more than one dye to a protein, resulting in *reduced brightness* for some protein-conjugates [12]. **Cy3**-labeled antibodies fluoresce very well, even at high dye-to-protein ratios, including when bound to (strept)avidin with up to four **Cy3** molecules. On the other hand, antibodies with six covalently bound **Cy5** labels are almost nonfluorescent, and only at 2–3 **Cy5** labels/IgG, a moderate fluorescence yield is obtained.

Shifting the carboxyalkyl group from *position 1* to *position 5* of the indolenine system does not seem to have a noticeable effect on the spectral and photophysical properties of **Cy5** [13].

The self-quenching behavior of cyanine dyes could be improved by increasing the number of sulfo-groups in the heterocyclic end-groups in **Cy3/Cy5**. These additional sulfo-groups were introduced in *position 1* while the reactive carboxylic functionality was shifted to *position 3* in the indolenine ring, a concept that was first introduced by Licha et al. [14]. Since then many other groups and companies have developed water-soluble cyanine labeling reagents based on modifications at the *3-position* in the indolenine ring. Most of the recently developed red and NIR dyes of the **Alexa Fluor**, **DyLight**, **HyLight**, and the **Seta** dye series of fluorescent labels are all based on such 3-modified indocyanine dyes.

The excitation and emission maxima of **Alexa Fluor 647**, **680**, and **750** (the numbers indicate approximate absorption maxima in nanometers), and the molar absorptivities are in the same range as those for **Cy5**, **Cy5.5**, and **Cy7**, but **Alexa Fluor** dyes are reported to have major advantages in terms of their photostability,

Fig. 2 Photostability of **Cy5**, oxo-squaraine **13b** (part 3 of this chapter) hio- (**41g**), and dicyanomethylene squaraine **41j** (part 3 of this chapter) in water (pH 7.4) measured by decay of the fluorescence intensity when exposed to light from a halogen lamp (200 W) [17, 18]

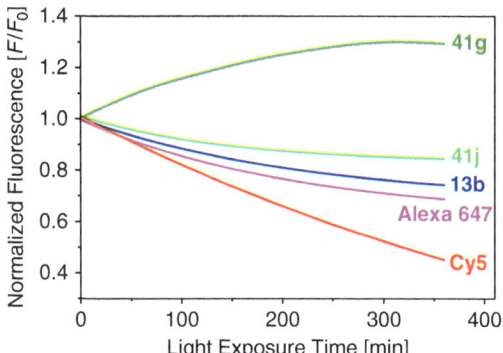

lower tendency to aggregate, and most importantly their ability to produce brighter fluorescent protein conjugates [15, 16]. However, our own investigations show that the increase in photostability for **Alexa Fluor 647** and its protein conjugates compared to **Cy5** is *not* very pronounced (Fig. 2) [17–19]. Both **Cy** and **Alexa Fluor** dyes have small Stokes shifts (20–30 nm).

The long-wavelength **DY** dyes from *Dyomics* are also based on cyanines but with quite different heterocyclic end-groups [20]. These dyes contain an indolenine moiety on one side of the polymethine chain and a benzopyrylium group at the other. Despite their highly unsymmetrical structure, the *Stokes shifts* of **DY** dyes are also small (17–28 nm). These dyes have molar absorptivities in the range of 100,000–200,000 $M^{-1}cm^{-1}$ and the *quantum yields* and *lifetimes* are dependent on the environment, e.g., the initial fluorescence lifetimes of **DY 630** and **DY 635** increases when bound to BSA and DNA. *Dyomics* also offers *large Stokes shift* dyes based on substituted pyridinium and coumarin end-groups. As expected, these dyes do have lower molar absorptivities ($\sim$50,000 $M^{-1}cm^{-1}$). Dyes with an increased number of sulfo-groups, showing reduced aggregation tendencies, are also available from this vendor. Binding of biotinylated **DY-635** and **DY-647** to streptavidin has a significant influence on both the intensity and anisotropy decays of the dyes and, in particular, the fluorescence anisotropy significantly changes upon binding to protein [21].

Red and NIR *bridged* cyanines, such as **HiLight** dyes, which are commercially available from *Anaspec Inc.,* are based on a concept that was described in the following cited patent applications [22–24].

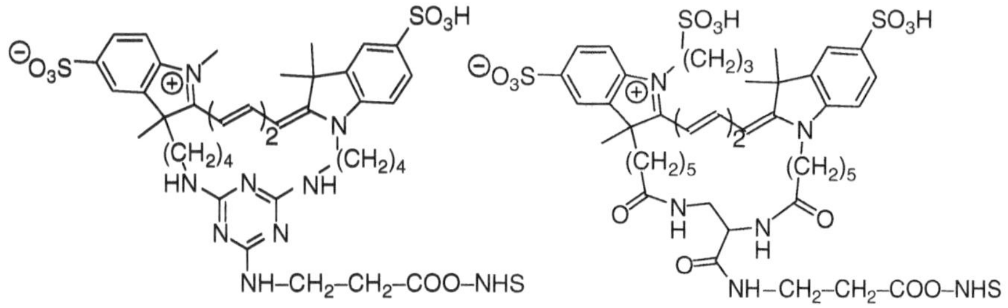

WO 03087052   **Bridged cyanine dyes**   **HyLight Dyes**

The fluorescent properties of another series of *bridged* dyes – symmetrical, cationic cyanines, and merocyanines containing a trimethylene bridge wherein the N-atom of the indolenine residue is bridged with the α-position of the polymethine chain are described in [25]. The constraining group influences their fluorescent properties both by steric and electronic effects leading to either an essential decrease or increase of the fluorescence quantum yields depending on the type of cyanine dye.

Cyanine dyes containing one or two *phosphate* groups attached directly to the aromatic indolenine residues were recently reported by M. Reddington [26].

A comprehensive study of the spectroscopic characteristics (absorption, emission, and fluorescence lifetime) of 13 commercially available red-absorbing fluorescent dyes (**Alexa Fluor 647, ATTO 655, ATTO 680, BODIPY 630/650, Cy5, Cy5.5, DiD, DY 630, DY 635, DY 640, DY 650, DY 655**, and **EVOblue30**) was provided by Sauer et al in 2003 [27]. The influences of polarity, viscosity, and the addition of detergent (Tween 20) on the spectroscopic properties were also investigated, and fluorescence correlation spectroscopy (FCS) data was utilized to assess the photophysical properties of these dyes under extreme excitation conditions. The study revealed that these dyes can be classified into groups based on the results presented: the fluorescence quantum yield of some of these dyes is primarily controlled by the polarity of the surrounding medium, and more hydrophobic and structurally flexible dyes of the **DY**-family are strongly influenced by the viscosity of the medium and the addition of detergents. Studies of biotin-conjugates and subsequent titration with streptavidin are also discussed.

A different type of cyanines of general formula **1** containing a *bridging group in the polymethine chain* were developed by Patonay et al. [28] at Georgia State University. Heptamethine cyanine dyes **1b,c** with extremely large Stokes shifts in the order of 150 nm were reported by X. Peng et al. [29]. These new dyes are very different from conventional cyanine dyes in their spectral properties: They have broader spectra and much stronger fluorescence than the parent dye **1a**. The maximum absorption wavelengths of the dyes (**1b,c**) exhibit a large blue shift and a large Stokes shift (>140 nm). These properties are usually not found in other heptamethine dyes. Nevertheless, the quantum yields and molar absorptivities of these dyes in water are *low* ($\Phi_F \sim 0.07$, $\varepsilon_M \sim 50–70,000$ M^{-1}cm^{-1}). Using the same principle, *Li-Cor Biosciences* developed a series of NIR cyanine dyes **IRDye 800** for their *Odyssey* imaging system [30].

An interesting imaging probe **1d** that can *selectively target bacteria* was recently reported by Smith et al. [31] also based on a heptamethine chromophore. The probe is composed of a bacterial affinity group, which is a synthetic zinc (II) coordination complex that targets the anionic surfaces of bacterial cells and a near infrared dye. The probe allowed detection of *Staphylococcus aureus* in a mouse leg infection model using whole animal near-infrared fluorescence imaging.

Various *hybrid* compounds comprised of two types of nitroxide radicals and either a pentamethine (**Cy5**) or trimethine cyanine (**Cy3**) were synthesized by Sato and co-workers [32]. These compounds seem to be promising fluorescent *chemosensors* for the measurement of reducing species such as Fe^{2+}, ascorbic acid, and hydroxyl radicals.

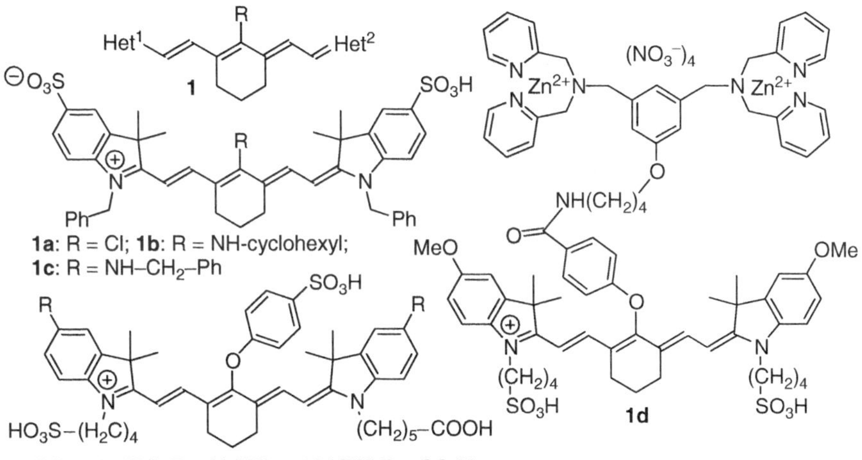

1a: R = Cl; **1b**: R = NH-cyclohexyl;
1c: R = NH–CH$_2$–Ph

IRDye 800RS: R = H; **IRDye 800CW**: R = SO$_3$H

Reference [33] describes recent progress on cyanine probes that bind *noncovalently* to DNA, with a special emphasis on the relationship between the dye structure and the DNA binding mode. Some of the featured dyes form well-defined helical aggregates using DNA as a template. This reference also includes spectroscopic data for characterizing these supramolecular assemblies as well as the monomeric complexes.

The synthesis and characterization of a somatostatin *receptor-specific peptide* H$_2$N-(DPhe)-cyclo[Cys-Phe-(D-Trp)-Lys-Thr-Cys]-Thr-OH, labeled with an indodicarbo- and an indotricarbocyanine dye at the *N*-terminal amino group were described in [34]. The ability of these fluorescent *contrast agents* to target the somatostatin receptor was demonstrated by flow cytometry in vitro, wherein the indotricarbocyanine conjugate led to elevated cell-associated fluorescence on somatostatin receptor-expressing tumor cells. The intracellular localization was visualized using NIR fluorescence microscopy.

Berezin et al. [35] show that the *fluorescence lifetime* of NIR polymethine dyes are highly sensitive to the *polarity* of solvents and biological mediums and further found that the fluorescence decays of polymethine dyes are monoexponential in a wide variety of solvents, except where they are not completely soluble or they form aggregates. A fluorescence lifetime based polarity index was developed for predicting the polarity of complex systems. This method overcomes the *poor solvatochromism* of these biocompatible NIR dyes and provides a strategy to translate *in vitro* studies to *in vivo*. They also attempt to correlate fluorescence lifetime with solvent orientation polarizability and aim to develop a *lifetime polarity index* for determining the polarity of complex systems, including micelles and albumin binding sites.

Recent developments on the *synthesis* and *applications* of cyanine dyes are reviewed in [36].

3   Oxo-Squaraines: Microenvironment-Sensitive Dyes

Squaraines (conventional *oxo-squaraines* and related compounds) are one of the most promising classes of dyes for use as long-wavelength fluorescent probes and labels [1, 17, 37]. With respect to the nature of their chromophoric system one can differentiate between three general types of oxo-squaraines (Fig. 3): (a) Squaraine ring is conjugated to two nitrogen-containing heterocycles via a methylene group ("*heterocyclic*" squaraines) [38, 39], (b) squaraine ring directly conjugated with two aromatic or heteroaromatic rings ("*aromatic*" squaraines) [40, 41], and (c) "*mixed*" squaraines, wherein the squaraine ring is conjugated with one methylene-heterocyclic and one aromatic end-group [42]. Another way to differentiate these dyes is according to their symmetry: *symmetrical* and *unsymmetrical* squaraines.

Symmetrical pyrrole- and phloroglucinol-based oxo-squaraines **2** and **3** were first synthesized by Treibs and Jacob in 1965 [43, 44] and aniline-based squaraines **4**

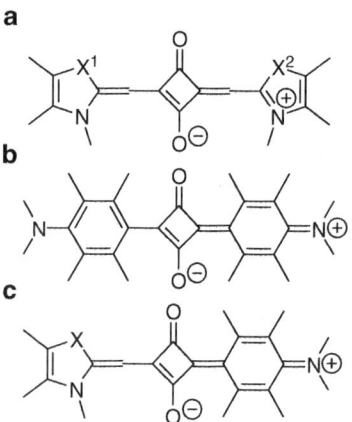

Fig. 3 Examples for different squaraine chromophores

by Sprenger and Ziegenbein in 1966 [38]. Symmetrical oxo-squaraines **5** with heterocyclic end-groups were also described by these authors in 1966–1967 [39, 40].

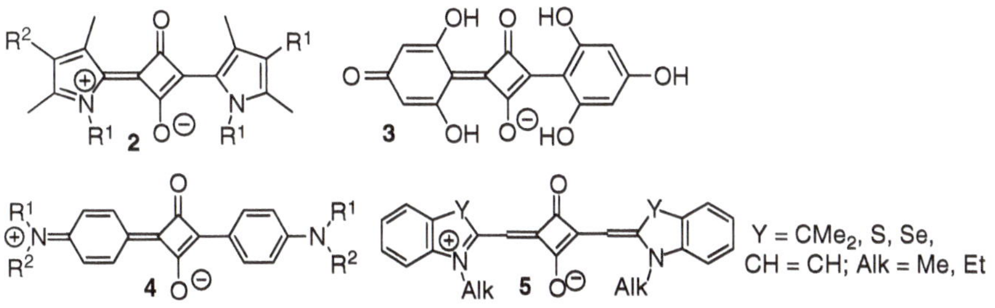

In general, *symmetrical* oxo-squaraines having the same end-groups are synthesized by reacting squaric acid with two equivalents of quaternized indolenine, 2-methyl-substituted benzothiazole, benzoselenazole, pyridine, quinoline [39, 45, 46] (Fig. 4) in a mixture of 1–butanol – toluene or 1–butanol – benzene with azeotropic removal of water in presence [39, 45] or absence [47] of quinoline as a catalyst. Other reported solvent systems include 1–butanol – pyridine [48], 1–propanol – chlorobenzene, or a mixture of acetic acid with pyridine and acetic anhydride [49]. Low CH-acidic, heterocyclic compounds such as quaternized aryl-azoles and benzoxazole do not react, and the corresponding oxo-squaraines cannot be obtained using this method [23, 50].

The key intermediates for the synthesis of *unsymmetrical*, "heterocyclic" oxo-squaraines are the *mono-squaraines* (semi-squaraines) shown in Fig. 5. These intermediates can be synthesized via condensation of dialkylsquarate with an equimolar amount of methylene base [51]. The obtained alkoxy-mono-squaraines are then reacted with the second methylene base to yield unsymmetrical oxo-squaraines. These mono-squaraine intermediates display a higher reactivity compared to squaric acid or its esters: they allow the synthesis of the corresponding

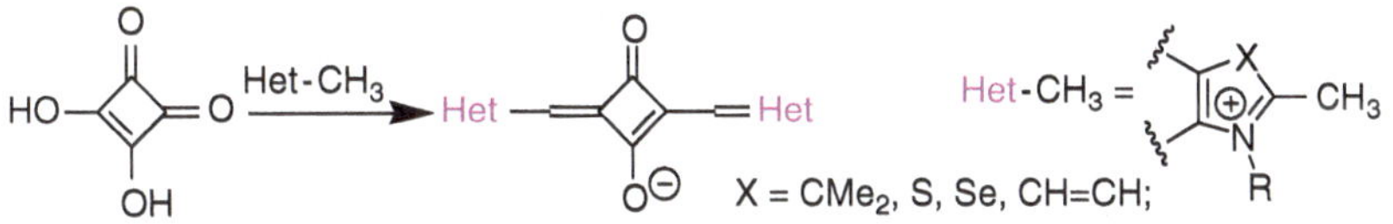

Fig. 4 Synthesis of symmetrical "heterocyclic" oxo-squaraines

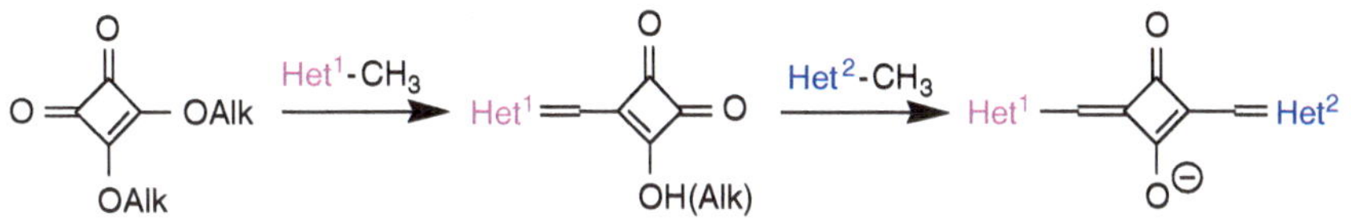

Fig. 5 Synthesis of unsymmetrical "heterocyclic" oxo-squaraines

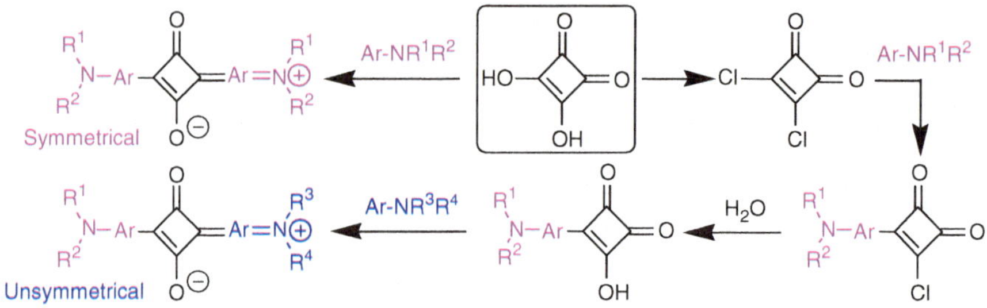

Fig. 6 Synthesis of symmetrical and unsymmetrical aniline-based oxo-squaraines

unsymmetrical squaraines based on quaternized *aryl-azoles* and *benzoxazoles* which cannot be obtained by reaction with squaric acid [50]. Unsymmetrical oxo-squaraines can also be synthesized by *cross-reacting* squaric acid with two different heterocyclic compounds [52]. Nevertheless, the yield of the unsymmetrical product will strongly depend on the relative reactivities of the two heterocycles used in the reaction.

Symmetrical, *aniline*-based, and *aromatic* oxo-squaraines are synthesized via a one-step reaction by heating two equivalents of the appropriate *N,N*-dialkylaniline or other reactive aromatic or heteroaromatic derivatives with squaric acid (Fig. 6) [38, 41]. *Unsymmetrical aniline*-type squaraines can be synthesized in two steps: first one component is reacted with squaric acid *dichloride* to yield a mono-squaraine intermediate, which in a subsequent step is then reacted with the second component to yield the unsymmetrical squaraine dye [53].

"*Mixed*" squaraines can be synthesized via "heterocyclic" or "aromatic" mono-squaraines [49]. Symmetrical and unsymmetrical oxo-squaraines and mono-squar-aine intermediates can be obtained also by *formation of squaraine ring* [42].

Due to the fact that the conventional, *open-chain cyanine* chromophore has an *ionic* structure with a delocalized positive charge, it is slightly soluble in aqueous solutions even without additional hydrophilic groups, while the *zwitter-ionic squar-aine* chromophore is neutral and therefore water-insoluble. As compared to the relatively nonpolar, open-chain cyanines, the squaraine chromophore exhibits a more *polar* structure due to the negatively charged squaraine oxygens. The higher polarity of squaraine dyes is the reason for their lower fluorescence quantum yields in aqueous solutions. Because of these differences, the quantum yields of squaraine dyes substantially *increase* in hydrophobic media or upon binding to high-molecu-lar-weight species (e.g., proteins), while for open-chain cyanines, they remain almost unchanged or are sometimes even reduced [18]. Squaraine dyes are there-fore well suited as fluorescent *probes* while open-chain cyanines such as **Cy5** or **Alexa 647** are better suited for *labeling* of small molecules such as peptides, nucleotides, and amino acids. However, increasing the number of hydrophilic functionalities such as sulfo groups in the squaraine chromophoric system helps to increase the quantum yield in aqueous media, thus making these dyes also useful labels for small molecules [18, 54, 55].

The synthesis of oxo-squaraines and related compounds, including their spectral properties and applications as biomedical probes, photoconducting materials, and photosensitizers are provided in a recent review [56].

3.1 Oxo-Squaraine Probes

3.1.1 Probes for Cellular Imaging, Proteins, and Other High-Molecular-Weight Analytes

The synthesis, spectral properties, and applications of symmetrical as well as unsymmetrical, hydrophobic oxo-squaraine *probes* for *noncovalent* interaction with proteins, lipids, cells, and other high-molecular-weight analytes are described in numerous publications and patents [52, 57, 58].

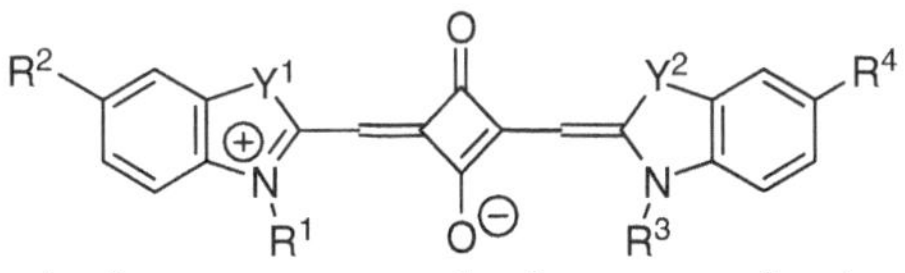

6: Y^1, Y^2 = CMe_2, S, Se; R^1, R^3 = Me, Et; R^2, R^4 = H, Cl

For the purpose of identifying long-wavelength fluorescent probes with good *quantum yields*, reasonably *long lifetimes* and *high photostabilities* for use in *fluorescence-based assays* and/or *imaging*, Terpetschnig et al. synthesized and tested a series of symmetrical and unsymmetrical squaraines 6 and investigated their absorption and fluorescence properties in different organic solvents and in water in presence of BSA [37, 45]. In general, when used as noncovalent probes, squaraine dyes exhibit increased quantum yields and fluorescence lifetimes when bound to proteins such as bovine serum albumin (BSA) [45, 59]. Like normal open-chain cyanines, squaraine dyes were found to have *negative solvatochromism* [45, 60]. Based on their investigation, the authors concluded that the most suitable probes for use in a biological application are the *symmetrical indolenine* squaraines, which also displayed the *highest photostability*. Their quantum yields and fluorescence lifetimes increase significantly upon binding to BSA, suggesting that a conjugatable derivative of these indolenine squaraines would be suitable for use as covalent labels of proteins. Importantly, the absorption maxima of these squaraines between 630 and 650 nm are well-suited for excitation with the same excitation sources and filter sets as **Cy5**.

A series of unsymmetrical oxo-squaraines 7, containing both heterocyclic and aniline-based end-groups, was synthesized and their absorption spectra were investigated [61]. Some of these dyes, absorbing between 680–820 nm with high molar absorptivities (log ε_M = 4.92–5.18) in chloroform, have the potential to be used as fluorescent probes.

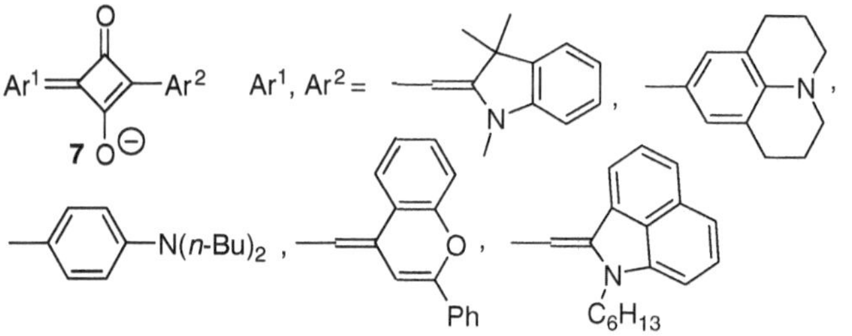

The *sulfo*-group containing squaraine–taurine probe **8** [62] displayed a very high affinity for BSA and other blood proteins. This probe exhibits low quantum yields and short fluorescence lifetimes in water and a significant increase of these characteristics upon binding to proteins.

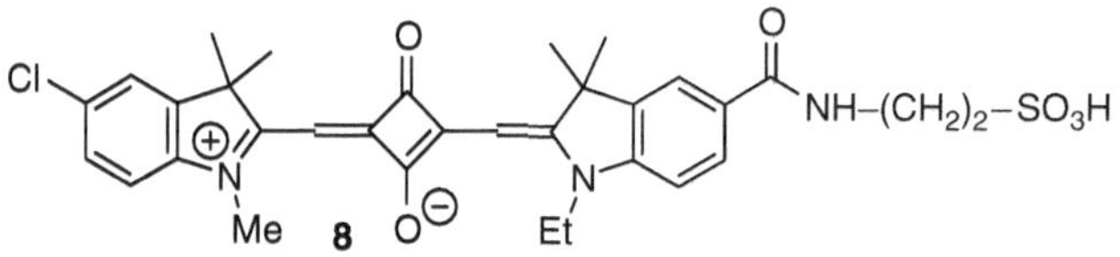

The noncovalent binding of a series of oxo-squaraine dyes **9a–e** to BSA was evaluated by measurement of absorption, emission, and circular dichroism [63]. The magnitude of the association constants (K_S) for the dye–BSA complexes depended on the nature of the side chains and ranged from 34×10^3 to 1×10^7 M^{-1}. Depending on the side chains, the K_S increase in the order: [$R^1 = R^2$ = butyl-phthalimide] <[$R^1 = R^2$ = cetyl] <[$R^1 = R^2$ = ethyl] <<[R^1 = butyl-phthalimide, R^2 = butyl-sulfonate] <<[$R^1 = R^2$ = butyl-sulfonate]. These dyes seem to interact mainly with a hydrophobic cavity on BSA. However, the association constants K_S increase substantially when the side chains are selected from *butyl sulfonate*.

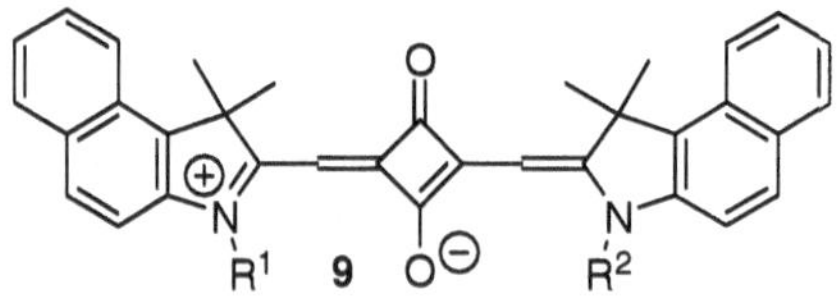

9: R^1, R^2 = Et (**a**), –(CH$_2$)$_4$–SO$_3$H (**b**), *N*-(*n*-butyl)phtalimide (**c**), cetyl (**d**), *N*-(*n*-butyl)phtalimide, –(CH$_2$)$_4$–SO$_3$H(**e**), –(CH$_2$)$_5$–COOH, –(CH$_2$)$_4$–SO$_3$H (**f**), butyl (**g**)

Hydrophobic (**10a**) and hydrophilic (**10b**) squaraines show a noticeable increase in fluorescence intensity in presence of HSA and importantly dye **10b**, containing a *sulfo* group, exhibits a large intensity increase when bound to avidin, a protein well-known to quench many fluorescent dyes [58].

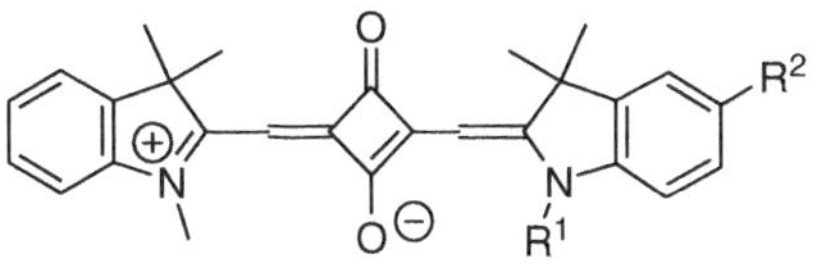

R[2] **10**: R[1] = Me, R[2] = H (**a**);
R[1] = (CH$_2$)$_5$–COOH, R[2] = SO$_3$H (**b**)
R[1] = n–Pr, R[2] = H (**c**)

A series of squaraine dyes was investigated by Nakazumi et al. as fluorescent probes for *capillary electrophoresis* (CE) based analysis of proteins [64, 65]. The fluorescence emissions of symmetrical (**9a–d**) and unsymmetrical (**9e–f, 10c, 11,** and **12**) squaraines were shown to be significantly enhanced (up to 97%) upon binding to proteins such as BSA, HSA, β-lactoglobulin A, and trypsinogen [65]. The nature of the interaction between these dyes and proteins was studied using different buffer systems, and it was found that electrostatic interactions are involved but not dominant. Dye-to-protein stoichiometries in the noncovalent complexes were found to be 1:1 for **10c, 11a–d,** and **12,** although various possible stoichiometries were found for **9a** depending upon pH and the nature of the protein. Association constants in the order of 10^5 and 10^7 were found for noncovalent complexes of **9a** and **11a** with HSA, indicating stronger interactions of **9a** with proteins. Dye **9a** was found to be a promising reagent capable of fluorescently labeling proteins for analysis by *capillary electrophoresis* (CE) with laser-induced fluorescence detection (CE–LIF).

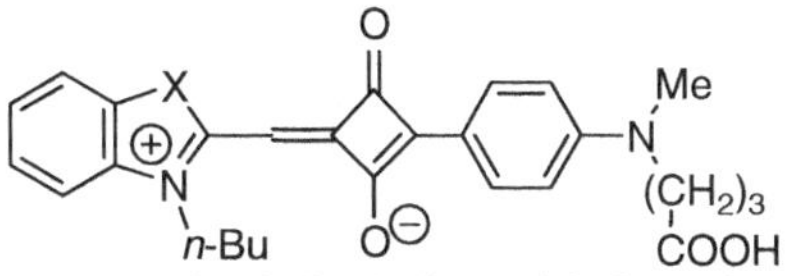

11: X = S (**a**), O (**b**), CMe$_2$ (**c**), CH = CH (**d**)   **12**

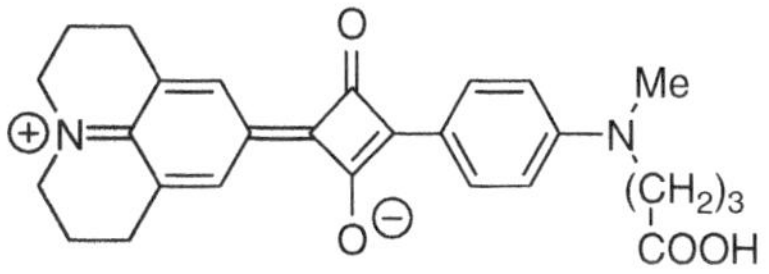

The *unsymmetrical* squaraine dye **11c** was also successfully utilized as a non-covalent probe for a variety of proteins [64]. This dye resulted in a *lager* fluorescence enhancement when complexed with BSA than the more symmetrical dye **10c**. Dye **11c** demonstrated a greater affinity toward BSA relative to β-lactoglobulin A in *capillary electrophoresis-frontal analysis* [66]. The ability to quantify the strength of noncovalent interactions between a specific fluorescent probe **11c** and protein analytes of interest (BSA and β-lactoglobulin A) can be used to optimize the separation conditions in regards to separation efficiency and sensitivity. Successful precolumn labeling experiments using **11c** as a tag for proteins were conducted by microchip and conventional CE [67].

The squaraine probe **9g** was tested for its sensitivity to trace the formation of *protein–lipid* complexes [57]. The binding of dye **9g** to model membranes composed of zwitter-ionic lipid phosphatidylcholine (PC) and its mixtures with anionic lipid cardiolipin (CL) in different molar ratios was found to be controlled mainly by hydrophobic interactions. Lysozyme (Lz) and ribonuclease A (RNase) influenced the association of **9g** with lipid vesicles. The magnitude of this effect was much higher for Lz as for RNase. Varying the membrane composition provided evidence for the dye's sensitivity toward both hydrophobic and electrostatic protein–lipid interactions. Fluorescence anisotropy studies were used to measure the restriction of **9g**'s rotational mobility in the lipid environment in the presence of Lz and RNase. The results of the binding, fluorescence quenching, and kinetic experiments suggest a lysozyme-induced, local lipid demixing upon protein association with negatively charged membranes with threshold concentrations of CL for lipid demixing of 10 mol%.

The influence of the *substitution of the indolenine-nitrogen* on the *spectral properties* and *photostability* was investigated on a series of water-soluble squaraine dyes **13** [60]. The maxima of the absorption and emission wavelengths of the dyes in different solvents (water, methanol, ethanol, DMF, and DMSO) were in the range 628–670 nm (Fig. 1) and exhibited *negative solvatochromism*. All these dyes show reasonable quantum yields (up to 0.22) in organic solvents. Electron-withdrawing substituents such as carboxyl or fluoro on the *N*-benzyl group help to improve the photostability in aqueous solution.

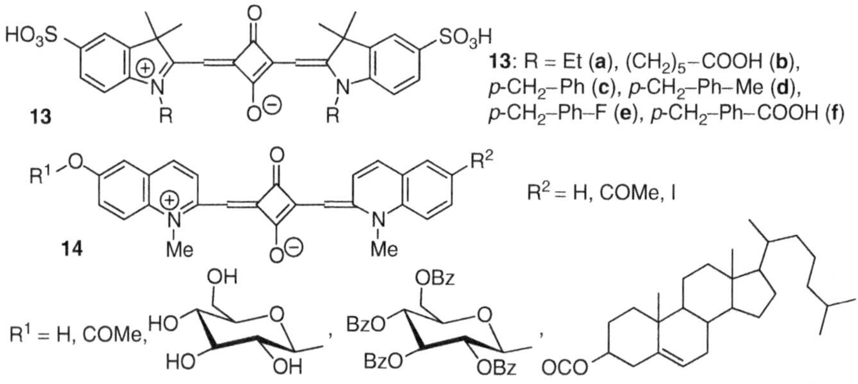

Symmetrical and unsymmetrical quinaldine-based squaraines **14** linked to *cellular recognition* elements that exhibit near-infrared absorption (>740 nm) could have potential biological and *photodynamic* therapeutical applications [68].

Energy transfer systems based on squaraines dye *"cassettes"* **15** were described in [69]. These dyes contain carbazole, phenothiazine, and phenoxazine donor-components that absorb around 300–320 nm, and squaraine acceptor-parts that fluoresce in the range 650–700 nm. The cassettes transfer energy very efficiently to their acceptors when irradiated in the donor absorption region (ca. 300 nm). With respect to fluorescence emission, the cassettes behave like dyes with a huge Stokes shift, absorbing around 300 nm and fluorescing in the near IR region (637–693 nm). Emission at such long wavelengths is of advantage for probing biological systems. The quantum yields of the acceptors in the cassettes are high. Despite these attributes, cassettes **15** have some disadvantages with respect to tagging biological molecules: firstly, the donor fragments display low molar absorptivities and secondly, the molecules have poor water solubility, which makes their conjugation to biomolecules relatively difficult.

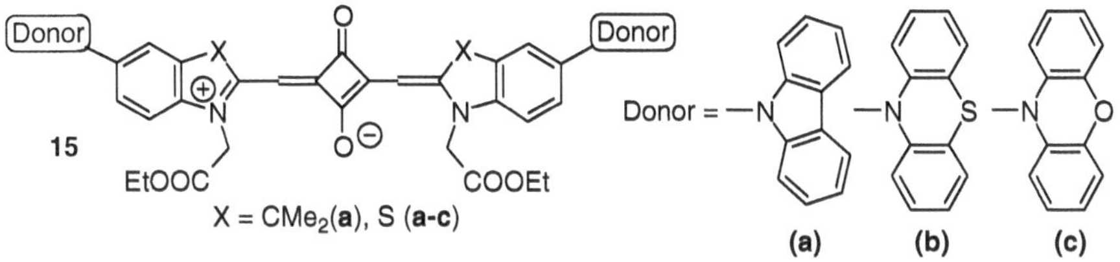

15

X = CMe₂(a), S (a-c)

Donor = (a) (b) (c)

Red and NIR *bis-squaraine* dyes **16**, wherein two squaraine moieties, are linked to an aromatic bridge –Ar–, have been synthesized and used by Nakazumi et al. for protein detection [70]. Depending on the nature of the bridging group, the absorption and emission spectra of **16** can be shifted in comparison to the analogous conventional squaraines. The combination of the pyrene spacer with the benzoindoline heterocycles results in significantly *red-shifted* absorption at 770 nm. Thiophene-bridged dyes display absorption in the NIR between 717 and 807 nm. An increase in the number of thiophene rings in the spacer leads to a *blue-shift* of the absorption maxima due to disruption of π-electron conjugation. Although the initial fluorescence quantum yields for these NIR dyes are low, all dyes exhibit a *quantum yield increase* upon binding to BSA or HSA. Selected dyes of this series containing a carboxyethyl group instead of an *n*-butyl group at the heterocyclic nitrogen were used for *labeling* of HSA or BSA.

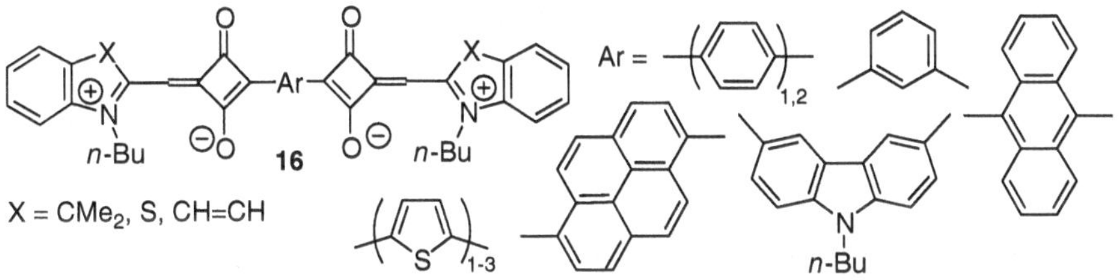

X = CMe₂, S, CH=CH

16

The *aniline-based* squaraines **4** with hydroxyalkyl, carboxypropyl, and ethylene glycol groups exhibit absorption maxima between 640–649 nm ($\varepsilon_M = 100{,}000$–$300{,}000$ $M^{-1}cm^{-1}$) and emission maxima between 663 and 675 nm in aqueous solutions [71]. The quantum yields are 0.15–0.21 in ethanol and 0.01–0.02 in an aqueous medium, but in *micelles*, the quantum yields are five to tenfold increased. The aggregation of these dyes was studied in [53]. The amphiphilic squaraines **4** combine favorable photophysical properties and good solubility in aqueous media and in addition interact efficiently with micelles, and therefore have the potential to be used as NIR fluorescent *sensors*. However, our own investigations show that *aniline-based squaraines lack chemical and photochemical stability* when compared to oxo-squaraines with heterocyclic end-groups.

It has been recently demonstrated that squaraine dye **3**, first synthesized by Treibs and Jacob [40], acts as a *dual mode recognition system* for serum albumin [72]. The significant enhancement in emission quantum yields and lifetimes suggests that **3** can be used as a noncovalent probe in *immunochemical assay* and biophysical studies. This dye is quite soluble in aqueous buffer and interacts with

HSA selectively at site II and signals the event by a visual color change and turning on fluorescence. The less hydroxylated version of **3** without *ortho*-hydroxy groups is used as a *protein-binding indicator*. It noncovalently interacts with proteins and provides a dramatic color change from orange to deep purple upon binding to proteins [73]. This indicator responds to various proteins, and the detection of the proteins is not affected by the presence of contaminants. This dye was successfully used to stain proteins after electrophoresis on one-dimensional SDS-PAGE mini-gels. The staining had high sensitivity and was easy to conduct [73].

Dialkylanthracene-containing squaraine dyes **17** show intense absorption and emission in the NIR region (720–810 nm) [74]. They are compatible with aqueous environments and show substantial enhancement of quantum yields and fluorescence lifetimes in hydrophobic and micellar media, suggesting that these dyes can be potentially useful as fluorescent probes in biological applications, e.g., for imaging of hydrophobic domains such as *cell membranes*.

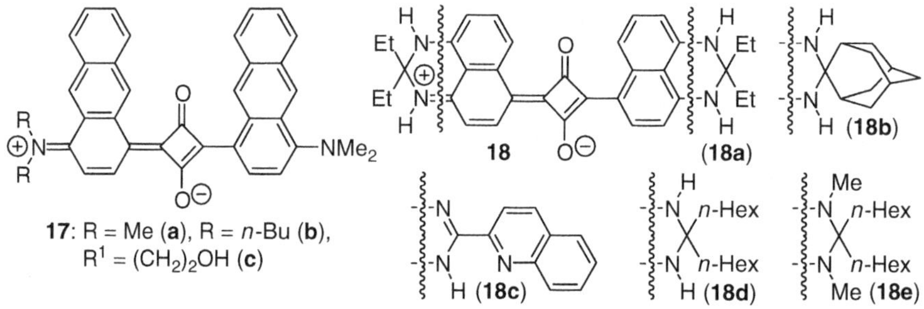

17: R = Me (**a**), R = *n*-Bu (**b**),
R^1 = (CH$_2$)$_2$OH (**c**)

Symmetrical 2,3-dihydroperimidine-based squaraines **18** absorb with high molar absorptivity at even longer wavelengths (>820 nm) [75]. Their photostability increases in the order: **18b** < **18a** < **18c**. The aggregation tendencies were studied in DMSO/water solutions. The *N*-methyl substituted perimidine-squaraine **18e** shows certain advantages compared to **18d**: it exhibits a higher molar absorptivity (200,000 M^{-1}cm^{-1}), a *sharper emission band*, a *higher quantum yield* (0.56 in cyclohexane), and *positive* solvatochromism [76]. These properties suggest that **18e** can function as NIR-emitting probe for *proteins*, *lipids*, and *cells*.

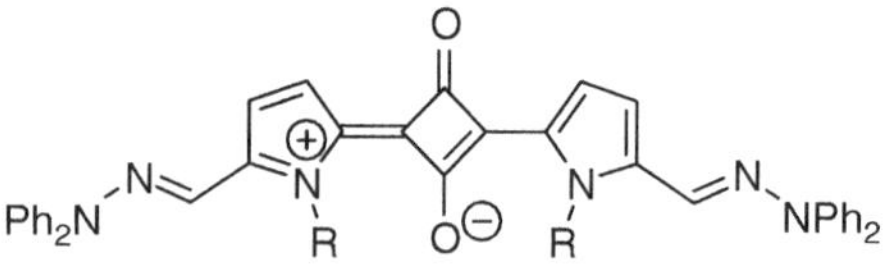

19: R = –(CH$_2$)$_{10}$Me, –CH$_2$O(CH$_2$CH$_2$O)$_2$Me

Also, the recently introduced *hydrazone* end-capped squaraines **19** [77], absorbing in the 600–800-nm range, could be promising candidates for the development of fluorescent probes.

3.1.2 Metal Cation Chemosensors

A summary of squaraine-based *metal cation probes* up to the year 2004 is provided in the review article by Ajayaghosh [78]. The described probes are mostly based on bridged bis-squaraines such as **20–22** or crown-esters such as **23a** and **23b**, although other principles were also used. The methine-bridged squaraine probes **20** and **21**, absorbing between 600–850 nm, are capable of detecting trace amounts of *toxic* and environmentally *hazardous* metal cations such as Cu^{2+}, Hg^{2+}, Pb^{2+}, and Mn^{2+} as well as several *lanthanide* metal ions in aqueous media, even in the presence of alkali and alkaline earth metal ions [79]. The synthesis and complexation of methine-bridged bisquaraine dyes **22a–c** with transition metal cations was described in [80]. These dyes exhibit an intense light absorption at 815–825 nm ($\varepsilon_M = 251{,}000{-}392{,}000$ $M^{-1}cm^{-1}$) in chloroform. Complexation with Cr^{3+}, Mn^{2+}, Fe^{2+}, Co^{2+}, and Cu^{2+} results in large hypsochromic shifts of absorption maxima demonstrating that these dyes are potentially applicable as the molecular sensors for *transition metal cations*.

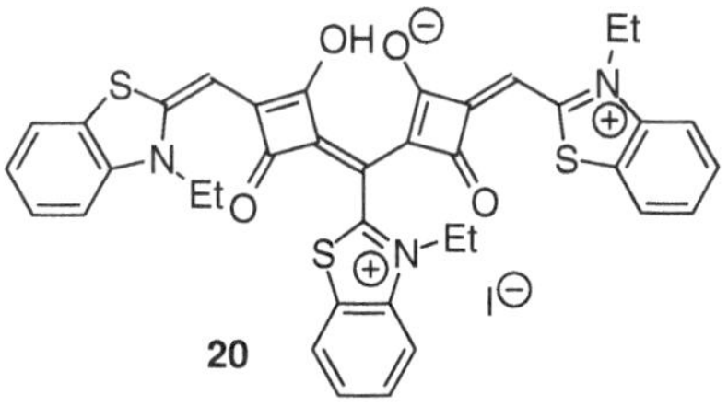

20

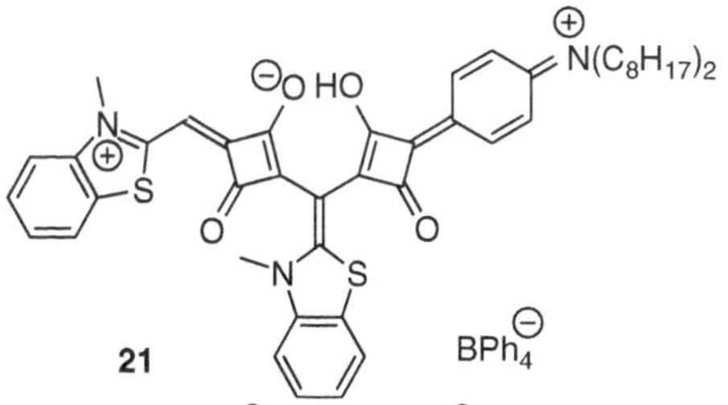

21

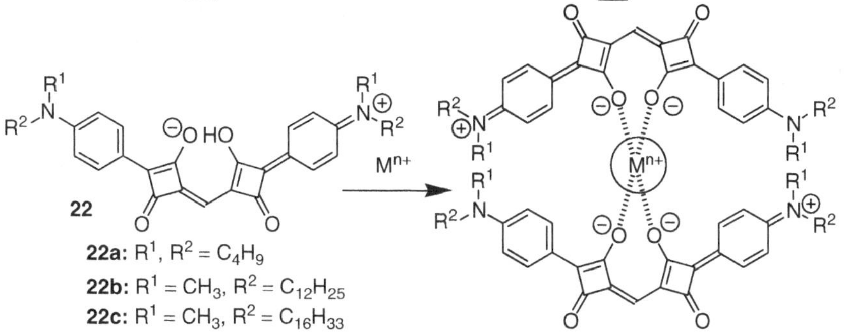

22

22a: R^1, $R^2 = C_4H_9$
22b: $R^1 = CH_3$, $R^2 = C_{12}H_{25}$
22c: $R^1 = CH_3$, $R^2 = C_{16}H_{33}$

An *azacrown*-squaraine-based chemosensor **23a** absorbing at 635 nm ($\varepsilon_M = 260{,}000$ $M^{-1}cm^{-1}$) and emitting at 665 nm signals *alkaline* and *earth-alkaline metal ions* in millimolar concentrations in acetonitrile [81]. In presence of Na^+ ions, the fluorescence signal weakly increases while it significantly decreases in presence of Ca^{2+}, Mg^{2+}, and Ba^{2+} and does not change substantially upon addition of K^+ ions. The same squaraine **23a** and the azacrown-squaraine **23b** [82] were used for Na^+ and K^+ sensing in a plasticized PVC matrix [83]. The squaraine derivatives exhibited fluorescence emission based optical responses to

Na^+ and K^+ with a detection limit of 10^{-9} M. The sensor compositions exhibited wide response ranges between 10^{-9} and 10^{-5} M Na^+ or K^+, and, therefore, may be an alternative method to flame emission spectroscopy. The sensor is fully *reversible* within the dynamic range and the response time is 3 min under batch conditions. Cross sensitivity to pH is negligible in the pH range of 6.2–7.3.

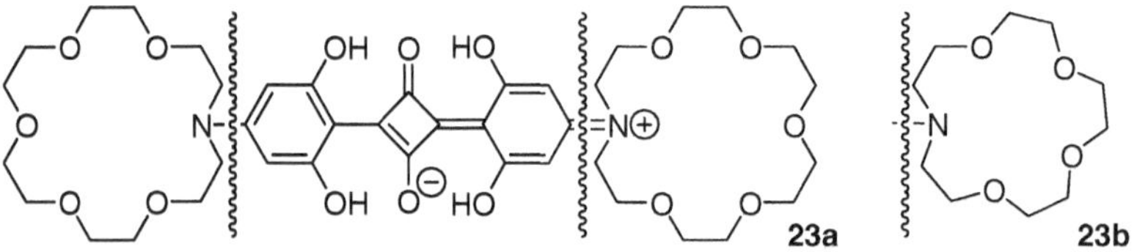

23a   23b

A new class of NIR-emitting squaraine probes **24** with *tetrahydroquinoxaline* moieties as the donor group show an intense absorption at 700 nm with a molar absorptivity ($>100,000$ $M^{-1}cm^{-1}$) [84]. Theses probes can be used for selective detection of Cu^{2+} ions in hydrophobic media such as THF.

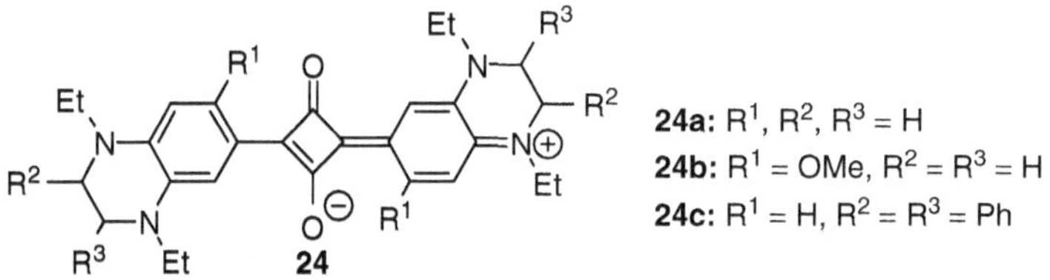

24a: R^1, R^2, R^3 = H
24b: R^1 = OMe, R^2 = R^3 = H
24c: R^1 = H, R^2 = R^3 = Ph

A Ca^{2+}-specific fluorescent chemosensor **25** in aqueous buffer signals Ca^{2+} via a *decrease* in fluorescence intensity, whereas excess of Mg^{2+} ions has no effect on the emission [85]. This probe has limited solubility in aqueous solution after binding to Ca^{2+}. A Zn^{2+} sensitive probe **26** showing different fluorescence responses depending on the complexation stoichiometry is described in [86].

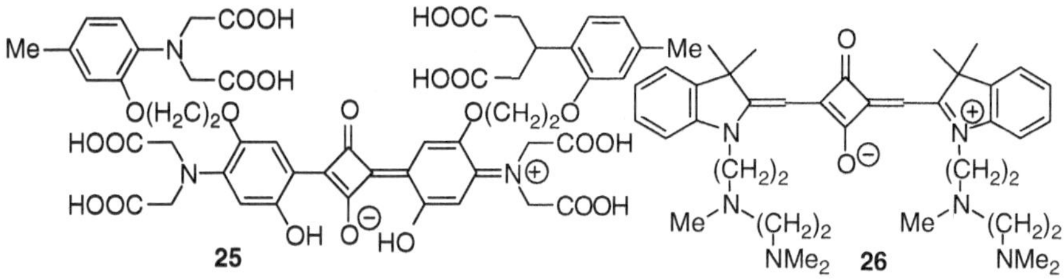

25   26

A Ca^{2+}-ion selective *rigid – flexible – rigid* type bichromophoric sensors based on the conformation liable bis-squaraine dyes **27** works on the principle of Ca^{2+}-ion-steered *folding*, which leads to dramatic perturbations in the optical properties as a result of exciton interactions [87].

The tripodal squaraine probe **28** displays a decrease in intensity in the presence of Ca^{2+} or Mg^{2+} ions; however, the quantum yields are approximately 25–30 times lower when compared to those of "monomeric" squaraines (e.g., **4**) [88].

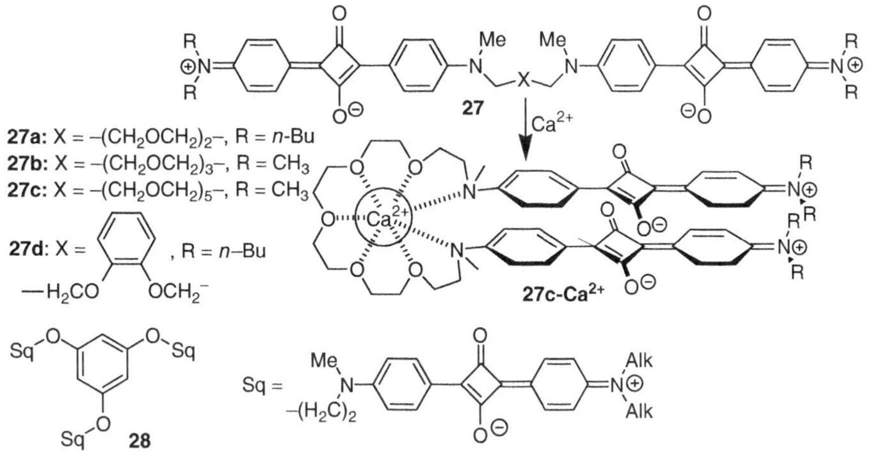
27
27a: X = –(CH2OCH2)2–, R = n-Bu
27b: X = –(CH2OCH2)3–, R = CH3
27c: X = –(CH2OCH2)5–, R = CH3
27d: X = , R = n–Bu
—H2CO OCH2⁻
Ca2+
27c-Ca2+
28
Sq =
Alk

3.1.3 Chemosensor for Carbohydrates

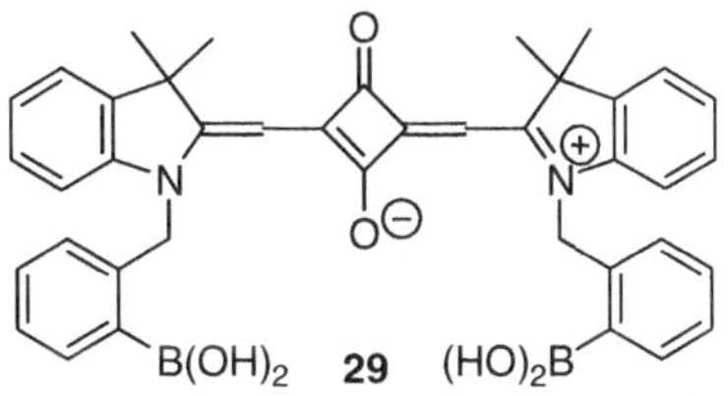
B(OH)2
29
(HO)2B

A long-wavelength probe **29** signaling *carbohydrates* in aqueous solutions by increasing of fluorescence was developed by Akkaya and Kukre on the basis of a symmetrical squaraine dye containing two phenylboronic acid functions [89]. The emission maximum of this probe is at 645 nm. A maximal response of about 25% was found for fructose.

3.1.4 pH Sensor

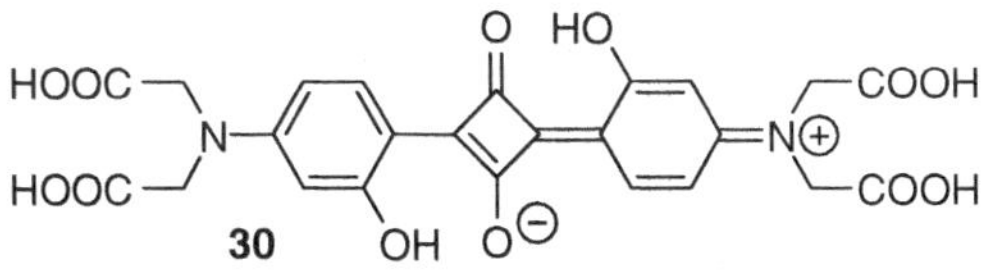
HOOC
HOOC
COOH
COOH
30

The dihydroxyaniline-squaraine chromophore was used by Akkaya and Isgor in the fluorescent chemosensor **30** for the measurement of pH [90]. This chemosensor, having the molar absorptivity about 200,000 $M^{-1}cm^{-1}$ and quantum yield 0.2, signals the pH from 10 to 7 by a 14-fold increase in the emission intensity at 651 nm, when excited at the isosbestic point (614 nm). The pK$_a$ of 8.8 for the probe is unfortunately unsuitable for biological measurements.

3.1.5 Chemosensors for Thiols

Because squaraines are sensitive toward nucleophilic attack resulting in decoloration and fluorescence quenching, they can be used as specific chemodosimeters for *thiol*-containing compounds. Squaraines **4** have been successfully applied as sensors for low-molecular-weight aminothiols like *cysteine* in a complex multicomponent mixture (e.g., human plasma) [91].

R = –CH$_2$O(CH$_2$)$_2$OMe, –CH$_2$OC(O)NHBu

A similar strategy for the detection of low-molecular-weight aminothiols is used in squaraine dye **31** absorbing and emitting in the NIR/Vis spectral region [92] (Fig. 7). Dye **31** is a *ratiometric* probe and its detection principle is based on the generation of a new fluorophore that emits at a different wavelength through an analyte-induced break up of π-conjugation. The emission properties of **31** show a dramatic change in presence of *cysteine*. When excited at 410 nm, the emission spectrum of the **31**–cysteine adduct exhibits a new emission peak at 592 nm with bright orange fluorescence, accompanied by a decrease in the intensity of the weak NIR emission at 800 nm (λ_{ex} = 730 nm). The probe was successfully used for the detection of aminothiols in human blood plasma and may have further

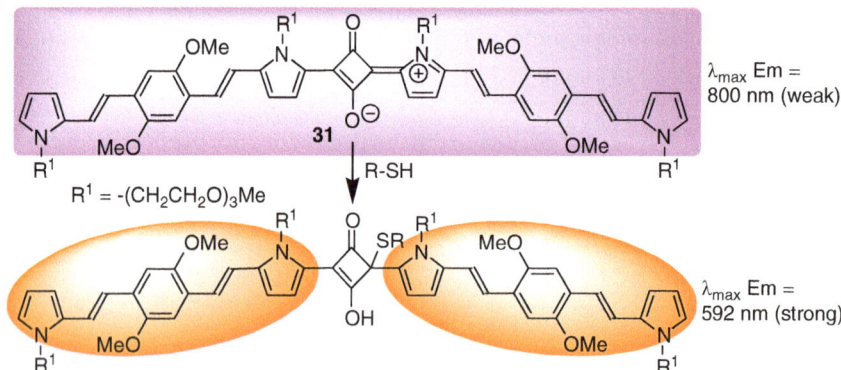

Fig. 7 Transformation of weak fluorescent probe **31** to the strongly fluorescent product upon reaction with thiol

applications for fluorescent labeling and imaging of thiol-containing proteins in biological fluids.

A reverse cross-coupling reactions mediated by palladium was used to develop a *colorimetric* sensitive chemodosimeter for the detection of trace *palladium (II)* salts [93]. The decolorization of **4** is produced by a nucleophilic attack of ethanethiol in basic DMSO solutions. Palladium detection is done via thiol scavenging from the **4**–ethanethiol complex leading to a color "turn-on" of the parent squaraine. "Naked-eye" detection of $Pd(NO_3)_2$ is as sensitive as 0.5 ppm in solution, and the instrument-based detection can go as low as 0.1 ppm.

3.2   Oxo-Squaraine Labels

Fluorescent labels based on oxo-squaraines were described in numerous articles and patents [45, 52]. *Mono-reactive* hydrophobic (**32a**) and hydrophilic (**32b**) squaraine labels containing one NHS ester group were synthesized by Terpetschnig et al. [62]. The initially low *quantum yields* and short *fluorescence lifetime* of **32b** in aqueous solutions significantly *increase* after covalent binding to proteins.

Wolfbeis, Terpetschnig, and co-workers synthesized and investigated the spectral properties of two reactive long-wavelength squaraine labels **Sq635-m** and **Sq635–b (RG-634)** with either one or two NHS ester groups [94, 95]. For improvement of the water solubility, two sulfonic acid groups were introduced into the heterocyclic ring systems. These squaraine labels exhibit low quantum yields in water ($\Phi_F = 0.15$) and high quantum yields ($\Phi_F = 0.6$–0.7) when bound to proteins. The absorption maxima at 634 nm in water and at approximately 645 nm when bound to proteins allow excitation with commercially available diode lasers. The lower detection limit of a representative squaraine dye in blood was estimated to be half that of a commonly used cyanine label **Cy5**.

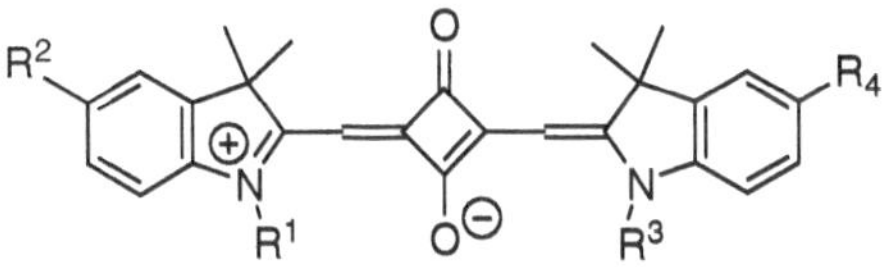

32a: R^1 = Me, R^2 = Cl,

R^3 = Et, R^4 = –COO–NHS

32b: R^1 = Me, R^2 = Cl,

R^3 = –(CH$_2$)$_4$SO$_3$H, R^4 = –COO–NHS

Sq635-m: R^1 = H, R^2 = SO$_3$H, R^3 = –(CH$_2$)$_5$COO–NHS, R^4 = SO$_3$H

λ_{max} Ab = 634 nm, λ_{max} Em = 646 nm

Sq635-b (RG-634): R^1 = R^3 = –(CH$_2$)$_5$COO–NHS, R^2 = R^4 = SO$_3$H

λ_{max} Ab = 634 nm, λ_{max} Em = 648 nm, Φ_F = 0.15

Sq635-b-HSA: λ_{max} Ab = 642 nm, λ_{max} Em = 653 nm, Φ_F = 0.68 (D/P = 0.8)

Sq635-b-*anti*-HSA: λ_{max} Ab = 647 nm, λ_{max} Em = 664 nm, Φ_F = 0.39 (D/P = 0.6)

Sq635–b and **Sq660** were also utilized as *donor-acceptor pairs* in combination with an HSA/anti-HSA system, in a *fluorescence energy transfer* (FRET)-based immunoassay [95, 96].

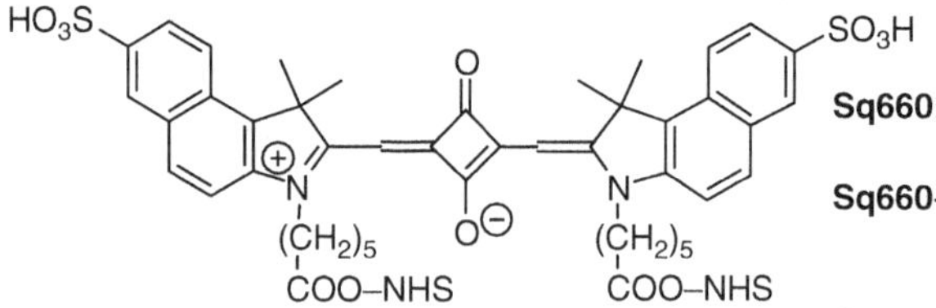

Sq660: λ_{max} Ab = 660 nm,
            λ_{max} Em = 680 nm, Φ_F = 0.05
Sq660-HSA: λ_{max} Ab = 672 nm,
            λ_{max} Em = 686 nm
            Φ_F = 0.43 (D / P = 2.2)
Sq660-*anti*-HSA: λ_{max} Ab = 668 nm,
            λ_{max} Em = 681 nm

One of the shortcomings of the above squaraine labels is the lack of water-solubility, which results in quenching of fluorescence. Highly improved fluorescent labels **33** for proteins showing *increased water-solubility*, and highly decreased tendencies to aggregate in aqueous solutions including much *higher sensitivity* were developed by Patsenker et al. [97, 98]. These dyes are based on symmetrical and asymmetrical oxo-squaraine dyes with reactive carboxyalkyl or sulfoalkyl groups at the *3-position* in one or both indolenine moieties. These dyes exhibit similar absorption and emission maxima as the conventional 3,3-dimethyl indolenine-analogs but exhibit *highly increased quantum yields* free in aqueous solutions and after binding to proteins or low-molecular-weight species.

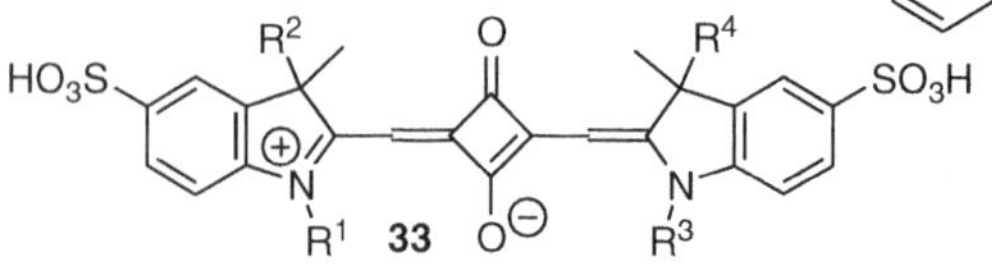

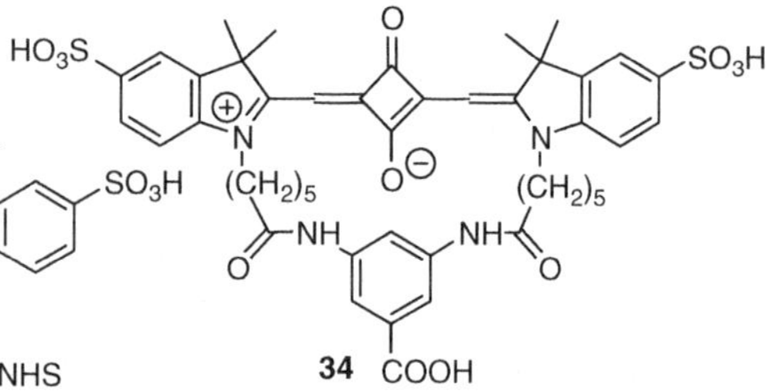

R^1, R^3 = –(CH$_2$)$_4$SO$_3$H, –(CH$_2$)$_5$COONHS
R^2, R^4 = Me, –(CH$_2$)$_5$COONHS, –(CH$_2$)$_4$SO$_3$H

Bridged squaraine labels such as **34** that absorb at 632 nm (ε_M = 250,000 M^{-1}cm^{-1}) and emit at 646 nm in PBS are reported in a patent [22]. A similar approach but linking the spacer in the *3-position* of one indolenine with the spacer group attached to the indolenine-nitrogen in the second heterocycle was proposed in [23, 24].

The NIR fluorescent 2,3-dihydroperimidine-squaraines **35a,b** were synthesized and used as biological labels [99]. Dye **35b** including four sulfopropyl groups exhibits high water-solubility, an emission maximum at 812 nm (Φ_F = 0.08) after conjugation with BSA (at dye-to-protein ratio D/P = 1), and 817 nm in the presence of BSA, which indicates compatibility with commercially available NIR laser diodes.

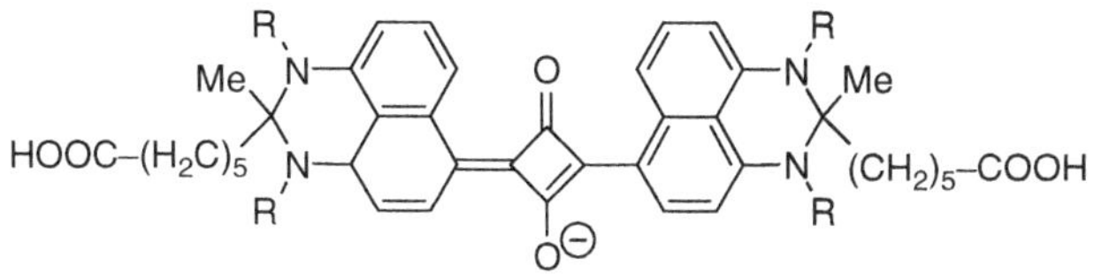

35a: R = Me;
35b: R = –(CH$_2$)$_3$–SO$_3$H

The nonfluorescent monofunctional azulenyl squaraine labels **NIRQ$_{700}$** and **NIRQ$_{750}$** with absorption maxima between 600 and 750 nm are capable of quenching a number of NIR fluorochromes. They were specifically developed to extend the spectrum for FRET-based assays [100, 101].

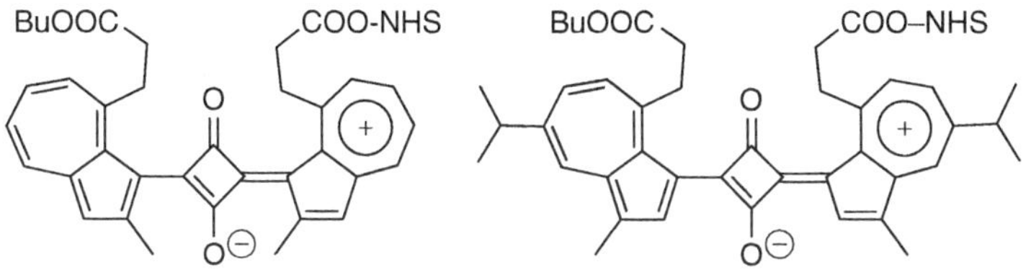

NIRQ$_{700}$ **NIRQ$_{750}$**

A thiol-reactive squaraine **36** (iodoacetamide) that displays fluorescence emission above 650 nm was used to develop a reagentless glucose monitoring assay [102].

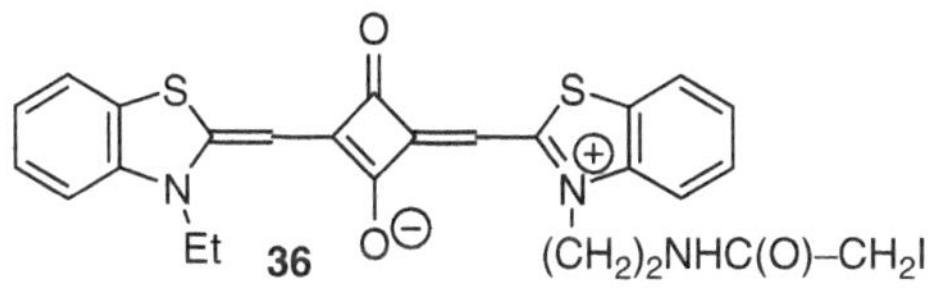

4 Ring-Substituted Squaraines: Probes and Labels with High Sensitivity and Photostability

Ring-substituted squaraines, wherein the oxygens in the squaraine moiety are substituted by *carbon, nitrogen,* or *sulphur,* have been described but are not as well investigated than oxo-squaraines. Nevertheless, the available literature data evidence the potential of these dyes as fluorescent probes and labels. They show very favorable properties such as *high sensitivity to the microenvironment* and *high quantum yields* when bound to high-molecular-weight analytes (e.g., proteins and cells), including *high photostability* [18, 58].

Dithio-squaraines **37** can be synthesized *via* thionation of the correspondent oxo-squaraines **5** with phosphor pentasulphide or *Lawesson* reagent (Fig. 8) [103, 104]. However, dithio-squaraines containing carboxylic groups cannot be obtained by this method. Symmetrical indolenine- and benzothiazole-based monothio- (**38**) and dithio- (**37**) squaraines containing carboxy and sulfo groups are synthesized using 1,2-dithio- and tetrathiosquaric acid, respectively, while the reaction of these

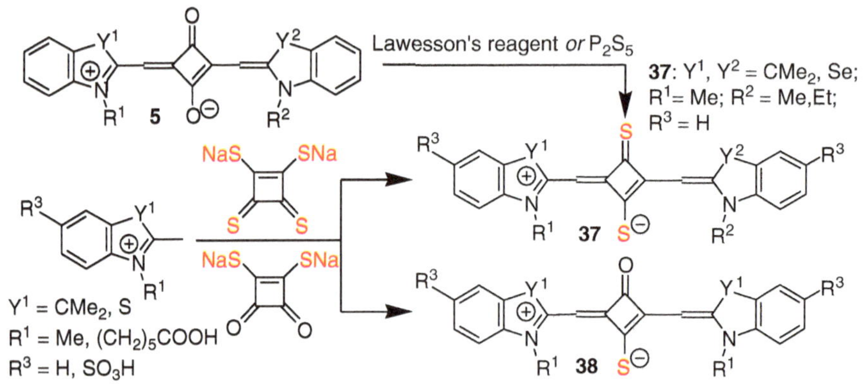

Fig. 8 Synthetic route to thio-squaraines

heterocyclic bases with 1,3-dithiosquaric acid primarily leads to the formation of conventional oxo-squaraines [98].

The direct substitution of the squaraine oxygen with CH or NH acidic compounds was unsuccessful. A general route to symmetrical and unsymmetrical ring-substituted squaraines, wherein the squaraine oxygen is replaced by *carbon, nitrogen,* or *sulfur,* is shown in Fig. 9 [23, 45, 50, 98]. The synthesis of water-soluble derivatives of these ring-substituted squaraine dyes containing sulfo groups and reactive functionalities was also reported [18, 105].

Amino-squaraines **39** ($Z^1 = O$, $Z^2 = NR^1R^2$) are synthesized by etherification of the oxo-squaraine oxygen using dimethylsulfate or methyl triflate followed by substitution of the methoxy group with amines [52] (Fig. 10).

Aniline-based amino-squaraines **40** can be obtained *via* reduction of oxo-squaraines with $NaBH_4$ in methanol followed by substitution and oxidation with chloranil or lead dioxide [106] (Fig. 11).

4.1 Probes

A series of symmetrical and unsymmetrical, hydrophobic and hydrophilic squaraine probes such as **41** (Table 1) with *substituted* squaraine ring oxygen was developed and compared to conventional oxo-squaraines **10a,b** and **13b** [18, 50, 98, 107]. The substituent on the squaraine ring have a strong influence on the spectral properties. Substitution of the squaraine oxygen by S, $C(CN)_2$, $C(CN)COOR$, $N(CN)$, $N(OH)$, $C(CN)[PO(OEt)_2]$, indanedione, barbituric, and thiobarbituric acid causes *red-shifted* absorption and emission spectra [50].

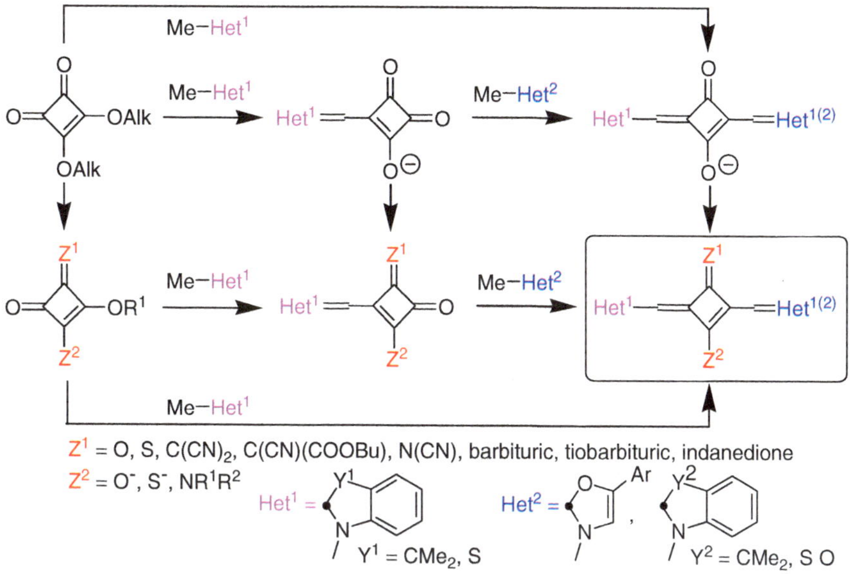

Fig. 9 Synthesis of symmetrical and unsymmetrical ring-substituted squaraines

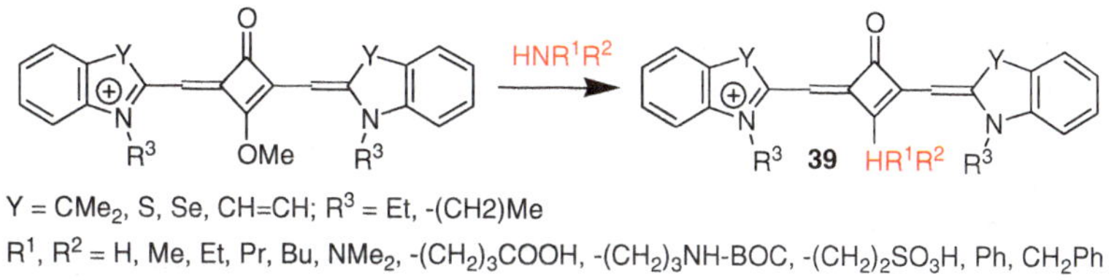

Y = CMe$_2$, S, Se, CH=CH; R^3 = Et, -(CH2)Me

R^1, R^2 = H, Me, Et, Pr, Bu, NMe$_2$, -(CH$_2$)$_3$COOH, -(CH$_2$)$_3$NH-BOC, -(CH$_2$)$_2$SO$_3$H, Ph, CH$_2$Ph

Fig. 10 Synthetic route to amino-squaraines

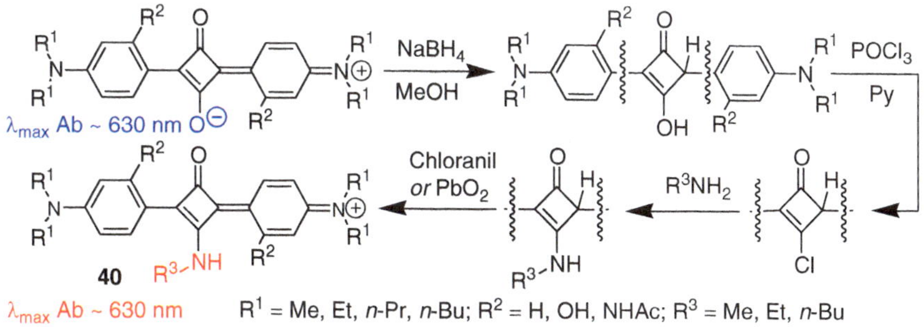

R^1 = Me, Et, n-Pr, n-Bu; R^2 = H, OH, NHAc; R^3 = Me, Et, n-Bu

Fig. 11 Synthesis of aniline-based amino-squaraines

Table 1 Selected substituents for squaraine dyes **10**, **13**, and **41**

Dye	Z	Y^1	Y^2	R^1	R^2	R^3	R^4
10b	O	CMe_2	CMe_2	$(CH_2)_5COOH$	Me	SO_3H	H
13b	O	CMe_2	CMe_2	$(CH_2)_5COOH$	$(CH_2)_5COOH$	SO_3H	SO_3H
41a	$C(CN)_2$	CMe_2	CMe_2	Me	Me	H	H
41b	$N(CN)$	CMe_2	CMe_2	Me	Me	H	H
41c	$C(CN)_2$	O	CMe_2	Me	Me	H	H
41d	$C(CN)_2$	CMe_2	CMe_2	$(CH_2)_5COOH$	Me	H	H
41e	$C(CN)_2$	CMe_2	CMe_2	$(CH_2)_5COOH$	Me	SO_3H	H
41f	S	CMe_2	CMe_2	$(CH_2)_5COOH$	Me	SO_3H	H
41g	S	CMe_2	CMe_2	$(CH_2)_5COOH$	$(CH_2)_5COOH$	SO_3H	SO_3H
41h	$C(CN)_2$	CMe_2	CMe_2	$(CH_2)_5COOH$	Me	SO_3H	H
41i	$C(CN)_2$	CMe_2	CMe_2	$(CH_2)_5COOH$	Et	H	SO_3H
41j	$C(CN)_2$	CMe_2	CMe_2	$(CH_2)_5COOH$	$(CH_2)_5COOH$	SO_3H	SO_3H

Table 2 Spectral properties of squaraine dyes in absence and presence of 6 mg/mL BSA and when covalently bound to BSA (phosphate buffer, pH 7.4)* [18]

Dye	λ_{max} Ab [nm]	ε_M [$M^{-1}cm^{-1}$]	λ_{max} Em [nm]	Φ_F
Cy5	647/647/653	250,000	664/664/672	0.27/0.27/0.20
10b	624/639/638	245,000	635/648/650	0.02/0.40/0.19
13b	632/645/642	265,000	642/654/655	0.06/0.38/0.27
41f	628 (373)/649/646	163,000 (21,500)	642/661/662	0.03/0.35/0.09
41g	636 (370)/655/648	216,000 (22,000)	648/666/666	0.06/0.28/0.04
41h	659 (385)/673/671	207,000 (44,500)	676/693/694	0.02/0.30/0.03
41i	658 (385)/679/676	182,000 (34,000)	677/695/695	0.03/0.45/0.13
41j	667(380)/683/681	188,000 (26,500)	685/699/702	0.07/0.44/0.24

*λ_{max} Ab, λ_{max} Em and Φ_F are given for *free dye/complex with BSA/conjugate with BSA (D/P = 1)*. λ_{max} Ab and ε_M for short-wavelength absorption bands are given in brackets

39a: R^1 = H, R^2 =*n*–Bu ; **39b**: R^1 = R^2 = Et   **10, 13, 41**

Similar to cyanines and oxo-squaraines, the terminal heterocyclic moieties in the ring-substituted squaraines cause a *red-shift* of the absorption and emission maxima in the order: 5-aryl-1,3,4-oxadiazole <5-aryl-1,3-oxazole < benzoxazole < indolenine < benzothiazole < 5-nitro-indolenine [98]. Ring-substituted squaraines exhibit long-wavelength absorption (628–728 nm) and emission maxima (642–757 nm), high molar absorptivities (ε_M = 100,000–216,000 $M^{-1}cm^{-1}$), low quantum yields, and fluorescence lifetimes in aqueous solutions but high quantum yields and lifetimes in hydrophobic media and in presence of proteins (Table 2). Unlike the conventional cyanines such as **Cy5** and **Alexa Fluor 647** and oxo-squaraines, ring-substituted squaraines (with exception of cyanamide **41b**) exhibit *additional absorption bands* in the 380–470 nm spectral region (ε_M = 16,000 and

44,500 $M^{-1}cm^{-1}$), and therefore, these dyes are excitable not only with red (635 or 670-nm) but also with blue (380, 405, and 470 nm) diode lasers or LEDs (Fig. 1). Carbonyl containing substituents such as 1,3-indanedione, cyanoacetic ester, barbituric, and thiobarbituric acid form *intramolecular H-bond* with the polymethine hydrogens of the squaraine bridge. As a result, the molar absorptivities and quantum yields of these dyes are substantially decreased.

Due to their higher polarity compared to oxo-squaraines and cyanines, *ring-substituted* squaraine dyes exhibit more pronounced *solvation effects*: the absorption and emission spectra measured in methanol are blue-shifted by about 10–30 nm compared to chloroform, and the quantum yields are somewhat lower. These dyes form nonfluorescent *aggregates* in aqueous media, and as a result, the long-wavelength absorption band broadens and a new maximum appears [58]. The quantum yields in water are low, but in hydrophobic media and upon binding to proteins, the *quantum yields* increase noticeably and can become as high as 95% (for **41c**). The quantum yield increase upon binding to proteins is generally higher for ring-substituted squaraines than for oxo-squaraines and cyanines [18]. The affinity of these dyes for BSA increases in the same order as the polarity of their molecules: **Cy5** $<<$ oxo-squaraines $<$ monothio-squaraines $<$ dicyanomethylene-squaraines [18]. The binding constants (K_S) for squaraine dyes are comparable to some common drugs, which suggest that these probes could be used for determining of *drug-binding constants* by the dye-displacement method [108]. Because of their high sensitivity toward the microenvironment, ring-substituted squaraines are used as fluorescent probes for *proteins* [18, 58], *lipids* [109, 110], *cells* [17], and other biological species with high molecular weight. These dyes are perfectly suited for fluorescence-based *quantification of proteins* and in-situ biological *imaging* as they readily stain different types of cells [17]. Ring-substituted squaraines are also used as stains for *gel electrophoresis* applications.

Squaraine dyes **10b**, **39a**, **39b**, **41a**, **41c**, **41d**, and **41e** were used to measure different *proteins* such as BSA, HSA, ovalbumin, avidin from hen egg white, lysozyme, and trypsin (Fig. 12) [58]. It is difficult to predict correlations between the dyes' structures and the affinity or sensitivity of the dyes for different proteins. *All* squaraine probes exhibit considerable *fluorescence increases in the presence of BSA*. Dicyanomethylene-squaraine **41c** is the brightest fluorescent probe and demonstrates the most pronounced intensity increase (up to 190 times) in presence of *BSA*. At the same time, the fluorescent response of the dyes **10b**, **39a**, **39b**, **41a**, **41c**, **41d**, and **41e** in presence of other albumins (HSA and ovalbumin) is, in general, significantly lower (intensity increases up to 24 times). Dicyanomethylene-squaraine **41a** and amino-squaraines **39a** and **39b** are the most sensitive probes for *ovalbumin*. Dyes **41d**, **10b**, and **41e** containing an *N*-carboxyalkyl-group demonstrate sufficient enhancement (up to 16 times) in the presence of *avidin*. Nevertheless, the presence of hydrolases like lysozyme or trypsin has only minor effects on the fluorescence intensity of squaraine dyes.

The dicyanomethylene-squaraine dye **41e** was found to be highly sensitive to trace *protein-lipid interactions* [109]. *Lysozyme* association with the lipid bilayer leads to a noticeable decrease in the fluorescence intensity of **41e**. In a separate

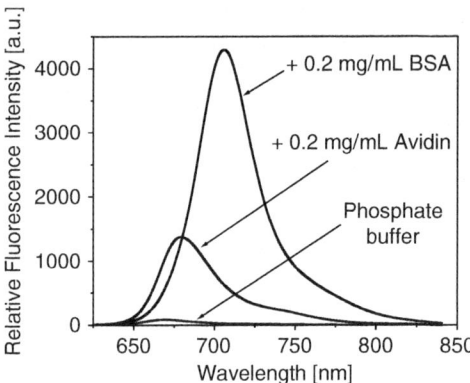

study, the dye **41e** was used to trace the changes of lipid bilayer properties, and its
performance was compared to the oxo-squaraine dye **13b** [110]. Binding of dye **41e**
to the lipid bilayer was accompanied by a significant increase of its fluorescence
intensity and a red-shift of its emission maximum, while the fluorescence intensity
of dye **13b** increased only slightly upon binding to lipid environment. This suggested that dye **41e** can be considered a sensitive probe for examining *membrane-related* processes.

Ring-substituted squaraines such as **42** and **43** [98, 111] also can be used as
fluorescent probes; however, the quantum yields for *dithio*-squaraines **42** are lower
compared to *monothio*-squaraines or *oxo*-squaraines.

42a : Y = CMe₂
42b : Y = Se, λ_max Ab = 687 nm,
        λ_max Em = 724 nm (CHCl₃)

43a : λ_max Ab = 696 nm (MeOH)

Z =   (b),   (c)

43b : λ_max Ab = 682 nm (MeOH)
43c : λ_max Ab = 677 nm (MeOH)

The *photostability* of ring-substituted squaraines, especially *dicyanomethylene-*
and *thio*-squaraines is, in general, higher compared to oxo-squaraines and cyanines
(Fig. 2). Increasing of the number of sulfo groups further improves the photostability of these dyes: the photostability for hydrophilic dyes increases in the order
10b < **13b** < **Cy5** < **41h** < **41i** ≈ **41j** < **41f** < **41g** [18]. *Thio-squaraine* dyes

such as **41g** demonstrate an *increase of the absorbance and emission intensity during light exposure* (Fig. 2), which is due to the *photo-induced hydrolysis* of thio-squaraine (**41g**) into oxo-squaraine (**13b**), which is more photostable and has a higher molar absorptivity (brightness) than **41g**.

Compared to oxo-squaraines or other ring-substituted squaraines, *amino-squaraines* **39** [45, 52, 112] have ionic character, similar to open-chain cyanine dyes, and due to the positive net charge, these dyes are to some extend *water-soluble*. Amino-squaraines absorb and emit at *longer* wavelength than the corresponding oxo-squaraines: the absorption maxima are between 650–710 nm (ε_M = 85,000–300,000 $M^{-1}cm^{-1}$) [45, 112]. The increase in solvent polarity is accompanied by a *hypsochromic* shift of the absorption. Amino-squaraine dyes are potentially used as fluorescent probes but because their photostability is inferior to those of oxo-squaraines and other ring substituted squaraines of similar structure, their applications are rather limited.

4.2 Labels

Water-soluble, ring-substituted squaraines such as **41f–41j** (Table 1) were converted to amine-reactive NHS esters and labeled to proteins including BSA, HSA, and anti-HSA [18, 96, 98]. Covalent attachment of these labels to protein causes a *red-shift* in the absorption and emission spectra, an *increase* of the *quantum yields* and *fluorescence lifetimes* (Table 2). However, when compared to noncovalently-bound dyes, the absorption maxima of the dye–BSA conjugates are slightly blue-shifted by about 2–7 nm, while the emission maxima are unchanged or slightly red-shifted by about 3 nm. After binding to BSA, the quantum yields increase up to 0.24 (D/P = 1), which is in between those for the noncovalently bound dyes and those for free dyes, while the fluorescence lifetimes increase up to 3.3 ns. The most pronounced lifetime increase was found for *dicyanomethylene*-squaraine labels. *Monothio*- and *dicyanomethylene*-squaraine labels and their protein conjugates are *more photostable* compared to **Cy5**, **Alexa 647**, including oxo-squaraines in their conjugated as well as nonconjugated forms (Fig. 2) [18]. Due to the favorable spectral and photophysical properties, these dyes can be used as fluorescent labels for intensity- and *fluorescence lifetime*-based applications.

The commercially available dicyanomethylene squaraine dye **Seta-670-mono-NHS** showed *extremely low blinking effects* and *good photostability* when used in *single-molecule* studies of multiple-fluorophore labeled antibodies [113]. **Seta-670-mono-NHS** and **Seta-635-NH-mono-NHS** were covalently labeled to antibodies and used in *a surface-enhanced immunoassay* [114]. From the fluorescence intensity and lifetime changes determined for a surface that had been coated with silver nanoparticles, both labeled compounds exhibited a 15- to 20-fold fluorescence enhancement and dramatic lifetime changes on the silver-coated surface compared to the noncoated surface.

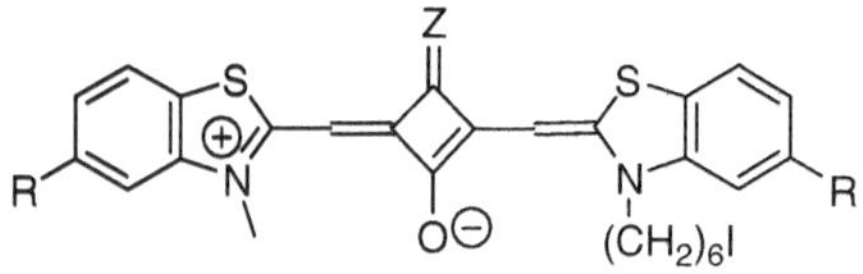

44a: R = H, Z = C(CN)$_2$; **44b**: R = H, Z = O; **44c**: R = F, Z = O

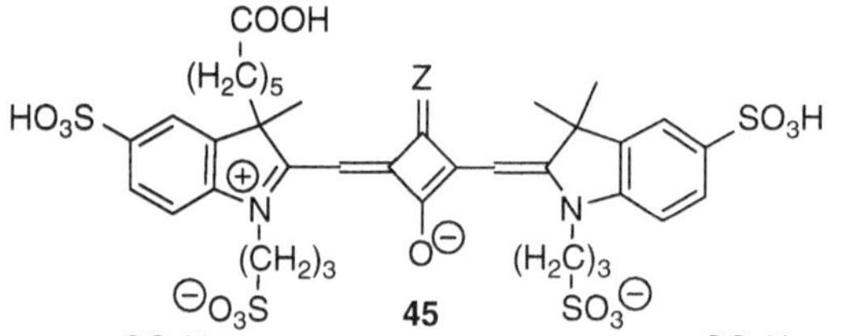

Z = S, C(CN)$_2$, N(CN), N(OH), barbituric

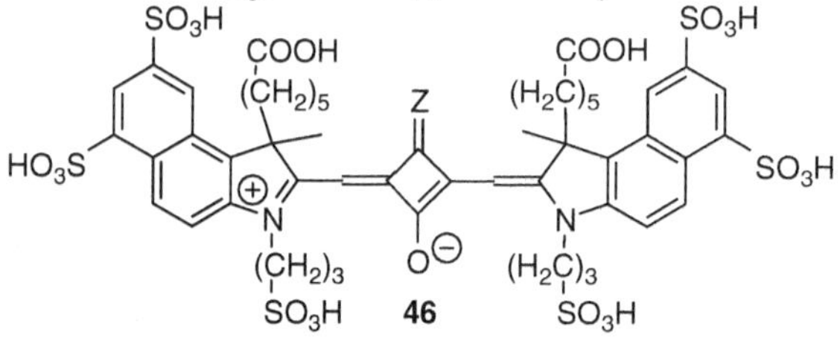

A *benzothiazole* dicyanomethylene-squaraine label **44a** containing a thiol-reactive iodohexyl function was compared to the relevant *fluorinated* (**44c**) and nonfluorinated (**44b**) oxo-squaraines and investigated for labeling of a series of oligonucleotides with various sequences, lengths, and chemistries [105]. While oxo-squaraine **44c** was successfully used for *oligonucleotide* labeling, the conjugation of the dicyanomethylene-squaraine **44a** failed due to the formation of aggregates and lack of stability under the reaction conditions. These conjugates exhibit emission wavelength >670 nm and high quantum yields (0.27–0.39) in phosphate buffer pH 7. The fluorinated oxo-squaraine label showed *improved* photophysical properties and chemical stability as compared to the corresponding nonfluorinated squaraines. Nevertheless, fluorinated versions of dicyanomethylene-squaraines are expected to have even better properties than oxo-squaraines 44c.

SETA BioMedicals further improved the fluorescent properties of squaraine dyes used for protein labeling and developed highly water-soluble, symmetrical, and asymmetrical *ring-substituted* squaraine dyes **45** and **46** containing reactive NHS ester or maleimide functions in the *3-position* of one or both indolenine moieties [97, 98]. Due to the increased water-solubility and lower aggregation tendencies, these dyes exhibit highly increased quantum yields free in aqueous solution and after binding to proteins or low-molecular-weight molecules.

5 Fluorescence Lifetime Dyes

Because the *fluorescence lifetime* (FLT) of certain *squaraine* dyes is strongly dependent on their environment, it can be used as a parameter to report on molecular-binding events such as receptor–ligand, oligonucleotide–oligonucleotide, oligonucleotide–aptamer, or antigen–antibody interactions. FLT has several advantages as compared to conventional methods. In particular in *high throughput screening* (HTS), where the assay *robustness* is of utmost importance, FLT promises great potential to reduce the number of undesired false positives and false negatives results [1, 2]. Despite its advantages, fluorescence lifetime is a parameter that has only been moderately explored for use in biomedical assays. This is probably due to the limited number of fluorophores particularly in the long-wavelength region that exhibit distinct lifetime changes upon interaction of the fluorescently labeled binding partner with its counterpart.

While *open-chain cyanines* such as **Cy5**, **Cy5.5**, or **Alexa 647** typically exhibit *small* fluorescence lifetime changes upon binding to biomolecules, *squaraine* dyes, in general, demonstrate *pronounced lifetime changes* upon binding to large-molecular-weight analytes [17, 18], which can be used for FLT-based assays.

Recently, *SETA BioMedicals* has developed a new near-infrared squaraine-based label **Seta-633**, which can be used to study the interaction between low-molecular-weight analytes and proteins using fluorescence lifetime as the readout parameter [19]. This label exhibits lower quantum yields and shorter fluorescence lifetimes when free in solution, but these values substantially increase upon interaction with proteins, which is contrary to tracers like **Cy5** or **Alexa 647**. It was demonstrated in a model assay that a biotinylated **Seta-633** binds to anti-biotin with high specificity. Importantly, the lifetime of **Seta-633**-biotin increases about 2.76 fold upon binding to a specific antibody (anti-biotin, MW = 160 kDa), while the titration with BSA or nonspecific antibody does not result in a noticeable change in lifetime (Fig. 13). The label is compatible with readily available light sources (635 nm or 640 nm lasers) and filter sets (as for **Cy5** or **Alexa 647**) and its

Fig. 13 Average fluorescence lifetime changes of **Seta-633** – biotin upon titration with antibiotin (specific), (nonspecific) human IgG, or BSA [19]

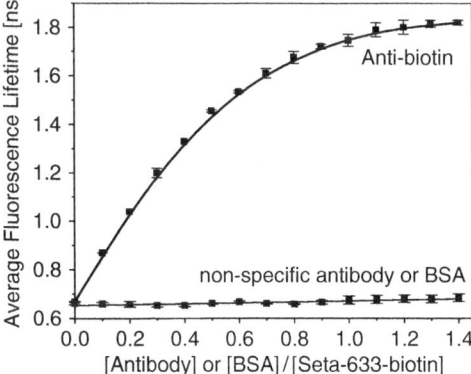

Table 3 Spectral characteristics and fluorescence lifetimes of free and BSA-bound forms of selected **Square** and **Seta** dyes (*SETA BioMedicals*) in phosphate buffer pH 7.4 [19, 115]

Label	λ_{max} Ab [nm]	ε_M [M^{-1}cm^{-1}]	λ_{max} Em [nm]	τ_{mean} [ns]	
				Free	Conjugated
Square-635-di-NHS	636	216,000	648	0.27	2.44
Square-670-di-NHS	667	188,000	685	0.54	3.26
Square-660-mono-NHS	658	182,000	677	0.26	3.32
Seta-633-mono-NHS	633	250,000	644	0.25	2.81

conjugates are *more photostable* as compared to **Cy5** and **Alexa 647**. Other squaraine-type labels that show large, distinct lifetime changes after binding to proteins are listed in the Table 3 [115].

6 Norcyanines and Norsquaraines: pH-Sensitive Probes and Labels

Cyanine and squaraine dyes with hydrogen substituents on the indolenine-nitrogen in one or both of the heterocyclic end-groups, the so-called norcyanines and norsquaraines, are useful as fluorescent pH-indicators due to the reversible equilibrium between their protonated and deprotonated forms:

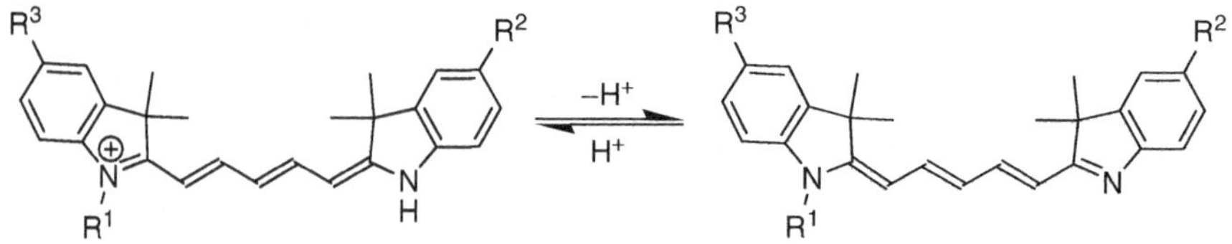

A series of functionalized, water-soluble, pH-sensitive *pentamethine norcyanines* **47**, the so-called **CypHer** dyes, was developed by Cooper et al. [114]. These dyes are *fluorescent* when protonated and become *nonfluorescent* upon proton abstraction and have a pK_a between 6.1 and 7.5 with observable changes in its fluorescent emission across a pH in physiological range between 6.0–8.0. Subtle changes to the structure of these dyes lead to pronounced changes in their pK_a (Table 4). **CypHer** dyes have been utilized for *cell-based assays*. **CypHer 5** is the commercially available, pH-sensitive label from this series and mainly used for pH-sensing applications in cells [117, 118]. Due to their ionic nature, **CypHer** dyes do NOT passively penetrate through the cell membrane. Open-chain cyanines like **CypHer** dyes are therefore almost exclusively used for *covalent* labeling of the outer cell membranes or cell surface receptors, for example, G-protein-coupled receptors. Agonist activation of this class of receptor results in the internalization of

Table 4 Spectral characteristics and pK_a's of **CypHer** dyes

Dye	R^1	R^2	R^3	R^4	X	λ_{max} Ab* [nm]	λ_{max} Em [nm]	pK_a
47a	H	$(CH_2)_4SO_3H$	SO_3H	CH_2COOH	CMe_2	645/484	665	7.5
47b	H	Et	SO_3H	SO_3H	CMe_2	650/515	665	6.4
47c	$(CH_2)_5COOH$	H	H	SO_3H	S	640/531	670	6.85
47d	H	$(CH_2)_5COOH$	SO_3H	SO_3H	CMe_2	653/501	660	7.5
CypHer 5	H	H	COOH	SO_3H	CMe_2	655/500	665	6.1

*At pH 4.67/9.00

the receptor from the plasma membrane (pH 7.4) to the endosomal pathway, which is accompanied by a change in pH [116].

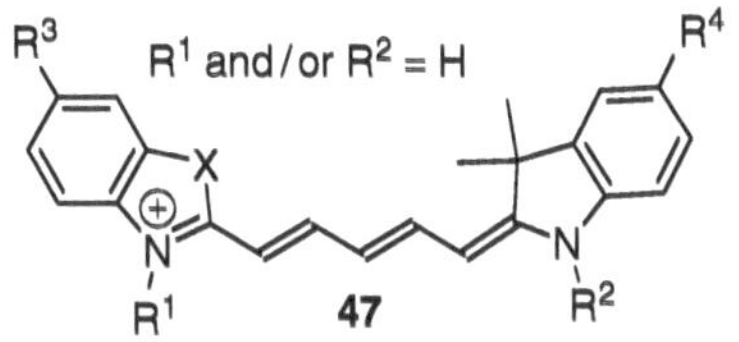

A series of amine-reactive, long-wavelength fluorescent labels **48a–48c** was recently developed on the basis of a *norsquaraine* chromophore [97, 98]. Importantly, norsquaraine dyes show shifted pK_a values as compared to norcyanines with the same heterocyclic end-groups by several pH units: e.g., while the unsymmetrical norsquaraine **48b** has a pK_a around 11.2, the norcyanine dye **47d** carrying the same substituents has a pK_a of 7.5 [118]. Due to this pK_a-shift to the alkaline pH range, some of these norsquaraines are also useful as fluorescent *labels* as their spectral properties are not affected in the pH range below 7.5.

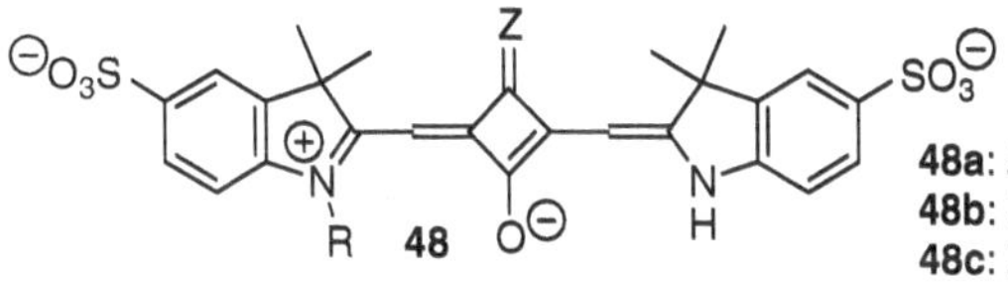

48a: Z = O, R = H, pK_a = 10.4
48b: Z = O, R = –$(CH_2)_5COOH$, pK_a = 11.2
48c: Z = $C(CN_2)$, R = H, pK_a = 8.8

Square-650-pH having a pK_a in the physiological pH range ($pK_a = 7.1$ for free dye and the $pK_a \sim 6.1$ when labeled to an antibody) was recently introduced by *SETA BioMedicals* [119]. This dye is commercially available as a free carboxylic acid and a mono-NHS ester. **Square-650-pH** has spectral properties similar to those of the **CypHer** dyes but is *fluorescent in both the protonated and deprotonated forms*. This dye displays reasonable molar absorptivities (135,000 and 48,000 $M^{-1}cm^{-1}$) and quantum yields (16% and 9%) for the protonated and deprotonated forms, an extremely *large Stokes shift* of more than 100 nm for the deprotonated form, and enables excitation and emission *ratiometric measurement*

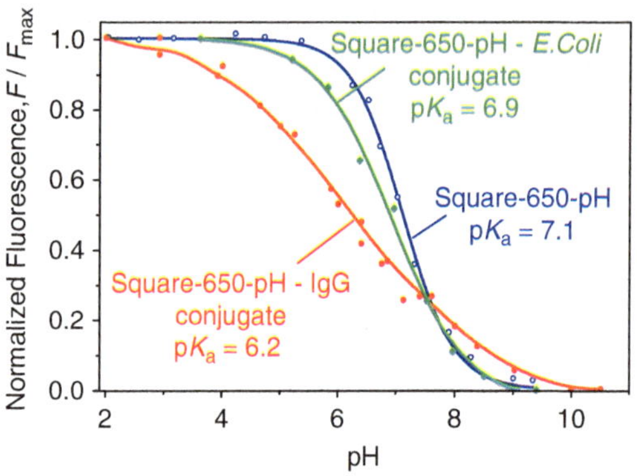

Fig. 14 pH-titration curves: normalized fluorescence intensity versus pH (λ_{exc} = 630 nm) of **Square-650-pH**, **Square-650-pH**–IgG, and **Square-650-pH**–*E. coli* conjugates

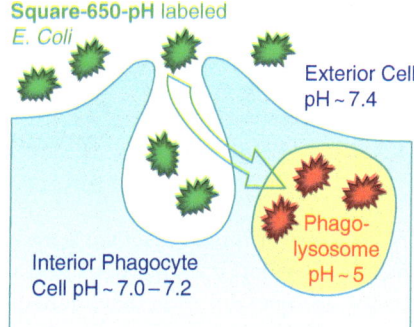

Fig. 15 Scheme for **Square-650-pH**–*E. coli*-based detection of phagocytic events

of pH. **Square-650-pH** exhibits a pronounced change of its fluorescence lifetime (τ_F = 0.53 ns at pH = 9.0 to τ_F = 1.17 ns at pH = 5.3), and, therefore, it is also suitable for *FLT-based* pH sensing and *fluorescence lifetime imaging* (FLIM) applications. The pH-dependent properties of the **Square-650-pH** were also measured after conjugation to an antibody (IgG) and *E.coli*. Importantly, the pK_a of the pH probe (pK_a = 7.1) only slightly changes after conjugation to *E. coli* (pK_a = 6.9, Fig. 14), and, therefore, the fluorescence intensity increases dramatically as the pH of its environment becomes more acidic (e.g., upon internalization in vesicles during phagocytic events) (Fig. 15) [119].

7   Concluding Remarks

Due to their low sensitivity toward the environment, cyanine dyes are perfect candidates as fluorescent labels. Squaraine dyes on the other hand display a highly environment-sensitive response and are therefore not only useful as fluorescent probes and labels but also, in particular, well-suited for lifetime-based applications.

Due to their displayed higher photostability, indolenine-based cyanines and squaraines are preferred for the design of fluorescent tracers. Dicyanomethylene- and thio-squaraines with substituted squaraine oxygens also show potential as highly photostable fluorescent probes.

The search for brighter, more sensitive fluorescent probes and labels with increased quantum yields, lifetimes and photostability including new ways to protect these chromophores from unfavorable environmental impact and methods to synthesize these advanced tracers is still ongoing.

References

1. Lakowicz JR (1999) Principles of fluorescence spectroscopy, 2nd edn. Kluwer Academic/ Plenum Publishers, New York
2. Demchenko AP (2010) Comparative analysis of fluorescence reporter signals based on intensity, anisotropy, time-resolution and wavelength-ratiometry. In: Demchenko AP (ed) Advanced Fluorescence Reporters in Chemistry and Biology I. Springer Ser Fluoresc 8:3–24
3. Miller JN (2008) Long-wavelength and near-infrared fluorescence: state of the art, future applications, and standards. Springer Ser Fluoresc 5:147–162
4. Mishra A, Behera RK, Behera PK, Mishra BK, Behera GB (2000) Cyanines during the 1990s: a review. Chem Rev 100:1973–2011
5. Gonçalves MST (2009) Fluorescent labeling of biomolecules with organic probes. Chem Rev 109:190–212
6. Hermanson GT (2008) Bioconjugate techniques, 2nd edn. Academic, New York
7. Gupta RR, Strekowski L (eds) (2008) Heterocyclic polymethine dyes. Topics in heterocyclic chemistry, vol 14. Springer-Verlag, Berlin, Heidelberg
8. Fei X, Gu Y (2009) Progress in modifications and applications of fluorescent dye probe. Prog Nat Sci 19:1–7
9. Williams CHG (1856) Trans Roy Soc Edinburgh 21:377
10. Sturmer DM (1977) Synthesis and properties – cyanine and related dyes. In: Weissberger A, Taylor EC (eds) The chemistry of heteroaromatic compounds, vol 30. Wiley, New York
11. Mujumdar RB, Ernst LA, Mujumdar SR, Lewis CJ, Waggoner AS (1993) Cyanine dye labeling reagents: sulfoindocyanine succinimidyl esters. Bioconjugate Chem 4:105–111
12. Gruber HJ, Hahn CD, Kada G, Riener CK, Harms GS, Ahrer W, Dax TG, Knaus HG (2000) Anomalous fluorescence enhancement of Cy3 and Cy3.5 versus anomalous fluorescence loss of Cy5 and Cy7 upon covalent linking to IgG and noncovalent binding to avidin. Bioconjugate Chem 11:696–704
13. Kuznetsova VE, Davydov AV, Vasiliskov VA, Stomakhin AA, Chudinov AV, Zasedatelev AS (2007) Novel asymmetric indodicarbocyanine dyes. Russ Chem Bull Int Edit 56:2263–2267
14. Licha K, Riefke B, Semmler W, Speck U, Hilger CS (2000) US Patent 6083485
15. Haugland RP (2002) Handbook of fluorescence probes and research products, 9th edn. Molecular Probes, Eugene
16. Leung WY, Cheung CY, Yue S (2002) US Patent Application 20020077487
17. Patsenker L, Tatarets A, Kolosova O, Obukhova O, Povrozin Y, Fedyunyayeva I, Yermolenko I, Terpetschnig E (2008) Fluorescent probes and labels for biomedical applications. Ann New York Acad Sci 1130:179–187
18. Tatarets AL, Fedyunyayeva IA, Dyubko TS, Povrozin YA, Doroshenko AO, Terpetschnig EA, Patsenker LD (2006) Ring-substituted squaraine dyes as probes and labels for fluorescence assays. Anal Chim Acta 570:214–223

19. Povrozin YA, Kolosova OS, Obukhova OM, Tatarets AL, Sidorov VI, Terpetschnig EA, Patsenker LD (2009) Seta-633 – a NIR fluorescence lifetime label for low-molecular-weight analytes. Bioconjugate Chem 20:1807–1812

20. Czerney P, Lehmann F, Wenzel M, Buschmann V, Dietrich A, Mohr GJ (2001) Tailor-made dyes for fluorescence correlation spectroscopy (FCS). Biol Chem 382:495–498

21. Luschtinetz F, Dosche C, Kumke MU (2009) Influence of streptavidin on the absorption and fluorescence properties of cyanine dyes. Bioconjugate Chem 20:576–582

22. Singh R, Gorski G, Frenzel G (2002) US Patent 6403807

23. Terpetschnig EA, Patsenker L, Tatarets A, Fedyunyaeva I, Borovoy I (2003) WO Patent 03087052

24. Diwu Z, Zhang J, Tang Y (2006) WO Patent 2006047452

25. Kulinich AV, Derevyanko NA, Ishchenko AA, Bondarev SL, Knyukshto VN (2008) Structure and fluorescence properties of indole cyanine and merocyanine dyes with partially locked polymethine chain. J Photochem Photobiol A: Chem 200:106–113

26. Reddington MV (2007) Synthesis and properties of phosphonic acid containing cyanine and squaraine dyes for use as fluorescent labels. Bioconjugate Chem 18:2178–2190

27. Buschmann V, Weston KD, Sauer M (2003) Spectroscopic study and evaluation of red-absorbing fluorescent dyes. Bioconjugate Chem 14:195–204

28. Narayanan N, Strekowski L, Lipowska M, Patonay G (1995) A new method for the synthesis of heptamethine cyanine dyes: synthesis of new near infrared fluorescent labels. J Org Chem 60:2391–2395

29. Peng X, Song F, Lu E, Wang Y, Zhou W, Fan J, Gao Y (2005) Heptamethine cyanine dyes with a large stokes shift and strong fluorescence: a paradigm for excited-state intramolecular charge transfer. J Am Chem Soc 127:4170–4171

30. Kovar J, Chen J, Draney DR, Olive MD, Volcheck WM, Xu X, Lugade AG, Narayanan N (2009) US Patent 7597878

31. DiVittorio KM, Leevy WM, O'Neil EJ, Johnson JR, Vakulenko S, Morris JD, Rosek KD, Serazin N, Hilkert S, Hurley S, Marquez M, Smith BD (2008) Zinc(II) coordination complexes as membrane-active fluorescent probes and antibiotics. Chembiochem 9:286–293

32. Sato S, Tsunoda M, Suzuki M, Kutsuna M, Takido-uchi K, Shindo M, Mizuguchi H, Obara H, Ohya H (2009) Synthesis and spectral properties of polymethine-cyanine dye–nitroxide radical hybrid compounds for use as fluorescence probes to monitor reducing species and radicals. Spectrochim Acta A 71:2030–2039

33. Armitage BA (2005) Cyanine dye–DNA groove binding, and aggregation. Top Curr Chem 253:55–76

34. Licha K, Hessenius C, Becker A, Henklein P, Bauer M, Wisniewski S, Wiedenmann B, Semmler W (2001) Synthesis, characterization, and biological properties of cyanine-labeled somatostatin analogues as receptor-targeted fluorescent probes. Bioconjugate Chem 12:44–50

35. Berezin MY, Lee H, Akers W, Achilefu S (2007) Near infrared dyes as lifetime solvatochromic probes for micropolarity measurements of biological systems. Biophysical J 93:2892–2899

36. Mojzych M, Henary M (2008) Synthesis of cyanine dyes. In: Gupta RR, Strekowski L (eds) Topics in heterocyclic chemistry, vol 14. Springer, Berlin, Heidelberg, pp 1–9

37. Terpetschnig E, Wolfbeis OS (1998) Luminescent probes for NIR sensing applications. In: Daehne S, Resch-Genger U, Wolfbeis OS (eds) Near-infrared dyes for high technology applications, NATO ASI Ser 3, vol 53. Kluwer Academic, Dordrecht (NL), pp 161–182

38. Sprenger HE, Ziegenbein W (1966) Condensation products of squaric acid and tertiary aromatic amines. Angew Chem Int Ed Engl 5:894

39. Sprenger HE, Ziegenbein W (1967) Das Cyclobuten-diylium-Kation, ein neuartiger Chromophor aus Quadratsäure. Angew Chem 79:581–582

40. Treibs A, Jacob K (1966) Über Tetracyclotrimethin-Farbstoffe. Cyclobutenderivate der Pyrrolreihe. Liebigs Ann Chem 699:153–167

41. Sprenger HE, Ziegenbein W (1968) Cyclobutenediylium dyes. Angew Chem Int Ed Engl 7:530–535
42. Law KY, Bailey FC (1992) Squaraine chemistry. Synthesis, characterization, and optical properties of a class of novel unsymmetrical squaraines: [4-(dymethylamino)phenyl](4'-methoxyphenyl)squaraine and its derivatives. J Org Chem 57:3278–3286
43. Treibs A, Jacob K (1965) Von der Quadratsäure abgeleitete Cyclotrimethinfarbstoffe. Angew Chem 77:680–681
44. Treibs A (1966) Über Pyrrolfarbstoffe. Chimia 20:329
45. Kim SH, Hwang SH, Kim JJ, Yoon CM, Keum SR (1998) Syntheses and properties of functional aminosquarylium dyes. Dyes Pigm 37:145–154
46. Terpetschnig E, Szmacinski H, Lakowicz JR (1993) Synthesis, spectral properties and photostabilities of symmetrical and unsymmetrical squaraines; a new class of fluorophores with long-wavelength excitation and emission. Anal Chim Acta 282:633–641
47. Nakazumi H, Natsukawa K, Nakai K, Isagawa K (1994) Synthesis and structure of new cationic squarylium dyes. Angew Chem Int Ed Engl 33:1001–1003
48. Lin T, Peng BX (1997) Synthesis and spectral characteristics of some highly soluble squarylium cyanine dyes. Dyes Pigm 35:331–338
49. Kim SH, Hwang SH (1997) Synthesis and photostability of functional squarylium dyes. Dyes Pigm 35:111–121
50. Tatarets AL, Fedyunyaeva IA, Terpetschnig E, Patsenker LD (2005) Synthesis of novel squaraine dyes and their intermediates. Dyes Pigm 64:125–134
51. Terpetschnig E, Lakowicz JR (1993) Synthesis and characterization of unsymmetrical squaraines: a new class of cyanine dyes. Dyes Pigm 21:227–229
52. Hamilton AL, West RM, Cummins WJ, Briggs MSJ, Bruce IE (2000) US Patent 6140494
53. Chen H, Farahat MS, Law KY, Whitten DG (1996) Aggregation of surfactant squaraine dyes in aqueous solution and microheterogeneous media: correlation of aggregation behavior with molecular structure. J Am Chem Soc 118:2584–2594
54. Deroover G, Missfeldt M, Simon L (2006) US Patent 6995262
55. Patsenker LD, Povrozin YA, Sidorov VI, Tatarets AL, Terpetschnig EA (2009) Fluorescence lifetime based hybridization assay using the new long-wavelength fluorescent label Seta-670. In: 24th International conference on photochemistry (ICP 2009). Book of Abstracts, p 430
56. Yagi S, Nakazumi H (2008) Squarylium dyes and related compounds. In: Gupta RR, Strekowski L (eds) Topics in heterocyclic chemistry, vol 14, Springer. Berlin, Heidelberg, pp 133–181
57. Ioffe VM, Gorbenko GP, Deligeorgiev T, Gadjev N, Vasilev A (2007) Fluorescence study of protein – lipid complexes with a new symmetric squarylium probe. Biophys Chem 128:75–86
58. Volkova KD, Kovalska VB, Tatarets AL, Patsenker LD, Kryvorotenko DV, Yarmoluk SM (2007) Spectroscopic study of squaraines as protein-sensitive fluorescent dyes. Dyes Pigm 72:285–292
59. Meadows F, Narayanan N, Patonay G (2000) Determination of protein–dye association by near infrared fluorescence-detected circular dichroism. Talanta 50:1149–1155
60. Song B, Zhang Q, Ma WH, Peng XJ, Fu XM, Wang BS (2009) The synthesis and photo-stability of novel squarylium indocyanine dyes. Dyes Pigm 82:396–400
61. Yagi S, Hyodo Y, Matsumoto S, Takahashi N, Kono H, Nakazumi H (2000) Synthesis of novel unsymmetrical squarylium dyes absorbing in the near-infrared region. J Chem Soc Perkin Trans: 599–603
62. Terpetschnig E, Szmacinski H, Ozinskas A, Lakowicz JR (1994) Synthesis of squaraine-N-hydroxysuccinimide esters and their biological application as long-wavelength fluorescent labels. Anal Biochem 217:197–204
63. Sophianopoulos AJ, Lipowski J, Narayanan N, Patonay G (1997) Association of near-infrared dyes with bovine serum albumin. Appl Spectrosc 51:1511–1515

64. Yan W, Sloat AL, Yagi S, Nakazumi H, Colyer CL (2006) Protein labeling with red squarylium dyes for analysis by capillary electrophoresis with laser-induced fluorescence detection. Electrophoresis 27:1347–1354

65. Nakazumi H, Colyer CL, Kaihara K, Yagi S, Hyodo Y (2003) Red luminescent squarylium dyes for noncovalent HSA labeling. Chem Lett 32:804–805

66. Yan W, Colyer CL (2006) Investigating noncovalent squarylium dye–protein interactions by capillary electrophoresis–frontal analysis. J Chromatogr A 1135:115–121

67. Sloat AL, Roper MG, Lin X, Ferrance JP, Landers JP, Colyer CL (2008) Protein determination by microchip capillary electrophoresis using an asymmetric squarylium dye: noncovalent labeling and nonequilibrium measurement of association constants. Electrophoresis 29:3446–3455

68. Jyothish K, Avirah RR, Ramaiah D (2006) Synthesis of new cholesterol- and sugar-anchored squaraine dyes: further evidence of how electronic factors influence dye formation. Org Lett 8:111–114

69. Jiao GS, Loudet A, Lee HB, Kalinin S, Johansson LBÅ, Burgess K (2003) Syntheses and spectroscopic properties of energy transfer systems based on squaraines. Tetrahedron 59:3109–3116

70. Yagi S, Ohta T, Akagi N, Nakazumi H (2008) The synthesis and optical properties of bis-squarylium dyes bearing arene and thiophene spacers. Dyes Pigm 77:525–536

71. Arun KT, Ramaiah D (2005) Near-infrared fluorescent probes: synthesis and spectroscopic investigations of a few amphiphilic squaraine dyes. J Phys Chem A 109:5571–5578

72. Jisha VS, Arun KT, Hariharan M, Ramaiah D (2006) Site-selective binding and dual mode recognition of serum albumin by a squaraine dye. J Am Chem Soc 128:6024–6025

73. Suzuki Y, Yokoyama K (2007) A protein-responsive chromophore based on squaraine and its application to visual protein detection on a gel for SDS-PAGE. Angew Chem Int Ed 46:4097–4099

74. Basheer MC, Santhosh U, Alex S, Thomas KG, Suresh CH, Das S (2007) Design and synthesis of squaraine based near infrared fluorescent probes. Tetrahedron 63:1617–1623

75. Kim SH, Kim JH, Cui JZ, Gal YS, Jin SH, Koh K (2002) Absorption spectra, aggregation and photofading behaviour of near-infrared absorbing squarylium dyes containing perimidine moiety. Dyes Pigm 55:1–7

76. Umezawa K, Citterio D, Suzuki K (2007) A squaraine-based near-infrared dye with bright fluorescence and solvatochromic property. Chem Lett 36:1424–1425

77. Binda M, Agostinelli T, Caironi M, Natali D, Sampietro M, Beverina L, Ruffo R, Silvestri F (2009) Fast and air stable near-infrared organic detector based on squaraine dyes. Org Electron 10:1314–1319

78. Ajayaghosh A (2005) Chemistry of squaraine-derived materials: near-IR dyes, low band gap systems, and cation sensors. Acc Chem Res 38:449–459

79. Basheer MC, Alex S, Thomas KG, Suresh CH, Das S (2006) A squaraine-based chemosensor for Hg^{2+} and Pb^{2+}. Tetrahedron 62:605–610

80. Yagi S, Fujie Y, Hyodo Y, Nakazumi H (2002) Synthesis, structure, and complexation properties with transition metal cations of a novel methine-bridged bisquarylium dye. Dyes Pigm 52:245–252

81. Oguz U, Akkaya EU (1997) One-pot synthesis of a red-fluorescent chemosensor from an azacrown, phloroglucinol and squaric acid: a simple in-solution construction of a functional molecular device. Tetrahedron Lett 38:4509–4512

82. Oguz U, Akkaya EU (1998) A squaraine-based sodium selective fluorescent chemosensor. Tetrahedron Lett 39:5857–5860

83. Ertekin K, Tepe M, Yenigü B, Akkaya EU, Henden E (2002) Fiber optic sodium and potassium sensing by using a newly synthesized squaraine dye in PVC matrix. Talanta 58:719–727

84. Chandrasekaran Y, Dutta G, Kanth RB, Patil S (2009) Tetrahydroquinoxaline based squaraines: Synthesis and photophysical properties. Dyes Pigm 83:162–167

85. Akkaya EU, Turkyilmaz S (1997) A squaraine-based near IR fluorescent chemosensor for calcium. Tetrahedron Lett 38:4513–4516
86. Dilek G, Akkaya EU (2000) Novel squaraine signalling Zn(II) ions: three-state fluorescence response to a single input. Tetrahedron Lett 41:3721–3724
87. Ajayaghosh A, Chithra P, Varghese R, Divya KP (2008) Controlled self-assembly of squaraines to 1D supramolecular architectures with high molar absorptivity. Chem Commun: 969–971
88. Chithra P, Varghese R, Divya KP, Ajayaghosh A (2008) Solvent-induced aggregation and cation-controlled self-assembly of tripodal squaraine dyes: optical, chiroptical and morphological properties. Chem Asian J 3:1365–1373
89. Kukrer B, Akkaya EU (1999) Red to near IR fluorescent signalling of carbohydrates. Tetrahedron Lett 40:9125–9128
90. Isgor YG, Akkaya EU (1997) Chemosensing in deep red: a squaraine-based fluorescent chemosensor for pH. Tetrahedron Lett 38:7417–7420
91. Ros-Lis JV, García B, Jiménez D, Martínez-Máñez R, Sancenón F, Soto J, Gonzalvo F, Valldecabres MC (2004) Squaraines as fluoro-chromogenic probes for thiol-containing compounds and their application to the detection of biorelevant thiols. J Am Chem Soc 126:4064–4065
92. Sreejith S, Divya KP, Ajayaghosh A (2008) A near-infrared squaraine dye as a latent ratiometric fluorophore for the detection of aminothiol content in blood plasma. Angew Chem 120:8001–8005
93. Houk RJT, Wallace KJ, Hewage HS, Anslyn EV (2008) A colorimetric chemodosimeter for Pd(II): a method for detecting residual palladium in cross-coupling reactions. Tetrahedron 64:8271–8278
94. Oswald B, Patsenker L, Duschl J, Szmacinski H, Wolfbeis OS, Terpetschnig E (1999) Synthesis, spectral properties, and detection limits of reactive squaraine dyes, a new class of diode laser compatible fluorescent protein labels. Bioconjugate Chem 10:925–931
95. Oswald B, Lehmann F, Simon L, Terpetschnig E, Wolfbeis OS (2000) Red laser-induced fluorescence energy transfer in an immunosystem. Anal Biochem 280:272–277
96. Oswald B, Gruber M, Böhmer M, Lehmann F, Probst M, Wolfbeis OS (2001) Novel diode laser-compatible fluorophores and their application to single molecule detection, protein labeling and fluorescence resonance energy transfer immunoassay. Photochem Photobiol 74:237–245
97. Terpetschnig EA, Tatarets A, Galkina O, Fedyunyayeva I, Patsenker L (2008) US Patent 7411068
98. Terpetschnig EA, Patsenker L, Tatarets A (2007) US Patent 7250517
99. Umezawa K, Citterio D, Suzuki K (2008) Water-soluble NIR fluorescent probes based on squaraine and their application for protein labeling. Anal Sci 24:213–217
100. Pham W, Weissleder R, Tung CH (2002) An azulene dimer as a near-infrared quencher. Angew Chem Int Ed Engl 41:3659–3662
101. Pham W, Weissleder R, Tung CH (2003) A practical approach for the preparation of monofunctional azulenyl squaraine dye. Tetrahedron Lett 44:3975–3978
102. Thomas J, Sherman DB, Amiss TJ, Andaluz SA, Pitner JB (2007) Synthesis and biosensor performance of a near-IR thiol-reactive fluorophore based on benzothiazolium squaraine. Bioconjugate Chem 18:1841–1846
103. Kim SH, Han SK, Park SH, Park LS (1998) A new dithiosquarylium dye for use as an electron transport material in an organic electroluminescent device having poly(p-phenylene vinylene) as an emitter. Dyes Pigm 38:49–56
104. Terpetschnig EA (2007) EP Patent 1849837
105. Renard BL, Aubert Y, Asseline U (2009) Fluorinated squaraine as near-IR label with improved properties for the labeling of oligonucleotides. Tetrahedron Lett 50:1897–1901
106. Griffiths J, Park S (2002) Facile preparative redox chemistry of bis(4-dialkylaminophenyl) squaraine dyes. Tetrahedron Lett 43:7669–7671

107. Nizomov N, Ismailov ZF, Nizamov SN, Salakhitdinova MK, Tatarets AL, Patsenker LD, Khodjayev G (2006) Spectral-luminescent study of interaction of squaraine dyes with biological substances. J Mol Struct 788:36–42

108. Patonay G, Salon J, Sowell J, Strekowski L (2004) Noncovalent labeling of biomolecules with red and near-infrared dyes. Molecules 9:40–49

109. Ioffe VM, Gorbenko GP, Tatarets AL, Patsenker LD, Terpechnig EA (2006) Examining protein-lipid interactions in model systems with a new squarylium fluorescent dye. J Fluoresc 16:547–554

110. Ioffe VM, Gorbenko GP, Domanov YA, Tatarets AL, Patsenker LD, Terpetsching EA, Dyubko TS (2006) A new fluorescent squaraine probe for the measurement of membrane polarity. J Fluoresc 16:47–52

111. Hiroo T, Masaki O (2000) JP Patent 2000285978

112. Reis LV, Serrano JP, Almeida P, Santos PF (2009) The synthesis and characterization of novel, aza-substituted squarylium cyanine dyes. Dyes Pigm 81:197–202

113. Luchowski R, Matveeva EG, Gryczynski I, Terpetschnig EA, Patsenker L, Laczko G, Borejdo J, Gryczynski Z (2008) Single molecule studies of multiple-fluorophore labeled antibodies. Effect of homo-FRET on the number of photons available before photobleaching. Curr Pharm Biotechnol 9:411–420

114. Matveeva EG, Terpetschnig EA, Stevens M, Patsenker L, Kolosova OS, Gryczynski Z, Gryczynski I (2009) Near-infrared squaraine dyes for fluorescence enhanced surface assay. Dyes Pigm 80:41–46

115. http://www.setabiomedicals.com/products/biomedical/flt-labels.htm

116. Cooper ME, Gregory S, Adie E, Kalinka S (2002) pH-Sensitive cyanine dyes for biological applications. J Fluoresc 12:425–429

117. Adie EJ, Kalinka S, Smith L, Francis MJ, Marenghi A, Cooper ME, Briggs M, Michael P, Milligan G, Game S (2002) A pH-sensitive fluor, CypHer 5, used to monitor agonist-induced G protein-coupled receptor internalization in live cells. BioTechniques 33:1152–1157

118. Cooper ME, Gregory SJ, Adie E, Kalinka S, Burns DD (2001) pH Sensitive cyanine dyes for biological applications. 7th Conference on Methods and Applications of Fluorescence: Spectroscopy, Imaging, and Probes, Amsterdam. http://www4.gelifesciences.com/aptrix/upp00919.nsf/Content /A29DEB85A10D6C84C1257628001CEE23/$file/MAFS_pH_dye.pdf

119. Povrozin YA, Markova LI, Tatarets AL, Sidorov VI, Terpetschnig EA, Patsenker LD (2009) Near-infrared, dual-ratiometric fluorescent label for measurement of pH. Anal Biochem 390:136–140

Two-Photon Absorption in Near-IR Conjugated Molecules: Design Strategy and Structure–Property Relations

Olga V. Przhonska, Scott Webster, Lazaro A. Padilha, Honghua Hu, Alexey D. Kachkovski, David J. Hagan, and Eric W. Van Stryland

Abstract In the past few years, applications built around two-photon absorption (2PA) have emerged, which require new materials to be designed and characterized in order to discover new applications and to advance the existing ones. This chapter describes the nonlinear optical processes and characterization techniques along with design strategies and structure–property relations of cyanine and cyanine-like molecular structures with the goal of enhancing 2PA in the near-IR for multiphoton fluorescence sensing applications. Specifically, a detailed analysis of the linear and nonlinear optical properties of several classes of polymethine dyes, which include symmetrical and asymmetrical combinations of π-conjugated bridges with electron donating (D) or electron accepting (A) terminal groups, are presented. These structures are: D–π–D, A–π–A, D–π–A, and a quadrupolar type arrangement of D–π–A–π–D.

The results of this research combined with the growing literature on structure–property relations in organic materials is moving us closer to the ultimate goal of developing a predictive capability for the nonlinear optical properties of molecules.

Keywords Cyanine dyes · Excited state absorption · Polymethine dyes · Pump-probe · Two-photon absorption · Z-scan

O.V. Przhonska (✉)
Institute of Physics, National Academy of Sciences of Ukraine, Prospect Nauki 46, 03028, Kyiv, Ukraine
CREOL, the College of Optics and Photonics, University of Central Florida, Orlando, FL, USA
e-mail: olga@creol.ucf.edu

S. Webster, L.A. Padilha, H. Hu, D.J. Hagan, and E.W.V. Van Stryland
CREOL, the College of Optics and Photonics, University of Central Florida, Orlando, FL, USA

A.D. Kachkovski
Institute of Organic Chemistry, National Academy of Sciences, Kyiv, Ukraine

A.P. Demchenko (ed.), *Advanced Fluorescence Reporters in Chemistry and Biology I:*   105
Fundamentals and Molecular Design, Springer Ser Fluoresc (2010) 8: 105–148,
DOI 10.1007/978-3-642-04702-2_4, © Springer-Verlag Berlin Heidelberg 2010

Contents

1   Introduction

Significant effort in the past decade has been given to the development of organic molecules and semiconductor quantum dots with large two-photon absorption (2PA) for applications in fluorescence sensing and biological imaging. Advances in several areas, including molecular and synthetic chemical designs, understanding of structure–property relations, and the wide availability of tunable femtosecond sources have enabled the development and discovery of molecules with larger 2PA cross sections, δ_{2PA}, and increased fluorescence quantum yields in the visible and near-infrared (NIR).

Organic molecules, with the capability of tailoring their linear and nonlinear optical properties by molecular structure modification, can be utilized in fluorescence sensing and biological imaging. Fluorescence imaging, where a dye molecule is attached to a particular component or species of a system under investigation, utilizes the difference in excitation energy and the emitted fluorophore's photon energy for increased signal to noise discrimination. This technique can be further enhanced by utilizing nonlinear excitation processes such as 2PA, where the excitation photon energy is less than the energy gap of the fluorophore where there is no linear absorption. Figure 1a shows a schematic of frequency degenerate (both excitation photons having the same frequency) 2PA (1) into the first allowed singlet state, (2) above the first allowed singlet state, and (3) a near double resonant condition, with a small intermediate state resonance, Δ, and a transition into an allowed final 2PA state.

Since 2PA is a third-order nonlinear susceptibility, $\chi^{(3)}$, process, large irradiances are needed to excite fluorophores into their excited state. Typically, femtosecond lasers are used in conjunction with tight focusing geometries to produce these large irradiances, thereby restricting 2PA to the focal volume, which can be as small as femtoliters. The one-photon excitation density in the focal region is proportional to the light intensity, whereas the two-photon excitation density is

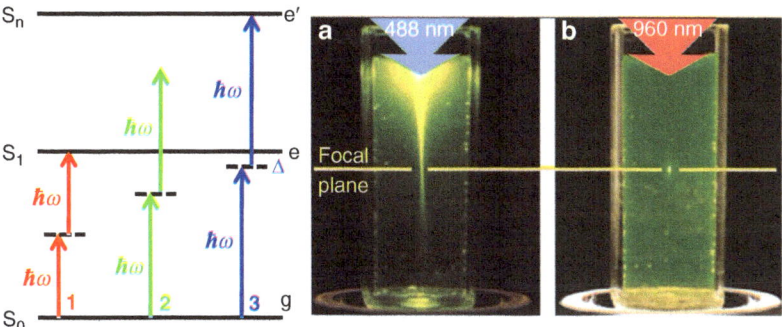

Fig. 1 (*Left*) Schematic of frequency degenerate 2PA (1) into the first allowed singlet state, (2) above the first allowed singlet state, and (3) a double resonant condition, with a small intermediate state resonance energy difference, Δ, and a transition into an allowed final 2PA state. (*Right*) Photograph illustrating the much sharper contrast of two-photon (**b**) versus one-photon excitation (**a**) (taken from [2])

proportional to the square of the irradiance and therefore falls rapidly away from the focus allowing decreased scattering. The 2PA review by M. Pawlicki et al [1] details the advantages and disadvantages of excitation by one- and two-photons. Figure 1b is a photograph of a cell containing a fluorophore that is excited with one-photon absorption (1PA) and 2PA, which illustrates the much sharper contrast in the excitation density for 2PA (taken from [2]).

Since the excitation wavelength is also less than the main transition energy, and usually in the NIR wavelength region, 2PA excitation is less likely to cause photochemical decomposition and decreases scattering due to longer wavelengths, thereby allowing applications involving in vivo imaging due to the low absorption and scattering of tissue in this wavelength range allowing for deep tissue imaging (up to 1 mm in commercially available systems) with high spatial resolution. Many excellent reviews for multiphoton microscopy and related techniques and applications are available [3–11].

This chapter summarizes our current knowledge and understanding of linear π-conjugated systems for NIR dyes: cyanines and cyanine-like molecules for one- and multiphoton applications, with a focus on the intramolecular spectroscopic properties and dynamics and the experimental methodologies used to characterize these particular organic systems and others.

1.1 Brief Historical Account of Two-Photon Absorption and π-Conjugated Systems

In the 1930s, Maria Göppert-Mayer [12], theoretically predicted the process of an atom being raised to an excited state of energy equal to the sum of two simultaneously absorbed photons [13], but it was not until 1961, shortly after the development of the laser, that the phenomenon was experimentally verified due to the large irradiances needed. With the development of pulsed lasers of decreasing pulse widths (nanosecond, picoseconds, and femtosecond pulse widths) and increasing output energies, larger peak irradiances were obtained in the 1980s and 1990s, and 2PA research grew leading to the development of two-photon induced fluorescence microscopy for enhanced imaging [14].

The many theoretical and experimental studies of 2PA in the past few decades have enabled a more complete understanding of molecular transitions. Selection rules for 2PA are different from those for 1PA, which has led to complementary spectroscopic techniques that are currently employed in discovering new structure–property relationships. Several reviews exist concerning the design, synthesis, and characterization of particular organic dyes [1, 15–19] with the desire to understand their structure–property relationships for 2PA-enabled applications such as 3D optical data storage [20, 21], multiphoton 3D microfabrication [22], imaging (both biological [23] and chemical [24] sensing), and two-photon photodynamic therapy [25–27]. This review focuses on describing the 1PA fluorescence, 2PA, and related linear and nonlinear optical properties in π-conjugated cyanine and cyanine-like molecules for applications involving two-photon fluorescence.

1.2 Nonlinear Mechanisms in π-Conjugated Molecules

To understand the response of materials upon light irradiation, we describe the macroscopic polarization P as a function of the electric field E as:

$$P = P^{(1)} + P^{(2)} + P^{(3)} + \ldots = \varepsilon_0(\chi^{(1)}E + \chi^{(2)}E^2 + \chi^{(3)}E^3 + \ldots), \quad (1)$$

where ε_0 is the dielectric constant in vacuum, $P^{(n)}$ denotes the nth order of polarization, and $\chi^{(n)}$ is the nth order optical susceptibility; $\chi^{(1)}$ describes the linear optical properties of the material; $\chi^{(2)}$ represents the second harmonic generation, sum frequency generation, optical rectification, parametric generation, and electro-optic effect, etc.; and $\chi^{(3)}$ is connected with third harmonic generation, nonlinear refraction (higher orders of $\chi^{(n)}$ for odd n can also contribute to nonlinear refraction), 2PA, stimulated Raman or Brillouin scattering, and four wave mixing, etc.

1.2.1 Two-Photon Absorption

Frequency degenerate 2PA is a third order, $\chi^{(3)}$, nonlinear optical process whereby two photons of equal energy are simultaneously absorbed to raise a system into an excited state of energy equal to that of the sum of the two photons. The propagation of light of irradiance I, through a material of thickness z with 1PA and 2PA is written as:

$$\frac{dI}{dz} = -\alpha_1 I - \alpha_2 I^2 - \ldots, \tag{2}$$

where α_1 and α_2 are the one- and two-photon absorption coefficients. Higher order nonlinear absorptions are not discussed here, but formulations for three- and multiphoton absorption can be found elsewhere [28]. In this brief review, we focus on degenerate 2PA, which corresponds to two photons of equal energy. Details concerning nondegenerate 2PA can be found elsewhere [29–32]. Degenerate 2PA at a specific frequency, ω, is proportional to the imaginary part of $\chi^{(3)}$, expressed in SI units as:

$$\alpha_2(\omega) = \frac{3\omega}{2n^2 c^2 \varepsilon_0} \text{Im}\left(\chi^3(-\omega; \omega, -\omega, \omega)\right), \tag{3}$$

where n is the refractive index and c is the speed of light. To characterize the 2PA of individual molecules, we define the 2PA cross section, $\delta_{2PA}(\omega)$, which is given in units of 1×10^{-50} cm^4 s photon^{-1} molecule^{-1}. This unit is called "Goppert-Mayer" or "GM" in honor of the author of [12]. $\delta_{2PA}(\omega)$ for an individual molecule can be expressed in SI units by:

$$\delta_{2PA}(\omega) = \frac{\hbar\omega\alpha_2}{N} = \frac{3\hbar\omega^2}{2n^2 c^2 \varepsilon_0 N} \text{Im}\left(\chi^3(-\omega; \omega, -\omega, \omega)\right), \tag{4}$$

where N is the number of molecules per unit volume.

In this chapter, we will discuss a simple quantum mechanical approach to model 2PA in molecular systems since it provides an accurate and comprehensive picture of the physics involved in the interaction. In 1971, Orr and Ward used a sum-over-states (SOS) model based on perturbation theory to derive $\chi^{(3)}$ [33]. This methodology sums all transitions between the ground state, and all possible excited states. Fortunately, a complete sum over all states is not necessary for modeling 2PA. Often, even as few as three states can give good insight. Therefore, a three-state model is developed to simplify the SOS approach [34–36]. In this model, the ground state (g), excited state as intermediate state (e), and another higher-lying excited state as a final two-photon state (e'), are considered (see Fig. 1). Transitions between ground state and excited state (g, e), and transitions between two excited states (e, e') are one-photon allowed so that their transition dipole moments are not zero: $\mu_{ge}, \mu_{ee'} \neq 0$, while transitions between the ground state and the higher-lying excited state (g, e') are forbidden: $\mu_{ge'} = 0$. All of the resonance terms are kept and the antiresonance terms are omitted. Considering only degenerate 2PA of linearly parallel polarized light, the third-order susceptibility tensor can be written in SI Units as:

$$
\chi_{xxxx}^{(3)}(2PA) = \frac{2N}{3!\varepsilon_0}
\begin{cases}
\dfrac{\mu_{ge}^x \Delta\mu^x \Delta\mu^x \mu_{eg}^x}{(\Omega_{eg}-\hbar\omega)(\Omega_{eg}-2\hbar\omega)(\Omega_{eg}-\hbar\omega)} & \text{D - term} \\[4mm]
+\dfrac{\mu_{ge}^x \Delta\mu^x \Delta\mu^x \mu_{eg}^x}{(\Omega_{eg}^*-\hbar\omega)(\Omega_{eg}-2\hbar\omega)(\Omega_{eg}-\hbar\omega)} & \text{D - term} \\[4mm]
+\dfrac{\mu_{ge}^x \bar\mu_{ee'}^x \bar\mu_{e'e}^x \mu_{eg}^x}{(\Omega_{eg}-\hbar\omega)(\Omega_{e'g}-2\hbar\omega)(\Omega_{eg}-\hbar\omega)} & \text{T - term} \\[4mm]
+\dfrac{\mu_{ge}^x \bar\mu_{ee'}^x \bar\mu_{e'e}^x \mu_{eg}^x}{(\Omega_{eg}^*-\hbar\omega)(\Omega_{e'g}-2\hbar\omega)(\Omega_{eg}-\hbar\omega)} & \text{T - term}
\end{cases}
\tag{5}
$$

$$
-\frac{2N}{3!\varepsilon_0}
\begin{cases}
\dfrac{\mu_{ge}^x \mu_{eg}^x \mu_{ge}^x \mu_{eg}^x}{(\Omega_{eg}-\hbar\omega)(\Omega_{eg}-\hbar\omega)(\Omega_{eg}-\hbar\omega)} & \text{N - term} \\[4mm]
\dfrac{\mu_{ge}^x \mu_{eg}^x \mu_{ge}^x \mu_{eg}^x}{(\Omega_{eg}-\hbar\omega)(\Omega_{eg}^*-\hbar\omega)(\Omega_{eg}-\hbar\omega)} & \text{N - term}
\end{cases}
$$

where Ω_{eg} is the energy difference between e and g states including the transition linewidth, Γ_{eg}: $\Omega_{eg} = \hbar\omega_{eg} - i\Gamma_{eg}, \Omega_{eg}^* = \hbar\omega_{eg} + i\Gamma_{eg}$; $\Delta\mu^x$ is the permanent dipole moment difference between the excited state e and ground state g : $\Delta\mu^x = \mu_{ee}^x - \mu_{gg}^x$.

The first two terms in (5) are called "D-terms" or "dipolar terms," which are nonzero only if $\Delta\mu^x \neq 0$. The two-photon resonance denominator, $(\Omega_{eg} - 2\hbar\omega)$, indicates that an electron is excited into the lower excited state e. If we consider a near resonance condition: $\hbar\omega = \hbar\omega_{eg}/2$, the imaginary part of the D-terms can be written in SI units as:

$$
\mathrm{Im}\left(\chi^3(2PA)_{\text{D - term}}\right) = \frac{N\mu_{ge}^2 \Delta\mu^2}{3!\varepsilon_0 \Gamma_{eg}} \frac{\left(\hbar\omega_{eg}\right)^2}{\left[\left(\hbar\omega_{eg}/2\right)^2 + \Gamma_{eg}^2\right]^2}
\tag{6}
$$

If the molecules possess different excited state permanent and ground state permanent dipole moments, the D-terms can contribute to 2PA. Centrosymmetric molecules do not have permanent dipole moments in both ground and excited states, so their D-terms are zero.

The second two terms in (5) are called "T-terms" or "two-photon terms", which have $(\Omega_{e'g} - 2\hbar\omega)$ in the denominator corresponding to the excitation of an electron into the higher-lying excited state e'. If we consider a resonance condition where $\hbar\omega = \hbar\omega_{e'g}/2$ and assume that the transition linewidth is narrow, $\Gamma_{eg} << [\hbar(\omega_{eg} - \omega_{e'g}/2)]/2$, the imaginary part of the T-term can then be expressed in SI units as:

$$
\mathrm{Im}[\chi_{xxxx}^{(3)}(2PA)_{\text{T - term}}] = \frac{4N\mu_{ge}^2 \mu_{ee'}^2}{3!\varepsilon_0 \Gamma_{e'g}} \frac{1}{\left(\hbar\omega_{eg} - \hbar\omega_{e'g}/2\right)^2}.
\tag{7}
$$

In both cases, μ_{ge} and $\Delta\mu$, as well as μ_{ge} and $\mu_{ee'}$ are assumed to be parallel to each other. If there is an angle θ between the corresponding dipole moments μ_{ge} and $\mu_{ee'}$, an effective excited state transition dipole moment $\mu_{ee'}^{\text{eff}}$ should be used instead of $\mu_{ee'}$ [37]: $\mu_{ee'}^{\mathrm{eff}} = \mu_{ee'}\left[\left(2\cos^2(\theta) + 1\right)/3\right]^{1/2}$. In isotropic media such as a solution, all D- and T-terms are averaged over the random orientations and should be divided by a factor of 5 [38].

The last two terms in (5) are called "N" or "negative terms" that do not contribute to two-photon absorption when the incident photon energy is far below the 1PA edge. However, very close to the one-photon edge, the N-terms in (5) may negatively contribute to the 2PA. This term is sometimes referred to as "virtual saturation" since it turns into real saturation at frequencies on or very near to resonance (within the linewidth). In semiconductors, this term is usually referred to as the AC-Stark or quadratic Stark effect and physically represents the shifting of the energy level with large optical fields [29].

The 2PA cross section spectrum, $\delta_{2\mathrm{PA}}(\omega)$, including D- and T- terms but assuming we are far enough below the 1PA edge to ignore the N terms, can be presented in SI units as [39, 40]:

$$\delta_{2\mathrm{PA}}(\omega) = \frac{1}{5c^2\hbar n^2\varepsilon_0^2}\frac{(\hbar\omega)^2}{\left(\hbar\omega_{eg} - \hbar\omega\right)^2 + \Gamma_{eg}^2}$$

$$\times\left[\frac{|\mu_{eg}|^2|\Delta\mu|^2\Gamma_{eg}}{\left(\hbar\omega_{eg} - 2\hbar\omega\right)^2 + \Gamma_{eg}^2} + \frac{|\mu_{eg}|^2|\mu_{ee'}|^2\Gamma_{e'g}}{\left(\hbar\omega_{e'g} - 2\hbar\omega\right)^2 + \Gamma_{e'g}^2}\right]. \tag{8}$$

In practice, for linear π-conjugated molecules, the 2PA spectra typically consist of several 2PA bands, corresponding to several final states (e') and the same intermediate state (e). Analysis of (8) identifies the main spectroscopic parameters responsible for $\delta_{2\mathrm{PA}}$ and formulates the following general trends in structure–property relations. As seen from (8), the main spectroscopic parameters are: change in the permanent dipole moment $\Delta\mu$; transition dipole moments μ_{ge}, $\mu_{ee'}$; angles between dipole moments; linewidth Γ, and detuning energies from intermediate and final states, $(\hbar\omega_{eg} - \hbar\omega)$ and $(\hbar\omega_{e'g} - 2\hbar\omega)$.

Factors that can enhance $\delta_{2\mathrm{PA}}(\omega)$ are:

1. *Increasing the transition dipole moments.* In molecular design, this can be realized by increasing the π-conjugation length, or by introducing electron donor/acceptor groups. In noncentrosymmetric molecules, increasing the difference of the ground and excited state permanent dipole moments can also increase $\delta_{2\mathrm{PA}}(\omega)$
2. *Maximizing resonance terms.* Decreasing the detuning energy between intermediate and ground states can significantly enhance $\delta_{2\mathrm{PA}}(\omega)$. This effect is called intermediate state resonance enhancement (ISRE). This is illustrated in Fig. 1a and is discussed in detail in [32]. If the intermediate state is located halfway between ground state and final state, a "double resonance" condition can be achieved, which can lead to a dramatic enhancement of $\delta_{2\mathrm{PA}}(\omega)$
3. *Reducing the linewidth of the lowest energy one-photon transition.* Minimizing Γ increases $\delta_{2\mathrm{PA}}(\omega)$, which allows for photons to closely approach the 1PA edge without one-photon losses

Additional details for enhancing 2PA in linear cyanine-like molecules will be discussed in Sect. 3.

1.2.2 Excited State Absorption (ESA)

Understanding the role of ESA in nonlinear absorbing systems is important for: (1) determining correct 2PA cross sections, illustrated in Sect. 1.2.3 by decoupling ESA from 2PA; (2) gaining insight into the nature of transitions from intermediate to final states in 2PA spectra; (3) determining the nature of higher-lying excited states that are not physically accessible through 2PA measurements; and (4) determining the intermediate transition dipole moments that can be additionally calculated by quantum chemical methods. Unlike 2PA, which is a $\chi^{(3)}$ process, ESA is a cascaded first-order susceptibility process, $\chi^{(1)}$, where two photons are sequentially absorbed to take the molecule to one excited and then a final state. This requires that the first absorbed photon has an energy equal to or larger than the lowest molecular transition energy.

The physical mechanisms involved in absorption and emission by the molecule can be illustrated by the 5-level energy model based on a Jablonski diagram [41] in Fig. 2a. The system can be excited by one-photon into vibrational sublevels of the first excited state (S_1), followed by a rapid vibronic relaxation to the lowest level in the S_1 state. There are several competing processes to depopulate the S_1 state: spontaneous decay or stimulated emission to the ground state (S_0); excitation into the higher-lying excited state (S_n) by absorbing another photon; or intersystem crossing leading to population of the first triplet state (T_1).

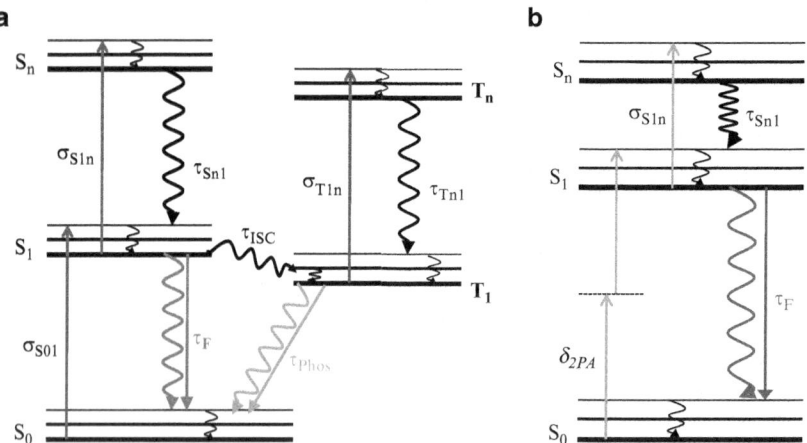

Fig. 2 (a) Energy level schematics for (a) a 5-level model including both singlet and triplet ESA and (b) 2PA-induced ESA, both including relaxation dynamics

Spontaneous decay from S_1 to S_0 can follow either a radiative pathway by emitting a photon (fluorescence), or a nonradiative pathway. The lifetime of the S_1 state is determined by the total decay rate of these two pathways, which is equal to the fluorescence lifetime τ_F. Stimulated emission occurs when there is spectral overlap between excitation and fluorescence; however, it may usually be neglected since excitation wavelengths are intentionally set out of the range of the fluorescence spectrum. ESA occurs when an S_1 electron absorbs another photon and is promoted to a higher lying state (S_n); if $\sigma_{S1n} > \sigma_{01}$, this is referred to as reverse saturable absorption (RSA). The decay rate from S_n to S_1 is normally much faster than the lifetime of the S_1 state ($\tau_{Sn1} \ll \tau_F$); so with small fluences (energy per area), the population of S_n can usually be assumed to be zero. But this approximation fails if the input fluence is very large, leading to significant population of the S_n state. In this case, another higher-lying singlet excited state $S_{n'}$ often needs to be included to take into account absorption from S_n to $S_{n'}$ (not shown in Fig. 2a). Intersystem crossing to the triplet state T_1 may occur when the excited electron undergoes a spin conversion, which is characterized by an intersystem crossing rate: $k_{ISC} = 1/\tau_{ISC}$. The decay from T_1 to S_0 can also follow a radiative pathway (phosphorescence) or nonradiative pathway; however, the lifetime of the T_1 state (τ_{Phos}) is long (usually 10^{-8} to 10^2 s) due to its spin-forbidden nature. Therefore, provided the presence of a triplet state in the molecular system and long-duration input laser pulse widths are used (usually larger than nanoseconds, but possibly as small as picoseconds [42]), ESA absorption from T_1 to T_n should also be considered when modeling results.

To mathematically describe the physical processes indicated above, the following propagation and rate equations are introduced in the form of differential equations as:

$$
\frac{dI}{dz} = -\sigma_{S01}N_0 I - \sigma_{S1n}N_{S1}I - \sigma_{T1n}N_{T1}I
$$

$$
\frac{dN_0}{dt} = -\frac{\sigma_{S01}N_0 I}{\hbar\omega} + \frac{N_{S1}}{\tau_F} + \frac{N_{T1}}{\tau_{Phos}}
$$

$$
\frac{dN_{S1}}{dt} = \frac{\sigma_{S01}N_0 I}{\hbar\omega} - \frac{N_{S1}}{\tau_F} - \frac{\sigma_{S1n}N_{S1}I}{\hbar\omega} + \frac{N_{Sn}}{\tau_{Sn1}} - \frac{N_{S1}}{\tau_{ISC}}
$$

$$
\frac{dN_{Sn}}{dt} = \frac{\sigma_{S1n}N_{S1}I}{\hbar\omega} - \frac{N_{Sn}}{\tau_{Sn1}}
\tag{9}
$$

$$
\frac{dN_{T1}}{dt} = -\frac{\sigma_{T1n}N_{T1}I}{\hbar\omega} + \frac{N_{Tn}}{\tau_{Tn1}} + \frac{N_{S1}}{\tau_{ISC}} - \frac{N_{T1}}{\tau_{Phos}}
$$

$$
\frac{dN_{Tn}}{dt} = \frac{\sigma_{T1n}N_{T1}I}{\hbar\omega} - \frac{N_{Tn}}{\tau_{Tn1}},
$$

where I is the irradiance, z is the sample thickness, σ_{Sij} and σ_{Tij} are the singlet and triplet cross sections from their respective ground and excited states, τ_{Sn1} and τ_{Tn1} are the nonradiative relaxation lifetimes from upper excited states S_n and T_n, and τ_F and τ_{Phos} are decay lifetimes from S_1 and T_1 to the singlet ground state S_0.

1.2.3 Excited State Absorption via Two-Photon Absorption

It is necessary to pay special attention to the role of ESA in 2PA measurements. ESA induced by 2PA in an organic system was first observed and explained in 1974 by Kleinschmidt et al [43]. Without separating ESA from 2PA, $\delta_{2PA}(\omega)$ measured by nanosecond pulses could be incorrectly interpreted as being two orders of magnitude larger than that obtained by femtosecond pulses [44]. In order to characterize 2PA induced ESA, pulsewidth dependent measurements are used in order to distinguish irradiance/fluence processes [45, 46]. ESA induced by 2PA, as shown in Fig. 2b, can be properly modeled by incorporating the 2PA term into the propagation and rate equations as [46, 47]:

$$
\begin{aligned}
\frac{dI}{dz} &- \frac{\delta_{2PA}N_0 I^2}{\hbar\omega} - \sigma_{S1n}N_1 I \\
\frac{dN_0}{dt} &= -\frac{\delta_{2PA}N_0 I^2}{2(\hbar\omega)^2} + \frac{N_1}{\tau_F} \\
\frac{dN_1}{dt} &= \frac{\delta_{2PA}N_0 I^2}{2(\hbar\omega)^2} - \frac{N_1}{\tau_F} - \frac{\sigma_{S1n}N_1 I}{\hbar\omega} + \frac{N_n}{\tau_{Sn1}} \\
\frac{dN_n}{dt} &= \frac{\sigma_{S1n}N_1 I}{\hbar\omega} - \frac{N_n}{\tau_{Sn1}}.
\end{aligned}
\tag{10}
$$

1.3 Linear π-Conjugated Molecular Systems

Cyanine and cyanine-like dyes have been known for more than a century and have found numerous applications as photosensitizers in photography and photodynamic therapy, fluorescent probes in chemistry and biology, active and passive laser media, materials for nonlinear optics and electroluminescence, memory devices etc. [48, 49]. They are among a particular class of organic compounds that exhibit large (with molar absorbance up to $3 \times 10^5 \ \text{M}^{-1} \ \text{cm}^{-1}$) and tunable absorption bands in the visible and NIR regions, which is important for the development of organic materials with large third-order nonlinearities for all-optical signal processing [50]. Based on the number of methine ($-CH=$) groups in the π-conjugation, linear conjugated dyes can be divided into two categories: *polymethine* and *polyene dyes*. *Polymethines* are compounds made up from an *odd number* of methine groups bound together by alternating single and double bonds, which form a π-conjugated chain bridging together two terminal groups R_1 and R_2 as shown in Fig. 3a.

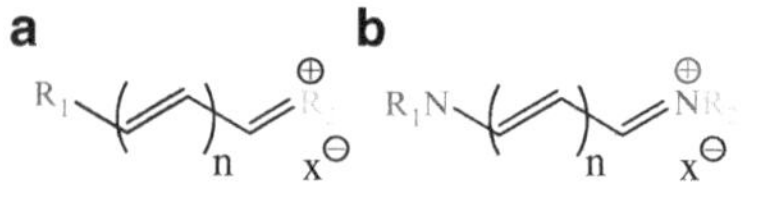

Fig. 3 (a) Typical *polymethine* and (b) *cyanine* molecular structures; n is the number of methine groups and X are counter ions

Polyenes are compounds made up from an *even number* of methine groups (not shown). These two classes of structures have different charge distribution and bond length alternation (BLA) along the main conjugation chain, thus leading to different electronic state configurations. Polymethine dyes are characterized by equalization of bond lengths and large charge alternation between neighboring carbon–carbon atoms in the π-conjugated chain. In contrast, polyene dyes have large BLA, while maintaining similar charges along the conjugated chain [51]. These two types of dyes typically have distinctive electronic structures, and thus differ by their linear and nonlinear optical properties. The influence of BLA to nonlinear optical properties is described in [52]. Simplified polymethine and cyanine dye structures are shown in Fig. 3. *Cyanine dyes*, which belong to the polymethine family, consist of *nitrogen atoms* at the end of the conjugated chain as shown in Fig 3b.

The electronic properties of these dyes can be tailored by changing the length of conjugation chain or by adding specific terminal groups R_1 and R_2. Due to their different electron affinities, these terminal groups can be classified into electron acceptor (A), and electron donor (D) groups.

Appending different terminal groups to the π-conjugated ends, the cyanine-like molecules may have the following molecular structures: D–π–D, D–π–A, and A–π–A. Additionally, electron acceptor/donor groups may be included into the main π-conjugation chain to form D–π–A–π–D or A–π–D–π–A quadrupolar structures. These basic structures are shown schematically in Fig. 4 and discussed in detail in Sect. 3.

Previously, 2PA properties of π-conjugated chromophores have been investigated primarily for molecules with absorption bands in the visible range (400–650 nm) [1, 15]. NIR cyanine-like dyes are now being developed due to their large optical nonlinearities, which makes them potentially applicable for fluorescence imaging, optical power regulation in the telecommunication wavelength range [53], and for all-optical signal processing [54]. A detailed analysis of the linear and nonlinear optical properties of the various symmetrical and asymmetrical cyanine and cyanine-like dyes will be discussed in Sect. 4.3. Throughout the text, we use the labeling for polymethine dyes as PD or PDs, squaraine dyes as SD or SDs, and tetraone dyes as TD or TDs.

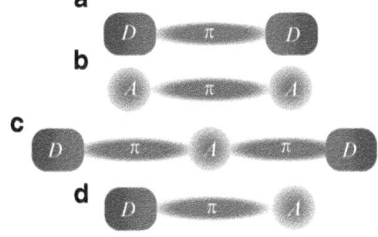

Fig. 4 Schematic of symmetrical (**a**) D–π–D, (**b**) A–π–A, (**c**) D–π–A–π–D, and (**d**) asymmetrical D–π–A molecular structures

2 Experimental Methodologies for Linear and Nonlinear Optical Characterization

To design organic conjugated molecules with optimal nonlinear optical properties, accurate and complete experimental analysis of their linear and nonlinear optical properties are needed. It is common to find in the literature reports of large nonlinear responses in different organic molecules; unfortunately, it is also not uncommon to encounter promising results due to erroneous or incomplete experimental analysis [55]. In the following section, we discuss different experimental techniques used for the nonlinear optical characterization of organic molecules.

Before being able to study the nonlinear optical properties of any material, it is necessary to have a complete understanding of its linear optical properties. Therefore, we start this section with a brief discussion of the techniques used to measure some of the most important linear properties, e.g., *linear absorption, fluorescence, anisotropy, and fluorescence quantum yield*.

2.1 Linear Optical Characterization

Two of the most important properties of any optical material are the linear absorption (one-photon) spectra and one-photon-excited fluorescence (1PF) spectra. In general, linear absorption spectra should be checked before and after each experiment to verify that the sample has not decomposed, especially after nonlinear optical measurement where photo-induced damage is often observed. All optical measurements, linear and nonlinear, should be performed at concentrations below the aggregation threshold, significant time should be given for the solute to fully dissolve, and the use of solution filters (filtration $\sim$200 nm) can be useful for removing large aggregates. The ratio between the number of emitted photons and absorbed photons is known as the fluorescence quantum yield, Φ_F, and is an important parameter for all fluorophores.

In most cases, the *linear absorption* is measured with standard spectrometers, and the fluorescence properties are obtained with commercially available spectrofluorometers using reference samples with well-known Φ_F for calibration of the *fluorescence quantum yield*. In the ultraviolet and visible range, there are many well-known fluorescence quantum yield standards. Anthracene in ethanol ($\Phi_F = 0.27$) [56], POPOP in cyclohexane ($\Phi_F = 0.97$), Rhodamine 6G in ethanol ($\Phi_F = 0.95$), and Cresyl Violet in methanol ($\Phi_F = 0.54$) are among the most commonly used reference samples for wavelengths of 350–650 nm. For wavelengths longer than 650 nm, there is a lack of fluorescence references. Recently, a photochemically stable, D–π–D polymethine molecule has been proposed as a fluorescence standard near 800 nm [57]. This molecule, PD 2631 (chemical structure shown in Fig. 5) in ethanol, has $\Phi_F = 0.11$ and has its fluorescence peak at 809 nm. Fig. 5 compares the linear absorption and fluorescence spectra of the reference PD 2631 in ethanol to

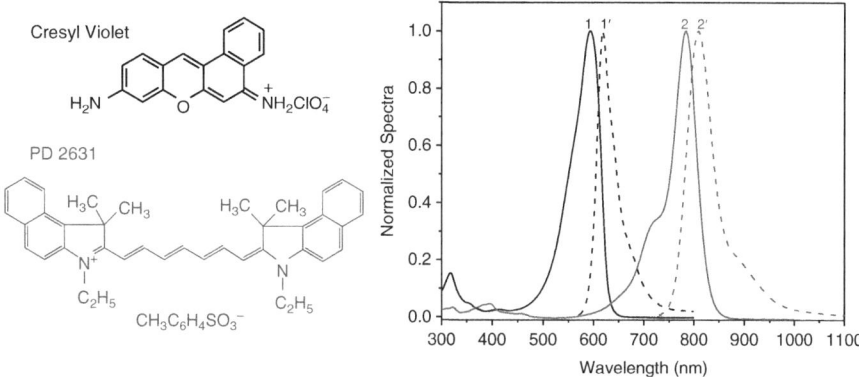

Fig. 5 Linear absorption (1, 2) and one-photon-excited fluorescence (1′, 2′) for the quantum yield standard Cresyl Violet (1, 1′) and the proposed standard PD 2631 (2, 2′) for NIR wavelengths. Molecular structures are shown to the *left*

the well-known standard Cresyl Violet in methanol. Additionally, a series of A–π–A dyes have been synthesized and show significant potential for fluorescence applications [58, 59]. The three shortest dyes (G37, G38, and G74) in Fig. 20 of Sect. 3.2.3 show good photochemical stability and surprisingly large one-photon fluorescence quantum yields of ∼0.90, ∼0.66, and ∼0.18 at the red to NIR region of ∼640 nm, ∼730 nm, and ∼840 nm, respectively.

Another important linear parameter is the *excitation anisotropy* function, which is used to determine the spectral positions of the optical transitions and the relative orientation of the transition dipole moments. These measurements can be provided in most commercially available spectrofluorometers and require the use of viscous solvents and low concentrations ($c_M \sim 1\ \mu M$) to avoid depolarization of the fluorescence due to molecular reorientations and reabsorption. The anisotropy value for a given excitation wavelength λ can be calculated as

$$r(\lambda) = \frac{I_{\parallel}(\lambda) - I_{\perp}(\lambda)}{I_{\parallel}(\lambda) + 2I_{\perp}(\lambda)}, \tag{11}$$

where $I_{\parallel}(\lambda)$ and $I_{\perp}(\lambda)$ are the intensities of the fluorescence signal (typically measured near the fluorescence maximum) polarized parallel and perpendicular to the excitation light, respectively [41].

The anisotropy value $r(\lambda)$ ranges between −0.2 and 0.4 in correspondence with the angle γ between the absorption and fluorescence dipole moments, which can range from 90° to 0° in accordance to:

$$r(\lambda) = \frac{2}{5}\left(\frac{3\cos^2(\gamma) - 1}{2}\right). \tag{12}$$

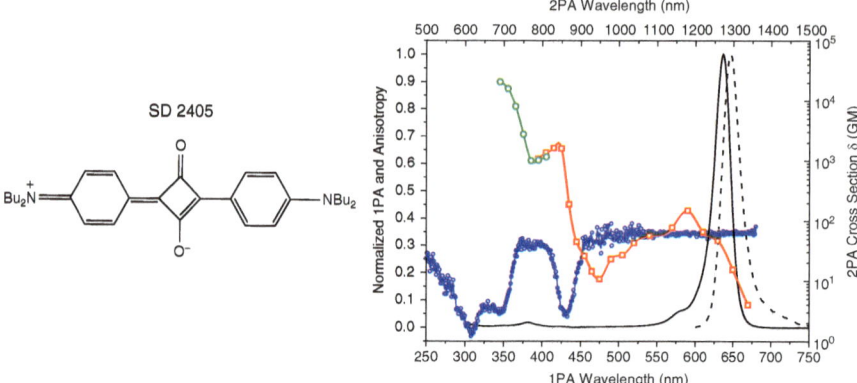

Fig. 6 Normalized linear absorption (*solid black line*), anisotropy (*blue circles*), and corresponding 2PA spectrum measured by two-photon fluorescence (*red squares*) and Z-scan (*green circles*) for SD 2405. Molecular structure is shown to the *left*

As seen from (12) and Fig. 6, the peaks in the excitation anisotropy spectrum indicate a small angle between the absorption and emission transition dipoles suggesting allowed 1PA transitions; while valleys indicate large angles between these two dipoles, suggesting a forbidden 1PA transition. Due to selection rules for symmetrical cyanine-like dyes, the valleys in the anisotropy spectrum could indicate an allowed 2PA transition as demonstrated in Fig. 6. Thus, an excitation anisotropy spectrum can serve as a useful guide to suggest the positions of the final states in the 2PA spectra.

From spectroscopic measurements, we can estimate the *fluorescence lifetime*, $\tau_F = \Phi_F \tau_N$, where the natural lifetime, τ_N, can be calculated from the Strickler–Berg equation in CGS units [60]:

$$\frac{1}{\tau_N} = 2.88 \times 10^{-9} n^2 \varepsilon^{max} \left[\frac{\int F(v) dv \times \int \frac{\varepsilon(v)}{v} dv}{\int \frac{F(v)}{v^3} dv} \right), \tag{13}$$

where $F(v)$ and $\varepsilon(v)$ are the normalized fluorescence and absorption spectra and ε^{max} (M^{-1} cm^{-1}) is the molar absorbance at the peak of the absorption band, and n is the refractive index of the solvent. For many cyanine-like molecules with spectral mirror symmetry between absorption and fluorescence spectra and small changes in excited state geometry, (14) gives reasonably good agreement with directly measured lifetimes. Thus, linear spectroscopic measurements allow the calculation of the values for the *transition dipole moments*, μ_{01}, an important parameter for 2PA, in CGS units as:

$$\mu_{01} = \sqrt{\frac{1,500(\hbar c)^2 \ln(10)}{\pi N_A E_{01}}} \int \varepsilon_{01}(v) dv, \tag{14}$$

where $\varepsilon_{01}(v)$ is the molar absorbance, N_A is Avogadro's number, and E_{01} is the energy at the absorption peak [41]. Calculations indicate that all cyanine-like dyes have similar μ_{01} values ranging from 10 to 17 D.

Summarizing, the linear optical characterization not only reveals important properties of organic molecules but also provides a necessary background for the nonlinear optical characterization, which will be discussed in the next section.

2.2   Nonlinear Optical Characterization

The term nonlinear optical property refers to an optical property, which can be modified by exposing the material to intense light irradiation. In this section, we focus on the cascaded first- ($\chi^{(1)}$) and third-order ($\chi^{(3)}$) susceptibilities describing nonlinear absorption (ESA and 2PA) and nonlinear refraction (n_2) processes. Z-scan, pump-probe, and two-photon upconverted fluorescence techniques are among the most used experimental methods for determining optical nonlinearities.

2.2.1   Pump-Probe Technique

The *pump-probe technique* is a method that can be used for determining lifetimes of excited states and their anisotropy. This method is the most common technique for time-resolved studies. A strong laser pulse (pump) is used to change the optical properties of the sample and a much weaker pulse (probe, with irradiance usually less than 10% of the pump irradiance) is used to study the magnitude and time evolution of the induced changes. The time evolution is investigated by delaying the probe pulse with respect to the pump. In this way, the pump-probe method can be applied to measure many nonlinear optical mechanisms, like nondegenerate 2PA and ESA. The temporal accuracy of the measurements is defined mainly by the pulse width of the laser beam used.

For 2PA measurements using the pump-probe technique, it is preferable to use femtosecond excitation due to the large irradiance to energy ratio. However, for ESA studies, both picosecond and femtosecond excitation can be used, since for most organic molecules the ESA lifetime is on the order of picoseconds to nanoseconds. The ESA spectrum can easily be obtained by the pump-probe technique, pumping the sample at the peak linear absorption (or slightly blue shifted from the main transition) and probing at a wavelength where ESA is expected, typically at shorter wavelengths compared to the pump, but longer probe wavelengths can give information about other, typically less intensive, ESA transitions. Currently, it is common to use a femtosecond white-light continuum, WLC, as the probe to obtain a spectrum. These can be generated in wide-bandgap crystalline materials, like CaF_2 or Sapphire, due to the broad WLC produced. In principle, a complete ESA spectrum can be obtained in a single laser shot experiment [61].

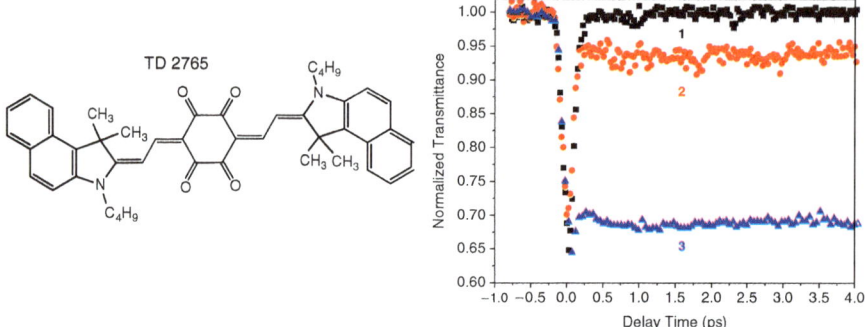

Fig. 7 Femtosecond pump-probe data for TD 2765 (molecular structure shown *left*). The probe wavelength is set at 670 nm and the pump wavelengths are (1) 710 nm, (2) 700 nm, and (3) 690 nm. See [62] for additional details

As discussed in Sect. 1.2.3, it is usually not possible to distinguish ESA from 2PA with Z-scan experiments if they are performed with only one excitation pulsewidth. However, since ESA is not an instantaneous process as is 2PA, the pump-probe technique can be successfully used to verify the origin of the nonlinearity for the spectral regions close to the main absorption band. Figure 7 illustrates how the influence of the ESA can be distinguished from the 2PA with pump-probe experiments. The curve labeled (1) shows an instantaneous 2PA response without ESA and the long-lived components of the transmittance change seen in (2) and (3) are due to ESA.

The pump-probe method can be also applied to measure the population decays or fluorescence lifetimes. For fluorescence lifetimes shorter than 100 ps, the relative polarization between pump and probe beams does not typically interfere with the decays. However, for longer lived excited states, the reorientation of the molecule, which typically takes hundreds of picoseconds, can play an essential role in decay kinetics and can affect the pump-probe results. When the reorientation time of the molecule is shorter or comparable to the fluorescence lifetime, the relative polarization of the pump and probe beams is important and the reorientation time has to be taken into account during the data analysis. To be able to eliminate the influence of the molecular reorientation on the pump-probe results, the angle between the pump and probe polarization is fixed at 54.7°, which is known as the "magic angle" (the 3-D analogy of 45 degrees) [63], which represents the angle at which the effects of the reorientation of the molecules cancel each other. Polarization-resolved pump-probe data can be fit using:

$$
\begin{aligned}
\Delta T_{\parallel}(t) &= \Delta T_{\text{magic}}(0) \exp(-t/\tau_{\text{F}})[1 + 2r(0)\exp(-t/\tau_{\text{Rot}})] \\
\Delta T_{\text{magic}}(t) &= \Delta T_{\text{magic}}(0) \exp(-t/\tau_{\text{F}}) \\
\Delta T_{\perp}(t) &= \Delta T_{\text{magic}}(0) \exp(-t/\tau_{\text{F}})[1 - r(0)\exp(-t/\tau_{\text{Rot}})],
\end{aligned}
\tag{15}
$$

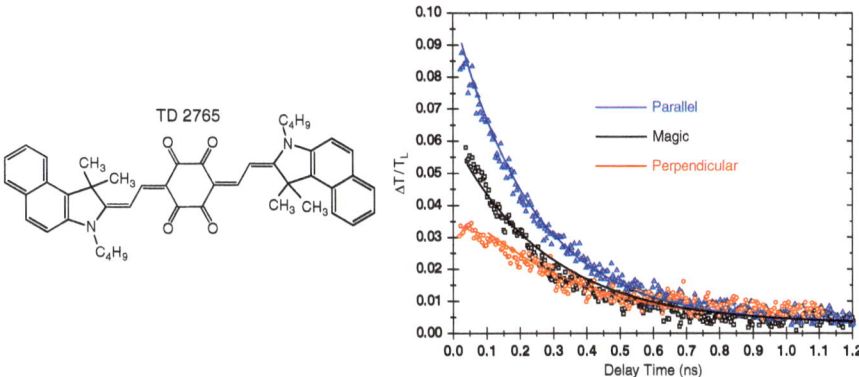

Fig. 8 Polarization-resolved picosecond pump-probe data for TD 2765 in ethanol (molecular structure shown *left*). The orientations of the probe beam relative to the pump are: perpendicular (*red*), "magic angle" (*black*), and parallel (*blue*). Data is modeled using (15) to obtain $\tau_F = 280$ ps, $\tau_{Rot} = 550$ ps, and $r = 0.35$. See [62] for additional details

where $\Delta T_{\parallel}(t)$, $\Delta T_{\text{magic}}(t)$, and $\Delta T_{\perp}(t)$ are the change in transmittance measured for parallel, magic, and perpendicular probe polarizations with respect to the pump polarization, respectively, τ_F and τ_{Rot} are the fluorescence lifetime and molecular rotational lifetime, and $r(0)$ is the anisotropy at $t = 0$. Figure 8 and (15) show how molecular reorientation can influence the pump-probe results. If the magic angle is not used in the experimental setup, an over- or underestimation of the decay lifetime, τ_F, will be obtained for parallel and perpendicular polarizations respectively.

2.2.2 Z-Scan Technique

The *Z-scan technique*, first introduced in 1989 [64, 65], is a sensitive single-beam technique to determine the nonlinear absorption and nonlinear refraction of materials independently from their fluorescence properties. The simplicity of separating the real and imaginary parts of the nonlinearity, corresponding to nonlinear refraction and absorption processes, makes the Z-scan the most widely used technique to measure these nonlinear properties; however, it does not automatically differentiate the physical processes leading to the nonlinear responses.

The Z-scan technique is performed by scanning the sample through the focus of a Gaussian beam (the technique can be performed with any beam shape, but in practice, modeling of a Gaussian beam is preferred). The transmission through the sample is measured as a function of the Z position (with respect to the focal point $Z = 0$) and, consequently, of the beam waist. The change in transmission as a function of the beam waist (which corresponds to a function of fluence and/or irradiance) gives information pertaining to the nonlinear absorption (so-called

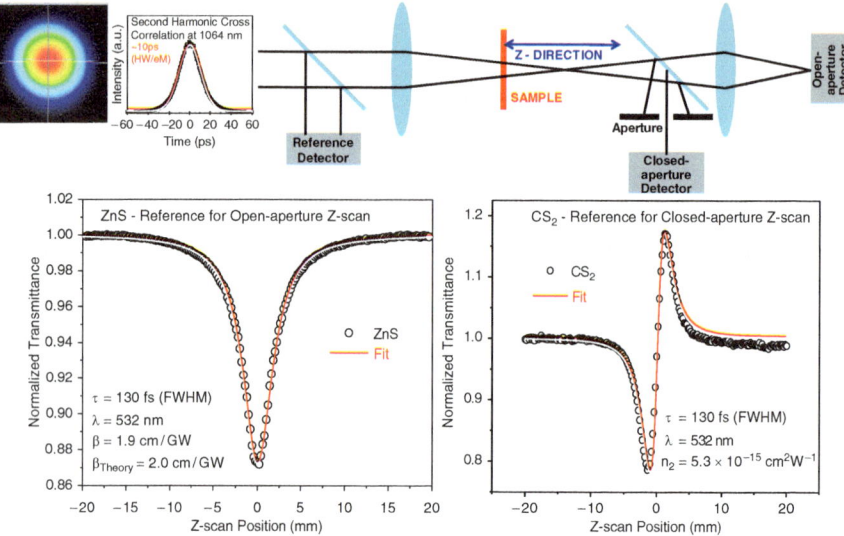

Fig. 9 Schematic of simplified Z-scan setup, and examples of open- and closed-aperture measurements of the common reference materials zinc sulfide (ZnS) and carbon disulfide (CS$_2$) with fitting using the theory proposed by [66]

open-aperture Z-scan). Adding an aperture after the sample, before any focusing element and linearly transmitting ~30% [66], the nonlinear phase change can be detected by changes in transmittance through the aperture onto the detector. When the nonlinear absorbance has been taken into account, this corresponds to the nonlinear refraction. Figure 9 demonstrates a simplified Z-scan setup together with examples of both the open- and closed-aperture experiments and their fitting using the theory proposed by Sheik-Bahae et.al. [66]. The fitting is done assuming a perfectly Gaussian laser beam and applying the "thin sample" approximation [66]. Knowledge of the beam, both spatially, by use of beam profiling cameras and knife-edge scans [67, 68] to determine focused spot sizes and propagation constants, and temporally, by use of fast detectors and autocorrelation techniques [69, 70], are needed to accurately analyze results.

For 2PA or ESA spectral measurements, it is necessary to use tunable laser sources where optical parametric oscillators/amplifiers (OPOs/OPAs) are extensively used for nonlinear optical measurements. An alternative approach, which overcomes the need of expensive and misalignment prone OPO/OPA sources, is the use of an intense femtosecond white-light continuum (WLC) for Z-scan measurements [71, 72]. Balu et al. have developed the WLC Z-scan technique by generating a strong WLC in krypton gas, allowing for a rapid characterization of the nonlinear absorption and refraction spectra in the range of 400–800 nm [72].

The main advantage of the usual Z-scan technique is that this method allows for a direct measurement of the nonlinearity, not requiring the use of reference samples, and thus minimizing experimental errors. On the other hand, the necessity of having a fully characterized (spatially and temporally) Gaussian laser beams requires perfect alignment at each wavelength, making spectral determination time consuming. The WLC Z-scan is a promising technique to reduce the time needed to obtain spectra using the traditional Z-scan. From the open-aperture Z-scan trace, and the dependence of the signal on the energy per pulse, it is possible to verify the order of the nonlinear absorption process involved in the measurement. Note that in the spectral regions where there is linear absorption, 1PA followed by ESA (an RSA mechanism) can be confused with 2PA since both processes involve two photons. In this case, single pulsewidth Z-scan measurements cannot determine which physical process is occurring. However, one can separate these two processes by performing Z-scans with different pulse widths (ESA is fluence dependent while 2PA is irradiance dependent), or using time-resolved experiments such as pump-probe, which was discussed in Sect. 2.2.1.

For 2PA spectra measurements, it is preferable to perform Z-scans with femtosecond excitation due to the large laser irradiance with relatively low pulse energy. On the other hand, for ESA characterization, picosecond lasers are typically used due to their larger energy per pulse while maintaining an impulse excitation (as opposed to a nanosecond laser where the temporal pulse width is comparable to the excited state lifetimes). For this regime, picosecond Z-scans can be taken at different pulse energies (for example, 5 energies spanning at least one order of magnitude), and the ESA cross section can be calculated by simplifying the propagation and rate equations in (9) to only include singlet transitions (decay to the triplet states normally takes nanoseconds, but can be as short as picoseconds [42]). From these equations, a 3-level singlet model can be used assuming that the fluorescence lifetime is known (measured by pump-probe, or time-resolved fluorescence) and higher excited state lifetimes are much shorter than the excitation pulse width. For larger pump fluences, where the population of S_2 becomes significant, a fourth singlet level may need to be added to describe the complete process.

Shown in Fig. 10 are examples of fitting Z-scan results for a squaraine molecule where triplet states are significantly populated during picosecond excitation. In this case, femtosecond excitation was needed to determine the singlet parameters, shown in Fig. 10a. Picosecond excitation was required to model the full five-level system, shown in Fig. 10b [73], to determine the triplet parameters. To account for the triplet states, it is necessary to consider at a minimum the 5-level system shown Fig. 2 (three singlet and two triplet states). Typically, PDs do not populate triplet states during picosecond excitation [74, 75].

2.2.3   Two-Photon Fluorescence (2PF) Technique

The *2PF technique* measures the upconverted fluorescence induced by 2PA. This technique allows for the measurement of 2PA cross sections that are less than tens of GMs when the fluorescence quantum yield is large. The technique was first proposed by Kaiser et al. [76] in 1961 and represents an indirect way to measure the 2PA cross section, which can be calculated by comparing the integrated fluorescence signal

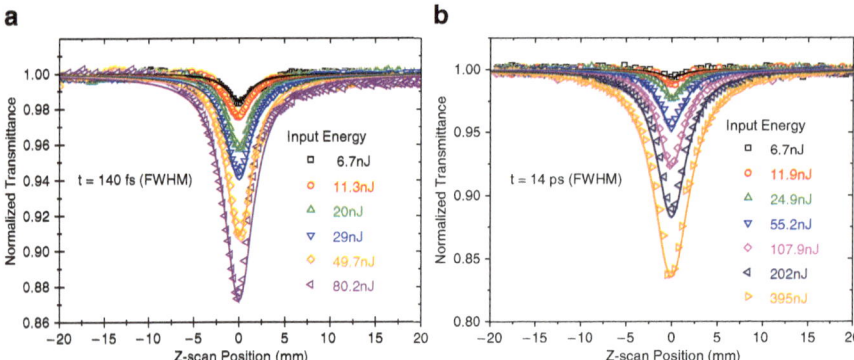

Fig. 10 Example of fitting Z-scan data for a squaraine molecule for (**a**) a 3-level model (singlet states only) with femtosecond excitation and (**b**) a 5-level model (both singlet and triplet states) with picoseconds excitation. The schematic is shown in Fig. 2a and the propagation and rate (9) are used for modeling

from an unknown molecule to a reference molecule with known 2PA cross section measured under identical conditions. The most common reference molecules are Rhodamine B in Methanol and Fluorescein in H_2O (pH = 11) [77]. Assuming that one knows the fluorescence quantum yields, Φ_F, (calculated as described in Sect. 2.1) and the concentrations, c_M, of the sample and reference and the 2PA cross section for the reference, Φ_{FRef}, the 2PA cross section for the sample can be calculated from:

$$\delta_{2PAsample} = \frac{\langle F \rangle_{sample}\, \Phi_{Fref}\, {}^c M_{ref} \langle P \rangle_{ref}^2}{\langle F \rangle_{ref}\, \Phi_{Fsample}\, {}^c M_{sample} \langle P \rangle_{sample}^2}\, \delta_{2PAref} \tag{16}$$

where $\langle F \rangle$ is the integrated fluorescence and $\langle P \rangle$ is the average pump power.

The experiment is performed with a spectrofluorometer similar to the ones used for linear fluorescence and quantum yield measurements (Sect. 2.1). The excitation, instead of a regular lamp, is done using femtosecond pulses, and the detector (usually a photomultiplier tube or an avalanche photodiode) must either have a very low dark current (usually true for UV–VIS detectors but not for the NIR), or to be gated at the laser repetition rate. Figure 11 shows a simplified schematic for the 2PF technique.

The 2PA cross section is a molecular parameter and, therefore, both techniques of Z-scan and 2PF should yield identical results when performed properly. From extensive investigations of many organic molecules, Z-scan and 2PF experiments complement each other, filling different spectral ranges and providing a double-check of the results. Figure 12 shows the 2PA spectrum measured for an extended dithiolene molecule [72], shown using open-aperture single-wavelength and WLC Z-scans, and the 2PF method. Figures 19, 23, and 24 also show excellent agreement between independently measured Z-scan and 2PF.

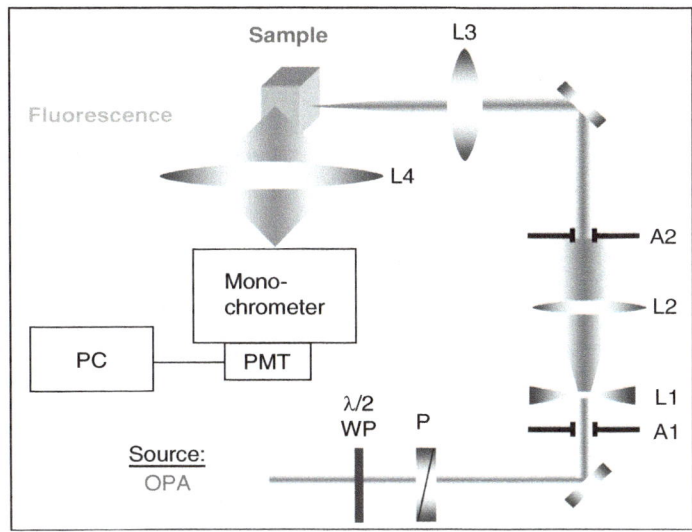

Fig. 11 Experimental setup for 2PF measurements. L are lenses, A are apertures, P is a polarizer, and WP is a half waveplate

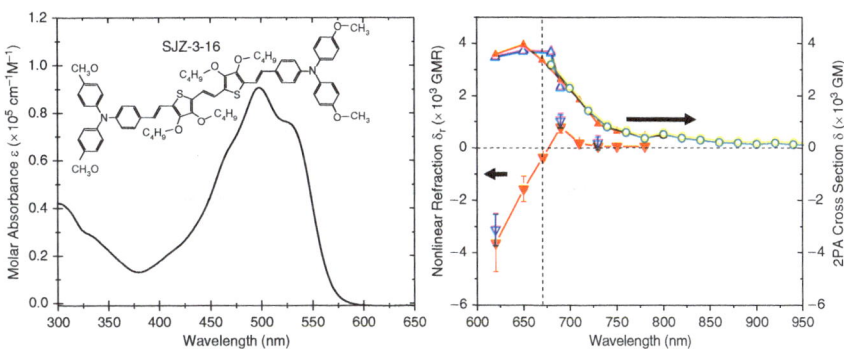

Fig. 12 (**a**) Molar absorbance and molecular structure (inset) for SJZ-3-16. (**b**) Comparison of the 2PA measured by single wavelength open-aperture Z-scan (*blue upward triangles*), WLC open-aperture Z-scan (*red upward triangles*), and 2PF (*green circles*). The nonlinear refraction measured by closed-aperture single wavelength (*blue downward triangles*), and WLC (*red downward*).

In conclusion, we stress that the complementary NLO characterization techniques of pump-probe, Z-scan, and 2PF allow for the unambiguous determination of nonlinear optical processes in organic materials. The important molecular parameters of 2PA cross section, fluorescence efficiency, reorientation lifetimes, excited state cross sections, etc. can be determined.

3 Trends in Dye Design and Structure–Property Relations

In order to design the best molecules for nonlinear optical applications, a link between molecular structure and two-photon absorption properties must be developed. To provide this link, one must start from an understanding of the formation of one-and two-photon absorption spectra in a series of molecules with systematic changes in structure. Detailed experimental characterization combined with quantum-chemical calculations and modeling can give the necessary information for the development of a design strategy. In this section, we show the connection between efficient 2PA and various elements of molecular structure, such as the length of the conjugated chromophore, the types of substitutions, including symmetrical and asymmetrical combinations of electron donor and acceptor terminal groups, and the addition of such groups in the middle of the chromophore.

3.1 Extending Absorption into the NIR

Before discussing the ways to enhance the 2PA cross section, we consider one of the most important and useful properties of cyanine-like dyes – the possibility to tune their absorption bands from the visible to NIR region up to ~1,600 nm. The shift of absorption spectra into the NIR region can be accomplished using two methods: lengthening the polymethine chromophore (polymethine chain) or by introducing specific terminal groups with their own extended conjugation system, which can strongly interact with the main chromophore and extend the total effective length of conjugation in the molecule [57, 78]. The first method typically decreases photochemical stability of the molecules, which can be partially improved by the introduction of bridge units within the chain. However, the second method allows significant shifting of the absorption bands without a substantial decrease of the photochemical stability. This method is described in [57] and is demonstrated in Fig. 13 for two sets of polymethine, squaraine, and tetraone dyes having similar lengths of the chain and different terminal groups.

The first set of dyes, so called "visible set", is presented by polymethine dye PD 2630, squaraine dye SD 2243, and tetraone dye TD 2765, all with benzo[e]indolium terminal groups. The second set of dyes, so called "NIR set", is presented by polymethine dye PD 2658, squaraine dye SD 2878, and tetraone dye TD 2824, all with 5-butyl-7,8-dihydrobenzo[cd]furo[2,3-f]indolium terminal groups. A distinguishing feature seen from this figure is a remarkably large, ≈300 nm, red shift of the absorption bands for PD 2658 and SD 2878 as compared to PD 2630 and SD 2243. The absorption spectrum of TD 2824 is red-shifted by ≈200 nm as compared to TD 2765. Thus, the effect of the 5-butyl-7,8-dihydrobenzo[cd]furo[2, 3-f]indolium terminal groups is equivalent to the extension of the chain to three vinylene groups.

Introduction of the acceptor squaraine and tetraone bridges to the conjugated chain causes BLA in the bridges resulting in a blue shift of the main absorption

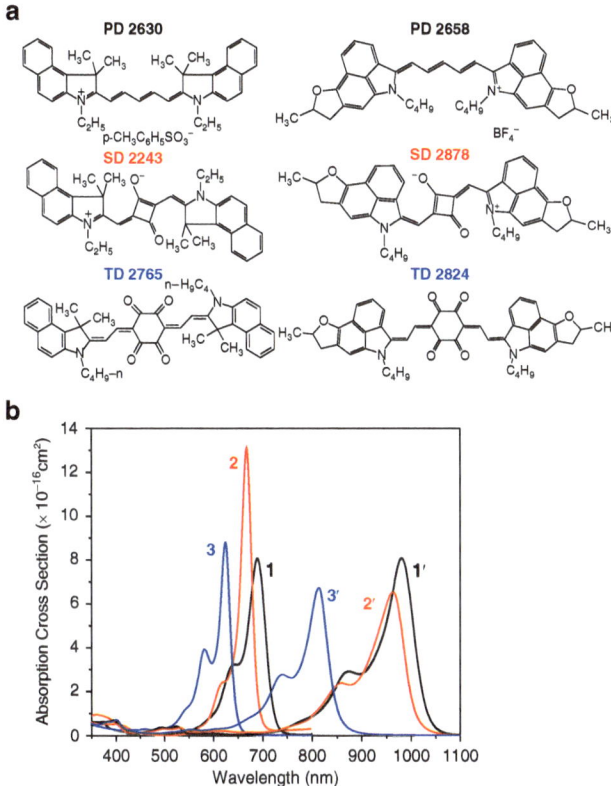

Fig. 13 (**a**) Molecular structures and (**b**) 1PA cross section of two sets of PDs, SDs, and TDs with similar chain lengths and different terminal groups. The first set of dyes (visible set) consists of PD 2630 (1), SD 2243 (2), and TD 2765 (3), all with benzoindolium terminal groups and the second (NIR set) is comprised of PD 2658 (1'), SD 2878 (2'), and TD 2824 (3'), all with benzofuroindolium terminal groups

bands. Additionally, incorporation of the stronger tetraone acceptor units leads to a larger BLA within the bridge and changes the nature of the molecules from polymethine to a polyenic type of conjugation with an even number of carbon atoms in the chain. Therefore, BLA for tetraone dyes are observed within the whole conjugated system, and their absorption spectra are more blue-shifted, especially for the NIR TD 2824 as compared to the corresponding PD 2658. The large red shifts for the "NIR set" of molecules can be explained by the extended π-system within the terminal groups and its strong conjugation with the π-system of the chain.

Linear absorption and fluorescence spectra for the series of symmetrical cationic polymethines with 5-butyl-7,8-dihydrobenzo[cd]furo[2,3-f]indolium terminal groups are shown in Fig. 14 for solvents of different polarity. It is known that the polarity of solvents can be characterized by their orientational polarizability, which is given by $\Delta f = (\varepsilon - 1)/(2\varepsilon + 1) - (n^2 - 1)/(2n^2 + 1)$, where ε is the static dielectric constant and n is the refractive index of the solvent [41]. Calculated Δf values

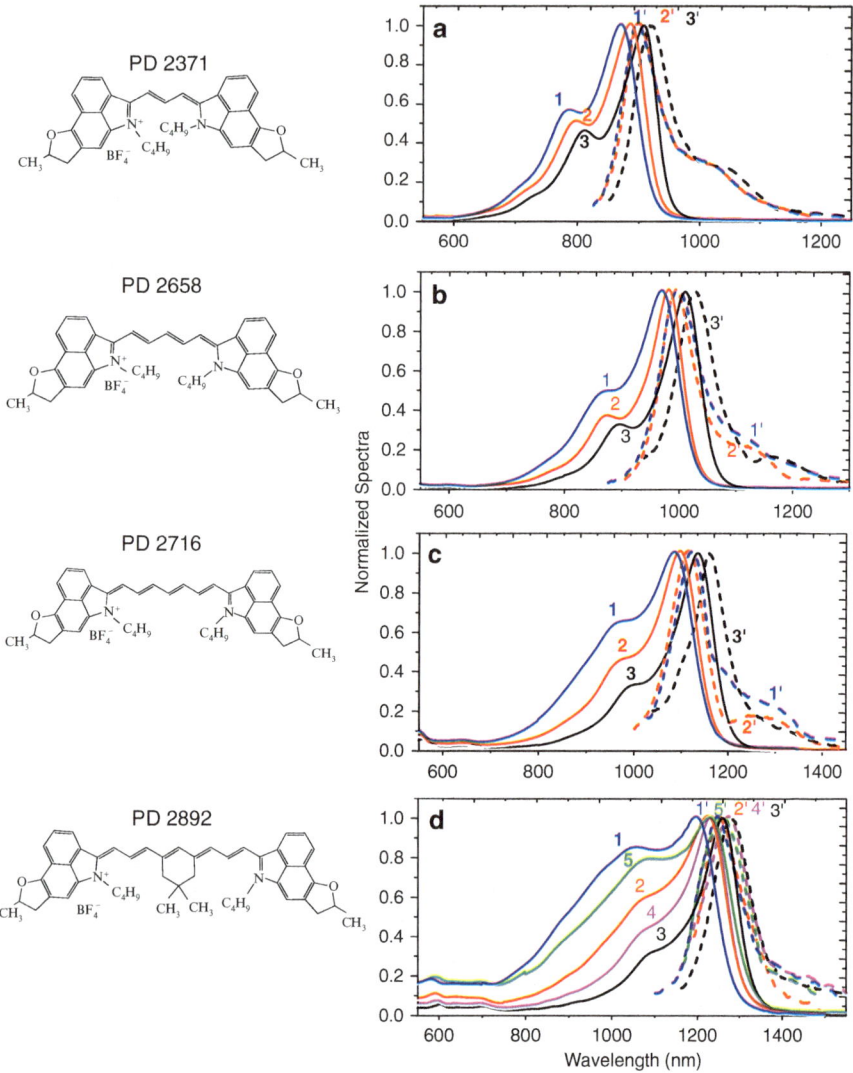

Fig. 14 (*left*) Molecular structures and (*right*) 1PA (*solid lines* 1, 2, 3, 4, 5) and 1PF spectra (*dashed lines* 1′, 2′, 3′, 4′, 5′) for PD 2371 (**a**), PD 2658 (**b**), PD 2716 (**c**), and PD 2892 (**d**) in acetonitrile (1, 1′), butanol (2, 2′), dichlorobenzene (3, 3′), methylene chloride (4, 4′), and dimethyl sulfoxide (5, 5′), respectively

range from the smallest polarity of 0.208 for dichlorobenzene (cationic dyes cannot be dissolved in solvents of lower polarity) to the largest in this series of 0.306 for acetonitrile (ACN).

The absorption spectra for all these dyes are composed of intense cyanine-like bands attributed to the $S_0 \rightarrow S_1$ absorption, with the main absorption peaks shifted by ≈ 100 nm to longer wavelengths upon lengthening of the main conjugation chain, and weak linear absorption in the visible and UV region corresponding to absorption to higher excited states $S_0 \rightarrow S_n$. Relatively short wavelength absorbing PDs exhibit classic nonpolar solvatochromism, i.e., a red shift of the absorption peak with an increase in solvent polarity, which correlates with a decrease of the refractive index. This is consistent with a symmetrical ground and excited state charge distribution and small permanent dipole moments, 1–2 D, oriented perpendicular to the polymethine chromophore [41]. In contrast, absorption spectra of PDs absorbing in the range of $\approx 1,000$ nm demonstrate a strong dependence on solvent polarity, see Fig. 14. An increase in solvent polarity leads to a substantial band broadening represented by the growth of the short wavelength shoulder. This is a strong indication of polar solvatochromism, which is typical for dyes that exhibit charge localization and a large ground state permanent dipole moment. This effect was investigated earlier theoretically [79, 80] and experimentally [81] and explained by a *symmetry breaking effect*, leading to the appearance of a ground state structural form with asymmetrical charge distribution and, as a result, with an asymmetrical bond-length alternation. Our explanation is based on the previously proposed theoretical concept of the formation of charge density waves (or solitonic waves) in the linear conjugated chromophores [82]. In the theoretical paper [80], we show that the minimum number of vinylene groups, n, in the conjugated chain necessary to break the symmetry of the simple streptocyanine molecule, is eight in the gas phase and six in nonpolar cyclohexane. As the symmetry breaks, an additional absorption band with large oscillator strength appears in the electronic spectrum. Charge localization is additionally stabilized by the solvent, which increases the energy barrier between symmetrical and asymmetrical forms and results in ground state symmetry broken geometries occurring at a shorter (than in the gas phase) length of the conjugated system. An additional absorption peak at the shorter wavelength region corresponds to a molecular geometry with charge localized at one of the molecular terminal groups that is additionally stabilized by the solvent. This finding demonstrates the possibility of coexistence of the two forms in polar solutions. Thus, our results suggest that a solvatochromic effect may be an important factor in absorption band broadening.

Quantum-chemical calculations show that the symmetry of molecular geometry is conserved for polymethines with 5-butyl-7,8-dihydrobenzo[cd]furo[2,3-f]indolium terminal groups at $n = 1$–3. However, for longer molecules, starting from $n = 4$, calculations show an inequality of the corresponding bond lengths in the chain. We note that, experimentally, a substantial broadening of the absorption band in polar ACN is observed even for tricarbocyanine PD 2716 ($n = 3$), as shown in Fig. 14. Therefore, we suggest that the existence of the asymmetrical form in the polar solvent may be observed in this series of polymethines starting from $n = 3$. An increase in solvent polarity shifts the equilibrium between these two forms to favor the asymmetrical form. Experiments show that the less polar solvent primarily stabilizes the symmetrical form.

In contrast to the absorption spectra, fluorescence spectra for all polymethine molecules are similarly narrow, independent of the solvent polarity, indicating that emission originates from the symmetrical form only. The symmetry breaking effect

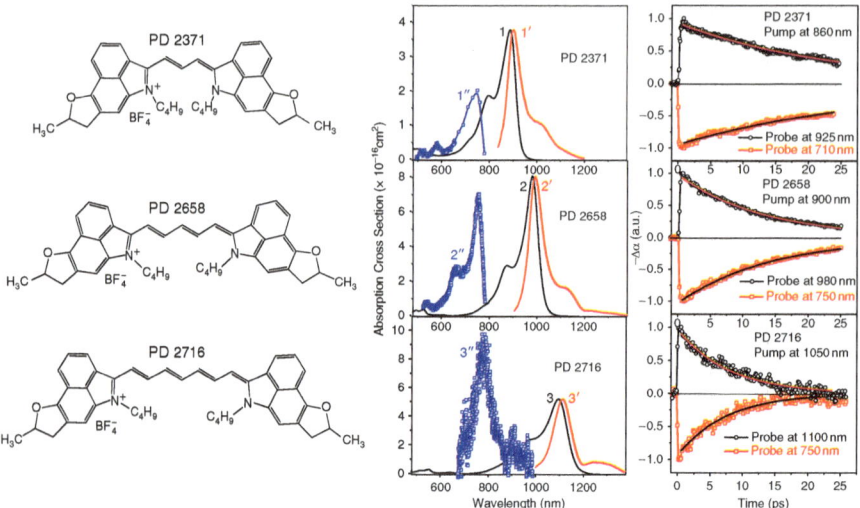

Fig. 15 (*left*) molecular structures; (*center*) 1PA (1, 2, 3), 1PF (1′, 2′, 3′), and ESA cross sections (1″, 2″, 3″) of PD 2371 (1, 1′, 1″), PD 2658 (2, 2′, 2″), and PD 2716 (3, 3′, 3″); (*right*) decay kinetics (Δα is the change of absorption of probe)

can also be responsible for the small quantum yields (typically less than 1%) and short lifetimes (typically less than 100 ps) for dyes absorbing in the range of ≈1,000 nm. We suggest that the most efficient route of energy deactivation is via formation of the asymmetrical excited state molecular geometry, which is strongly coupled to the ground state geometry.

The ESA spectra and decay kinetics of this series of polymethines are shown in Fig. 15. A small red shift of the ESA spectra (comparing to a large red shift of the 1PA main band) is observed as the conjugation length increases. Lengthening of the conjugation chain also leads to an increase of both the ESA cross section peak values and the ratio of the total ESA to total 1PA main transition bands, which are proportional to their oscillator strengths. These totals are defined as the integrated area in a plot of absorbance versus photon energy. For PD 2716 with $n = 3$, the oscillator strength for ESA can be as large as that of the ground state, which is favorable for some nonlinear optical applications [78]. To obtain the decay kinetics of the series of polymethines shown in Fig. 15, the samples are pumped at their 1PA peaks and probed first, near the 1PA peaks where saturable absorption occurs and second, in the region of reverse saturable absorption. Both saturable and reverse saturable absorption decays show the same lifetimes (corresponding to the lifetime of the S_1 state) for all polymethine dyes. This confirms that no other intermediate states are involved with the ESA process. The lifetime of the S_1 state, shown in the decay kinetics, decreases as the conjugation length increases, which is also in accord with the decrease of their fluorescence quantum yields [78].

From these studies, we find that the dyes with dihydrobenzo[cd]furo[2,3-f] indolium terminal groups are characterized by a remarkably large shift of their linear absorption bands to the red region (300 nm for PDs and 200 nm for SDs). These large red shifts for the "NIR" set of molecules can be explained by the extended π-system within the terminal groups and their strong connection with the π-system of the chain resulting in a significant extension of the total effective conjugation length. The effect of these terminal groups is equivalent to the extension of the chain to three vinylene groups. This is specifically true for PDs with benzo[e]indolium terminal groups.

In the next sections, we systematically describe structure–property relations in symmetrical (Sect. 3.2) and asymmetrical (Sect. 3.3) series of cyanine-like molecules.

3.2   Symmetrical π-Conjugated Cyanine-Like Systems

3.2.1   Cationic D–π–D Dyes

Molecular structures of a series of cationic D–π–D dyes, and their 1PA and 2PA spectra are shown in Fig. 16. These dyes have increasing lengths of conjugation and share the same indolium terminal groups, which are electron donors. *An increase of conjugation length* by one unit leads to a red shift of the one-photon main absorption band by $\sim$100 nm [83]. 2PA spectra for all these D–π–D molecules show one weak band corresponding to two-photon excitations into the vibrational shoulder of the main $S_0 \rightarrow S_1$ absorption band, which is weakly allowed due to vibronic coupling; and a second much stronger band, corresponding to two-photon

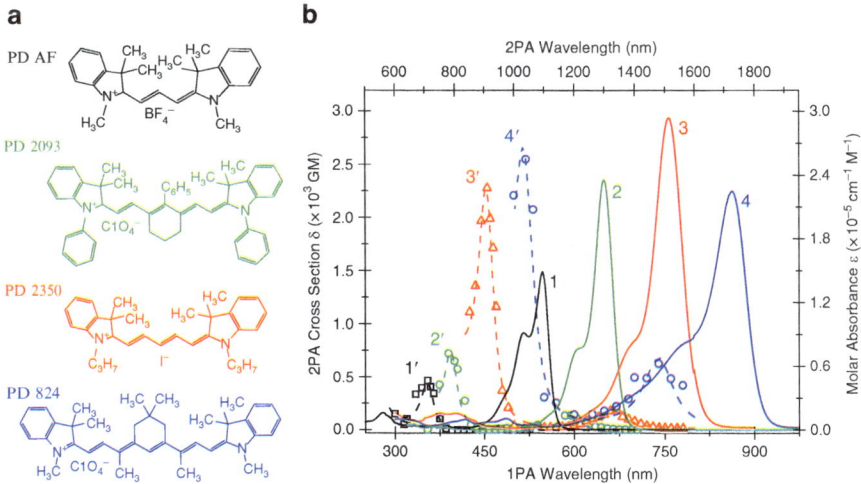

Fig. 16 (a) Molecular structures of PD AF ($n = 1$), PD 2093 ($n = 2$), PD 2350 ($n = 3$), and PD 824 ($n = 4$); (b) 1PA and 2PA bands for of PD AF (1, 1$'$), PD 2093 (2, 2$'$), PD 2350 (3, 3$'$), and PD 824 (4, 4$'$), respectively

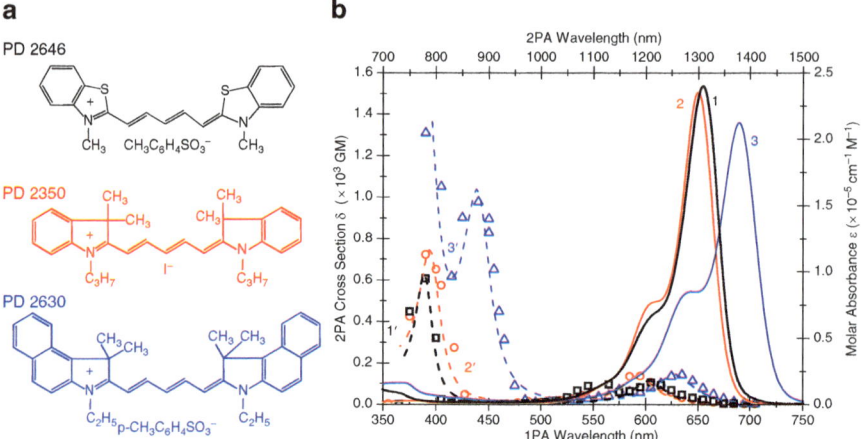

Fig. 17 (**a**) Molecular structures and (**b**) 1PA and 2PA of PD 2646 (1, 1′), PD 2350 (2, 2′), and PD 2630 (3, 3′), respectively

excitation into the S_2 state. In correspondence with (8), lengthening of the conjugation chain leads to increase in magnitude of 2PA cross sections. This can be explained by an increase of the ground state transition dipole moment and a decrease in detuning energy. Lengthening the chain results in an increase of δ_{2PA} for the first 2PA band from 10 to 600 GM and for the second 2PA band from 470 to 2,550 GM (see Fig. 16). However, it is worth noting that an increase of conjugation length could also lead to symmetry breaking for long wavelength absorbing dyes (seen for PD 824), resulting in an asymmetrical charge distribution within the conjugated chain [83], which affects 2PA cross sections.

To investigate the *effect of the donor properties* of the terminal groups, we compare a series of molecules with the same conjugation length ($n = 2$), but different terminal groups, see molecular structures in Fig. 17a. The electron donor strength increases from thiazolium (PD 2646), to indolium (PD 2350), and to benzoindolium (PD 2630) terminal groups. From Fig. 17b, it is clearly seen that an increase in the donor strength leads to an increase of the 2PA cross section. A more detailed description of structure–property trends in a series of cationic D–π–D dyes is presented in [83].

3.2.2 Neutral D–π–A–π–D Dyes

The addition of an electron acceptor group into the D–π–D can further enhance the 2PA cross sections. A series of dyes, cationic PD 2630, and two neutral dyes, SD 2243 and TD 2765, are shown in Fig. 18a. Compared with the polymethine dye PD 2630, SD 2243 has a squaraine fragment and a strong electron acceptor inserted into the main conjugation chain, while TD 2765 contains a tetraone fragment and an

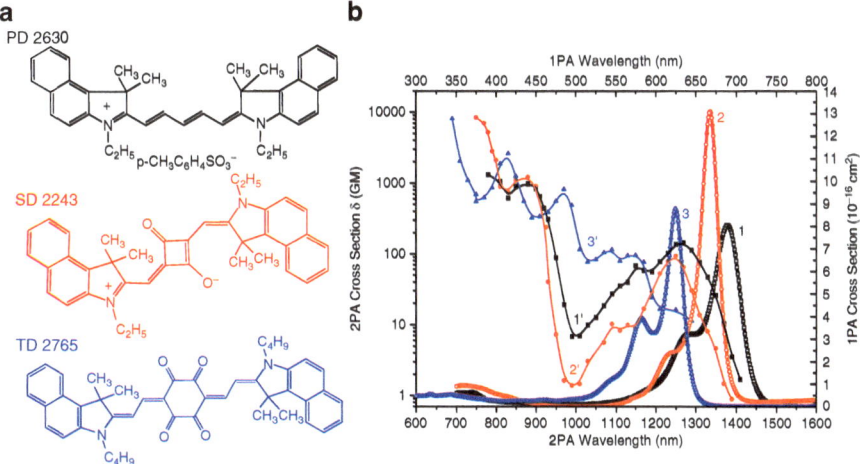

Fig. 18 (**a**) Molecular structures of PD 2630, SD 2243, TD 2765; (**b**) 1PA and 2PA of PD 2630 (1, 1′), SD 2243 (2, 2′), TD 2765 (3, 3′), respectively

even stronger electron acceptor group. As seen from Fig. 18b, the introduction of acceptors leads to a blue shift of the 1PA main transition band. The 2PA cross section of SD 2243 increases by ~6 times compared to PD 2630 at its 750 nm 2PA wavelength; while TD 2765 shows broader 2PA bands with a nearly monotonic increase of δ_{2PA} (ω) towards the 1PA edge. The difference of 2PA between D–π–A–π–D dyes and D–π–D dyes can be explained by the intermediate state resonance enhancement (ISRE) due to a narrowing of the 1PA main transition band, and by an increase in the density of final states, which is supported by quantum-chemical calculations.

A detailed experimental investigation and quantum-chemical analysis of 2PA spectra for quadrupolar D–π–A–π–D structures in cyanine-like molecules are presented in [62].

3.2.3 Anionic A–π–A Dyes

A series of anionic A–π–A dyes with different conjugated chains (G37, G38, G74, and G152) is obtained by connecting two diethylamino-coumarin-dioxaborine acceptors (A) via a π-conjugation (see Fig. 20a for molecular structures) [58]. The linear absorption, fluorescence, anisotropy, and 2PA spectra are shown in Fig. 19. 2PA spectra for all these A–π–A molecules show one weak band corresponding to two-photon excitations into the vibrational shoulder of the 1PA main transition band and one or two much stronger bands, corresponding to two-photon excitations into S_2 and higher electronic states. Similarly to that for symmetrical cationic D–π–D dyes, an increase of conjugation length leads to both a red shift of the 2PA transition bands and an increase of δ_{2PA} (ω). The longest dye, G152, has a large δ_{2PA} (ω) of

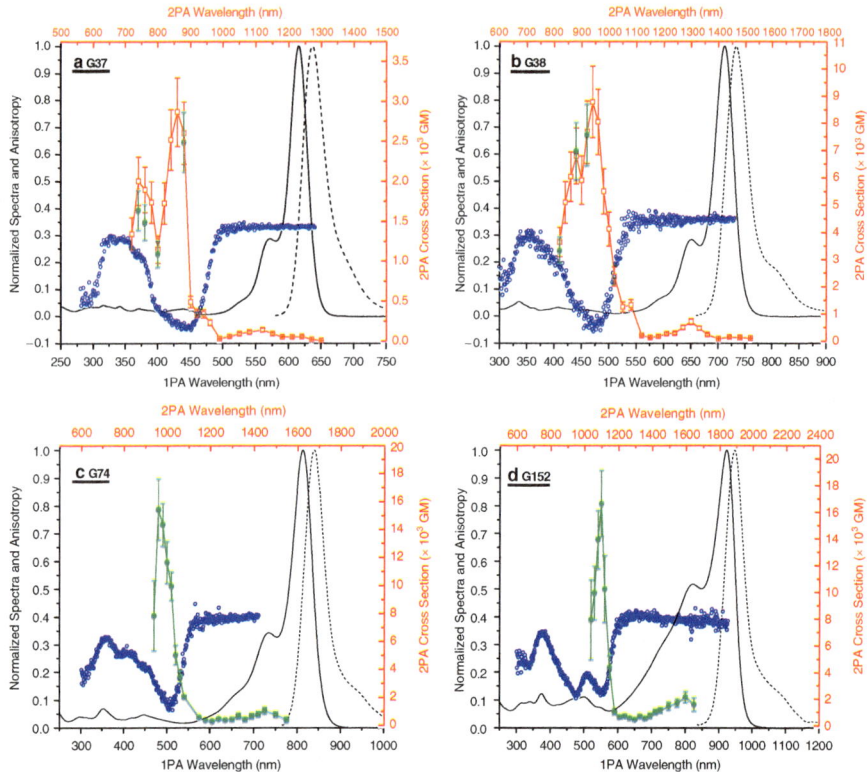

Fig. 19 1PA (*solid black lines*), 1PF (*dashed black lines*), anisotropy (*blue circles*), and 2PA spectra measured by 2PF (*red squares*) and Z-scan (*green circles*) of G37, G38, G74, and G152 in ACN. *Red axes* correspond to 2PA measurements

$\approx$16,000 GM at its peak and a δ_{2PA} (ω) of $\approx$2,000 GM at 1,600 nm (near telecommunication wavelengths of 1,300–1,600 nm) which is one of the largest reported 2PA cross section for organic molecules in this wavelength range. These large 2PA cross sections are explained by the combination of very large transition dipole moments of μ_{01} $\sim$13–20 D and μ_{1F} $\sim$9–13 D, which is confirmed by quantum chemical calculations.

As seen from Fig. 19, fluorescence excitation anisotropy is a very useful tool to predict the positions of 2PA bands as was concluded in Sect. 2.1.

The ESA spectra of this series of A–π–A dyes are shown in Fig. 20. They exhibit broad and intense bands in the visible range (400–600 nm for G37, 400–630 nm for G38, 450–630 nm for G74, and 450–700 nm for G152) and weak bands in the NIR as revealed in Fig. 20 for G38. We observe that lengthening of the conjugation chain leads to a $\approx$30–40 nm red shift of the ESA peaks, which is similar to the behavior of D–π–D polymethine dyes. This red shift ($\sim$30–40 nm) is much smaller than for the linear absorption bands ($\sim$100 nm). Another experimental feature is connected with the redistribution of the ESA magnitude from the shorter to the

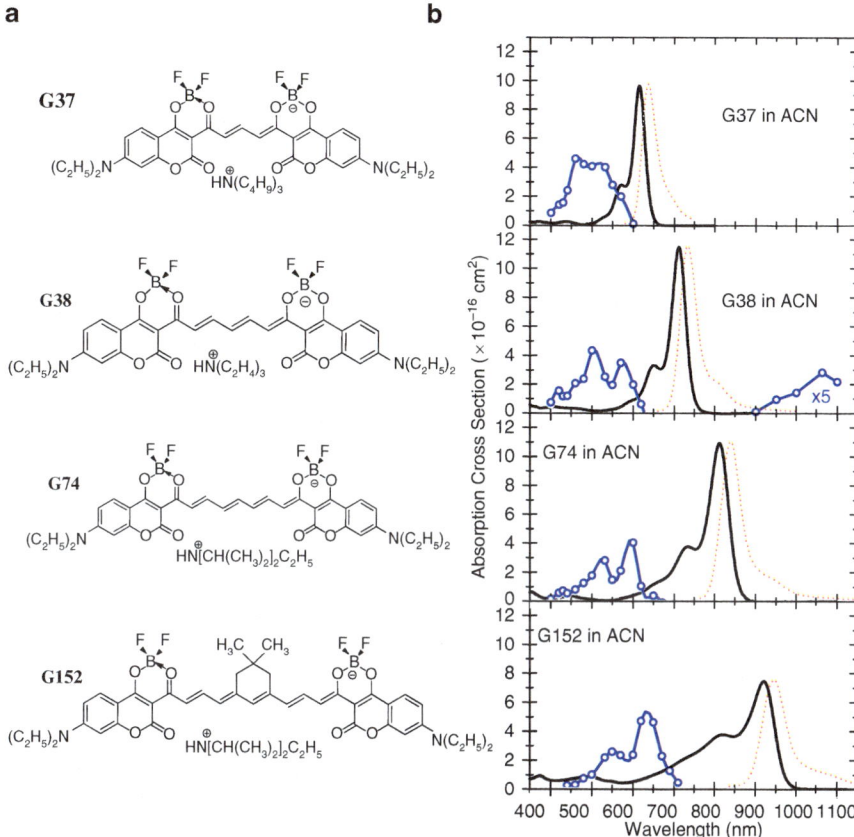

Fig. 20 (a) Molecular structures of G37 ($n = 1$), G38 ($n = 2$), G74 ($n = 3$), and G152 ($n = 4$). (b) 1PA (*solid black lines*), 1PF (*dashed black lines*), and ESA (*blue circles*) spectra

longer wavelength band. This redistribution is enhanced as the conjugation length increases and clearly observed for G152, which is probably a result of the involvement of a large number of excited state transitions.

A more detailed description of structure–property trends in a series of anionic A–π–A dyes is presented in [59].

3.3 Asymmetrical π-Conjugated Cyanine-Like Systems

The asymmetrical D–π–A dyes, often referred to as push–pull polyenes, are an additional class of cyanine-like molecules of interest. Due to their dipolar nature, the linear and nonlinear optical properties of this series of dyes can be strongly influenced by solvent polarity [84]. The structures of a series of such dyes (G19,

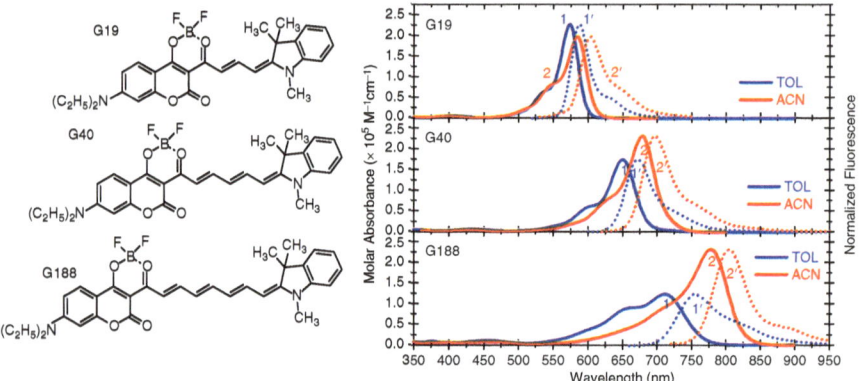

Fig. 21 (*left*) Molecular structures and (*right*) 1PA (1, 2) and 1PF (1′, 2′) spectra of G19, G40 and G188 in toluene (1, 1′) and ACN (2, 2′)

G40, and G188) with different conjugation lengths ($n = 1$–3) are shown in Fig. 21 with linear absorption and fluorescence spectra measured in two different solvents, toluene and ACN. All three dyes contain the same trimethylindolin donor group and diethylamino-coumarin-dioxaborine acceptor group. Significant solvent-dependent absorption and fluorescence spectra are observed for all three dyes. The main properties of these compounds can be explained based on the well-developed two-state model for push–pull polyenes as quasi-one-dimensional molecules containing an electron-donating group (D) and electron-accepting group (A) interacting via a π-conjugated chromophore [52, 85]. The structure of these molecules can be presented in two resonance forms: neutral D–π–A and ionic (or zwitterionic) with the separated charges D^+–π–A^-. Using this model in conjunction with extensive experimental data and quantum-chemical analysis, valuable insights may be gained for the explanation of the linear and nonlinear properties of G19, G40, and G188. Our understanding is that for the molecule G19 with the shortest conjugated chain, the donor–acceptor properties of the terminal groups in both solvents dominate the properties of the polyenic chain, and the ground state can be represented by a "polymethine-like" structure with almost equalized bond lengths within the conjugated chromophore and with the charges alternating at carbon atoms.

For the dye G188 with the longest conjugated chain, we suppose that the ground state represents a mixture of a "polymethine-like" structure, connected with the donor-acceptor properties of the terminal groups, and a "polyene-like" structure, mainly determined by a polyenic-type of conjugated chain with strong BLA. The relative contribution of these two resonance structures to the ground state is controlled by the polarity of the solvent: a more polar solvent can increase the ground state polarization and make the charge-separated form dominant. The neutral polyenic form dominates in less polar toluene resulting in a change of the absorption shape (growth of the short wavelength shoulder), clearly seen for G188 in Fig. 21. Dye G40 presumably represents an intermediate case between the shortest G19 and the longest G188 based on linear absorption data and quantum-chemical analysis. Analyzing the shift of the absorption peaks with lengthening of the chromophore, we note that an increase in the conjugation length from G19 to G188 leads to a shift of ≈100 nm in ACN and ≈70 nm in toluene, which is in accord with the model presented of a co-existence of "polyenic-like" and "polymethine-like" forms. Asymmetrical D–π–A dyes show a bathochromic shift of the main absorption peaks, ≈11 nm for G19, ≈28 nm for G40, and ≈65 nm for G188, with increasing solvent polarity from toluene to ACN.

The solvatochromic behavior of these dyes in solution can be explained by the comparison of their permanent dipole moments. If the excited state exhibits a larger dipole moment (μ_1) than the ground state (μ_0), it is preferentially stabilized by the more polar solvent, and the energy between these two states decreases, that is, the absorption and emission spectra both shift to the red region.

Additionally, note that the polarity of the solvent significantly affects not only the positions of absorption and fluorescence spectra but also the fluorescence quantum yields. The largest difference in quantum yield is observed for G19 (eight times larger in toluene) [86]. The effect of solvent polarity on quantum yield and fluorescence lifetime was investigated in mixtures of toluene and ACN (polarity range 0.013–0.306). Polarity dependent quantum yield and lifetime measurements are presented in Fig. 22.

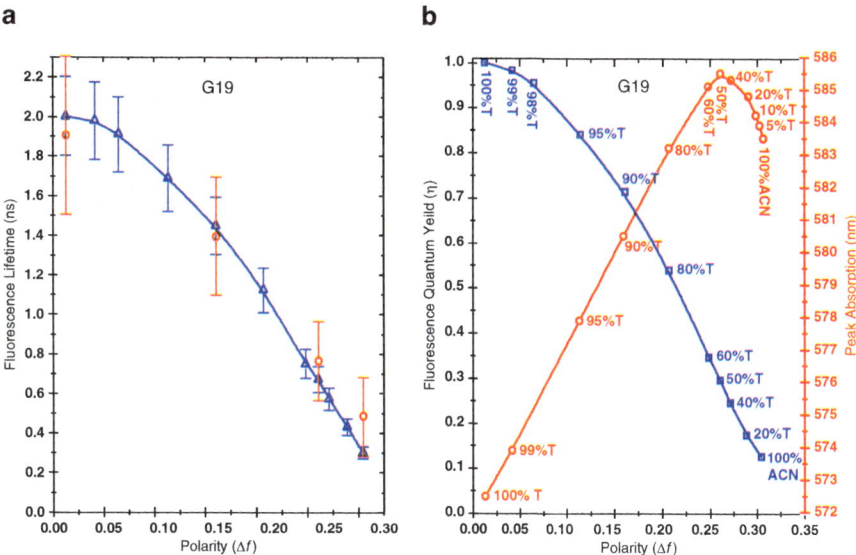

Fig. 22 (**a**) Comparison of fluorescence lifetime (*blue triangles*), calculated from (13), and measured by time-resolved fluorescence (*red circles*) as a function of solvent polarity for G19. (**b**) Fluorescence quantum yield (*blue squares*) and peak ground state absorption wavelength (*red circles*) as a function of solvent polarity given by the percentage of toluene (T) in toluene-ACN mixtures for G19

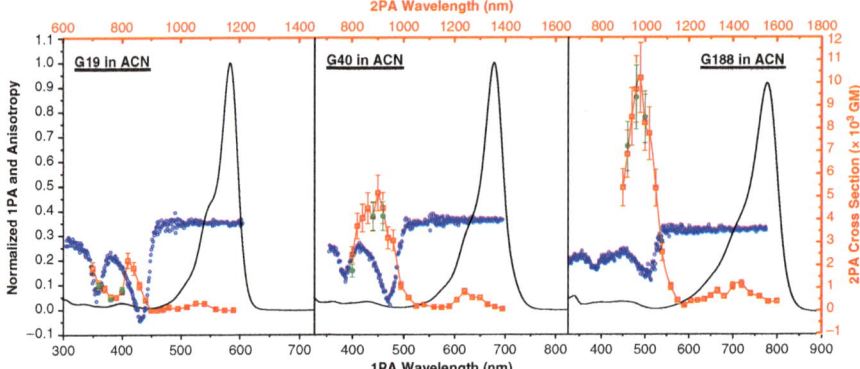

Fig. 23 1PA (*solid black lines*), anisotropy (*blue circles*), and 2PA spectra measured by 2PF (*red squares*) and Z-scan (*green circles*) of G19, G40, and G188 in ACN. *Red axes* correspond to 2PA measurements

It is seen that the fluorescence quantum yield and lifetime of G19 gradually decreases with increasing solvent polarity. For example, the insertion of 20% ACN by volume into toluene leads to a decrease of a factor of two. Based on these results we can conclude that G19 is very sensitive to solvent polarity and can be used as an efficient probe to test the polarity of its microenvironment. A reverse trend of the absorption peak at 1:1 mixture of ACN and toluene (50%T in Fig. 22b) corresponds to a change of the sign of $\mu_0 - \mu_1$ due to a transition from a "polyene-like" structure in nonpolar toluene to a "polymethine-like" structure in polar ACN.

These spectroscopic studies have advanced our knowledge of the structure–property relations, which are extremely important for understanding the nonlinear optical behavior of these dyes, and specifically for their 2PA properties discussed below.

The 2PA spectra in ACN are shown in Fig. 23 in comparison to fluorescence excitation anisotropy. The increase of conjugation length leads to an increase of 2PA cross section as well as to a red shift of the peak the absorption. Interestingly, there is no significant indication of 2PA under the 1PA main peak, which is different from observation for typical asymmetrical dyes [86]. This is explained by quantum-chemical calculations due to the large angle between transition dipole moment μ_{01} and the change of permanent dipole moment, $\Delta\mu$, under excitation, see Sect. 1.2.1. The first weakly allowed 2PA band, similar to symmetrical dyes, can be attributed to the coupling between the first excited electronic state S_1 and its vibrational modes. The second, strongly allowed 2PA band is connected with two-photon excitation into S_2 and higher electronic states and corresponds approximately to the anisotropy valley as seen in Fig. 23. The peak of the third 2PA band for G19 could not be resolved due to the presence of 1PA edge.

In order to investigate the δ_{2PA} (ω) solvent dependence of the D–π–A dyes, the 2PA spectra of G188 in toluene and ACN are presented in Fig. 24. As large as two-fold enhancement of the second 2PA band is observed in ACN (10,000 GM)

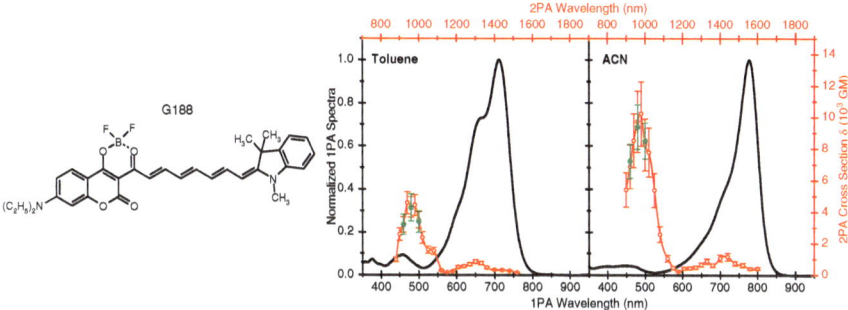

Fig. 24 1PA (*solid black lines*) and 2PA spectra measured by 2PF (*red squares*) and Z-scan (*green circles*) of G188 in ACN and toluene. *Red axes* correspond to 2PA measurements

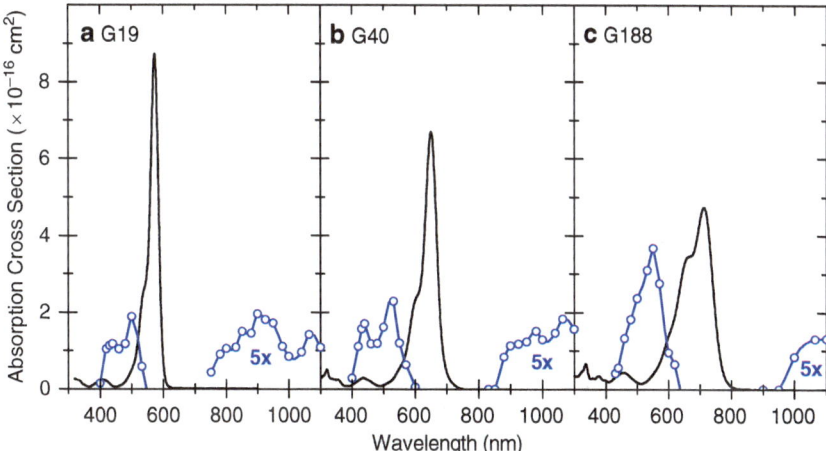

Fig. 25 1PA (*black*) and ESA (*blue*) of G19, G40, and G188 in toluene

comparing to toluene (4,600 GM), while the position of 2PA band does not shift in these two solvents. Since a large bathochromic effect is observed in the 1PA main transition band, this enhancement can be explained by ISRE due to the decrease of the energy difference between S_1 and S_0 in ACN [32, 86].

The ESA spectra of asymmetrical dyes in toluene are shown in Fig. 25. They show broad structureless bands in the NIR region (750–1,100 nm for G19, 850–1,100 nm for G40, and 950–1,100 nm for G188) and more intense transitions in the visible range (400–550 nm for G19, 400–600 nm for G40, and 450–650 nm for G188). Similarly to symmetrical anionic polymethine dyes (Fig. 20), the increase of conjugation length leads to a small red shift of ESA spectra, and to an enhancement of ESA cross sections and the ratio between the ESA and linear absorption oscillator strengths by approximately a factor of two. More detailed experimental description and quantum-chemical analysis can be found in [86].

3.4 Enhancement of 2PA Cross Sections

Increasing 2PA and ESA cross sections is important for multiphoton imaging and for many other fields. Ongoing work is focused on (1) intelligent molecular design strategies to enhance $\delta_{2PA}(\omega)$; (2) increasing the spectral range of the largest 2PA (2PA for most cyanine-like dyes have large cross sections within a relatively narrow spectral range); and (3) obtaining large ESA cross sections while maintaining long-lived excited state lifetimes, perhaps through triplet generation.

In order to determine the structural factors maximizing 2PA cross section values, we analyze (8) from Sect. 1.2.1. For all cyanine-like molecules, symmetrical and asymmetrical, several distinct 2PA bands can be measured. *First*, the less intensive 2PA band is always connected with two-photon excitation into the main absorption band. The character of this 2PA band involves at least two dipole moments, μ_{01} and $\Delta\mu$. It is well-known that 2PA into the S_1 band is symmetry forbidden for centrosymmetrical molecules, such as squaraines with C_i symmetry due to $\Delta\mu = 0$, and only slightly allowed for polymethine dyes with C_{2v} symmetry ($\Delta\mu$ is small and oriented nearly perpendicular to μ_{01}). It is important to note that a change in the permanent dipole moment under two-photon excitation into the linear absorption peak, even for asymmetrical D–π–A molecules, typically does not lead to the appearance of a 2PA band. 2PA bands under the main absorption peak are typically observed only for strongly asymmetrical molecules, for example, Styryl 1 [83], whose $S_0 \rightarrow S_1$ transitions are considerably different from the corresponding transitions in symmetrical dyes and represent much broader, less intense, and blue-shifted bands. Thus, for typical cyanine-like molecules, both symmetrical and asymmetrical, with strong and relatively narrow $S_0 \rightarrow S_1$ transitions, we observe the first 2PA band occurring at an energy shifted to the "blue" range at $\approx 1,000\text{--}1,200$ cm^{-1} as compared to the peak of the $S_0 \rightarrow S_1$ transition. The 2PA final state in this case corresponds to vibrational levels of S_1. In conclusion, the nature of the first 2PA band can be attributed to the coupling between the first excited electronic state S_1 and its vibrational modes.

Second, we analyze the nature of the next, strong 2PA bands. The positions of their final states correspond to one-photon symmetry forbidden bands and can be found from excitation anisotropy measurements, as illustrated in Figs. 6, 19, and 23. Excitation anisotropy spectra for all cyanine-like molecules typically reveal a large alternation of maximum and minimum features suggesting the positions of the 1PA and 2PA transitions. Two-photon excitation into final states involves two dipole moments, μ_{01} and μ_{1fi}.

Finally, we formulate the following general trends in structure–property relations:

1. An increase of 2PA cross section can be achieved by increasing μ_{01}, which can be realized upon lengthening of the polymethine chromophore. For linear conjugated molecules, molar absorbance can be as large as 100,000–300,000 M^{-1} cm^{-1}, which corresponds to $S_0 \rightarrow S_1$ transition dipole moments of $\mu_{01} = 12\text{--}18$ D. Note that lengthening of the conjugation chain leads to an increase of δ_{2PA} for all 2PA bands. The limitation of this factor is connected with the saturation of μ_{01}

values for the long molecules absorbing in the range $\approx 1,000$ nm due to the ground state symmetry breaking effect discussed in Sect. 3.1. An example is shown in Fig. 14 for a series of symmetrical D–π–D polymethine dyes with trimethylindolin donor terminal groups.

2. An increase of 2PA cross section can be achieved by increasing μ_{1f}, which can be realized by introduction of a strong acceptor group into the conjugated bridge leading to a quadrupolar type arrangement D–π–A–π–D (squaraine and tetraone structures). This arrangement stimulates an effective charge transfer processes over a large distance (extended conjugation bridge) resulting in large 2PA cross section values (up to 30,000 GM [87]). Examples are shown in Fig. 18 for PD 2630, SD 2243, and TD 2765 with the same D terminal groups and similar conjugation length. As seen, considerably larger δ_{2PA} can be accessed in SD and especially in TD molecules due to the effective charge transfer transitions allowed in 2PA processes.

3. An increase of 2PA cross section can be achieved by decreasing the detuning energy $(\hbar\omega_{eg}-\hbar\omega)$ leading to ISRE (see Sect. 1.2.1). This effect can be realized in dyes with a relatively narrow absorption band with a steep edge (for example, SD, see Fig. 18) allowing the use of optical pumping at frequencies closer to the linear absorption resonance.

4. An increase of 2PA cross section can be obtained by specific arrangement of the molecular energy levels allowing for an increase of the density of final states and reaching the final states fi at the smallest detuning energy $(\hbar\omega_{eg}-\hbar\omega)$. This case corresponds to the so called "double resonance" condition: $\hbar\omega \rightarrow \hbar\omega_{eg}$ and $2\hbar\omega \rightarrow \omega_{e'g}$ (see Sect. 1.2.1), and is realized in SD molecules due to the increased density of final states [88].

5. An increase of 2PA cross section can be achieved in asymmetrical molecules with the proper choice of solvent polarity [86]. The main idea here is connected with the strong effect of solvent polarity on the position of the linear absorption peak $S_0 \rightarrow S_1$ but much smaller influence on the position of the 2PA band. This case is illustrated in Fig. 24 for the D–π–A dye G188. The position of the second 2PA band for this dye remains unshifted in solvents of different polarity, ACN and toluene, in spite of the large solvatochomic shift of the linear absorption band $S_0 \rightarrow S_1$. This effect leads to a large ISRE in ACN, allowing tuning closer to resonance and results in more than a two times larger δ_{2PA} in this solvent, δ_{2PA} $\approx 10,000$ GM ($\approx 4,700$ GM in toluene). The first 2PA band follows the solvato-chromic shift of the linear absorption peak and exhibits a smaller difference in 2PA cross sections: $\delta_{2PA} \approx 1,150$ GM in ACN and ≈ 850 GM in toluene.

4 Conclusions and Future Directions

Much of the work presented in this chapter represents a synergistic effort from several complementary research fields: quantum-chemical theory, chemical synthesis, and nonlinear optical materials characterization. This combination of expertise is essential in order to make progress in a meaningful and directed way. Obtaining the nonlinear spectrum of a single molecule has limited value. What is needed is a large database of molecular nonlinear optical studies on several series of molecules having well-controlled differences. These differences, often small, can give trends in properties that are crucial to understanding the underlying physics that produces nonlinear absorption. Having the knowledge to be able to separate various nonlinear effects and their temporal response is also essential so as to not confuse trends. Understanding the quantum mechanical molecular states that ultimately lead to the observed nonlinear response is the final test of success in predicting nonlinear properties.

The discussion in this chapter is limited to cyanine-like NIR conjugated molecules, and further, is limited to discussing their two-photon absorption spectra with little emphasis on their excited state absorption properties. In principle, if the quantum mechanical states are known, the ultrafast nonlinear refraction may also be determined, but that is outside the scope of this chapter. The extent to which the results discussed here can be transferred to describe the nonlinear optical properties of other classes of molecules is debatable, but there are certain results that are clear. *Designing molecules with large transition dipole moments that take advantage of* intermediate state resonance *and* "double resonance" *enhancements are definitely important approaches to obtain large two-photon absorption cross sections.*

The results of this research combined with a growing literature on structure–property relations in organic materials is moving us closer to the ultimate goal of developing a predictive capability for the nonlinear optical properties of molecules; however, there is still a long way to go. Progress is slow due to the difficulties of synthesis (and the time it takes), the inadequacies of models (and the capacity of computers), and the nearly infinite variety of possible molecular structures. Of course, this last difficulty is what makes organic materials so interesting! If we do obtain a predictive capability for molecular properties, there will still be questions of what happens in the solid-state/neat materials due to strong molecular interactions, although progress is also being made in this field [89].

When we do have a good grasp of the ultimate possibilities and limits on the two-photon absorption along with the nonlinear refraction, we should have a much better understanding on what is possible to do with these nonlinear materials, i.e., what possible applications are practical and what devices can be made. On the other hand, as has been found with linear optical properties, it is often that other properties including processability, longevity, cost, and extrinsic properties that ultimately demand what material is used for a given application. Although the goal of practical molecular-based nonlinear optical devices remains mostly a hope, the continued progress in this field brings it closer to realization.

References

1. Pawlicki M, Collins HA, Denning RG, Anderson HL (2009) Two-photon absorption and the design of two-photon dyes. Angew Chem Int Ed Engl 48:3244–3266
2. Zipfel WR, Williams RM, Webb WW (2003) Nonlinear magic: multiphoton microscopy in the biosciences. Nat Biotechnol 21:1369–1377

3. Williams RM, Piston DW, Webb WW (1994) Two-photon molecular excitation provides intrinsic 3-dimensional resolution for laser-based microscopy and microphotochemistry. FASEB J 8:804–813
4. So PT, Dong CY, Masters BR, Berland KM (2000) Two-photon excitation fluorescence microscopy. Annu Rev Biomed Eng 2:399–429
5. Piston DW (1999) Imaging living cells and tissues by two-photon excitation microscopy. Trends Cell Biol 9:66–69
6. König K (2000) Multiphoton microscopy in life sciences. J Microsc 200:83–104
7. Diaspro A, Robello M (2000) Two-photon excitation of fluorescence for three-dimensional optical imaging of biological structures. J Photochem Photobiol B 55:1–8
8. Rubart M (2004) Two-photon microscopy of cells and tissue. Circ Res 95:1154–1166
9. Diaspro A, Chirico G, Federici F, Cannone F, Beretta S, Robello M (2001) Two-photon microscopy and spectroscopy based on a compact confocal scanning head. J Biomed Opt 6:300–310
10. Scherschel JA, Rubart M (2008) Cardiovascular imaging using two-photon microscopy. Microsc Microanal 14:492–506
11. Bates M, Huang B, Zhuang X (2008) Super-resolution microscopy by nanoscale localization of photo-switchable fluorescent probes. Curr Opin Chem Biol 12:505–514
12. Maria GoeppertMayer Biography http://nobelprize.org/nobel_prizes/physics/laureates/1963/mayer-bio.html
13. Goeppert-Mayer M (1931) Uber elementarakte mit zwei quantensprungen. Ann Phys 401:273–294
14. Denk W, Strickler JH, Webb WW (1990) Two-photon laser scanning fluorescence microscopy. Science 248:73–76
15. He GS, Tan LS, Zheng Q, Prasad PN (2008) Multiphoton absorbing materials: molecular designs, characterizations, and applications. Chem Rev 108:1245–1330
16. Reinhardt BA, Brott LL, Clarson SJ, Dillard AG, Bhatt JC, Kannan R, Yuan L, He GS, Prasad PN (1998) Highly active two-photon dyes: design, synthesis, and characterization toward application. Chem Mater 10:1863–1874
17. Terenziani F, D'Avino G, Painelli A (2007) Multichromophores for nonlinear optics: designing the material properties by electrostatic interactions. Chemphyschem 8:2433–2444
18. Bhawalkar JD, He GS, Prasad PN (1996) Nonlinear multiphoton processes in organic and polymeric materials. Rep Prog Phys 59:1041–1070
19. Albota M, Beljonne D, Bredas JL, Ehrlich JE, Fu JF, Heikal AA, Hess SE, Kogej T, Levin MD, Marder SR, McCord-Maughon D, Perry JW, Röckel H, Rumi M, Subramaniam G, Webb WW, Wu XL, Xu C (1998) Design of organic molecules with large two-photon absorption cross sections. Science 281:1653–1656
20. Corredor CC, Huang Z, Belfield KD (2006) Two-photon 3D optical data storage via fluorescence modulation of an efficient fluorene dye by a photochromic diarylethene. Adv Mater 18:2910–2914
21. Cumpston BH, Ananthavel SP, Barlow S, Dyer DL, Ehrlich JE, Erskine LL, Heikal AA, Kuebler SM, Lee IS, McCord-Maughon D, Qin J, Röckel H, Rumi M, Wu XL, Marder SR, Perry JW (1999) Two-photon polymerization initiators for three-dimensional optical data storage and microfabrication. Nature 398:51–54
22. Kuebler SM, Rumi M (2004) Nonlinear optics - applications: three-dimensional microfabrication. In: Guenther RD, Steel DG, Bayvel L (eds) Encyclopedia of modern optics. Oxford, Elsevier
23. Picot A, D'Aléo A, Baldeck PL, Grichine A, Duperray A, Andraud C, Maury O (2008) Long-lived two-photon excited luminescence of water-soluble europium complex: applications in biological imaging using two-photon scanning microscopy. J Am Chem Soc 130:1532–1533
24. Briñas RP, Troxler T, Hochstrasser RM, Vinogradov SA (2005) Phosphorescent oxygen sensor with dendritic protection and two-photon absorbing antenna. J Am Chem Soc 127:11851–11862

25. Kim S, Ohulchanskyy TY, Pudavar HE, Pandey RK, Prasad PN (2007) Organically modified silica nanoparticles co-encapsulating photosensitizing drug and aggregation-enhanced two-photon absorbing fluorescent dye aggregates for two-photon photodynamic therapy. J Am Chem Soc 129:2669–2675

26. Ogawa K, Kobuke Y (2008) Recent advances in two-photon photodynamic therapy. Anticancer Agents Med Chem 8:269–279

27. Ogawa K, Kobuke Y (2009) Design of two-photon absorbing materials for molecular optical memory and photodynamic therapy. Org Biomol Chem 7:2241–2246

28. Sutherland RL (1996) Handbook of nonlinear optics. Marcel Dekker, New York

29. Sheik-Bahae M, Hutchings DC, Hagan DJ, Van Stryland EW (1991) Dispersion of bound electronic nonlinear refraction in solids. IEEE J Quantum Electron 27:1296–1309

30. Hutchings DC, Van Stryland EW (1992) Nondegenerate 2-photon absorption in zinc-blend semiconductors. J Opt Soc Am B 9:2065–2074

31. Sheik-Bahae M, Wang J, Van Stryland EW (1994) Nondegenerate optical Kerr-effect in semiconductors. IEEE J Quantum Electron 30:249–255

32. Hales JM, Hagan DJ, Van Stryland EW, Schafer KJ, Morales AR, Belfield KD, Pacher P, Kwon O, Bredas JL (2004) Resonant enhancement of two-photon absorption in substituted fluorine molecules. J Chem Phys 121:3152–3160

33. Orr BJ, Ward JF (1971) Perturbation theory of the non-linear optical polarization of an isolated system. Mol Phys 20:513–526

34. Dirk CW, Cheng L, Kuzyk MG (1992) A simplified three-level model describing the molecular third-order nonlinear optical susceptibility. Int J Quantum Chem 43:27–36

35. Kuzyk MG, Dirk CW (1990) Effects of centrosymmetry on the nonresonant electronic third-order nonlinear optical susceptibility. Phys Rev A 41:5098–5109

36. Birge RR, Pierce BM (1979) A theoretical analysis of two-photon properties of linear polyenes and the visual chromophores. J Chem Phys 70:165–178

37. Cronstrand P, Luo Y, Agren H (2002) Generated few-state models for two-photon absorption of conjugated molecules. Chem Phys Lett 352:262–269

38. Monson PR, McClain WM (1970) Polarization dependence of the two-photon absorption of tumbling molecules with application to liquid 1-chloronaphthalene and benzene. J Chem Phys 53:29–37

39. Kamada K, Ohta K, Iwase Y, Kondo K (2003) Two-photon absorption properties of symmetric substituted diacetylene: drastic enhancement of the cross section near the one-photon absorption peak. Chem Phys Lett 372:386–393

40. Ohta K, Kamada K (2006) Theoretical investigation of two-photon absorption allowed excited states in symmetrically substituted diacetylenes by ab initio molecular-orbital method. J Chem Phys 124:124303

41. Lakowicz JR (1999) Principle of fluorescence spectroscopy, 2nd edn. Kluwer Academic/Plenum, New York

42. Ramakrishna G, Goodson T III, Rogers-Haley JE, Cooper TM, McLean DG, Urbas A (2009) Ultrafast intersystem crossing: excited state dynamics of platinum acetylide complexes. J Phys Chem C 113:1060–1066

43. Kleinschmidt J, Rentsch S, Tottleben W, Wilhelmi B (1974) Measurement of strong nonlinear absorption in stilbene-chloroform solutions, explained by the superposition of two-photon absorption and one-photon absorption from the excited state. Chem Phys Lett 24:133–135

44. Kannan R, He GS, Lin TC, Prasad PN, Vaia RA, Tan LS (2004) Toward highly active two-photon absorbing liquids. Synthesis and characterization of 1, 3, 5-triazine-based octupolar molecules. Chem Mater 16:185–194

45. Ehrlich JE, Wu XL, Lee IS, Hu ZY, Röckel H, Marder SR, Perry JW (1997) Two-photon absorption and broadband optical limiting with bis-donor stilbenes. Opt Lett 22:1843–1845

46. Webster S, Odom SA, Padilha LA, Przhonska OV, Peceli D, Hu H, Nootz G, Kachkovski AD, Matichak J, Barlow S, Anderson HL, Marder SR, Hagan DJ, Van Stryland EW (2009) Linear and nonlinear spectroscopy of a porphyrin-squaraine-porphyrin conjugated system. J Phys Chem B 113:14854–14867

47. Sutherland RL, Brant MC, Heinrichs J, Rogers JE, Slagle JE, McLean DG, Fleitz PA (2005) Excited-state characterization and effective three-photon absorption model of two-photon-induced excited-state absorption in organic push-pull charge-transfer chromophores. J Opt Soc Am B 22:1939–1948

48. Mishra A, Behera RK, Behera PK, Mishra BK, Behera GB (2000) Cyanines during the 1990s: a review. Chem Rev 100:1973–2012

49. Fabian J, Nakazumi H, Matsuoka M (1992) Near-infrared absorbing dyes. Chem Rev 92:1197–1226

50. Peyghambarian N, Dalton L, Jen A, Kippelen B, Marder S, Norwood R, Perry J (2006) Technological advances brighten horizons for organic nonlinear optics. Laser Focus World 42:85–94

51. Daehne S (1978) Color and constitution: one hundred years of research. Science 199:1163–1167

52. Meyers F, Marder SR, Pierce BM, Bredas JL (1994) Electric field modulated nonlinear optical properties of donor-acceptor polyenes: sum-over-states investigation of the relationship between molecular polarizabilities (α, β, and γ) and bond length alternation. J Am Chem Soc 116:10703–10714

53. Beverina L, Fu J, Leclercq A, Zojer E, Pacher P, Barlow S, Van Stryland EW, Hagan DJ, Brédas JL, Marder SR (2005) Two-photon absorption at telecommunications wavelengths in a dipolar chromophore with a pyrrole auxiliary donor and thiazole auxiliary acceptor. J Am Chem Soc 127:7282–7283

54. Hales JM, Zheng S, Barlow S, Marder SR, Perry JW (2006) Bisdioxaborine polymethines with large third-order nonlinearities for all-optical signal processing. J Am Chem Soc 128:11362–11363

55. Fisher JAN, Susumu K, Therien MJ, Yodh AG (2009) One- and two-photon absorption of highly conjugated multiporphyrin systems in the two-photon Soret transition region. J Chem Phys 130:134506

56. Eaton DF (1988) Reference materials for fluorescence measurement. Pure Appl Chem 60:1107–1114

57. Webster S, Padilha LA, Hu H, Przhonska OV, Hagan DJ, Van Stryland EW, Bondar MV, Davydenko IG, Slominsky YL, Kachkovski AD (2008) Structure and linear spectroscopic properties of near IR polymethine dyes. J Lumin 128:1927–1936

58. Gerasov AO, Shandura MP, Kovtun YP (2008) Series of polymethine dyes derived from 2, 2-difluoro-1, 3, 2-(2H)-dioxaborine of 3-acetyl-7-diethylamino-4-hydroxycoumarin. Dyes Pigm 77:598–607

59. Padilha LA, Webster S, Przhonska OV, Hu H, Peceli D, Ensley TR, Bondar MV, Gerasov AO, Kovtun YP, Shandura MP, Kachkovski AD, Hagan DJ, Van Stryland EW (2010) Efficient two-photon absorbing acceptor-π-acceptor polymethine dyes. J Phys Chem A, Submitted

60. Strickler SJ, Berg RA (1962) Relationship between absorption intensity and fluorescence lifetime of molecules. J Chem Phys 37:814–822

61. Negres RA, Przhonska OV, Hagan DJ, Van Stryland EW, Bondar MV, Slominsky YL, Kachkovski AD (2001) The nature of excited-state absorption in polymethine and squarylium molecules. IEEE J Sel Top Quantum Electron 7:849–863

62. Webster S, Fu J, Padilha LA, Przhonska OV, Hagan DJ, Van Stryland EW, Bondar MV, Slominsky YL, Kachkovski AD (2008) Comparison of nonlinear absorption in three similar dyes: polymethine, squaraine, and tetraone. Chem Phys 348:143–151

63. Lessing HE, Von Jena A (1976) Separation of rotational diffusion and level kinetics in transient absorption spectroscopy. Chem Phys Lett 42:213–217

64. Sheik Bahae M, Said AA, Van Stryland EW (1989) High-sensitivity, single beam n2 measurements. Opt Lett 14:955–957

65. Sheik-Bahae M, Said AA, Wei TH, Hagan DJ, Van Stryland EW (2007) Special 30th anniversary feature: sensitive measurement of optical nonlinearities using a single beam. IEEE LEOS Newslett 21:17–35

66. Sheik-Bahae M, Said AA, Hagan DJ, Van Stryland EW (1990) Sensitive measurement of optical nonlinearities using a single beam. IEEE J Quantum Electron 26:760–769

67. Johnston TF (1998) Beam propagation (M^2) measurement made as easy as it gets: the four-cuts method. Appl Opt 37:4840–4850

68. Firester AH, Heller ME, Sheng P (1977) Knife-edge scanning measurements of subwavelength focused light beams. Appl Opt 16:1971–1974

69. Weber HP (1967) Method for pulsewidth measurement of ultrashort light pulses generated by phase-locked lasers using nonlinear optics. J Appl Phys 38:2231–2234

70. Diels JC, Rudolph W (1996) Ultrashort laser pulse phenomena: fundamentals, techniques, and applications on a femtosecond time scale. Academic, San Diego CA, pp 365–399

71. Balu M, Hales J, Hagan DJ, Van Stryland EW (2005) Dispersion of nonlinear refraction and two-photon absorption using a white-light continuum Z-scan. Opt Express 13: 3594–3599

72. Balu M, Padilha LA, Hagan DJ, Van Stryland EW, Yao S, Belfield K, Zheng S, Barlow S, Marder S (2008) Broadband Z-scan characterization using a high-spectral-irradiance, high-quality supercontinuum. J Opt Soc Am B 25:159–165

73. Webster S, Padilha L, Przhonska O, Peceli D, Hu H, Slominsky Y, Kachkovski A, Tolmachov A, Kurdyukov V, Hagan D, Van Stryland E (2009) Enhancement of triplet yields in cyanine-like molecules. Laser Science XXV, OSA Technical Digest (CD) paper: LSTuG2

74. Santos PF, Reis LV, Duarte I, Serrano JP, Almeida P, Oliveira AS, Vieira Ferreira LF (2005) Synthesis and photochemical evaluation of iodinated squarylium cyanine dyes. Helv Chim Acta 88:1135–1143

75. Lim JH, Przhonska OV, Khodia S, Yang S, Ross TS, Hagan DJ, Van Stryland EW, Bondar MV, Slominsky YL (1999) Polymethine and squarylium molecules with large excited-state absorption. Chem Phys 245:79–97

76. Kaiser W, Garrett CGB (1961) Two-photon excitation in $CaF_2:Eu^{2+}$. Phys Rev Lett 7:229–231

77. Xu C, Webb WW (1996) Measurement of two-photon excitation cross sections of molecular fluorophores with data from 690 to 1050 nm. J Opt Soc Am B 13:481–491

78. Padilha LA, Webster S, Hu H, Przhonska OV, Hagan DJ, Van Stryland EW, Bondar MV, Davydenko IG, Slominsky YL, Kachkovski AD (2008) Excited state absorption and decay kinetics of near IR polymethine dyes. Chem Phys 352:97–105

79. Iordanov TD, Davis JL, Masunov AE, Levenson A, Przhonska OV, Kachkovski AD (2009) Symmetry breaking in cationic polymethine dyes, part 1: ground state potential energy surfaces and solvent effects on electronic spectra of streptocyanines. Int J Quantum Chem 109:3592–3601

80. Ryabitsky AB, Kachkovski AD, Przhonska OV (2007) Symmetry breaking in cationic and anionic polymethine dyes. J Mol Struct-Theochem 802:75–83

81. Lepkowicz RS, Przhonska OV, Hales JM, Fu J, Hagan DJ, Van Stryland EW, Bondar MV, Slominsky YL, Kachkovski AD (2004) Nature of the electronic transitions in thiacarbocyanines with a long polymethine chain. Chem Phys 305:259–270

82. Su WP, Schrieffer JR, Heeger AJ (1979) Soliton in polyacetylene. Phys Rev Lett 42:1698–1701

83. Fu J, Padilha LA, Hagan DJ, Van Stryland EW, Przhonska OV, Bondar MV, Slominsky YL, Kachkovski AD (2007) Molecular structure – two-photon absorption property relations in polymethine dyes. J Opt Soc Am B 24:56–66

84. Luo Y, Norman P, Macak P, Ågren H (2000) Solvent-induced two-photon absorption of a push – pull molecule. J Phys Chem A 104:4718–4722

85. Terenziani F, Katan C, Badaeva E, Tretiak S, Blanchard-Desce M (2008) Enhanced two-photon absorption of organic chromophores: theoretical and experimental assessments. Adv Mater 20:4641–4678

86. Padilha LA, Webster S, Przhonska OV, Hu H, Peceli D, Rosch JL, Bondar MV, Gerasov AO, Kovtun YP, Shandura MP, Kachkovski AD, Hagan DJ, Van Stryland EW (2009) Nonlinear

absorption in a series of donor–π–acceptor cyanines with different conjugation lengths. J Mater Chem 19:7503–7513

87. Chung SJ, Zheng S, Odani T, Beverina L, Fu J, Padilha LA, Biesso A, Hales JM, Zhan X, Schmidt K, Ye A, Zojer E, Barlow S, Hagan DJ, Van Stryland EW, Yi Y, Shuai Z, Pagani GA, Brédas JL, Perry JW, Marder SR (2006) Extended squaraine dyes with large two-photon absorption cross-sections. J Am Chem Soc 128:14444–14445

88. Fu J, Padilha LA, Hagan DJ, Van Stryland EW, Przhonska OV, Bondar MV, Slominsky YL, Kachkovski AD (2007) Experimental and theoretical approaches to understanding two-photon absorption spectra in polymethine and squaraine molecules. J Opt Soc Am B 24:67–76

89. Hu H, Gerasov AO, Padilha LA, Przhonska OV, Webster S, Shandura MP, Kovtun YP, Masunov AE, Hagan DJ, Van Stryland EW (2010) Two-photon absorption in single crystals of cyanine-like dye. CLEO/QELS 2010, San Jose CA, Submitted

Discovery of New Fluorescent Dyes: Targeted Synthesis or Combinatorial Approach?

Eunha Kim and Seung Bum Park

Abstract Because of their advantageous properties, fluorescent organic molecules have been used extensively in various applications. Despite the high demand for fluorophores, there are only a limited number of fluorescent organic core skeletons with sufficient flexibility in their synthetic strategies. The rational design of fluorescent probes having desirable photophysical properties is still far from being achieved because of the complexity of underlying photophysical phenomena of fluorescent small molecules. To study and improve the photophysical properties of fluorescent core skeletons, many efforts have been pursued. Traditionally, targeted approach had been used to rationally study the photophysical properties of known fluorescent core skeleton. On the other hand, recent studies showed that combinatorial approach can act as a powerful tool to discover or develop novel fluorescent probes. Through a comparison between targeted synthesis and combinatorial approach in the study of fluorescent organic compounds, we can conclude that the targeted synthesis with rational design and systematic modification using combinatorial approach are complementary to each other for the discovery of novel, small molecules with desired photophysical properties.

Keywords Combinatorial approach · Design-based approach · Organic fluorophores · Photophysical property · Tunability

Contents

E. Kim and S.B. Park (✉)
Department of Chemistry and Department of Biophysics and Chemical Biology, Seoul National University, 599 Gwanak-ro, Gwanak-gu, Seoul 151-747, Korea
e-mail: sbpark@snu.ac.kr

A.P. Demchenko (ed.), *Advanced Fluorescence Reporters in Chemistry and Biology I:*   149
Fundamentals and Molecular Design, Springer Ser Fluoresc (2010) 8: 149–186,
DOI 10.1007/978-3-642-04702-2_5, © Springer-Verlag Berlin Heidelberg 2010

1   Introduction

Fluorescent organic small molecules have been used extensively in various applications in material science as well as in biological science because of their advantages, such as excellent sensitivity, good specificity, a large linear range of analysis, ease of handling, and so on [1]. Furthermore, many studies have focused on discovering novel fluorescent small molecules as well as tuning their photophysical properties, such as quantum yield Φ_F, Stokes shift, absorption, excitation, and emission wavelength [2]. Despite the high demand for fluorophores, however, there are only a limited number of fluorescent organic core skeletons with sufficient flexibility in their synthetic strategies. The extensive studies using targeted approach have been used to understand and manipulate the photophysical properties of photoluminescent materials. In addition, recent series of studies show that combinatorial approach is useful not only to improve the photophysical properties of known dyes but also to develop novel fluorescent probes. This chapter is subcategorized by the fluorescent core skeletons for the side-by-side comparison of targeted synthesis and combinatorial approach for the discovery of new fluorescent small molecule with desired photophysical properties.

2   Coumarin Dyes

The name of "Coumarins," 1-benzopyran-2-one, comes from a Caribbean word "*coumarou*" for the tonka bean [3]. Although, as we can expect based on the etymology of a word, they are in high concentration in tonka bean, coumarins are also found in many plants, such as vanilla grass, woodruff, mullein, sweet grass, etc. Coumarin is now the accepted trivial name for the compound possessing naturally occurring lactones as a fundamental structural unit (Fig. 1a). Depending on their structures, there are several types of coumarins; *dihydrofuranocoumarins, furanocoumarins, dihydropyranocoumarins, pyranocoumarins, and miscellaneous coumarins* (Fig. 1b). Because of their sweet scent, readily recognized as the scent of newly-mown hay, coumarins have been used in perfumes since 1882. In addition to

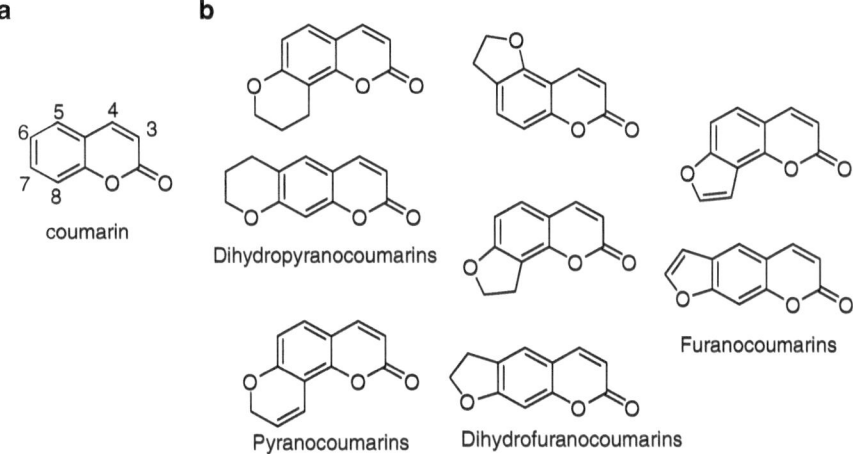

Fig. 1 Coumarin (**a**) and its derivatives (**b**)

their use in perfumes, coumarins also constitute an important class of natural and synthetic products [4, 5]. For instance, warfarin, a synthetic derivative of coumarin and also known under the brand names "Coumadin," "Jantoven," "Marevan," "Lawarin," and "Waran" is commonly used as an anticoagulant.

2.1 Coumarin-Based Targeted Approach

The history of coumarins' synthesis began in the mid-19th century by W. H. Perkin in his famous first synthesis of a vegetable perfume. As shown in Fig. 2a, he made a coumarin by treating the sodium salt of *ortho*-hydroxybenzaldehyde (also known as salicylic aldehyde) with acetic anhydride [6–8]. Since the report of famous coumarin synthesis, many other novel methods have been developed by Pechmann [9–11], Claisen [12], Knoevenagel [13, 14], Reformatsky [15], and Wittig [16] for the synthesis of pyrone-ring moieties in coumarins. Among them, Pechmann reaction has been widely used for the preparation of coumarins, especially for 4-substituted coumarins, since it proceeds from very simple starting materials, such as phenols and β-ketoesters or α,β-unsaturated carboxylic acids in the presence of acid catalysts (see Fig. 2b). This reaction yields coumarins with substituents at one or both rings in good yields. Not only conventional homogenous acid catalysts, such as H_2SO_4, P_2O_5, $FeCl_3$, $ZnCl_2$, $POCl_3$, $AlCl_3$, HCl, and phosphoric acid, and trifluoroacetic acid [17–20], but also cation exchange resins [21] and solid acid catalysts [22–26] have been used as an acid catalyst in Pechmann condensation to synthesize coumarin derivatives.

The Knoevenagel-type condensation of salicylaldehydes with malonic acid [14, 27], malonic ester [28], cyanoacetic ester [29], or Meldrum's acid [30], is

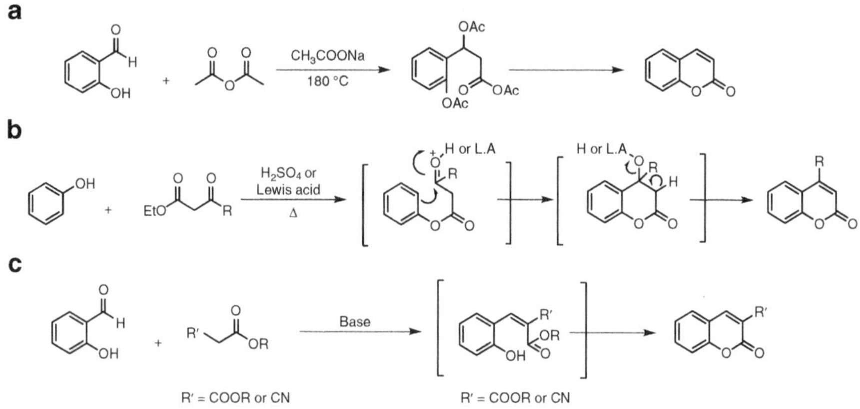

a

b

c

R' = COOR or CN   R' = COOR or CN

Fig. 2 Representative traditional methods for coumarins synthesis. (**a**) Perkin synthesis; (**b**) Pechmann-type condensation; (**c**) Knoevenagel-type condensation

another traditional method for the preparation of the 3-substituted coumarins, especially substituted with electron-withdrawing groups (EWG) at the C-3 position of coumarins (see Fig. 2c). The reaction is catalyzed by weak bases or by suitable combinations of amines and carboxylic or Lewis acids under homogeneous reaction conditions.

Besides their biological activities, coumarins have interesting optical properties. Due to their inherent physicochemical and photophysical characteristics, such as reasonable stability, usually good water solubility, and relative ease of synthesis, coumarin derivatives have been extensively investigated for electronic and photonic applications such as charge-transfer complexes, laser dyes, fluorescent whiteners, solar energy collectors, nonlinear optical materials, and so on [31–34]. In fact, coumarins are well known as the largest class of laser dyes in the "blue-green" region, and the fluorescence of coumarin derivatives changes drastically with electronic characteristics and regiochemical positions of their substituents. For instance, while coumarin itself is nonfluorescent, coumarin derivatives with electron-donating groups (EDG) at the C-7 position develop the intense fluorescence. This fluorescence emission mechanism was interpreted later by the charge-transfer from electron-rich substituents (electron donor) at the C-7 position to a lactone carbonyl group (electron acceptor) on coumarin rings, which is based on the intramolecular charge-transfer (ICT) mechanism arising from, namely, a push–pull electron relay [35]. After discovery of above-mentioned phenomena, a series of studies followed to understand the relationship between structural features and photophysical properties of coumarins – the structure–photophysical property relationship (SPR). These studies revealed that electron-withdrawing groups (EWG) at the C-3 position and electron-donating groups (EDG) at the C-7 position of coumarin synergistically enhance the fluorescence intensity of coumarins, first found in 1941 [36–38]. The charge-transfer from a donor (EDG) at the C-6 position to an acceptor (EWG) at the C-3 position also showed similar phenomena, again explained by ICT mechanism. In addition, the absence of electron withdrawing moieties in coumarins deteriorates their fluorescence properties. For instance, chromene-type compound **3l** (without carbonyl group on lactone ring of coumarin) is nonfluorescent compared to original chromenone-type fluorescent compound (**3k**).

Based on the knowledge accumulated through the targeted approach in coumarins, the systematic derivatization of EDG at the C-7 or C-6 position was pursued to induce the ICT toward either lactone carbonyl group or EWG at the C-3 position for brighter fluorescence of coumarins (see Fig. 3b). Based on a series of studies, we can summarize that ICT between EDG at the C-6 or C-7 position of coumarins to lactone carbonyl group is the most effective for their brightness of fluorescence, and ICT between EDG at the C-6 or C-7-position and EWG at the C-3 position is also helpful for the enhancement of fluorescent brightness. In addition, substituents at the C-7 and C-3 position of coumarins are important for the tuning of their emission wavelength (Fig. 3a) [35, 39–43]. As shown in Fig. 3b, the introduction of either EDG, such as methoxy (**3b, 3f**), hydroxy (**3c, 3g**), and diethylamino group (**3h**), at the C-7 position or EDG (**3i**) at the C-6 position of coumarin cause the bathochromic shift compared to coumarin (**3a**) or 4-methylcoumarin (**3e**). The introduction of EDG at the C-6 position of coumarins can induce more effective bathochromic shift than that at the C-7 position. For example, the bathochromic shift (98 nm) of the emission wavelength in compound (**3n**) with methoxy group at the C-6 position, compared to coumarin (**3a**), is much greater than that (30 nm) in compound (**3b**) with methoxy group at the C-7 position. Furthermore, several 6-methoxy coumarins (**3n, 3o** and **3r**) exhibit the more red-shifted emission wavelength (more than 50 nm) than all of corresponding 7-methoxy coumarins (**3b, 3p**, and **3s**). The comparison between coumarin **3l** and **3q** reveals that the introduction of EWG at the C-3 position can induce the more effective ICT from the electron donors at the C-6 or C-7 position than that with lactone carbonyl group. Therefore, the structure–photophysical properties of coumarin can be summarized as following: the ICT from EDG at the C-6 or C-7 position of coumarin to lactone carbonyl group or EWG at the C-3 position of coumarin causes a bathochromic shift of emission maxima and the C-6 position of coumarin is more sensitive toward the electronic changes by EDG for the effective bathochromic shift than the C-7 position.

2.2 Coumarin-Based Combinatorial Approach

Based on the preliminary understanding of ICT mechanism of coumarin core skeleton, the first systematic application of combinatorial approach toward the field of fluorescence chemistry was reported by Bäuerle and co-workers in 2001 [44]. In their study, the structure–photophysical property relationships (SPR) of coumarin fluorophore were revealed by means of a combinatorial approach.

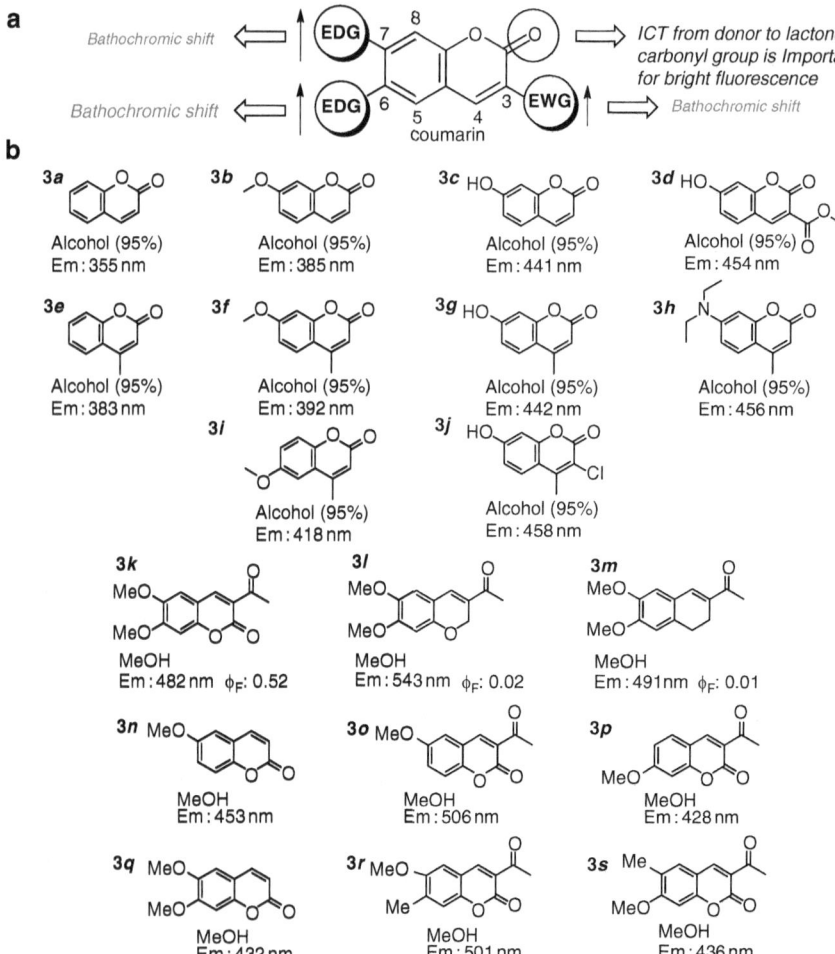

Fig. 3 Structure–photophysical property relationship of coumarin derivatives. (**a**) Schematic representation of the correlation between electronic effect of substituents at the C-3 and C-7 position and photophysical properties; (**b**) Structure and their emission maxima of various coumarins

Using Pd-mediated cross-coupling reactions, such as Suzuki, Heck, and Sonoga-shira– Hagihara reaction, researchers efficiently constructed a library of 151 coumarin derivatives from eight 3-bromocoumarins cross-coupled with ten aryl/heteroaryl boronic acids, ten alkenes, and ten alkynes (Fig. 4).

The resulting coumarin derivatives were isolated with >99% purity using the automated LC/MS purification system. Although there were neither rational explanation nor deduced structure–photophysical property relationships, this study successfully demonstrated the application of a combinatorial approach for the development of novel fluorescent compounds. The systematic transformation of

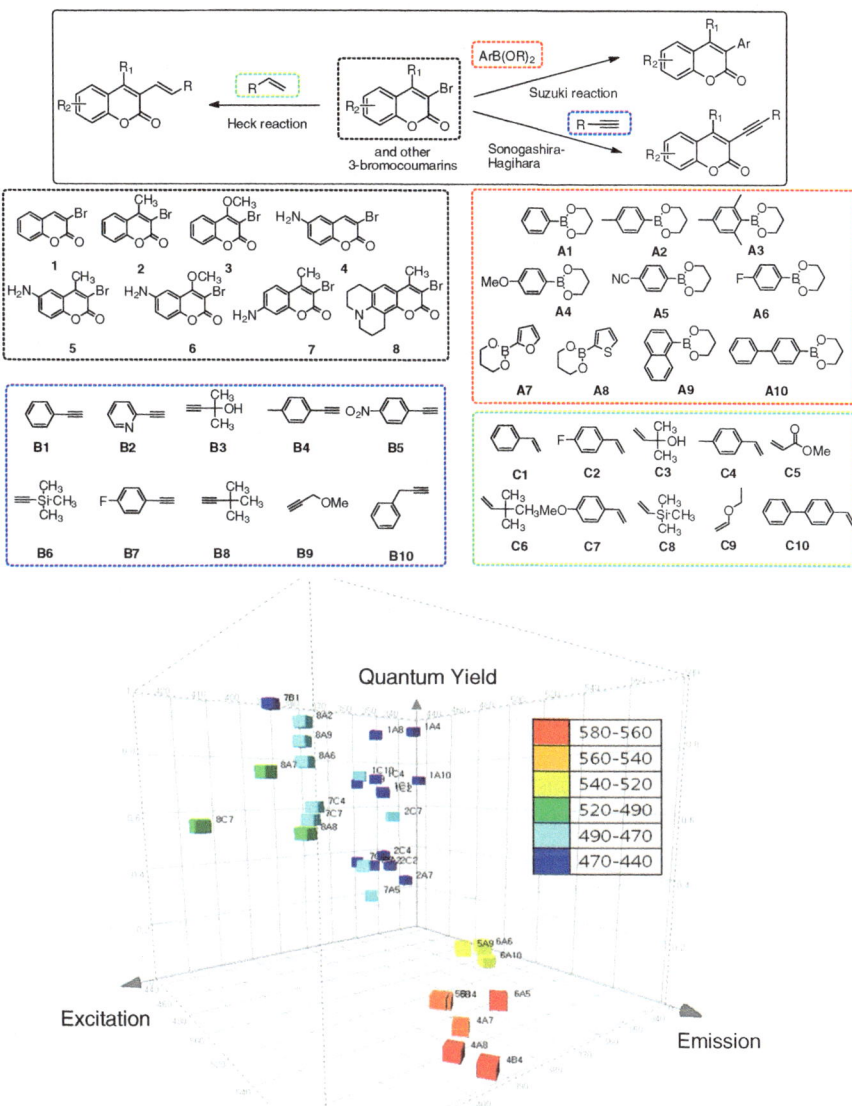

Fig. 4 Combinatorial synthesis of coumarin library using Pd-mediated cross-coupling reaction and 3D scatter plot of representative coumarin derivatives according to their excitation, emission wavelength, and quantum yield. All the photophysical properties were measured in ethanol

their structures not only led to the notable bathochromic shift of their emission maximum (ranging from 400 to 570 nm) but also significantly increased quantum yield Φ_F (ranging from zero to near 1). It is noteworthy that a large diversity of coumarin derivatives having various photophysical properties are achieved by the combinatorial approach in the field of fluorescent chemistry without complicated calculation and laborious synthetic work in relatively short time period.

Wang and co-workers also utilized coumarin as a key core skeleton for the systematic construction of fluorescent library in combinatorial fashion (see Fig. 5) [45]. In order to discover highly fluorescent triazolylcoumarin for the biological application, they synthesized 192-membered triazolylcoumarin library using Cu(I)-catalyzed 1,3-dipolar cycloaddition reaction, namely click chemistry, between nine 3-azidocoumarins and twenty-four alkyne building blocks. The resulting triazolyl-coumarin derivatives cover a relatively wide range of emission maxima between 388 and 521 nm, and the quantum yields of selected compounds ($\phi_F = 0.6$–0.7) are high on the background of parent 3-azidocoumarin ($\phi_F = 0$). Their combinatorial study using triazolylcoumarin as a core skeleton corroborates the previous conclusion regarding the structure–photophysical property relationship in coumarin, that is, the EDG at the C-7 position of coumarins increases the quantum yield of coumarin derivatives. More importantly, the combinatorial approach revealed that photophysical properties of triazolylcoumarin derivatives are affected not only by the substituents on coumarin skeleton itself, observed in above-mentioned series of studies, but also by substituents on the triazolyl moiety. It is not an easy task to figure out the structure–photophysical property relationship (SPR) in triazolylcou-marin derivatives, because it is difficult to compare the structure and electronic states of various substituents on triazolyl ring (see Fig. 5). In fact, there is no direct correlation between structures and photophysical properties in triazolylcoumarin library. However, this triazolylcoumarin library constructed in combinatorial fash-ion successfully achieved a large diversity in photophysical properties without established structure–photophysical property relationships and without the rational design with complicated calculation. Therefore, this study successfully demon-strated that the systematic construction with maximized diversity using combinato-rial approach can provide a powerful tool to access a novel small molecule with desired physical, biological, and photophysical properties, which might be very difficult to be predicted or designed by the rational target-based approach.

3 Xanthene Dyes

Xanthene, from *xanthos*, meaning yellow in Greek, is a yellow organic heterocy-clic compound. Xanthene derivatives are commonly referred to collectively as xanthenes. Although xanthenes have been used as fungicidal, antiviral, antibac-terial, and antiinflammatory agents [46, 47] due to their biological activities, they are the most well-known and one of the most important class of fluorescent dyes including fluorescein, eosins, and rhodamins (see Fig. 6). Xanthene dyes have been employed in various fields as following: (1) tracing agents in water pollution and aerial pesticide spraying studies; (2) colorant agents in drugs, cosmetics, textiles and inks; (3) laser dyes [48, 49]. Especially, fluorescein has been used extensively as a diagnostic tool in the field of ophthalmology and optometry, where topical fluorescein is used in the diagnosis of corneal abrasions, corneal ulcers, and hepatic corneal infections. Fluorescein is a highly fluorescent

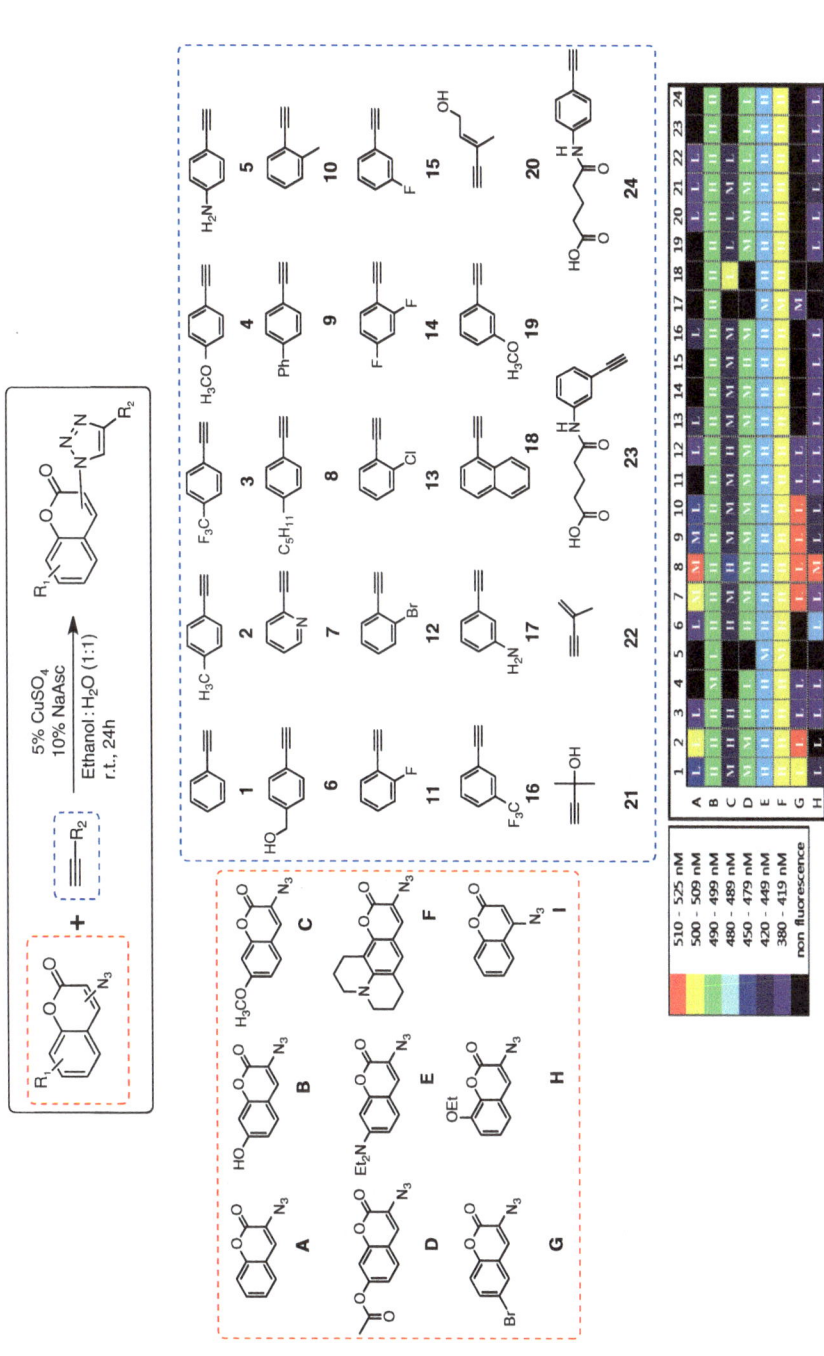

Fig. 5 Combinatorial synthesis of triazolylcoumarin library using Cu(I)-catalyzed 1,3-dipolar cycloaddition reaction. Each letter represents the relative fluorescence intensities; *H* High, *M* Middle, *L* Low. The excitation and emission spectra were recorded in DMF. Reproduced with permission from reference [45]

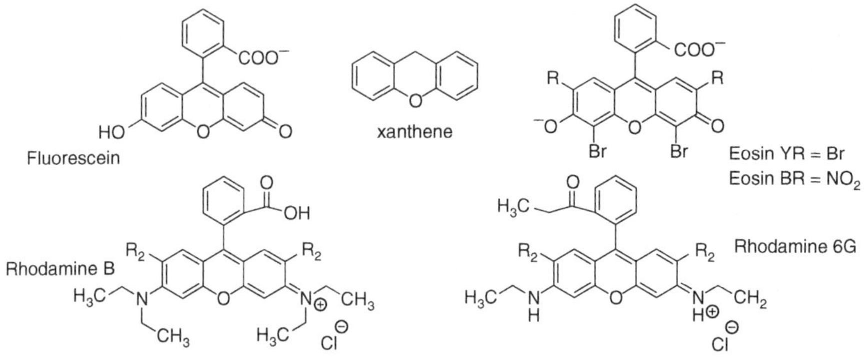

Fig. 6 General structure of xanthene dyes containing fluorescein, eosin, and rhodamine

molecule that absorbs light at 492 nm and emits 517 nm in water with a quantum yield of 0.92 at pH>8 [1].

3.1 Xanthene-Based Targeted Approach

Owing to its favorable photophysical properties, such as high molar absorbance ($\varepsilon = 9.3 \times 10^4 M^{-1}$ cm^{-1}), good water solubility, and excellent quantum yield ($\phi_F = 0.95$), fluorescein is one of the most common labeling agents used in biological applications [50].

Since its first discovery in 1871, the preparation of fluorescein derivatives has been generally pursued by the condensation of resorcinol derivatives with phthalic anhydride in the presence of zinc chloride as a catalyst via Friedel–Crafts reaction (see Fig. 7a). Although Friedel–Crafts-type synthetic methodology allows the access to fluorescein derivatives, the relatively low yields and the formation of regioisomeric mixtures are the weak point of this synthetic procedure. To overcome the limitation of the traditional Fiedel–Crafts-type procedure, the recently developed synthetic method using Grignard reaction between xanthone and substituted bromobenzenes has been considered as an alternative method for the preparation of fluorescein and its derivatives (Fig. 7b) [51, 52]. Through this procedure, a single regioisomer of fluorescein derivatives can be achieved in high yields.

One of the interesting features in the structure–photophysical property relationship of fluorescein is that the quantum yield of fluorescein increases under the basic condition. Therefore, many of fluorescein derivatives have been used as pH sensors to measure intracellular pH due to their pH-responding photophysical property [53]. Although fluorescein itself is slightly fluorescent in alcoholic solutions, the addition of alkali (pH > 8) to the fluorescein solution produces the very intense fluorescent alkali salt. The salt form of fluorescein

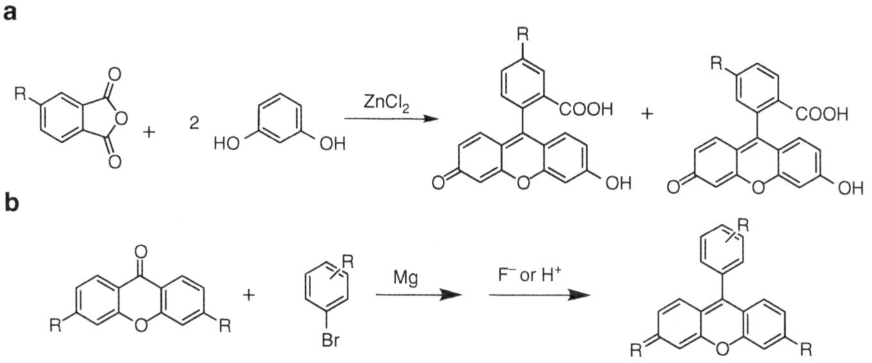

Fig. 7 General synthetic strategy for fluorescein derivatives

is readily soluble in water and exhibits the well-known yellowish green fluorescence – the characteristics of negative fluorescein ion [54]. In alkali condition, fluorescein is converted to dianionic form (**8c**) [55, 56]. Among the two anionic parts, the phenolate part of xanthenes ring is responsible for its intense fluorescence in alkali condition (see Fig. 8a). Another key feature of structure–photophysical property relationship in fluorescein is that the quantum yield of fluorescein is dominantly controlled by substituents on the external phenyl moiety of xanthene ring (see Fig. 8b). A plausible mechanism regarding the fluctuation on quantum yields (ϕ_F) of fluorescein derivatives has been suggested to be the photoinduced electron-transfer (PET) between benzoic acid moiety of fluorescein (EDG) and the singlet excited state of xanthenes moiety [51, 57, 58]. PET is a generally accepted mechanism for the fluorescence quenching, in which the photoinduced electron-transfer from the PET donor to the excited fluorophores diminishes the fluorescence. In fact, there are only small differences observed in the absorbance among fluorescein derivatives. And their dihedral angle between benzoic acid moiety and fluorogenic xanthene ring is almost 90°, as found by X-ray analysis [59]. Both observations suggested that benzoic acid and xanthenes moieties in fluorescein are orthogonal to each other and there is no ground-state interaction between them. The deletion of carboxylic group at the *ortho* position, fluorine (**8d**), gives significantly decreased quantum yield. However, methyl substitution (**8b**) at the *ortho* position of external phenyl ring in fluorescein gives the identical quantum yield of fluorescein without any changes in other photophysical properties (namely, Tokyo green, **8b**). Therefore, the preservation of fluorescent quantum yields can be achieved by the maintenance of dihedral angle between external phenyl moieties and fluorogenic xanthenes, and the bulky residues at the *ortho* position of external phenyl moiety can prevent the rotation of phenyl group, which leads to the increase of quantum yields. Moreover, the oxidation potentials of external phenyl substituents and the HOMO energy levels under basic and acidic conditions are closely related to the quantum yield (ϕ_F) of corresponding

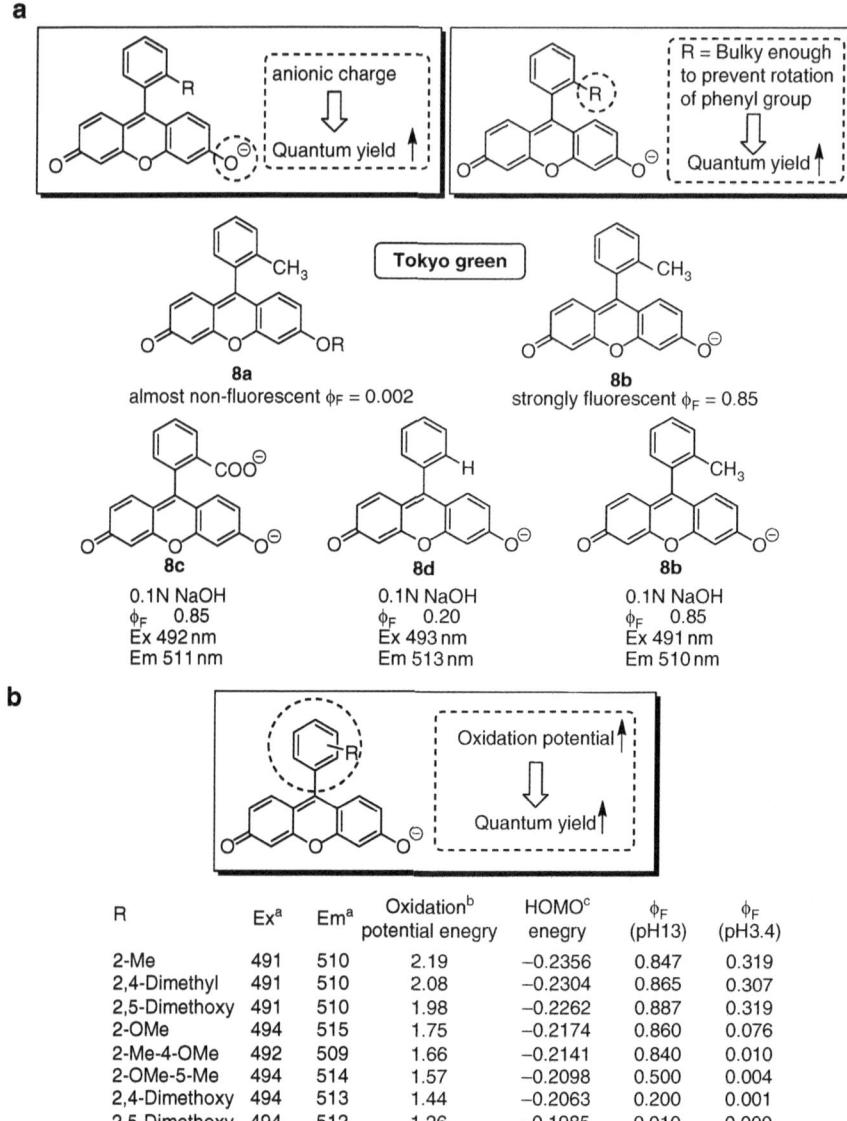

Fig. 8 Structure–photophysical properties relationship of fluorescein derivatives. [a]Measured in 0.1 N NaOH(aq). [b]Oxidation potential of corresponding benzene moiety, obtained in acetonitrile containing 0.1 M TBAP. [c]HOMO energy level of the corresponding benzene moiety, calculated with B3LYP/6-31G(d)//B3LYP/6-31G(d) by Gaussian 98 W

R	Ex[a]	Em[a]	Oxidation[b] potential enegry	HOMO[c] enegry	ϕ_F (pH13)	ϕ_F (pH3.4)
2-Me	491	510	2.19	−0.2356	0.847	0.319
2,4-Dimethyl	491	510	2.08	−0.2304	0.865	0.307
2,5-Dimethoxy	491	510	1.98	−0.2262	0.887	0.319
2-OMe	494	515	1.75	−0.2174	0.860	0.076
2-Me-4-OMe	492	509	1.66	−0.2141	0.840	0.010
2-OMe-5-Me	494	514	1.57	−0.2098	0.500	0.004
2,4-Dimethoxy	494	513	1.44	−0.2063	0.200	0.001
2,5-Dimethoxy	494	512	1.26	−0.1985	0.010	0.000

fluorescein derivatives (Fig. 8b). Based on this observation, it is reasonable to claim that external phenyl moiety can be the electron donor and fluorogenic xanthene moiety can be the electron acceptor in the photoinduced electron-transfer;

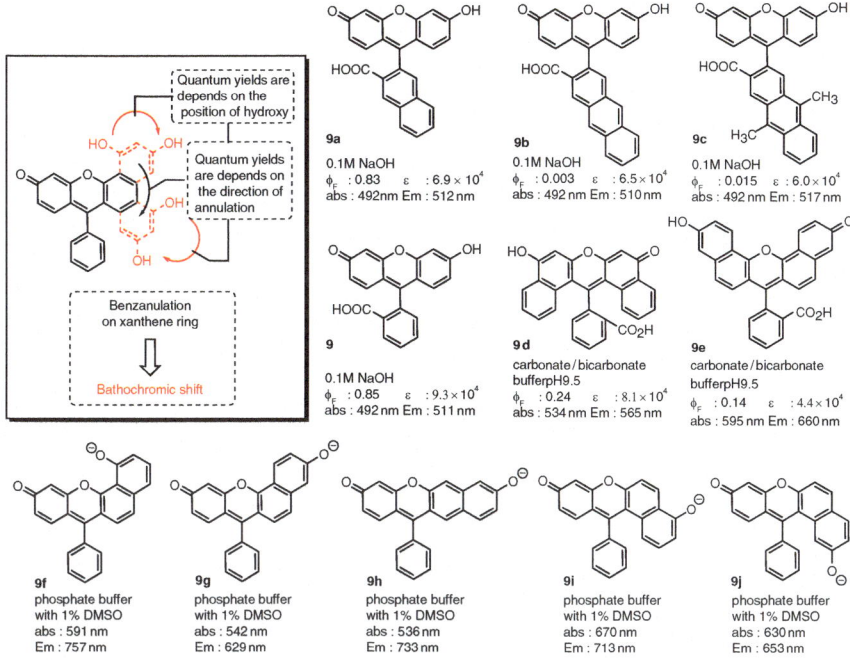

Fig. 9 Structure–photophysical properties relationship of benzannulated fluorescein derivatives

therefore, quantum yields of fluorescein derivatives can be tuned by controlling the oxidation potentials of substituents on phenyl moieties (Fig. 9).

Annulation of aryl moieties to a fluorescent core skeleton is one of the most effective strategies that have been utilized extensively to tune or red-shift the emission wavelength maximum through the extension of the conjugation lengths in various dye architectures [43, 60–62]. However, the extension of conjugation on phenyl moiety of fluorescein was not associated with the increase of their emission wavelengths (compounds **9a**, **9b**, and **9c**) compared to parent fluorescein (**9**) because of the orthogonality of external phenyl moieties with fluorogenic xanthene moiety [57]. On the other hand, the benzannulation on xanthene moiety – the fluorophore of fluorescein – can significantly perturb the emission maximum [63–66]. For example, benzannulated xanthene moieties, called as seminaphthofluoresceins (SNAFLs, **9d** and **9e**) and seminaphthofluorones (SNAFRs, **9f–9j**), generally have a bathochromic shift of emission wavelength. Further systematic investigation might be needed to rationalize the correlation of their photophysical properties with electronic states of the benzanulated architectural motif. It has been demonstrated that the direction of annulations as well as the location of hydroxyl group can influence the emission maximum and this annulations approach allows the discovery of near-IR fluorescent dyes (**9f**).

3.2   Xanthene-Based Combinatorial Approach

The rosamine library is one of excellent examples in the combinatorial construction of fluorescent compound library [52]. Chang and co-workers envisioned that rosamine dye without 2′-carboxylic group on the external phenyl moiety can be flexible enough to respond to the environmental changes through the specific interaction with various biological analytes. The absence of steric restrictions at the *ortho* position allows the free rotation of external phenyl ring, which might perturb the fluorescent quantum yields and emission properties of fluorogenic xanthenes core. Therefore, these series of rosamine fluorescent dyes could be a good candidate as a fluorescent probe, compared to rhodamine dyes containing nonrotatable bond between external phenyl moiety and xanthenes core. Starting from three different xanthone derivatives, they pursued the construction of rosamine library using solid-phase parallel synthesis. First, they synthesized twelve different nonsymmetrical xanthone derivatives through the modification of 3-amino group. Each xanthone was immobilized on the 2-chloro-trityl chloride polystyrene resins, followed by the second diversification with various Grignard reagents on solid supports. As a result, they constructed the 240-membered rosamine library having a wide range of emission wavelength (from 530 to 605 nm) and quantum yield (from 0.00,025 to 0.89). After the construction of rosamine library, they evaluated those library members as fluorescent probes against various analytes in the biological system. Interestingly, one of rosamine dyes exhibited a highly selective response toward reduced glutathione (GSH) – endogenous reducing agent in cellular system – compared to other biological analytes under the physiological conditions. Then, they successfully demonstrated the application of this fluorescent sensor to monitor GSH in living cells. A different fluorescent compound from this rosamine library was also identified as a fluorescent probe, which can distinguish the cell state between C2C12 myotubes and C2C12 myoblasts (Fig. 10) [67].

4   BODIPY (4,4-difluoro-4-bora-3a,4a-diaza-*s*-indacene) Dyes

Fluorophores containing 4,4-difluoro-4-bora-3a,4a-diaza-*s*-indacene as a core skeleton are commonly designated as BODIPY fluorophores. Due to their useful photophysical properties including high fluorescence quantum yields, high molar absorption coefficient, narrow absorption and emission band width, and their high photostability [50], BODIPY dyes are proven to be extremely versatile and useful in many biological applications Fig. 11 [68].

4.1   BODIPY-Based Targeted Approach

The first BODIPY fluorophore was synthesized by Treibs and Kreuzer in 1968 [69]. As described in recent review article [70], BODIPY dyes can be categorized into

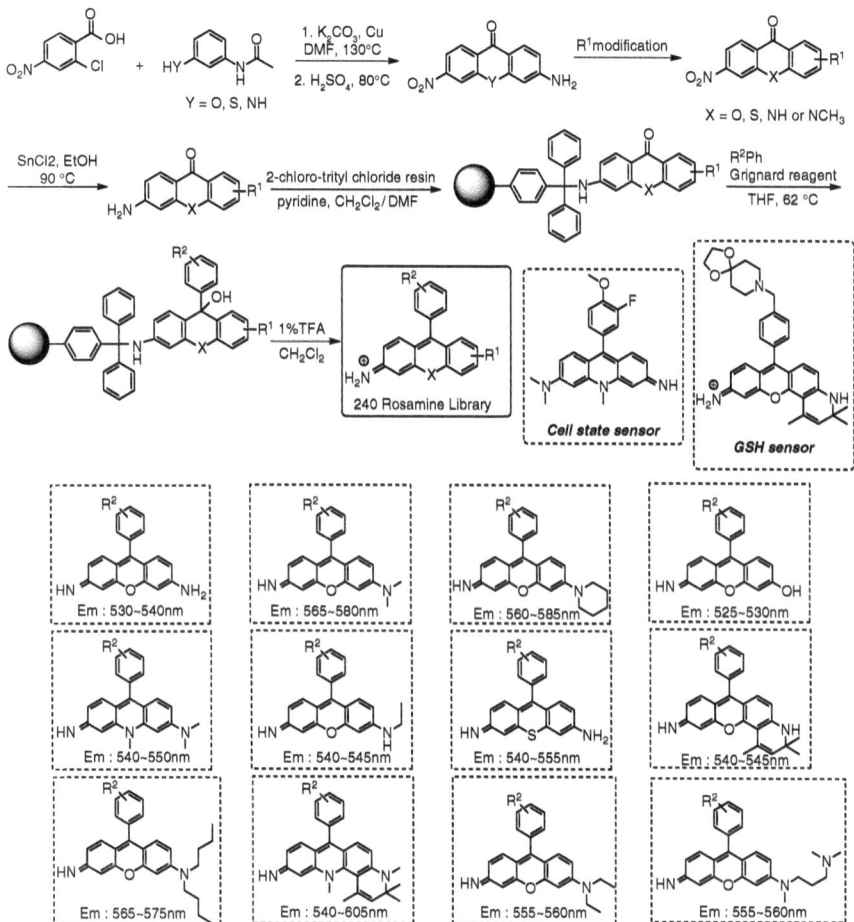

Fig. 10 Synthetic scheme for the construction of rosamine library using solid-phase combinatorial approach. Average emission values of fluorescent compounds derived from each rosamine moiety are presented below the structure of rosamine moieties. (All the emission wavelengths were obtained in PBS buffer)

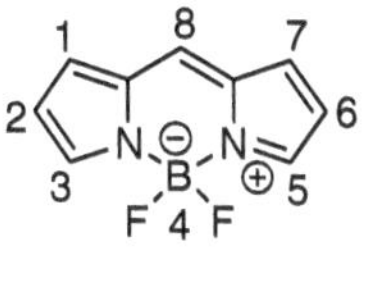

Fig. 11 General structure of BODIPY

two broad types: symmetrically substituted and nonsymmetrically substituted BODIPY dyes (see Fig. 12).

Symmetrically substituted BODIPY dyes are relatively easy to be synthesized via the condensation of pyrroles with carbonyl electrophiles, such as acyl chlorides,

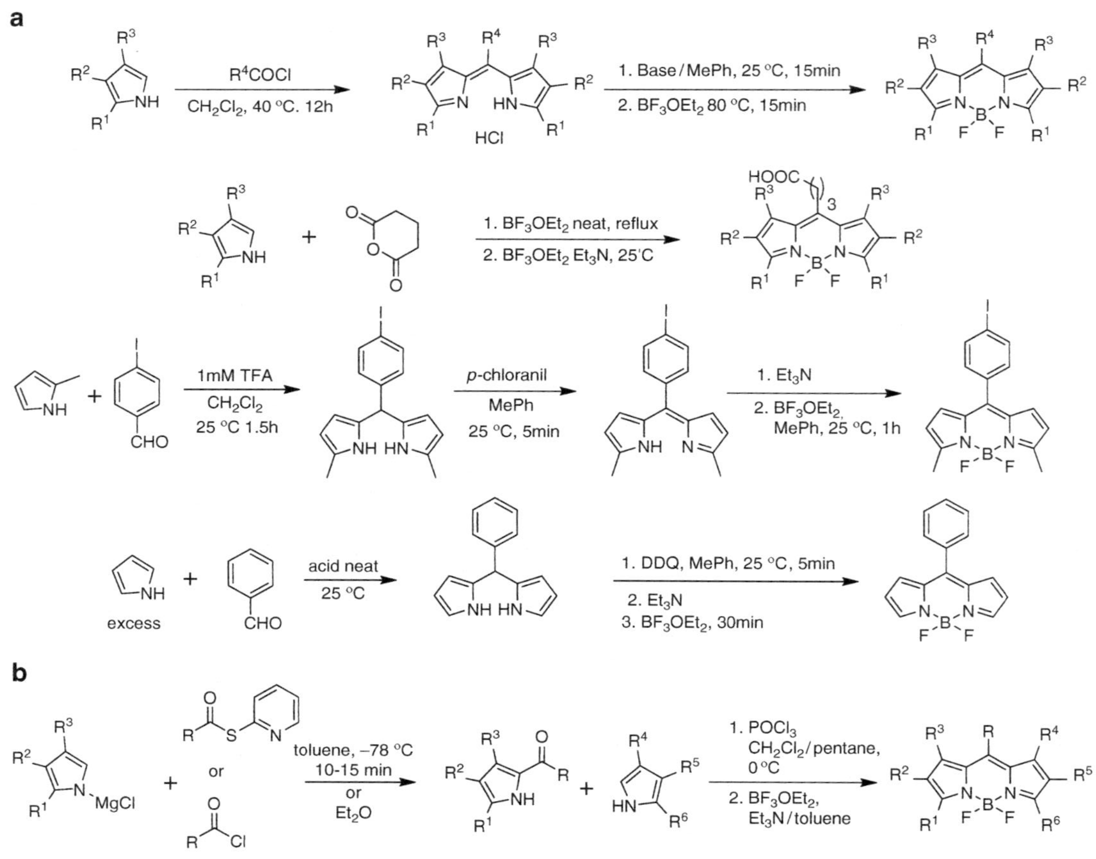

Fig. 12 Synthetic scheme for symmetrically substituted BODIPY dyes (**a**) and unsymmetrically substituted BODIPY dyes (**b**)

activated carboxylic acid derivatives, or alkyl/aryl aldehydes. The resulting unstable dipyrromethene can proceed to the next reaction for boron complexation using $BF_3 \cdot OEt_2$ without further purification, which lead to the formation of symmetrical BODIPY dyes. In the case of using aromatic aldehydes as carbonyl electrophiles, the further oxidation step either with p-chloranil or with 2,3-dichloro-5,6-dicyano-benzoquinone (DDQ) is needed to yield dipyrromethene intermediate for the subsequent complexation with boron trifluoride to acquire final BODIPY dyes (see Fig. 12a). On the other hand, unsymmetrically substituted BODIPY dyes can be obtained via the preparations of ketopyrrole intermediate using acyl chlorides or S-pyridin-2-yl arylthioate, followed by the Lewis-acid-mediated condensation of ketopyrrole intermediates with another pyrrole fragment in the presence of phosphoryl chloride ($POCl_3$). This series of chemical transformation can yield dipyrromethene intermediates for the subsequent complexation with boron trifluoride to acquire final nonsymmetrical BODIPY dyes (see Fig. 12b). The BODIPY core structure is robust enough to withstand a range of chemical transformations; therefore, a variety of BODIPY dyes has been synthesized via various direct chemical transformations of BODIPY core itself, using electrophilic reactions, nucleophilic aromatic substitution on halogenated BODIPYs, palladium-mediated Heck-type C-H functionalization, palladium-catalyzed cross-coupling reactions, and so on [70]. By virtue of recent review article, the structure–photophysical property relationships of BODIPY dyes were summarized with a focus on the position of its substituents. The substituents at the C-1 and C-7 position are important for the enhancement of quantum yields. The methyl substituent at the C-1 and C-7 positions of BODIPY dyes (**13d** and **13b**) increase quantum yields (ϕ_F) from 0.19 and 0.38 to 0.68 and 0.72, respectively, compared to parent BODIPY dyes (**13c** and **13a**) (see Fig. 13). The introduction of *ortho*-substituents on the external phenyl ring at the C-8 position (compound **13f**) also increases quantum yields of BODIPY dyes. This phenomenon is due to preventing the free rotation of external phenyl group by the introduction of steric elements either at the C-1 and C-7 position of BODIPY core skeleton or at the *ortho* position on the external phenyl rings, which lead to the reduction in the energy loss from the excited states via nonirradiative molecular motions. Thus, it was concluded that the restrictions on the free rotation of external phenyl moiety at the C-8 position can increase quantum yields of BODIPY dyes.

On the other hand, the introduction of halide substituents at the C-2 and C-6 position decreases fluorescence quantum yields and gives a bathochromic shift of emission maxima. For example, bromine at the C-2 and C-6 position in compound **14b** deteriorates fluorescence quantum yields from 0.95 (**14a**) to 0.45 and the emission maximum is red-shifted by 42 nm. Moreover, iodine at the C-2,6 position in compound **14d** gives the similar bathochromic shift to bromine (**14b**, 44 nm) and more dramatic reduction in quantum yields (almost nonfluorescent, $\phi_F = 0.02$). These changes in photophysical properties were interpreted as the heavy atom effect of halides on a BODIPY core skeleton. The bathochromic shift of BODIPY dyes without dramatic decrease in quantum yield was observed by the introduction of vinyl substituents at the C-2 and C-6 position. The extension of conjugation

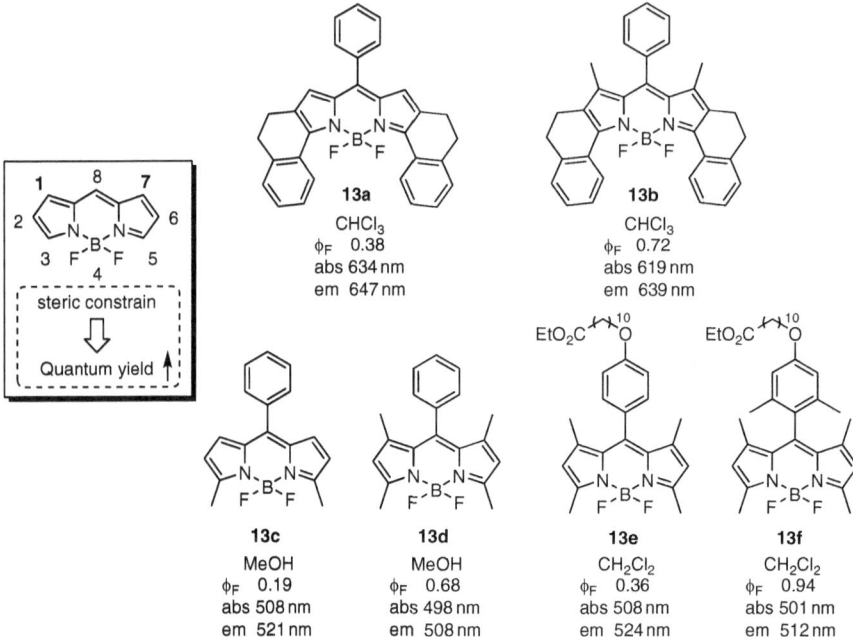

Fig. 13 Structure–photophysical property relationship at the C-1 and C-7 position of BODIPY derivatives

length at the C-2 and/or C-6 position with vinyl moieties containing electron withdrawing groups (EWG) triggers bathochromic shifts from 504 nm (compound **14a**) to 549, 551, and 580 nm (compound **14h**, **14i**, and **14j**, respectively), depending on its EWG or substitution pattern (mono- or disubstituted) (Fig. 14).

Although disubstituted BODIPY derivatives at the C-2 and C-6 position demonstrated decreased quantum yields, vinyl substitution at the C-2/C-6 position brings a bathochromic shift without significant decrease in quantum yield, which was observed in the case of halide substitution. In comparison, the direct substitution at the C-2 and C-6 position with EWG, such as sulfonate (**14f**) or nitrate group (**14g**), did not cause a bathochromic shift, as compared to ethyl-substituted BODIPY derivatives (compound **14e**). Therefore, the extension of conjugation length by the introduction of vinyl group at the C-2 and C-6 position is more important than the introduction of EWG for the bathochromic shift of BODIPY system. Interestingly, the increased quantum yield without significant red-shifts of emission wavelength was observed when alkene moiety was introduced with methylene spacer (compound **14l–14o**). In this case, the enhanced quantum yield is not associated with the direct extension of conjugation length.

The C-3 and C-5 position of BODIPY core skeleton is one of the most important points for the substitution to tune photophysical properties, including absorption and/or emission maxima as well as fluorescence quantum yields. First of all, the introduction of amine-based (**15d–15g**) or sulfur-based (**15b–15c**) nucleophiles can

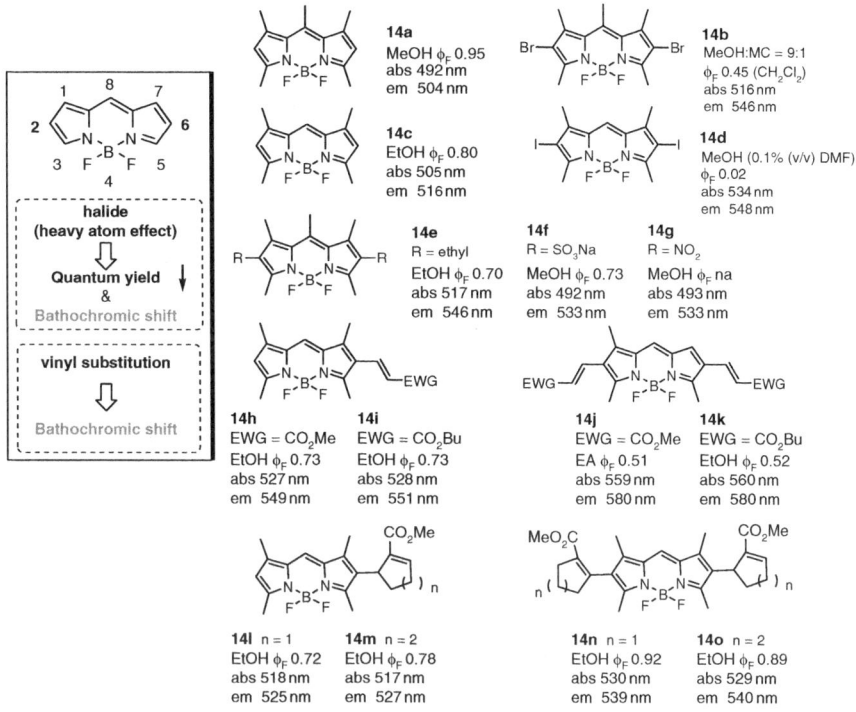

Fig. 14 Structure–photophysical property relationship at the C-2 and C-6 position of BODIPY derivatives

lead to a significant bathochromic shift (red shift) of both absorption and emission maxima along with reduced quantum yields compared to unmodified BODIPY derivatives (**15a**). The introduction of aryl substituents at the C-3 and C-5 position also caused bathochromic shifts with reduced quantum yields. For instance, phenyl (**15h**) or naphthyl (**15i**) disubstituents at the C-3 and C-5 position of BODIPY core skeletons brings a bathochromic shift around 80–90 nm compared to chlorine-substituted BODIPY dye (**15a**). Their photophysical properties were also perturbed depending on the electric nature of substituents as well as the substitution position on aryl group at the C-3 and C-5 position of BODIPY (**15j** derivatives). Their red-shifted emission maximum and reduced quantum yields are due to the extension of their conjugation length through substituted aryl groups and nonradiative loss of energy via rotation around the C–Ar bonds, respectively. Consistent with this rationalization, the extension of their conjugation through rigid bond, such as ethenyl- or ethynyl linker, provides the highest quantum yield with dramatic bathochromic shift (compound **15l**: ϕ_F 0.92, Em 639 nm; compound **15n**: ϕ_F 1.00, Em 622 nm) (Fig 15).

Compared to the introduction of various substituents at the BODIPY core skeleton, the extension of conjugation length can be achieved by the fusion of aryl group via furan or (hetero)carbocycles. As shown in Fig. 16, the furan-based

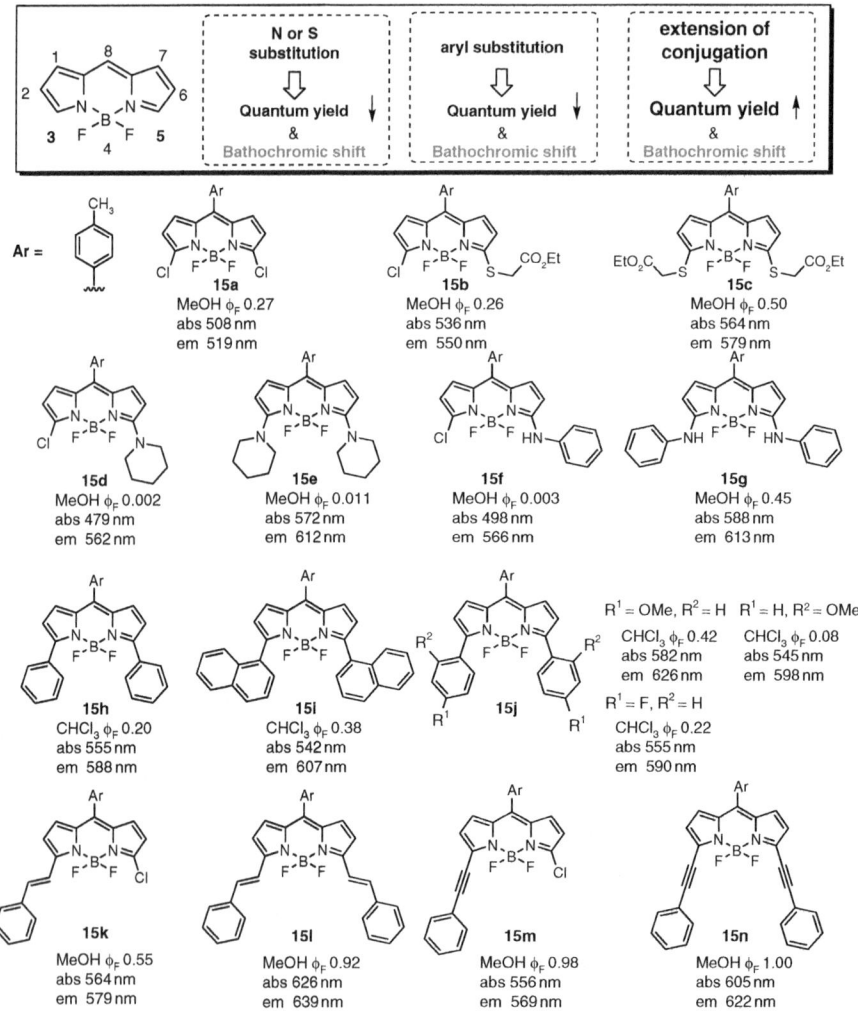

Fig. 15 Structure–photophysical property relationship at the C-3 and C-5 position of BODIPY derivatives

fusion of aryl-ring on a BODIPY core skeleton causes a bathochromic shift and enhanced molar absorbance. In fact, many research groups have been used aryl-ring fusion approach for the extension of its conjugation length with structural rigidity to induce bathochromic shifts of emission maxima in given fluorophores [43, 60–62, 66, 71, 72]. On the basis of this concept, furan-based ring fusion at the C-2/C-3 and C-5/C-6 positions significantly affects the photophysical property and causes bathochromic shifts in both absorption and emission maxima compared with alkyl-substituted counter parts. These types of structural modification often lead to dramatically enhanced fluorescence quantum yields and molar absorbance [73, 74].

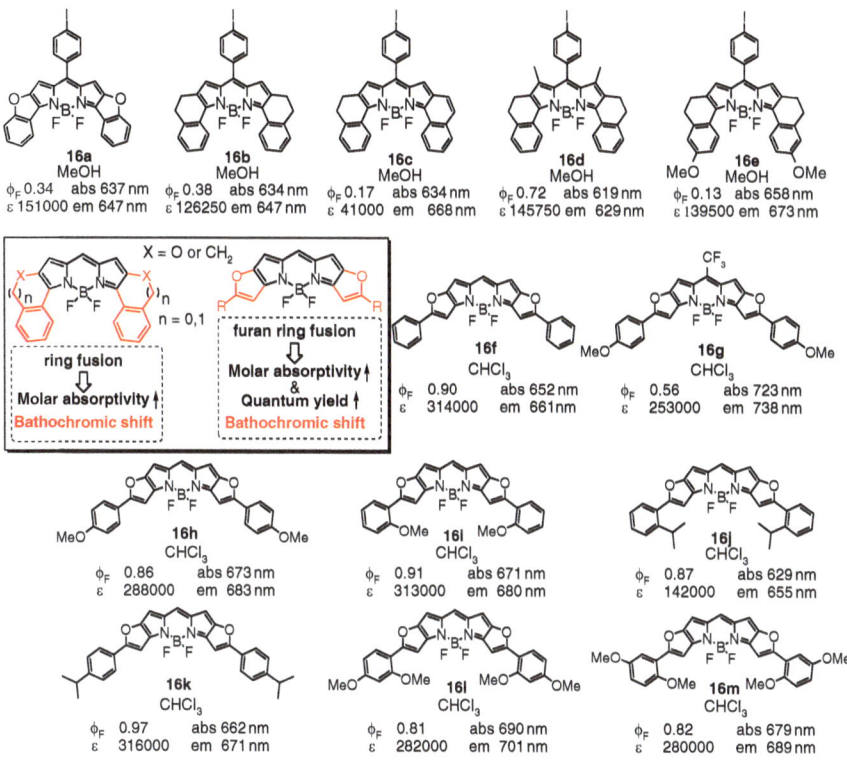

Fig. 16 Structural transformation through the introduction of aryl moiety at the C-8 position or through the extension of conjugation length via fused heteroaryl moiety on BODIPY core skeletons

Most of the furan-ring-fused BODIPY dyes exhibit the significantly high molar absorbance (ε from 2.53×10^5 to $3.16 \times 10^5 \mathrm{M}^{-1}\,\mathrm{cm}^{-1}$). For example, the molar absorbance of compound **16f** is about 4-times higher than tetramethyl substituted BODIPY dye (**14c**). More interestingly, furan-ring-fused BODIPY derivatives have high quantum yields ϕ_F (ranged from 0.81 to 0.97) and long emission wavelength (greater than 650 nm). Because of their excellent molar absorptivities, high quantum yields, and brightness [75], furan-ring-fused BODIPY dyes are significantly brighter than current commercial BODIPY dyes. In addition, the extension of conjugation wavelength using furan-ring fusion became a powerful method for the development of Vis/NIR-fluorescent dyes (**16g, 16l**).

4.2 BODIPY-Based Combinatorial Approach

It is quite difficult to predict or design a novel small molecule with desired photo-physical properties because a direct correlation between structures and photophysical properties is limited or not even observed. In addition, there is no report regarding the precise algorithm in computational or theoretical chemistry for the prediction of photophysical properties including emission wavelengths and quantum yields. Therefore, the systematic construction with maximized diversity using combinatorial approach can provide a powerful tool to access a novel small molecule with desired photophysical property, which might be very difficult to be predicted or designed by the rational target-based approach. Along with this research trends using combinatorial chemistry, the BODIPY-based fluorescent small molecule library was designed and synthesized by Chang and co-workers [76].

It is known that the methyl substituents at the C-3 and C-5 position on BODIPY core skeleton are acidic enough to participate as nucleophiles in Knoevenagel reaction [70], and they systematically pursued the Knoevenagel condensation of 1,3-dimethyl BODIPYs with aromatic aldehydes under the microwave irradiation in basic condition to append different styryl motifs to the BODIPY fluorophore. In order to make styryl-derivatized BODIPY library without complex purification steps, they designed unsymmetrically substituted BODIPY and synthesized a total of 160-membered monostyryl-containing BODIPY library (named as BD library) with emission wavelengths ranging from 560 to 730 nm, and the average quantum yield of 0.32. A notable bathochromic shift was observed when styryl moiety with amino substituents was introduced at the C-3 position of BODIPY core skeleton (from 710 to 730 nm). This bathochromic shift with lower quantum yield was rationalized by the photoinduced electron-transfer (PET) effect from amino substituents to the styryl moiety. As shown in Fig. 17, the resulting member of BODIPY-based combinatorial library has a wide coverage of emission wavelength and brightness ($\phi_F \times \varepsilon$). They also demonstrated the successful application of one of BODIPY dyes as a selective fluorescent sensor for cellular glucagon, confirmed in various cell lines.

5   Cyanine (Polymethine) Dyes

Polymethine fluorescent compounds can be generally categorized into two groups based on their structure; polymethine and meropolymethine. Donor–acceptor-substituted polymethines, containing both nitrogen and oxygen as a donor–acceptor pair, are categorized into meropolymethine dyes. On the other hand, Donor–acceptor-substituted polymethines containing same heteroatom (either nitrogen or oxygen) at their terminals are categorized into polymethine dyes [77]. As shown in Fig. 18, merocyanine dyes have both nonpolar and zwitterionic resonance forms, and the spectral properties of merocyanines are very sensitive to the solvent polarity. In comparison, polymethine dyes are charged molecules, and their spectral properties are not significantly influenced by the solvent polarity [78]. In fact, cyanine and merocyanine dyes have been extensively used for fluorescence-labeling of protein, nucleic acid, and many other-of-interests in various biological applications

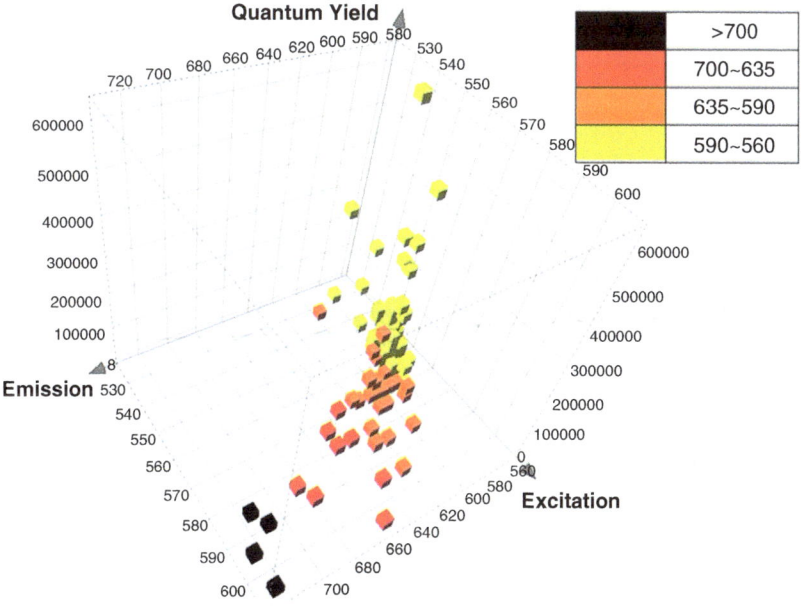

(a) POCl₃, DCM, −5 °C for 3 h, then 3 h at rt; (b) BF₃ OEt₂, DIPEA, rt for 3h
(overall yeild : 51%); (c) R-CHO (160 aromatic aldehydes), pyrrolidine, AcOH,
EtOH, microwave irradiation, 2–15 min

Fig. 17 Synthetic schemes of BODIPY-based combinatorial library and 3D scatter plot of synthesized library members according to excitation wavelength, emission maxima, and brightness ($\phi_F \times \varepsilon$). All the photophysical properties were measured in MeOH

because of their excellent photophysical properties, including superb fluorescence brightness (high quantum yields and molar absorbances), photostability, and long emission wavelengths without overlapping the cellular autofluorescence [79].

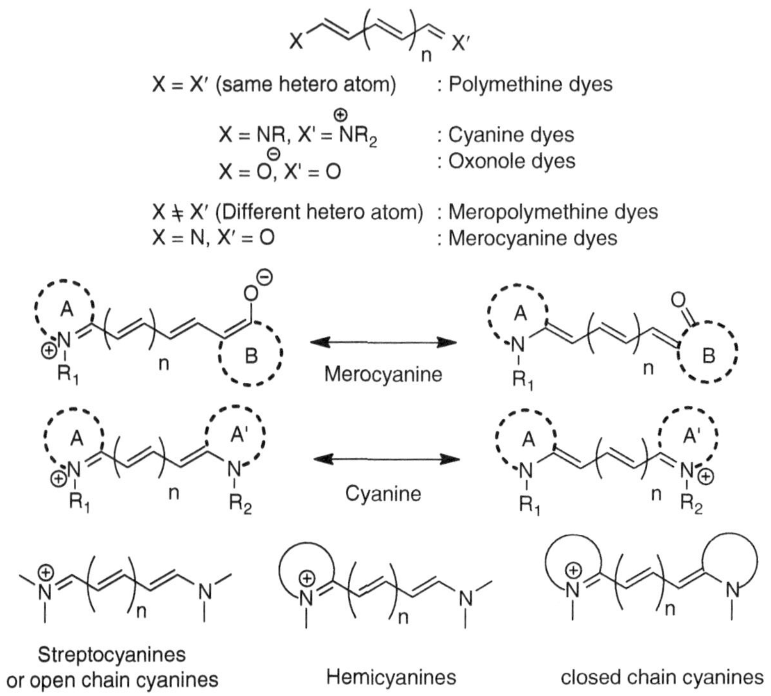

X = X′ (same hetero atom)   : Polymethine dyes

$$X = NR, X′ = \overset{\oplus}{N}R_2$$: Cyanine dyes
$$X = \overset{\ominus}{O}, X′ = O$$: Oxonole dyes

X ≠ X′ (Different hetero atom) : Meropolymethine dyes
X = N, X′ = O   : Merocyanine dyes

Merocyanine

Cyanine

Streptocyanines
or open chain cyanines

Hemicyanines

closed chain cyanines

Fig. 18 General structure of cyanine (polymethine) dyes

5.1   Cyanine-Based Targeted Approach

Among polymethine dyes, cyanine derivatives historically evoked the huge interest primarily because of their unrivaled ability to impart light sensitivity to silver halide emulsions in a region of the spectrum to which the silver halide is normally not sensitive [80]. Etymology of "cyanine" comes from the Greek word *Kyanos*, dark blue, because of the brilliant blue color in its solution. Cyanine is a nonsystematic name of a synthetic polymethine dye family. Polymethine/cyanine dyes have dominated the field of photography and other sophisticated arenas of dye application since 1856, and their numerous applications in various research areas have been reported in the literature every year [80]. Recently, cyanine dyes started to be used as fluorescent probes for biomolecular labeling, including genetic analysis, DNA sequencing, in vivo imaging, proteomics, and so on [81]. Depending on their structure, there are three types of cyanines: (1) streptocyanines (open chain cyanines); (2) hemicyanines; and (3) closed chain cyanines (see Fig. 18). The first synthetic endeavor was pursued by Williams to obtain the "corn fluor" blue cyanine dye in 1856 by treating impure quinoline (which contained some 4-methylquinoline) with amyl iodide followed by the treatment with ammonia [80]. A series of papers have been reported about the development of synthetic procedures for cyanine dyes [80]. A general method for the synthesis of stilbazolium dyes (hemicyanines) is the condensation reaction of heterocyclic bases (Fischer's base) containing activated methyl group with unsaturated bisaldehydes or their equivalents (Fig. 19a).

The facile synthetic method for heptamethine cyanines was also reported through the direct condensation of N-alkyl-substituted or 2,3,3-trimethylbenzindole with 2-chloro-1-formyl-3-(hydroxy methylene)cyclohex-1-ene under the refluxing condition in a solvent mixture of butanol and benzene (7:3) without any catalyst (Fig. 19b) [82]. This procedure is reproducible and robust enough to be utilized for the scale-up synthesis of both symmetric and nonsymmetric dyes. This methodology is also useful for the synthesis of cyanine intermediates, which can be easily derivatized for applications in fluorescent labeling of biopolymers or biologically important small molecules. The cyanine dyes have been utilized as a classic example of an organic fluorophore with a tunable emission wavelength for the development of multicolor fluorescent probes because their photophysical properties have a significant correlation with their structures (see Fig. 20) [75, 77, 83]. Through the alteration of heterocyclic nuclei and the extension of conjugation length in the polymethine chain, cyanine dyes can be tuned in their emission wavelengths. For instance, the extension of conjugation length by adding one vinylene unit in the cyanine bridge results in a bathochromic shift of 100 nm each (**20a** vs. **20b**, **20c** vs. **20d**, **20e–20 g**, and **20 h** vs. **20i**) [84, 85]. The annulations of benzene ring at the C-4 and C-5 position of indolenine nuclei is another option to tune the emission maximum of cyanine dyes, and the benzannulations on cyanines roughly causes 30–40 nm of bathochromic shift: red shift of emission wavelength from 660 nm (**20a**) to 693 nm (**20c**) and from 760 nm (**20b**) to 830 nm (**20d**).

The tunability on emission wavelength of cyanine derivatives is based on the understanding of structure–photophysical property relationships, which allows the development of near-IR fluorophores [80, 84–87]. Enhancement of the rigidity in

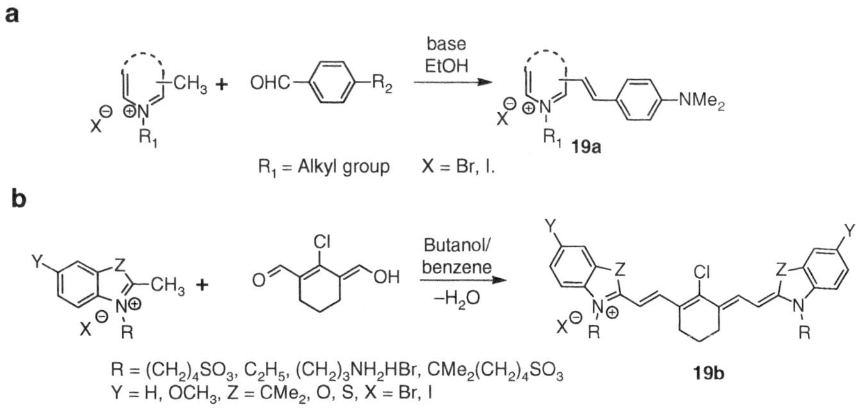

Fig. 19 General synthetic scheme for cyanine dyes

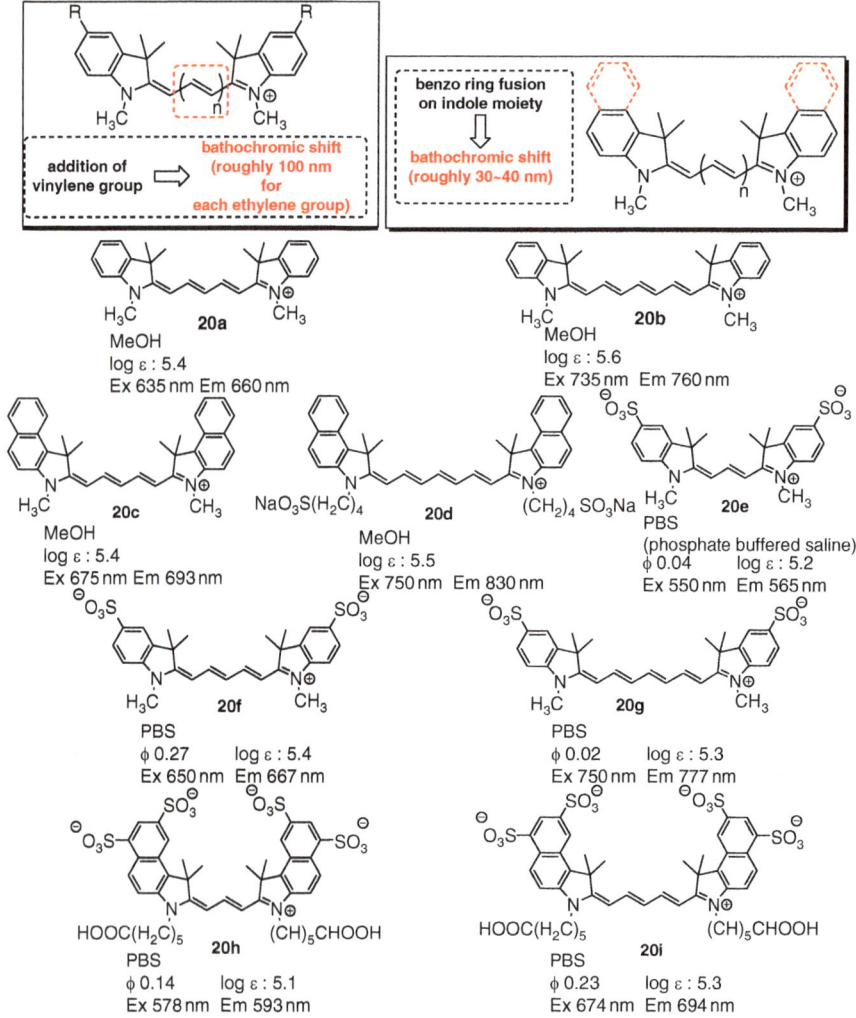

Fig. 20 Structure–photophysical property relationship of cyanine derivatives

the polymethine chain through alicyclic rings (compounds **19b**) is very effective in improving their photostability and applicability as near-IR sensitizing dyes [77].

5.2 Cyanine-Based Combinatorial Approach

Benzimidazolium is one of the typical scaffolds of the hemicyanine group, and cationic hemicyanine derivatives can be assumed as potential fluorescent sensors due to their electrostatic interactions. Based on this assumption, Chang and co-workers chose benzimidazolium motif as a key scaffold for the construction of fluorescent small molecule library Fig. 21 [88].

Starting from benzimidazolium ring containing two chloride substituents to achieve the longer wavelength of final fluorophores, they extended the conjugation length of benzimidazolium scaffold by condensation reaction with 96 aromatic aldehydes using solid-phase parallel synthesis. Due to their structural diversity, a variety of excitation/emission wavelengths was observed within this benzimidazolium library. To discover novel fluorescent sensors against biological analytes, they screened the fluorescent library members with various nucleotides and identified one of 94-membered library of benzimidazolium-based fluorophores as a specific sensor for GTP (Guanosine triphosphate) with excellent selectivity over other nucleotides. As a result, GTP sensor showed dramatically increased fluorescence (80-fold enhancement at 540 nm) upon binding with GTP, but not with other

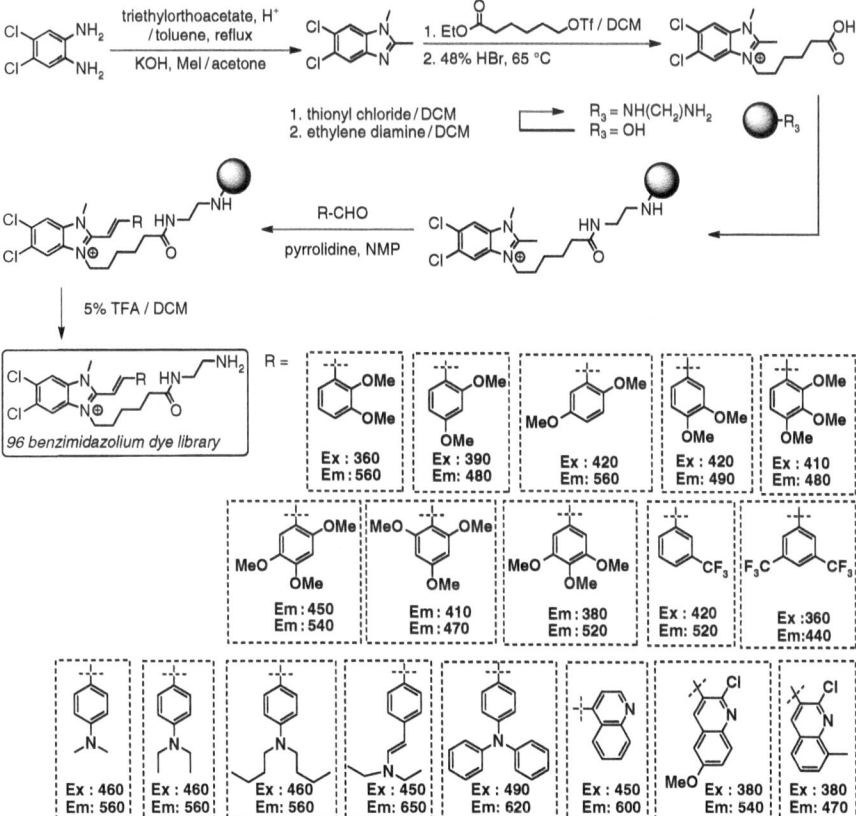

Fig. 21 Synthetic scheme of benzimidazolium scaffold-based combinatorial library and their general structures and photophysical properties of the representative examples. All the photophysical properties were measured in methanol

nucleotides. Interestingly, none of the other library members selectively causes any changes of their photophysical properties upon the treatment with GTP, which means that it might be almost impossible to develop a novel GTP fluorescent sensor using targeted synthesis guided by de novo design.

6 Combinatorial Approach for the Discovery of Novel Fluorescent Dyes

Because of the complexity of underlying photophysical phenomena of fluorescent small molecules, the rational design of fluorescent probes having desirable photophysical properties is still far from being achieved. Therefore, the rational discovery of new fluorescent probes with desired photophysical properties have been pursued through a huge amount of time-consuming effort and the most of fluorescent probes have been empirically developed with limited synthetic flexibility, one at a time. However, the birth of combinatorial chemistry in the early 1990s promised a new strategy to overcome the previous limitation, especially the difficulties in rational design [89]. Accordingly, researchers have used an empirical combinatorial approach to discover novel fluorophores as well as to improve the photophysical properties of existing fluorophores, and their fruitful outcomes provide the confidence to the scientific community that the systematic approach using combinatorial chemistry can be a powerful tool to develop novel fluorescent compounds.

One of the earliest embodiments is the combinatorial synthesis and subsequent evaluation of styryl-based fluorescent library [90]. Chang and co-workers conducted the diversification of known fluorophore, styryl scaffold [80], through a simple one-step condensation reaction of fourteen pyridinium salts with forty-one aldehydes, catalyzed by the microwave irradiation for 5 min. The constructed 276-membered fluorescent library covers a broad range of emission wavelengths, from 420 to 730 nm (see Fig. 22). Without a good deal of efforts for further purification, they directly utilized fluorescent compounds for the application in biological systems in conjunction with efficient cell-based screening and successfully demonstrated the discovery of novel fluorescent probes, which can specifically stain several organelles in various cell-lines with a wide spectrum of photophysical properties. Moreover, the unexpected dramatic changes in these properties have been achieved by simple structural changes in combinatorial fashion, which lead to the development of new DNA-sensitive dyes using extended styryl dyes (see Fig. 23) [91].

In spite of the great demand for fluorescent materials, the availability of novel molecular frameworks – critical components of new fluorescent probes – is quite limited. Recently, our research group reported the development of full-color-tunable fluorophores based on the novel core skeleton, 1,2-dihydropyrrolo[3,4-β] indolizin-3-one, using a combinatorial approach [92]. We developed a new

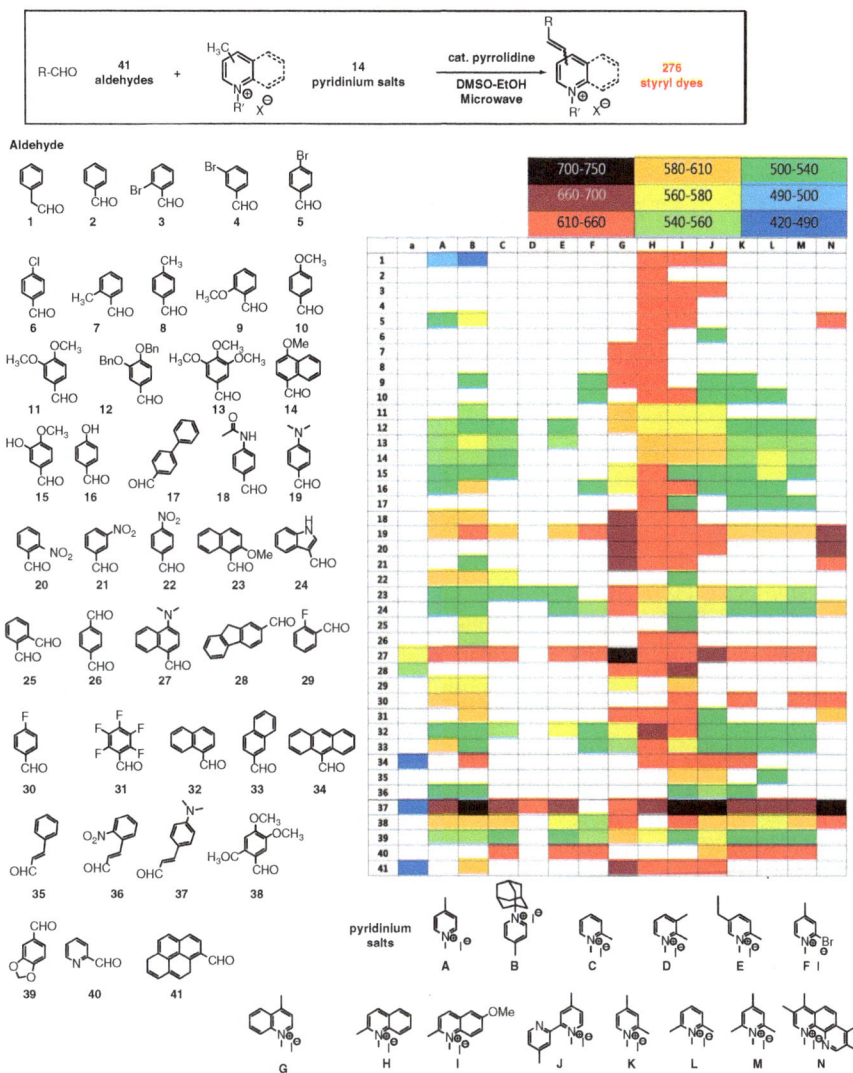

Fig. 22 Synthetic schemes styryl scaffold-based DOFLA approach and building blocks and table for photophysical properties of synthesized styryl dye derivatives. Reproduced with permission from [90]

synthetic method that was able to produce a novel fluorescent core skeleton by a complexity-generating one-pot reaction followed by oxidative aromatization using DDQ (Fig. 24).

After the integration of this synthetic method into a fully compatible protocol, the positions for electronic changes with various substituents were rationally predicted through the deliberation of computational studies; the R_1 position of the

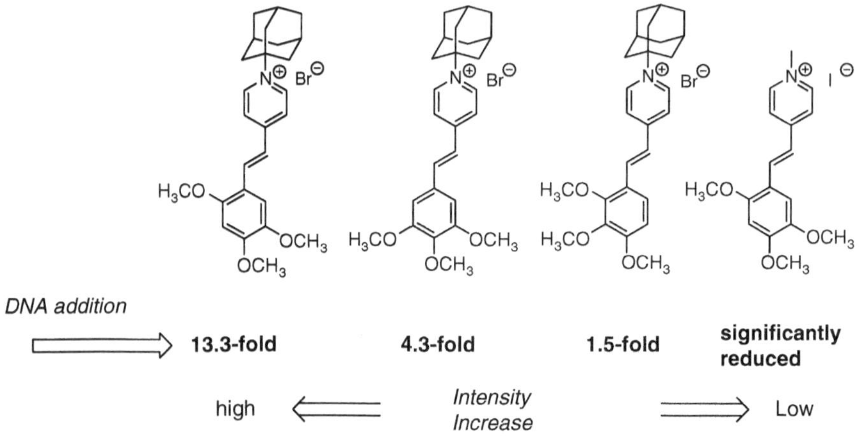

Fig. 23 Extended styryl fluorescent dyes as DNA sensitive probes

lowest unoccupied molecular orbital (LUMO) has a significantly smaller lobe than that of the highest occupied molecular orbital (HOMO) based on our calculation; therefore, we postulated that introduction of EDG at the R_1 position could trigger a bathochromic shift in our core skeleton [93]. The predicted positions (R_1 and R_2) were systematically modified with various substituents by the combination of five different aldehydes with five different pyridines having various electronic properties. Because of versatility of this approach, we accomplished the synthesis of 24 novel fluorescent compounds covering full-color emission wavelengths (from 420 to 613 nm). With this collection of fluorescent compounds with a single core skeleton, we rationalized substituents' effects on the photophysical properties. Interestingly, the emission wavelengths have significant correlations with the electronic properties of substituents, which is consistent with our original prediction. For example, a bathochromic shift was observed with the increments in electron-donating potentials of substituents at the R_1 position [from 471 to 613 nm].

As shown in Fig. 25b, the systematic tuning of emission wavelength was achieved by the combinatorial introduction of substitutents at the two diversity points on the fluorescent core skeleton. In addition to the synthetic versatility and predictability on emission wavelengths, these novel fluorophores were compatible with the modification of biopolymers and successfully applied in the immunofluorescence (see Fig. 25c).

Kool and co-workers recently reported a multicolor set of water-soluble dyes synthesized through the combination of three to five individual fluorophores assembled on a DNA-like backbone [94, 95]. As a continuation of their previous works on various DNA analogs [96–99], they synthesized the "oligodeoxyfluoroside" (ODF) with seven fluorescent monomers, such as pyrene, perylene, dimethylaminostilbene, and three stilbene derivatives, and they assembled these fluorescent DNA monomers into oligofluor chains using a DNA synthesizer (Fig. 26). Using

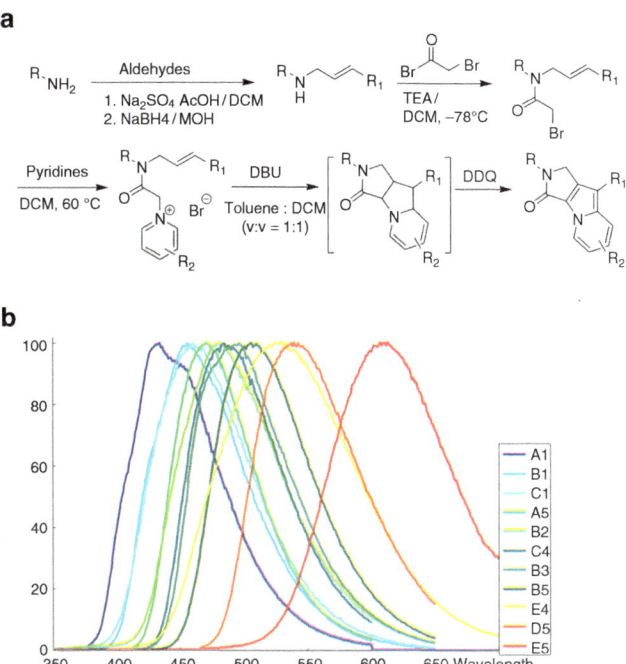

Cpd	R₁	R₂	clogP	λ_abs (nm)[a]	λ_em (nm)[b]	Gap(eV)[c]	Φ_f[d]
A1	Methyl	Hydrogen	1.59	326	434	2.92	0.41
A2	Methyl	Methoxy	1.61	–	–	2.71	–
A3	Methyl	Phenyl	3.48	349	433	2.55	0.19
A4	Methyl	Nitrile	1.06	342	460	2.53	0.76
A5	Methyl	Acetyl	1.12	396	471	2.27	0.82
B1	Phenyl	Hydrogen	2.97	298	420	2.80	0.57
B2	Phenyl	Methoxy	2.99	320	481	2.64	0.27
B3	Phenyl	Phenyl	4.87	350	490	2.47	0.83
B4	Phenyl	Nitrile	2.45	388	497	2.41	0.65
B5	Phenyl	Acetyl	2.50	403	507	2.17	0.74
C1	O–Methoxy phenyl	Hydrogen	2.34	320	461	2.82	0.37
C2	O–Methoxy phenyl	Methoxy	2.36	323	465	2.64	0.26
C3	O–Methoxy phenyl	Nitrile	4.23	381	489	2.46	0.69
C4	O–Methoxy phenyl	Phenyl	1.82	391	493	2.39	0.55
C5	O–Methoxy phenyl	Acetyl	1.87	404	508	2.15	0.71
D1	Thiophenyl	Hydrogen	2.84	322	481	2.55	0.08
D2	Thiophenyl	Methoxy	2.85	335	500	2.46	0.03
D3	Thiophenyl	Phenyl	4.72	283	515	2.30	0.11
D4	Thiophenyl	Nitrile	2.31	398	529	2.13	0.15
D5	Thiophenyl	Acetyl	2.36	349	540	1.92	0.35
E1	p–Dimethylaminophenyl	Hydrogen	3.15	298	495	2.53	0.10
E2	p–Dimethylaminophenyl	Methoxy	3.16	311	480	2.44	0.03
E3	p–Dimethylaminophenyl	Phenyl	5.04	403	530	2.17	0.21
E4	p–Dimethylaminophenyl	Nitrile	2.63	428	509	2.03	0.13
E5	p–Dimethylaminophenyl	Acetyl	2.67	440	613	1.78	0.15

[a]Only the longest absorption maxima are shown. [b]Excited at the maximum excitation wavelength. [c]Value of calculated energy gap between the HOMO and LUMO. [d]Absolute fluorescence quantum yield. Absolute quantum yields of known fluorescent dyes in various wavelengths were measured to confirm the reliability of the system. [anthracene: φ_F = 0.27 (reported: 0.27); fluorescein: φ_F = 0.76 (reported: 0.79); cresyl violet: φ_F = 0.48 (reported: 0.54)]

Fig. 24 Discovery of novel fluorophore, 1,2-dihydropyrrolo[3,4-β]indolizin-3-one, using a combinatorial approach; (**a**) Synthetic schemes of fluorescent core skeleton; (**b**) Collected emission spectra of selected compounds covering full-color emission wavelength; (**c**) Table of photophysical properties of all fluorescent compounds. All the photophysical properties were measured in DCM (dichloromethane). Reproduced with permission from [92]

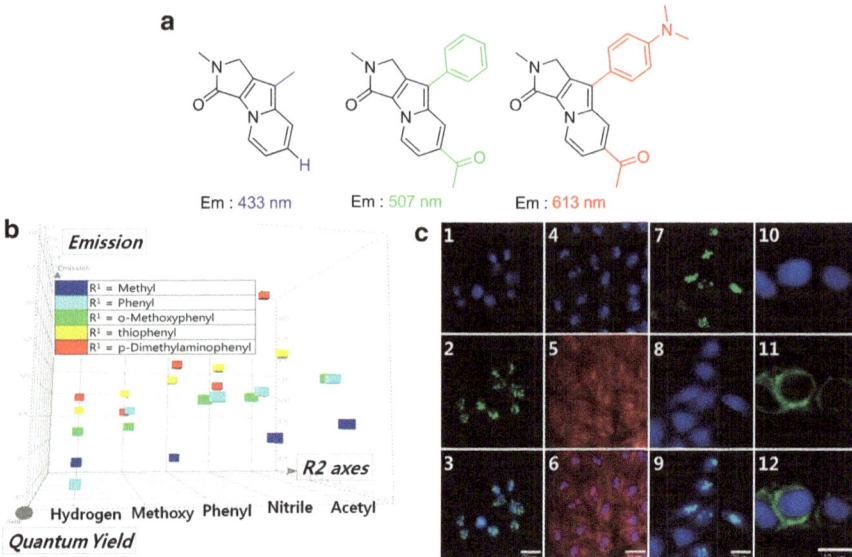

Fig. 25 Tunable photophysical properties of 1,2-dihydropyrrolo[3,4-β]indolizin-3-one fluoro-phores and their bio-application; (**a**) Structures and emission maxima of representative fluorescent compounds based on 1,2-dihydropyrrolo[3,4-β]indolizin-3-one; (**b**) 3D scatter plot of the whole member of fluorescent library according to quantum yield, emission maximum, and R^2 group. (each color represents the substituent at the R_1 position of fluorophore); (**c**) Immunofluorescence image of HeLa cell; Nucleus stained by Hoechst (**1, 10**); nucleol targeted by antinucleol 1 Ab and visualized by **B5**-labeled antimouse 2 Ab (**2, 7**) or **B1**-labeled antimouse 2 Ab (**4**); Actin visualized by TRITC-labeled phalloidin (**5**); EGFR targeted by Cetuximab and visualized by **B1**-labeled antihuman 2 Ab (**8**); EGFR visualized by **B5**-labeled Cetuximab (**11**). Panels **3, 6, 9**, and **12** are merged images of **1** and **2**, **4** and **5**, **7** and **8**, and **10** and **11**, respectively. The 1 Ab, 2 Ab, and EGFR represent the primary antibody, secondary antibody, and epidermal growth factor receptor, respectively. Cetuximab is chimeric (mouse/human) monoclonal antibody, targeting EGFR. Reproduced with permission from [92]

the split-and-pool synthesis in combination of seven ODF monomers and one nonfluorescent spacer, they initally constructed the 4096-membered tetrameric ODF library on polystyrene beads. Interestingly, synthesized ODF library exhibits the broad coverage of emission properties from 376 to 633 nm under the excitation with a single wavelength (354 nm). After the analysis of initial ODF tetramer library, they resynthesized the twenty-three ODF tetramers containing two, three, or four chromogenic ODF monomers and demonstrated the application of ODF tetramer as a bioimaging tool in human cell lines and vertebrate tissue (zebra fish embryo). Although it is still unclear whether the broad range of emission wavelengths in ODFs library is caused by the simple emission mixing of each ODF or the electronic interaction within ODFs, it is clear that this type of combinatorial approach can be valuable for the development of novel fluorescent compounds or tune the photophysical properties of various fluorophores.

a

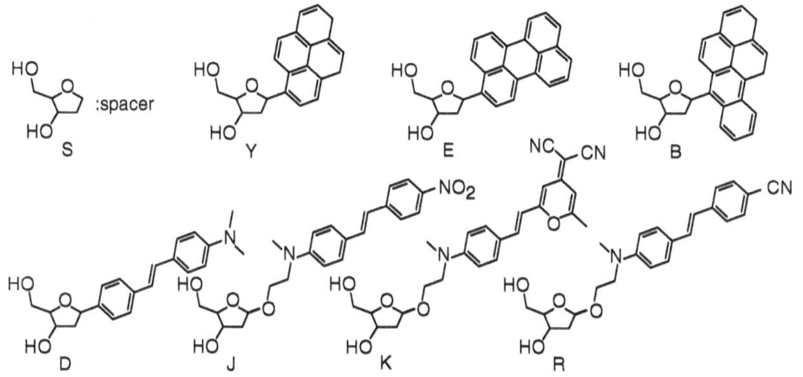

b

sequence (5'-XX..-3')	ε	λ_{max} (abs)	λ_{max} (em)	Φ_F	sequence (5'-XX..-3')	ε	λ_{max} (abs)	λ_{max} (em)	Φ_F
SY	46000	327,343	376,396	0.37	SSEE	43000	443,420	449,478,560	0.06
SB	28000	375,395	427	0.91	SSSBEK	53000	397,443	448,631	0.07
SE	39000	415,441	462	0.86	SSSKYY	89000	330,446,477	482,602	0.05
SSBY	49000	344,380,400	497	0.20	SSSEKB	53000	401,444	414,447,550	0.18
SSSYYY	67000	328,344	491	0.12	SSSYKY	81000	345,490	380,487,624	0.06
SSSBEY	50000	346,396,450	376,492	0.42	SSSBKBK	71000	381,444	415,625	0.09
SSEY	47000	346,448	464,490	0.65	SSSYKBK	92000	379	414,604	0.04
SSSBYE	50000	345,398,447	377,496	0.18	SSSEYBK	75000	346,398,448	414,624	0.028
SSSEBY	50000	344,397,447	520	0.29	SSBK	40000	396,480	412,633	0.10
SSEB	39000	396,448	412,519	0.26	SSSYKEB	72000	326,399,450	412,495,626	0.12
SSBB	48000	380	412,509	0.31	SSSYKYK	110000	346,477	380,482,629	0.09
SSSEYK	61000	326,425,450	488	0.17					

* Data were obtained in Phosphate buffer

Fig. 26 (a) Monomer structures of oligodeoxyfluorosides (ODF) library. (b) Photophysical properties of ODF fluorophores. Reproduced with permission from [94]

7   Conclusion

In this chapter, we have reviewed various fluorescent core skeletons through a side-by-side comparison between targeted synthesis and combinatorial approach for the discovery of novel fluorescent small molecules. Our comparison did not intend to make a conclusion whether one approach is superior to the other. In fact, we would like to emphasize that the targeted synthesis with rational design and systematic modification using combinatorial approach are complementary to each other for the discovery of novel small molecules with desired properties. Our survey demonstrates clearly that most of the knowledge regarding each fluorophore has been accumulated through a huge amount of time-consuming efforts on the targeted synthesis with rational design. However, it is quite difficult to achieve desired photophysical properties or to discover specific biosensors via de novo design of fluorescent probes because of the complexity in the fluorescence phenomena of organic small molecules. Therefore, combinatorial approach can provide a powerful tool to overcome this limitation, especially the difficulties in rational design, and the systematic modification of fluorophores can increase the possibility for the empirical discovery of novel fluorescent small molecules having desired biological sensing activity as well as a wide spectrum of photophysical properties. The further exploration of combinatorial approach in conjunction with rational design might address the huge demand of novel fluorophores for the application in biological science as well as in material science.

References

1. Guilbault GG (1999) Practical fluorescence. Revised and expanded, 2nd edn. Marcel Dekker, New York
2. Kim E, Park SB (2009) Chemistry as a prism: a review of light-emitting materials having tunable emission wavelengths. Chem Asian J 4:1646–1658
3. Murray RDH, Méndez J, Brown SA (1982) The natural coumarins: occurrence, chemistry, and biochemistry. Wiley, New York
4. Finn GJ, Kenealy E, Creaven BS, Egan DA (2002) In vitro cytotoxic potential and mechanism of action of selected coumarins, using human renal cell lines. Cancer Lett 183:61–68
5. Kirkiacharian S, Thuy DT, Sicsic S, Bakhchinian R, Kurkjian R, Tonnaire T (2002) Structure-activity relationships of some 3-substituted-4-hydroxycoumarins as HIV-1 protease inhibitors II. Farmaco 57:703–708
6. Perkin WH (1868) Artificial production of coumarin and formation of its homologues. J Chem Soc 21:53–63
7. Perkin WH (1868) Hydride of aceto-salicyl. J Chem Soc 21:181–186
8. Perkin WH (1877) Formation of coumarin and of cinnamic and of other analogous acids from the aromatic aldehydes. J Chem Soc 31:388–427
9. von Pechmann H, Duisberg C (1883) Ueber die Verbindungen der Phenol mit Acetessigäther. Chem Ber 16:2119–2128
10. von Pechmann H (1884) Neue Bildungsweise der Cumarine. Synthese des daphnetins. I. Chem Ber 17:929–979
11. Sethna S, Phadke C (1953) The Pechmann reaction. Org React 7:1–58
12. Cairns N, Harwood LM, Astles DP (1994) Tandem thermal claisen-cope rearrangements of coumarate derivatives. Total syntheses of the naturally occurring coumarins: suberosin, demethylsuberosin, ostruthin, balsamiferone and gravelliferone. J Chem Soc Perkin Trans 1:3101–3107
13. Brufola G, Fringuelli F, Piermatti O, Pizzo F (1996) Simple and efficient one-pot preparation of 3-substituted coumarins in water. Heterocycles 43:1257–1266
14. Bigi F, Chesini L, Maggi R, Sartori G (1999) Montmorillonite KSF as an inorganic, water stable, and reusable catalyst for the knoevenagel synthesis of coumarin-3-carboxylic acids. J Org Chem 64:1033–1035
15. Shriner RL (1942) The reformatsky reaction. Org React 1:15–18
16. Yavari I, Hekmat-Shoar R, Zonuzi A (1998) A new and efficient route to 4-carboxymethyl-coumarins mediated by vinyltriphenylphosphonium salt. Tetrahedron Lett 39:2391–2392
17. Appel H (1935) Improved method for the synthesis of coumarins by V Pechmann's method. J Chem Soc. Abstracts: 1031
18. Woods LL, Sapp J (1962) A new one-step synthesis of substituted coumarins. J Org Chem 27:3703–3705

19. Robinson R, Weygand F (1941) Experiments on the synthesis of substances related to the sterols, Part XXX. J Chem Soc:386–391
20. Nadkarni AJ, Kudav NA (1981) A convenient synthesis of 8-Methoxy-4-methylcoumarin. Indian J Chem Sect B 20:719–720
21. John EVO, Israelstam SS (1961) Use of cation exchange resins in organic reactions. I. The von Pechmann reaction. J Org Chem 26:240–242
22. Hoefnagel AJ, Gunnewegh EA, Downing RS, van Bekkum H (1995) Synthesis of 7-hydroxycoumarins catalysed by solid acid catalysts. J Chem Soc Chem Commun 1995:225–226
23. Biswas GK, Basu K, Barua AK, Bhattacharyya P (1992) Montmorillonite clay as condensing agent in Pechmann reaction for the synthesis of coumarin derivatives. Indian J Chem 31B: 628–628
24. Sabou R, Hoelderich WF, Ramprasad D, Weinand R (2005) Synthesis of 7-Hydroxy-4-methylcoumarin via the Pechmann reaction with Amberlyst ion-exchange resins as catalysts. J Catal 232:34–37
25. Palaniappan S, Shekhar RC (2004) Synthesis of 7-Hydroxy-4-methyl coumarin using polyaniline supported acid catalyst. J Mol Catal A Chem 209:117–124
26. Laufer MC, Hausmann H, Hölderich WF (2003) Synthesis of 7-hydroxycoumarins by Pechmann reaction using Nafion resin/silica nanocomposites as catalysts. J Catal 218:315–320
27. Jones G (1967) Organic reactions. Wiley, New York
28. Watson BT, Christiansen GE (1998) Solid phase synthesis of substituted coumarin-3-carboxylic acids via the knoevenagel condensation. Tetrahedron Lett 39:6087–6090
29. Wiener C, Schroeder CH, Link KP (1957) The synthesis of various 3-substituted-4-alkylcoumarins. J Am Chem Soc 79:5301–5303
30. Song A, Wang X, Lam KS (2003) A convenient synthesis of coumarin-3-carboxylic acids via Knoevenagel condensation of Meldrum's acid with *ortho*-hydroxyaryl aldehydes or ketones. Tetrahedron Lett 44:1755–1758
31. Christie RM, Lui CH (2000) Studies of fluorescent dyes: part 2. An investigation of the synthesis and electronic spectral properties of substituted 3-(2'-benzimidazolyl)coumarins. Dyes Pigm 47:79–89
32. Ayyangar NR, Srinivasan KV, Daniel T (1991) Polycyclic compounds Part VII. Synthesis, laser characteristics and dyeing behaviour of 7-diethylamino-2H-1-benzopyran-2-ones. Dyes Pigm 16:197–204
33. Moylan CR (1994) Molecular hyperpolarizabilities of coumarin dyes. J Phys Chem 98: 13513–13516
34. Fischer A, Cremer C, Stelzer EHK (1995) Fluorescence of coumarins and xanthenes after two-photon absorption with a pulsed titanium-sapphire laser. Appl Opt 34:1989–2003
35. Takadate A, Masuda T, Murata C, Tanaka T, Irikura M, Goya S (1995) Fluorescence characteristics of methoxycoumarins as novel fluorophores. Anal Sci 11:97–101
36. Rangaswami S, Seshadri TR (1940) A Note on certain constitutional factors controlling visible fluorescence in compounds of the benzo-pyrone group. Proc Ind Acad Sci 12A: 375–380
37. Rangaswami S, Seshadri TR, Venkateswarlu V (1941) The remarkable fluorescence of certain coumarin derivatives. Proc Ind Acad Sci 13A:316–322
38. Balaiah V, Seshadri TR, Venkateswarlu V (1942) Visible fluorescence and chemical constitution of compounds of the benzopyrone group. Part III. Further study of structural influences in coumarins. Proc Ind Acad Sci 16A:68–82
39. Wheelock CE (1959) The fluorescence of some coumarins. J Am Chem Soc 81:1348–1352
40. Atkins RL, Bliss DE (1978) Substituted coumarins and azacoumarins. Synthesis and fluorescent properties. J Org Chem 43:1975–1980
41. Sherman WR, Robins E (1968) Fluorescence of substituted 7-hydroxycoumarins. Anal Chem 40:803–805
42. Takadate A, Masuda T, Murata C, Shibuya M, Isobe A (2000) Structural features for fluorescing present in methoxycoumarin derivatives. Chem Pharm Bull 48:256–260

43. Murata C, Masuda T, Kamochi Y, Todoroki K, Yoshida H, Nohta H, Yamaguchi M, Takadate A (2005) Improvement of fluorescence characteristics of coumarins: syntheses and fluorescence properties of 6-methoxycoumarin and benzocoumarin derivatives as novel fluorophores emitting in the longer wavelength region and their application to analytical reagents. Chem Pharm Bull 53:750–758

44. Schiedel MS, Briehn CA, Bäuerle P (2001) Single-compound libraries of organic materials: parallel synthesis and screening of fluorescent dyes. Angew Chem Int Ed 40:4677–4680

45. Sivakumar K, Xie F, Cash BM, Long S, Barnhill HN, Wang Q (2004) A fluorogenic 1, 3-dipolar cycloaddition reaction of 3-azidocoumarins and acetylenes. Org Lett 6:4603–4606

46. Jamison JM, Krabill K, Hatwalkar A, Jamison E, Tsai CC (1990) Potentiation of the antiviral activity of poly r(A-U) by xanthene dyes. Cell Biol Int Rep 14:1075–1084

47. Chibale K, Visser M, van Schalkwyk D, Smith PJ, Saravanamuthu A, Fairlamb AH (2003) Exploring the potential of xanthene derivatives as trypanothione reductase inhibitors and chloroquine potentiating agents. Tetrahedron 59:2289–2296

48. Nestmann ER, Douglas GR, Matula TI, Grant CE, Kowbel DJ (1979) Mutagenic activity of rhodamine dyes and their impurities as detected by mutation induction in *Salmonella* and DNA damage in Chinese hamster ovary cells. Cancer Res 39:4412–4417

49. Shapiro HM (1988) Practical flow cytometry, 2nd edn. Alan R. Liss, New York

50. Haugland RP (2005) Handbook of fluorescence probes and research products, 10th edn. Molecular Probes, Eugene

51. Urano Y, Kamiya M, Kanda K, Ueno T, Hirose K, Nagano T (2005) Evolution of fluorescein as a platform for finely tunable fluorescence probes. J Am Chem Soc 127:4888–4894

52. Ahn YH, Lee JS, Chang YT (2007) Combinatorial rosamine library and application to in vivo glutathione probe. J Am Chem Soc 129:4510–4511

53. Han J, Burgess K (2009) Fluorescent indicators for intracellular pH. Chem Rev. doi:10.1021/cr900249z

54. Leonhardt H, Gordon L, Livingston R (1971) Acid-base equilibria of fluorescein and 2′, 7′-dichlorofluorescein. J Phys Chem 75:245–249

55. Zanker V, Peter W (1958) Die prototropen Formen des Fluoresceins. Chem Ber 91: 572–580

56. Diehl H, Horchak-Morris N (1987) Studies on fluorescein V. The absorbance of fluorescein in the ultraviolet, as a function of pH. Talanta 34:739–741

57. Tanaka K, Miura T, Umezawa N, Urano Y, Kikuchi K, Higuchi T, Nagano T (2001) Rational design of fluorescein-based fluorescence probes. Mechanism-based design of a maximum fluorescence probe for singlet oxygen. J Am Chem Soc 123:2530–2536

58. Miura T, Urano Y, Tanaka K, Nagano T, Ohkubo K, Fukuzumi S (2003) Rational design principle for modulating fluorescence properties of fluorescein-based probes by photoinduced electron transfer. J Am Chem Soc 125:8666–8671

59. Yamaguchi K, Tamura Z, Maeda M (1997) Disodium fluorescein octahydrate. Acta Cryst C53:284–285

60. Shen Z, Röhr H, Rurack K, Uno H, Spieles M, Schulz B, Reck G, Ono N (2004) Boron-diindomethene (BDI) dyes and their tetrahydrobicyclo precursors-en route to a new class of highly emissive fluorophores for the red spectral range. Chem Eur J 10:4853–4871

61. Goeb S, Ziessel R (2007) Convenient synthesis of green diisoindolodithienylpyrromethene-dialkynyl borane dyes. Org Lett 9:737–740

62. Wada M, Ito S, Uno H, Murashima T, Ono N, Urano T, Urano Y (2001) Synthesis and optical properties of a new class of pyrromethene-BF_2 complexes fused with rigid bicyclo rings and benzo derivatives. Tetrahedron Lett 42:6711–6713

63. Whitaker JE, Haugland RP, Prendergast FG (1991) Spectral and photophysical studies of benzo[c]xanthenes dyes: dual emission pH sensors. Anal Biochem 194:330–344

64. Lee LG, Berry GM, Chen CH (1989) Vita blue: a new 633-nm excitable fluorescent dye for cell analysis. Cytometry 10:151–164

65. Fabian WMF, Schuppler S, Wolfbeis OS (1996) Effects of annulation on absorption and fluorescence characteristics of fluorescein derivatives: a computational study. J Chem Soc Perkin Trans 2(5):853–856

66. Yang Y, Lowry M, Xu X, Escobedo JO, Sibrian-Vazquez M, Wong L, Schowalter CM, Jensen TJ, Fronczek FR, Warner IM, Strongin RM (2008) Seminaphthofluorones are a family of water-soluble, low molecular weight, NIR-emitting fluorophores. Proc Natl Acad Sci USA 105:8829–8834

67. Wagner BK, Carrinski HA, Ahn YH, Kim YK, Gilbert TJ, Fomina DA, Schreiber SL, Chang YT, Clemons PA (2008) Small-molecule fluorophores to detect cell-state switching in the context of high-throughput screening. J Am Chem Soc 130:4208–4209

68. Gonçalves MST (2009) Fluorescent labeling of biomolecules with organic probes. Chem Rev 109:190–212

69. Treibs A, Kreuzer FH (1968) Difluorboryl-Komplexe von Di- und tripyrrylmethenen. Justus Liebigs Ann Chem 718:208–223

70. Loudet A, Burgess K (2007) BODIPY dyes and their derivatives: syntheses and spectroscopic properties. Chem Rev 107:4891–4932

71. Yang Y, Lowry M, Schowalter CM, Fakayode SO, Escobedo JO, Xu X, Zhang H, Jensen TJ, Fronczek FR, Warner IM, Strongin RM (2006) An organic white light-emitting fluorophore. J Am Chem Soc 128:14081–14092

72. Yang Y, Lowry M, Schowalter CM, Fakayode SO, Escobedo JO, Xu X, Zhang H, Jensen TJ, Fronczek FR, Warner IM, Strongin RM (2007) An organic white light-emitting fluorophore. J Am Chem Soc 129:1008–1008

73. Umezawa K, Nakamura Y, Makino H, Citterio D, Suzuki K (2008) Bright, color-tunable fluorescent dyes in the visible–near-infrared region. J Am Chem Soc 130:1550–1551

74. Umezawa K, Matsui A, Nakamura Y, Citterio D, Suzuki K (2009) Bright, color-tunable fluorescent dyes in the Vis/NIR region: establishment of new "tailor-made" multicolor fluorophores based on borondipyrromethene. Chem Eur J 15:1096–1106

75. Lavis LD, Raines RT (2008) Bright ideas for chemical biology. ACS Chem Biol 3:142–155

76. Lee JS, Kang NY, Kim YK, Samanta A, Feng S, Kim HK, Vendrell M, Park JH, Chang YT (2009) Synthesis of a BODIPY library and its application to the development of live cell glucagon imaging probe. J Am Chem Soc 131:10077–10082

77. Fabian J, Nakazumi H, Matsuoka M (1992) Near-infrared absorbing dyes. Chem Rev 92:1197–1226

78. Ernst LA, Gupta RK, Mujumdar RB, Waggoner AS (1989) Cyanine dye labeling reagents for sulfhydryl groups. Cytometry 10:3–10

79. Sturmer DM (1977) Syntheses and properties of cyanine and related dyes. In: Weissberger A, Taylor EC (eds) The chemistry of heterocyclic compounds: special topics in heterocyclic chemistry. Wiley, New York

80. Mishra A, Behera RK, Behera PK, Mishra BK, Behera GB (2000) Cyanines during the 1990s: a review. Chem Rev 100:1973–2012

81. Gonçalves MST (2009) Fluorescent labeling of biomolecules with organic probes. Chem Rev 109:190–212

82. Narayanan N, Patonay G (1995) A new method for the synthesis of heptamethine cyanine dyes: synthesis of new near-infrared fluorescent labels. J Org Chem 60:2391–2395

83. Benson RC, Kues HA (1977) Absorption and fluorescence properties of cyanine dyes. J Chem Eng Data 22:379–383

84. Mujumdar SR, Mujumdar RB, Grant CM, Waggoner AS (1996) Cyanine-labeling reagents: sulfobenzindocyanine succinimidyl esters. Bioconjugate Chem 7:356–362

85. Kundu K, Knight SF, Willett N, Lee S, Taylor WR, Murthy N (2009) Hydrocyanines: a class of fluorescent sensors that can image reactive oxygen species in cell culture, tissue, and in vivo. Angew Chem Int Ed 48:299–303

86. Chen X, Conti PS, Moats RA (2004) In vivo near-infrared fluorescence imaging of integrin $\alpha_v\beta_3$ in brain tumor xenografts. Cancer Res 64:8009–8014

87. Lin Y, Weissleder R, Tung CH (2002) Novel near-infrared cyanine fluorochromes: synthesis, properties, and bioconjugation. Bioconjugate Chem 13:605–610
88. Wang S, Chang YT (2006) Combinatorial synthesis of benzimidazolium dyes and its diversity directed application toward GTP-selective fluorescent chemosensors. J Am Chem Soc 128: 10380–10381
89. Finney NS (2006) Combinatorial discovery of fluorophores and fluorescent probes. Curr Opin Chem Biol 10:238–245
90. Rosania GR, Lee JW, Ding L, Yoon HS, Chang YT (2003) Combinatorial approach to organelle-targeted fluorescent library based on the styryl scaffold. J Am Chem Soc 125: 1130–1131
91. Lee JW, Jung M, Rosania GR, Chang YT (2003) Development of novel cell-permeable DNA sensitive dyes using combinatorial synthesis and cell-based screening. Chem Commun 1852–1853
92. Kim E, Koh M, Ryu J, Park SB (2008) Combinatorial discovery of full-color-tunable emissive fluorescent probes using a single core skeleton, 1, 2-dihydropyrrolo[3, 4-β]indolizin-3-one. J Am Chem Soc 130:12206–12207
93. Higashiguchi K, Matsuda K, Asano Y, Murakami A, Nakamura S, Irie M (2005) Photochromism of dithienylethenes containing fluorinated thiophene rings. Eur J Org Chem:91–97
94. Teo YN, Wilson JN, Kool ET (2009) Polyfluorophores on a DNA backbone: a multicolor set of labels excited at one wavelength. J Am Chem Soc 131:3923–3933
95. Gao J, Strässler C, Tahmassebi D, Kool ET (2002) Libraries of composite polyfluors built from fluorescent deoxyribosides. J Am Chem Soc 124:11590–11591
96. Krueger AT, Kool ET (2008) Fluorescence of size-expanded DNA bases: reporting on DNA sequence and structure with an unnatural genetic set. J Am Chem Soc 130:3989–3999
97. Wilson JN, Teo YN, Kool ET (2007) Efficient quenching of oligomeric fluorophores on a DNA backbone. J Am Chem Soc 129:15426–15427
98. Gao J, Watanabe S, Kool ET (2004) Modified DNA analogues that sense light exposure with color changes. J Am Chem Soc 126:12748–12749
99. Ren RXF, Chaudhuri NC, Paris PL, Rumney S IV, Kool ET (1996) Naphthalene, phenanthrene, and pyrene as DNA base analogues: synthesis, structure, and fluorescence in DNA. J Am Chem Soc 118:7671–7678

Part III
Organic Dyes with Response Function

Physical Principles Behind Spectroscopic Response of Organic Fluorophores to Intermolecular Interactions

Vladimir I. Tomin

Abstract The main intermolecular processes and mechanisms affecting fluorescence properties of organic probes are overviewed. Among them are contact diffusion controlled reactions (static and dynamic quenching), formation of excimers and exciplexes, photoinduced proton transfer, and radiative and nonradiative energy transfer. Different methods of polarity determination, such as polarity empirical scales and single parameter and multiparameter approaches, are discussed. Solvation and solvatochromism are presented with the accounting of inhomogeneous broadening of electronic spectra due to thermal fluctuations, and their effect on the main spectroscopic properties of organic probes in polar and heterogeneous solutions. Effect of specific interactions and their contribution to electronic spectra are also discussed.

Keywords Dipole moment · Fluorescence · Hydrogen bond · Intermolecular interactions · Solvatochromism

Contents

V.I. Tomin
Institute of Physics, Pomeranian University in Slupsk, 76-200 Słupsk, Arciszewskiego str. 22B, Poland
e-mail: tomin@apsl.edu.pl

A.P. Demchenko (ed.), *Advanced Fluorescence Reporters in Chemistry and Biology I:* 189
Fundamentals and Molecular Design, Springer Ser Fluoresc (2010) 8: 189–224,
DOI 10.1007/978-3-642-04702-2_6, © Springer-Verlag Berlin Heidelberg 2010

1   Fluorescence and Main Intermolecular Processes Affecting Fluorescence Properties

1.1   Fluorescence and Fluorescence Probes

During the last decades, a permanent growth of interests and concentration of efforts of researcher teams in chemistry, laser physics, and spectroscopy have been observed on studying complex molecular systems including solutions of organic molecules and biological objects. From one side, this fact is due to scientific and practical importance of processes taking place in such systems, and from another side, the progress in improvement of traditional experimental techniques and the emergence of principally new ones have considerably broadened possibilities of researchers [1–10]. Fluorescence was first used as an analytical tool to identify or even determine the concentrations of various species (atoms, molecules, ions, and so on), when they are fluorescent. Fluorescence sensing is the method, which has a very high sensitivity reaching in proper conditions (laser excitation, confocal microscope configuration of excitation and registration, single photon detection) the level of single molecules. Fluorescence is also a powerful instrument for studying the intermolecular interactions, the structure and dynamics of matter or biological systems at a molecular level. The development of experimental techniques that use fluorescence and laser excitation has widened the horizons of physics, chemistry, biology, and medical applications.

   The Jablonski diagram of singlet and triplet energy levels of an organic molecule (see Fig. 1) is commonly used [1, 2] for interpretation of all main features of luminescence phenomenon. A set of vibrational and rotational levels of the fluorophore is associated with each electronic level of the diagram. Electronic transitions from the singlet ground S_o to the excited S_1 state are responsible for absorption spectra. After fast redistribution of energy excess in the excited molecule (10^{-12} s), the equilibrium over vibrational levels is established, and the molecule is ready for other intra- and intermolecular processes. The internal conversion from S_1 to S_0 state with the rate k_{nr} competes with spontaneous emission k_r that forms fluorescence spectrum, and occurs during fluorescence lifetime, τ. In agreement with the *Stokes rule*, the fluorescence maximum is always red-shifted (on wavelength scale) with respect to the absorption maximum (the *Stokes shift* is a gap between the maxima of absorption and fluorescence spectra denoted as Δv^{af} in Fig. 1) as a result

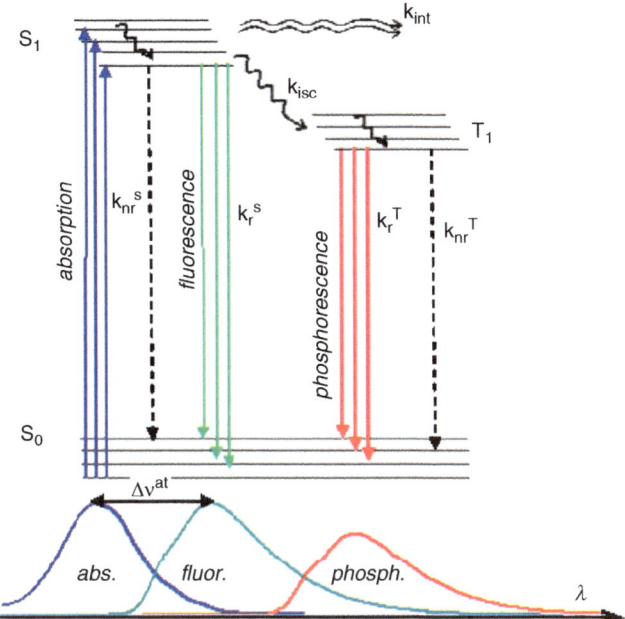

Fig. 1 Jablonski diagram of energy level for describing processes absorption, fluorescence and phosphorescence in complex molecules where k_r^S and k_{nr}^S are the radiative and nonradiative rates of fluorescence, respectively, k_r^T and k_{nr}^T are the radiative and nonradiative rates of phosphorescence, respectively, k_{isc} is the interconversion rate, and k_{int} is the rate of intermolecular processes; Δv^{af} denotes the Stokes shift of fluorescence

of energy losses due to vibrational relaxation and interaction with the solvent at the excited state.

A third possible channel of S_1 state deexcitation is the $S_1 \rightarrow T_1$ transition – nonradiative intersystem crossing k_{isc}. In principle, this process is spin forbidden, however, there are different intra- and intermolecular factors (spin–orbital coupling, heavy atom effect, and some others), which favor this process. With the rates $k_{isc} = 10^7 - 10^9 \ s^{-1}$, it can compete with other channels of S_1 state deactivation. At normal conditions in solutions, the nonradiative deexcitation of the triplet state T_1, k_{nr}^T, is predominant over *phosphorescence*, which is the radiative deactivation of the T_1 state. This transition is also spin-forbidden and its rate, k_r^T, is low. Therefore, normally, phosphorescence is observed at low temperatures or in rigid (polymers, crystals) matrices, and the lifetimes of triplet state τ_T at such conditions may be quite long, up to a few seconds. Obviously, the phosphorescence spectrum is located at wavelengths longer than the fluorescence spectrum (see the bottom of Fig. 1).

Many other processes in the S_1 singlet state are also possible; they may compete with fluorescence and affect directly the quantum yield, the lifetime, and the spectrum of emission. Among them are intramolecular charge transfer (ICT), dynamic quenching by other microscopic substances, absorption to higher singlet states, radiative and nonradiative transfer of excitation energy, processes of intermolecular relaxations, or change of solvation energy during the lifetime of fluorescence. Singlet states may serve as initial states for various photochemical reactions, the most important of them [1–3] are the changes of molecular geometry (conformational transformations), electron transfer, formation of exciplexes and excimers, and other more complicated photochemical transformations. These processes can be observed only during fluorescence lifetime (10^{-8}–10^{-10} s). Quite often, species appearing as a result of photoreaction also possess an ability to emit fluorescence or phosphorescence. Their contribution can be resolved in the recorded spectrum.

It is necessary to note that fluorescence characteristics demonstrate remarkable sensitivity to variations of physicochemical parameters of the environment. Therefore, such parameters as polarity, viscosity, temperature, electric potential, local electric field, pressure, pH, etc., can be registered successfully using the modern sensitive apparatus for fluorescence detection [1, 4–12]. As a consequence, fluorescent molecules are used successfully as molecular probes to study the local characteristics of physicochemical, biochemical and biological systems.

1.2 Main Intermolecular Processes Affecting Fluorescence Properties

Fluorescent molecules can participate in different intermolecular reactions starting from S_1 state with rate k_{int}; as a result, their properties, such as the quantum yield, Φ, and the fluorescence lifetime, τ, can change. The proper equations for τ and Φ, as illustrated in Jablonski diagram (Fig. 1), may be written as [1, 2]:

$$\tau^{-1} = \tau_0^{-1} + k_{int} \tag{1}$$

$$\Phi_F^{-1} = \Phi_{F0}^{-1} + \frac{k_{int}}{k_r} \tag{2}$$

In (1)–(2) all denotations are the same as in Fig. 1, τ_0 and Φ_{F0} are the lifetime and quantum yield in the simplest case, when there are no intermolecular reactions or their rate $k_{int} = 0$.

As seen from (1) and (2), intermolecular processes may reduce essentially the lifetime and the fluorescence quantum yield. Hence, controlling the changes of these characteristics, we can monitor their occurrence and determine some characteristics of intermolecular reactions. Such processes can involve other particles, when they interact directly with the fluorophore (bimolecular reactions) or participate (as energy acceptors) in deactivation of S_1 state, owing to nonradiative or radiative energy transfer. Table 1 gives the main known intermolecular reactions and interactions, which can be divided into four groups:

Table 1 Classification of the main intermolecular processes in solutions of organic molecules

Intermolecular processes			
Contact diffusion controlled reactions:	Reaction with a change of chemical nature of fluorophore:	Noncontact interactions:	Solvatochromism:
– dynamic and static quenching;	– electron transfer;	– radiative energy transfer;	– polarity effect;
– quenching by solvent	– proton transfer;	– nonradiative energy transfer	– universal or general interactions;
	– excimer formation;		– charge transfer;
	– exciplexes formation		– specific interactions

(1) Contact diffusion controlled reactions (without change of chemical nature of fluorophore)
(2) Reactions with the change of chemical nature of fluorophore in the excited state
(3) Noncontact interactions
(4) Solvatochromism

In this section, we discuss the first three groups (1)–(3); the forth one (4) will be treated in detail in Sect. 2.

Bimolecular reactions with paramagnetic species, heavy atoms, some molecules, compounds, or quantum dots refer to the first group (1). The second group (2) includes electron transfer reactions, exciplex and excimer formations, and proton transfer. To the last group (3), we ascribe the reactions, in which quenching of fluorescence occurs due to radiative and nonradiative transfer of excitation energy from the fluorescent donor to another particle – energy acceptor.

Let us consider in details all three groups (1)–(3).

1.3   Contact Diffusion Controlled Reactions

In this group, there are collisional interactions, which are responsible for quenching of excited states by molecular oxygen, paramagnetic species, heavy atoms, etc. [1, 2, 13–15]. Probability of such quenching can be calculated as:

$$k_{int} = k_Q[Q] \qquad (3)$$

where k_Q is a constant of bimolecular quenching and $[Q]$ is the concentration of the quencher particles. Molecular oxygen, which is ubiquitous in solvents at normal conditions, is a well known and efficient quencher of fluorescence. Molecular oxygen has a triplet ground state and two low-lying excited singlet states. Quenching of dye excited electronic states takes place via energy transfer resulting in production of singlet oxygen species.

The constant of bimolecular quenching for oxygen is very high and reaches $\sim 10^{10}$ L mol^{-1} s^{-1}; it means that at the atmospheric pressure, when oxygen concentration may be as high as $[Q] \sim 10^{-3}$ M, $k_{int} \sim 10^7$ s^{-1}. Therefore, for typical value of radiative constant $k_r \sim 10^8\ s^{-1}$ and $\Phi_0 \sim 1.0$, we have from (2) that $\Phi \sim 0.91$, i.e., noticeable quenching of emission takes place. Quenching of fluorescence by solvent is also associated with the collisional processes, which are usually controlled by diffusion, and their rates are time-dependent.

Besides the quenching in the excited state (*dynamic quenching*), there exists another type of quenching (*static*), which takes place in the ground state and occurs due to the formation of nonemitting complexes.

Paramagnetic species and heavy atoms through interaction with the solute may cause the solute to undergo intersystem crossing to the triplet state, which in liquids is completely quenched. Hence, an addition of some impurities to the solution with fluorescent solute may lead to a dynamic quenching of the excited state as a result of diffusion controlled process. The same effect gives formation of chemical complexes with different spectroscopic characteristics reducing number of emitters and also leading to a decrease of fluorescence intensity; such reactions will be briefly described in Sect. 1.4.

To study the dynamic quenching in steady state approach, the Stern–Volmer relations are commonly used:

$$\frac{\Phi_0}{\Phi} = \frac{I_0}{I} = 1 + k_{SV}[Q] \tag{4}$$

where Φ_0, Φ, I_0, and I are the steady state fluorescence quantum yields and intensities in the absence and the presence of the quencher, respectively,

$$k_{SV} = k_Q \tau_0 \tag{5}$$

is the Stern–Volmer constant. The slope of I_0/I versus $[Q]$ plot gives us the Stern–Volmer constant k_{SV}, and then the constant k_Q may be calculated from (5), if τ_0 is known. Further, applying different models for bimolecular processes and diffusion controlled reactions, we can get important information concerning solute–solvent system on microscopic level and the mechanisms leading to the contact of the quencher with the fluorophore.

As it was shown recently [16, 17], dynamic quenching is also a powerful method for determining the kinetic or thermodynamic characteristics of photoreactions in the excited sate, and for changing the rates [18, 19] of such reactions as, for example, ESIPT.

1.4   Reactions with a Change of Chemical Nature of Fluorophore

The second group of intermolecular reactions (2) includes [1, 2, 9, 10, 13, 14] electron transfer, exciplex and excimer formations, and proton transfer processes (Table 1). *Photoinduced electron transfer* (PET) is often responsible for fluorescence quenching. PET is involved in many photochemical reactions and plays a major role in photosynthesis of plants and in artificial systems for solar energy conversion on photoinduced charge separation. The oxidation and reductive properties of molecules can be enhanced in the excited state accordingly to the following schemes

$$^1D^* + A \rightarrow D^{\bullet+} + A^{\bullet-} \tag{6}$$

$$^1A^* + D \rightarrow A^{\bullet-} + D^{\bullet+} \tag{7}$$

where D^* and A^* are neutral molecules of donor and acceptor of electron in the excited states, respectively, $D^{\bullet+}$ and $A^{\bullet-}$ are ions of donor and acceptor, respectively. Examples of electron donors are naphthalene, anthracene, phenanthrene, pyrene, perylene, and as an electron acceptor may serve, for example, 9,10-dicyanoantracene [1, 2, 14, 15].

Excimers and exciplexes are formed in the excited states. *Excimers* are complexes of excited $^1M^*$ and unexcited ^1M molecules in the excited state:

$$^1M^* + {}^1M \leftrightarrow {}^1(MM)^* \tag{8}$$

Different aromatic hydrocarbons (naphthalene, pyrene and some others) can form excimers, and these reactions are accompanying by an appearance of the second emission band shifted to the red-edge of the spectrum. Pyrene in cyclohexane (CH) at small concentrations 10^{-5}–10^{-4} M has structured vibronic emission band near 430 nm. With the growth of concentration, the second smooth fluorescence band appears near 480 nm, and the intensity of this band increases with the pyrene concentration. At high pyrene concentration of 10^{-2} M, this band belonging to excimers dominates in the spectrum. After the act of emission, excimers disintegrate into two molecules as the ground state of such complex is unstable.

Exciplexes are complexes of the excited fluorophore molecule (which can be electron donor or acceptor) with the solvent molecule. Like many bimolecular processes, the formation of excimers and exciplexes are diffusion controlled processes. The fluorescence of these complexes is detected at relatively high concentrations of excited species, so a sufficient number of contacts should occur during the excited state lifetime and, hence, the characteristics of the dual emission depend strongly on the temperature and viscosity of solvents. A well-known example of exciplex is an excited state complex of anthracene and N,N-diethylaniline resulting from the transfer of an electron from an amine molecule to an excited anthracene. Molecules of anthracene in toluene fluoresce at 400 nm with contour having vibronic structure. An addition to the same solution of diethylaniline reveals quenching of anthracene accompanied by appearance of a broad, structureless fluorescence band of the exciplex near 500 nm (Fig. 2)

Photoinduced proton transfer reactions, undoubtedly, belong to the most important transformations in chemistry [20]. Proton transfers can take place both in the ground and excited states of organic compounds, and the most important for

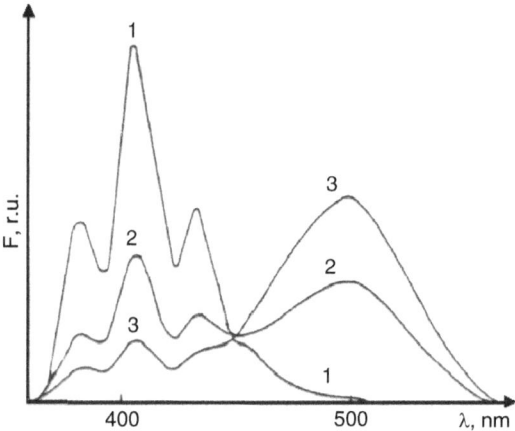

Fig. 2 Fluorescence spectra of anthracene $(2.4 \times 10^{-4}$ M) in hexane with various concentrations of diethylaniline: 1–0 M, 2–2.5 $\times 10^{-3}$ M; 3–0.1 M

fluorescence sensing is the excited state reaction. Excited state proton transfer (ESPT) reactions refer to relatively new classes of reactions in photochemistry, which more likely were observed for the first time by Weller, who discovered dual fluorescence of naphthols [21]. ESPT reactions can be divided into two large groups: intermolecular and intramolecular, which often are abbreviated in literature as ESIPT (Excited State Intramolecular Proton Transfer). In the first group, we can mention such well-known reactions as deprotonation with creating of anions in solutions (naphthols, methylumbelliferones, and some others). Acid–base properties of various molecules in electronically excited states, as well as multiple examples of intramolecular and intermolecular photoinduced proton transfer reactions, have been the subjects of numerous papers and reviews [20–23] (and references therein).

The proton transfer may occur rapidly after the excitation and form a tautomer, when either acidic or basic moieties of the same molecule become stronger acids or bases in the excited state. The majority of reactions of this type involve the proton transfer from an oxygen donor to an oxygen or nitrogen acceptor, although a few other cases are known, where a nitrogen atom can function as a donor and a carbon atom as the acceptor. Usually an intramolecular hydrogen bond between the two moieties of a molecule facilitates the proton transfer.

In 1979, Sengupta and Kasha [23] discovered and studied one of the most interesting, attractive, and, until now, widely studied ESIPT reaction that occurs in 3-hydroxyflavone and results in dual fluorescence in the violet and green spectral ranges. Chemical structure of 3-hydroxyflavone is shown in Fig. 3. Nowadays, ESIPT is often considered as a prototype of proton-transfer (or, more precisely, hydrogen atom transfer) reactions. A variety of organic fluorophores exhibiting ESIPT has been described in the literature [20, 22, 23], but 3-hydroxyflavones are unique in many respects. The two effective groups 3-OH (proton donor) and 4-carbonyl (proton acceptor) are attached to a rigid skeleton and are permanently connected by a hydrogen bond, which make possible the fastest pathway of ESIPT

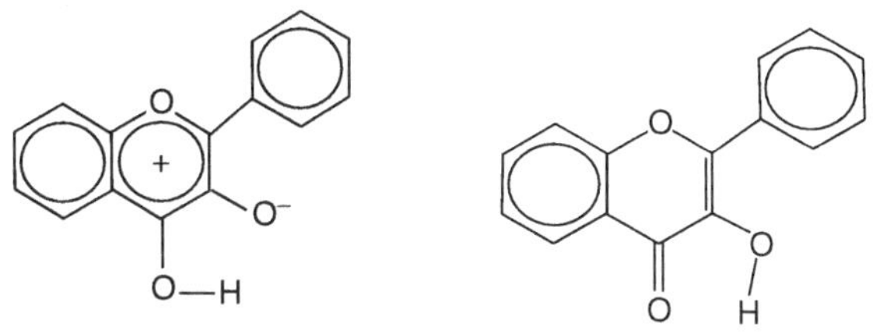

Fig. 3 Chemical structure of
3-hydroxyflavone tautomer
(T) and normal (N) form
(*right*)

reaction along this bond (see Fig. 3). Being isomeric to the initially excited normal (*N*) form, the tautomer (*T*) form is well fluorescing with emission spectrum strongly shifted to the red, up to 5000–6000 cm^{-1} in respect to the fluorescence of the *N* form having band maximum ≈390–400 nm. A reverse isomerization in the ground state occurs after decay of the excited tautomer. As a result, the entire excitation–emission process has the cyclic character.

In general, ESIPT reaction kinetics reveals itself in the rise and decay of the tautomer fluorescence. The rise of the tautomer fluorescence covers a broad time range from femtoseconds to nanoseconds, although the majority of the relevant processes occur in picoseconds range. The transfer of a proton, having positive charge, between groups in the molecule causes changes of molecular geometry, leading to large electronic and structural rearrangements, which are accompanied by significant changes in the dipole electric moments. In consequence, the dynamics of such processes can be essentially affected by the nature of the solvent, namely with respect to the formation of hydrogen bonds and dielectric properties. One of the best and efficient ways to monitor these changes is to register the ratio of the normal and tautomer bands' intensities; hence, molecules with dual fluorescence have one additional channel, ratiometric parameter, for studying and sensing of intermolecular interactions [7].

Detailed coverage of the topic can be found in the next chapter of this book [24].

1.5   Noncontact Interactions (Nonradiative and Radiative Energy Transfer)

The next group of bimolecular interactions (3) shown in Table 1, includes noncontact interactions, in which fluorescence quenching occurs due to radiative and nonradiative excitation energy transfer [1, 2, 13, 25, 26]. Energy transfer from an excited molecule (donor) to another molecule (acceptor), which is chemically different and is not in contact with the donor, may be presented according to the scheme:

$$D^* + A \rightarrow D + A^*$$

This process called *heterotransfer* is possible if the fluorescence spectrum of the donor overlaps the absorption spectrum of the acceptor. If the donor and the acceptor are identical chemically, we have *homotransfer*:

$$D^* + D \rightarrow D + D^*,$$

and this process can be observed in concentrated dye solutions.

Nonradiative transfer of excitation energy requires some interaction between donor and acceptor molecules and occurs if the emission spectrum of the donor overlaps the absorption spectrum of the acceptor, so that several vibronic transitions in the donor must have practically the same energy as the corresponding transitions in the acceptor. Such transitions are coupled, i.e., they are in *resonance*, and that is why the term *resonance energy transfer* (RET) or electronic energy transfer (EET) are often used.

Energy transfer can result from different interaction mechanisms. The interactions may be Coulombic and/or due to intermolecular orbital overlap. The Coulombic interactions consist of long-range dipole–dipole interactions (Förster's mechanism, and for this mechanism used in literature, acronym FRET may be denoted also as *Förster resonance energy transfer*) and short-range interactions. The interactions due to intermolecular orbital overlap, which include electron exchange (Dexter's mechanism) and charge resonance interactions, are, of course, only short-scale range. It should be noted that for singlet–singlet energy transfer ($^1D^* + {}^1A \rightarrow {}^1D + {}^1A^*$), all types of interactions are involved, whereas triplet–triplet energy transfer ($^3D^* + {}^1A \rightarrow {}^1D + {}^3A^*$) is only due to orbital overlap.

Förster derived the following expression for the transfer rate:

$$k_T^{dd} = k_D \left[\frac{R_0}{r}\right]^6 = \frac{1}{\tau_D^0}\left[\frac{R_0}{r}\right]^6 \tag{9}$$

where k_D is the emission rate constant of the donor and τ_D^0 its lifetime in the absence of transfer, r is the distance between the donor and the acceptor (which is assumed to remain unchanged during the lifetime of the donor), and R_0 is the critical distance or Förster radius, i.e., the distance at which transfer and spontaneous decay of the excited donor are equally probable ($k_T = k_D$). Parameter R_0 can be determined from the spectroscopic data, i.e., from the spectra of fluorescence and absorption, and is generally in the range of 15–60 Å [1, 2, 13, 26]. FRET is particularly widely used to determine the distances in membranes, biomolecules, and supramolecular associations and assembles. The sixth power dependence explains why RET is the most sensitive to donor–acceptor distance when this distance is comparable to the Förster critical radius.

Homotransfer between chemically identical molecules in polar ambience had attracted attention during the last years [27]. This transfer has the same mechanism as FRET, but due to the presence of inhomogeneous broadening of spectra (see below), it is not random as in homogeneous systems, where it is directed from structures with short-wavelength 0–0 (pure electronic) transitions to that with longer wavelength transitions. That is why such homotransfer may be called directed homotransfer (DHT). As a result of the latter, the fluorescence band of steady state spectra in concentrated solutions (from 10^{-3} M and up) in viscous or rigid environment are shifted to the long-wavelength range. The failure of FRET in rigid solutions at the red-edge excitation (the Weber effect [28]) is explained by site-selections of red centers from inhomogeneously broadened ensemble at such excitation [27, 29]. Hence, the rate of DHT depends on the energy of excitation and can be reduced to zero at quite long-wavelengths excitation.

Such a process can naturally be expected to play a certain part in the mechanism of directed energy transport in biological systems, in particular, in the transfer of absorbed energy from the antenna chlorophyll molecules to the reactive center in the photosynthetic system of plants. In Ref. [30], energy exchange between molecules of the photosynthetic pigments chlorophyll a and pheophytin a was studied experimentally with pigments introduced into the polar matrix.

DHT may occur over different tryptophan forms in proteins as they quite often have inhomogeneously broadened electronic spectra [31]. A very interesting case of DHT is described between two indole rings in bichromophoric solutes tryptophan dipeptide [32]. Such directed transport allows to correctly interpret spectral properties of dipeptide and other multichromophoric solutes. The theory of inductive-RET in solutions with inhomogeneous spectral broadening is given in Ref. [33]. In more detail, DHT mechanism will be explained in Sect. 2.2 (vide infra).

2 Solvation and Solvatochromism

2.1 Inhomogeneous Broadening of Electronic Spectra of Dye Molecules in Solutions

As well known, the electronic spectral bands show shifts as a whole in solvents of different nature. This phenomenon called solvatochromism is directly connected with the intermolecular interactions in the solute–solvent system.

In 1970, Galley and Purkey [34], and Rubinov and Tomin [35] independently demonstrated that fluorescence spectra of some organic molecules in frozen organic polar glasses shift to longer wavelengths when excited at the long wavelength slope, or "the red edge," of the absorption spectrum. The first team [34] dealt with solutes such as indole, tryptophan, β-naphthol, and proflavin, and the other one [35] with the derivatives of phthalimide, where all spectral shifts were essentially stronger. In nonpolar solvents, the effect failed to appear. It is obvious that these effects cannot originate from the violation of fundamental principles of the theory of complex molecules, but by neglecting some of the additional factors, which are stronger in polar solvents. Remember that for several decades it had been believed that the fluorescence spectra of organic molecules in solutions are independent on the frequency of the exciting light due to the fast energy exchange between the vibrational sublevels [1, 2, 13, 14].

The basic explanation of these effects was suggested in both papers [34, 35], where it was shown that apart from molecular vibrations, there is another cause of the substantial broadening of electronic spectra of organic molecules in solution, namely, the fluctuations of the structure of the solvation shell surrounding the molecule. The variation of the local electric field caused by the fluctuation of the shell structure leads to a statistical distribution of the frequencies of the electronic transitions of the molecules and, therefore, to the broadening of the dye spectrum. The character of this broadening is strongly controlled by molecular mobility, depends on physical conditions, and may be homogeneous in liquid solvents and inhomogeneous in rigid samples during steady state registration.

Later, Personov et al. [36] observed inhomogeneous broadening of electronic spectra of frozen solutions of organic molecules at lower (liquid helium) temperatures. It was shown that at liquid helium temperatures, discrete fluorescence spectra of organic molecules with the resolved vibrational structure could be obtained by eliminating the inhomogeneous broadening in the fluorescence spectrum by means of selective monochromatic excitation of the solution. This made possible to gain a deeper insight into the nature of the electronic spectra and led to new, more sophisticated methods for their study (*site-selection* spectroscopy). Subsequently, it was shown that inhomogeneous broadening of electronic spectra of organic molecules occurs not only in solid solutions but also in liquid solutions [29]. In the latter case, it has a dynamic nature and can only be detected by nano- or picosecond time-resolved spectroscopic techniques. Inhomogeneous spectral broadening in liquid solutions opens a door to new, interesting methods of studying molecular motions.

2.2   Solute Intermolecular Interactions and Inhomogeneous Broadening of Electronic Spectra in Solutions

Inherent peculiarities of liquid molecular systems are fluctuations of nanoscale density, polarity, and other characteristics of medium owing to thermal motion of molecules and their segments. Generally, some specific interactions like hydrogen bonding and preferential solvation in multicomponent solutions can also lead to differences in the structure of solvates. In polymeric and biological samples, together with the thermal fluctuations, there also exist gradients of polarity, density, viscosity, etc., the nature of which is connected with considerable spatial heterogeneity of the mentioned structures. Spatial and thermal fluctuations and structural heterogeneity in liquid molecular systems affect the spectra leading to new effects.

The spectroscopic properties of the solute–solvent system accounting fluctuations of solution structures can be analyzed using a simple model that includes a fluorescent solute and its immediate surrounding contributing to full potential of

Fig. 4 Polar solute
interacting with polarized
solvent

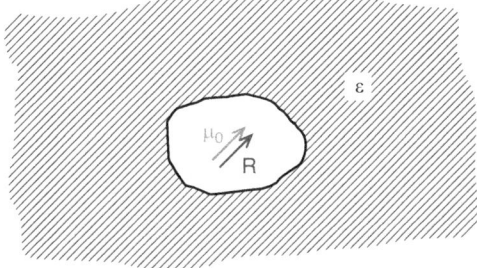

intermolecular interactions [37–39]. Due to the statistical variation of the solvate structure, different solvates may have different local electric fields.

Without specifying the dimensions and spatial configuration of the solvation shell, we will treat it in terms of its macroscopic characteristic, like of other dielectric materials. First, consider a polar solution, in which the solutes possess a constant dipole moment as is shown in Fig. 4. In each solvate of a solution, the immediate surroundings are polarized due to the dipole moment, μ_g, of the solute, thus giving rise to a reactive field, R, in the cell:

$$R = f \cdot \mu_g \qquad (10)$$

where function f is the polarizability (or factor of dielectric reaction) of the void, which is occupied by the solute. Physical sense of function f is the field intensity, which is produced by the unit dipole, $\mu = 1$ unit. Inhomogeneous broadening of spectra occurs as a result of the thermal fluctuations and, therefore, of existence of solvates having different reactive fields R, which differ somewhat from the mean value of R_I corresponding to the equilibrium electric configuration of all molecules in solvate. As shown in [29, 39], accounting fluctuations in solvate in the excited states over equilibrium field R_{II} and applying simple thermodynamic approaches, we came to the energetic diagram of solvates, where the reaction electric field R is chosen as a generalized coordinate (see Fig. 5).

Solvates with the different electric field strengths R are denoted by points on the axis R, and due to thermal fluctuations, there is a distribution of all solvates over R. The inhomogeneous broadening function characterizing absorption spectrum is found to be a symmetrical Gaussian shape $\rho(R)$ with the maximum for solvates with equilibrium field in the ground state R_I [29]. A distribution over different fields R, as a consequence, leads to a distribution over different energies of 0–0 transitions (see (15) vide infra). Half-width of this distribution in wavenumbers, as was shown in [29], is given by:

$$\Delta v_{0.5} = \Delta \mu (2f \cdot \kappa \cdot T)^{1/2} \qquad (11)$$

where $\Delta \mu = \mu_e - \mu_g$ is the difference of dipole moments of the solute in the ground and excited states. As seen, inhomogeneous broadening value $\Delta v_{0.5}$ depends on the

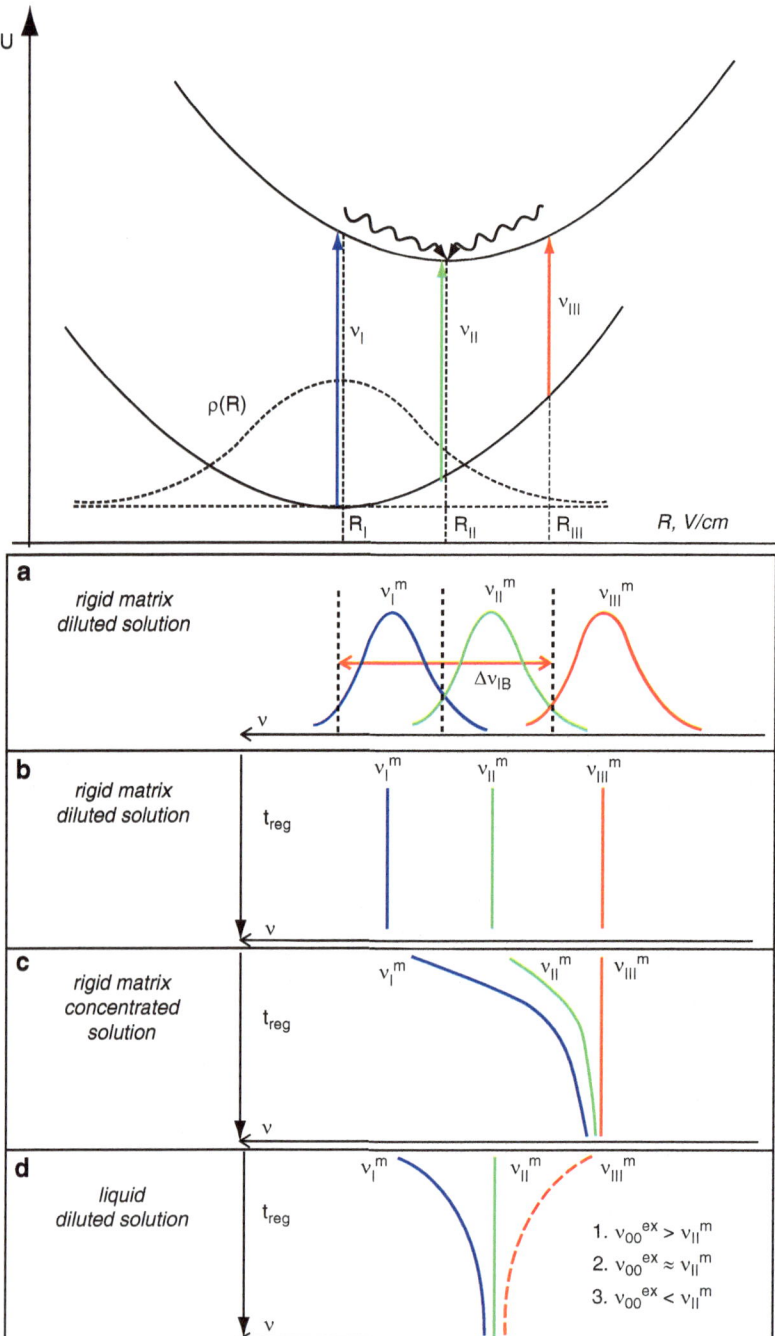

Fig. 5 Field diagram of full energy of solvate U in solution. Panels **a–d** illustrate main spectral features of polar solutions arising due to inhomogeneous broadening of their spectra: (**a**) Spectra of fluorescence appearing in the initial instant time $t = 0$ during time-resolved registration of

difference between the dipole moments of the dye molecule in its ground and excited states $\Delta\mu$ as well as on the solvent susceptibility f and temperature. The temperature dependence of $\Delta\nu_{0.5}$ exists until the freezing point of the solution is reached. Then, the spatial configurational fluctuations of the elementary cells become fixed, and the inhomogeneous broadening function becomes temperature-independent. This is a principal point, which shows an existence of considerable inhomogeneous broadening even at temperatures close to absolute zero.

As can be seen from (11), inhomogeneous broadening depends on the factor of reaction of the medium f. The most frequent and simple way for calculating function f is by applying the Onsager's model, which treats solvate like a cavity of spherical shape with a dipole μ inserted to the center [1, 4]. Outside of cavity is the solvent, which is considered as continuous medium with dielectric constant ε. Employing the Onsager model for a solvate, we account for inhomogeneous broadening, which then depends not only on the dipole moments of the dye molecule but also, and most essentially, on its dimensions. The smaller the molecule, the stronger is the inhomogeneous broadening effect. This agrees with experimental data: inhomogeneous broadening is the most pronounced for small solutes, for example, phthalimides. The estimates of $\Delta\nu_{0.5}$ for typical polar solutions, for example, 3-aminophthalimide ($\Delta\mu = 2.3$ D[1], radius of Onsager cavity a $= 0.35$ nm) in ethanol, show that at $T = 20°C$, the value $\Delta\nu_{0.5}$ reaches 300 cm^{-1} and decreases to 250 cm^{-1} at the freezing point of the solution (158°K). Estimation of $\Delta\nu_{0.5}$ for a nonpolar solvent (e.g., toluene) gives smaller values $\Delta\nu_{0.5} \approx 17$ cm^{-1}; hence, the broadening of electronic bands in toluene is an order of magnitude smaller than in polar solvents. By this reason, we treat further more common case of polar solutions, which are characterized by essentially larger spectral broadening.

Electronic transitions in a solute take place very fast, i.e., almost immediately in comparison with the movement of the molecules as a whole and vibrations of atoms in organic molecules. Hence, absorption and fluorescence can be denoted in Fig. 5 by vertical arrows, in accordance with Franck–Condon principle. Both these processes are separated by relaxations, which are intermolecular rearrangements of the solute–solvent system after the excitation.

Due to distribution of solvates over different intensities of electric field R, it is possible to excite the solvates with the different pure electronic, 0–0, frequencies on both branches of the excited parabola. Relaxation is accompanied by the spectral shifts of emission in time; when we excite solvates with fields $R \sim R_I$, we have

fluorescence in liquid solutions and in steady state spectra for rigid solutions for different ν_{ex}; (**b**) Maxima of fluorescence in rigid diluted solutions for different time of fluorescence registration observed upon various excitations ν_{ex}; (**c**) Maxima of fluorescence in rigid concentrated solutions for different time of fluorescence registration observed upon various excitations ν_{ex}; (**d**) Maxima of fluorescence in liquid diluted solutions for different time of fluorescence registration observed upon various excitations ν_{ex}

[1]in SI units 1 D (*Debye*) $= 3.33564 \times 10^{-30}$ C m

particles in the excited states with the storage of intermolecular energy on the left branch of the parabola and "red" shift of the fluorescence spectrum. Intermolecular relaxation takes place until excited solvates obtain equilibrium configurations with field R_{II} corresponding to minimum of energy in the excited state. Red spectral shift of fluorescence spectrum is well-known and usually called the *Time-Dependent Stokes Shift*. Important to note that, as seen from the diagram, there is one more interesting possibility – the "blue" time-dependent shift following excitation solvates with the electric fields $R > R_{II}$ and their relaxation to the equilibrium position with the gain of energy.

The inhomogeneous broadening effect will be apparent in practically all cases, and the character of this broadening may be both *stationary* (in rigid solutions or when time of relaxation τ_r is less than lifetime of fluorescence τ, or $\tau_r > \tau$) and *dynamic* in nature. Inhomogeneous broadening affects all spectral characteristics of organic molecules in solutions.

Thus, the discovery of inhomogeneous broadening of electronic spectra of organic molecules in solid and liquid solutions gave rise to a new field of research, namely, *the selective spectroscopy of organic solutions*. The main feature of this field is selective excitation of some members of an inhomogeneous ensemble of solutes, emitting in the narrower spectral range than overall spectrum, which enables them to be studied and treated selectively. Observation of these effects in steady state and time-resolved spectroscopy is different.

In steady state spectroscopy under the condition of slow reorientation of molecules in solution ($\tau_r > \tau$), the following peculiarities take place:

- Dependencies of luminescence bands (both fluorescence and phosphorescence), anisotropy of emission, and its lifetime on a frequency of excitation, when fluorescence is excited at the red edge of absorption spectrum. Panel **a** of Fig. 5 shows the fluorescence spectra at different excitations for the solutes with the 0–0 transitions close to ν_I, ν_{II}, and ν_{III} frequencies. Spectral location of all shown fluorescence bands is different and stable in time of experiment and during lifetime of fluorescence (panel **b**)
- Dependencies of fluorescence excitation spectra on the recorded emission wavelength.

A study of spectral properties in polar solutions in conditions of slow reorientations, $\tau_r > \tau$, was undertaken in a number of papers, for example, see [27, 29, 40] (and references herein). This condition can be realized either by increasing τ_r or decreasing the lifetime τ. The former can be achieved by freezing the solution or using polymer matrices, whereas the latter can be achieved by quenching the fluorescence by using admixtures or a powerful light flux [41–43]. Dozens of different solutes with considerable magnitude of dipole moment changes $\Delta\mu = \mu_e - \mu_g$ show such effect [29–36].

These experiments have firmly established that the red shift of the fluorescence band in polar solutions with the change of excitation quanta energy is caused by inhomogeneous configurational broadening of electronic energy levels. Later, it was found that not only singlet but also triplet states of dyes are broadened configurationally and, under the condition of slow reorientation of molecules in a solution, this broadening is inhomogeneous [44]. The term "red-edge effect" is applied commonly for these dependencies, the exploration of the red-edge effects are reviewed in detail by Demchenko in Ref. [40].

In concentrated dye solutions, when homotransfer of electronic excitation energy may take place due to inhomogeneous broadening, we observed [29, 40] the following interesting features:

- Red shift of fluorescence band with the increase of dye concentration due to *directed nonradiative energy homotransfer* (DHT) (owing to FRET mechanism) from the "blue" to the "red" centers of sample (see panel **b** in Fig. 5).
- Growth of the degree of fluorescence polarization (the Weber's effect) and a decrease of energy transfer efficiency while shifting the excitation wavelength to the red edge.

These features appear only when inhomogeneous broadening of electronic spectra takes place, i.e., in well polar matrices; the reason of DHT is a strong spectral overlap of "blue" centers with the absorption of the "red" centers. DHT was directly observed for the first time using time-resolved spectrofluorimetry [29, 45] as this technique makes possible to visualize the dynamics of the process and measure the required kinetic parameters. In [45–47] DHT was observed between the different configurational states of solvated molecules of 3-aminophthalimide in a rigid solution of polyvinyl alcohol. At high concentrations, $\sim 10^{-2}$ M, the instantaneous emission spectrum is shifting gradually toward "the red" region with time (illustration is in the panel **c** of Fig. 5); however, at a low dye concentration, the effect was absent. In this case, each instantaneous spectrum corresponds to a specific solvate state, whereas the entire sequence of spectra represents the energy transfer dynamics for a set of configurational states. The higher the dye molecule concentration, the faster is the energy transfer. Of particular interest is the fact that energy is transferred from the structures with a low degree of molecular orientation of dye surrounding those characterized by a high degree of orientation having higher intensities of electric field. It means that we have no random energy transfer in space of the sample as in the case of ordinary homotransfer. We conclude by noting that the theory of inductive-RET in solutions with inhomogeneous spectral broadening is given in Ref. [33].

The following remarkable features of fluorescence may be observed in polar solutions applying the method of *time-resolved spectroscopy*:

- Initial position of instant spectrum of fluorescence and character of spectral shifts in time depend on the excitation frequency, i.e., inhomogeneous broadening is of dynamic nature as a degree of broadening is maximal at the initial instants of time and decreases with time of emission registration (demonstration in panel **d** of Fig. 5).
- Different fluorescence lifetimes for particular states within inhomogeneously broadened population.

The relaxation of the fluorescence spectrum takes place as a result of rearrangement of all molecules in solvate (with characteristic times of orientational relaxation) and depends essentially on the exciting radiation frequency, that is, on the type of selectively excited solvates. The following three characteristic cases are possible:

1. Excitation of solutes with 0–0 transitions $v_{00} > v_{II}^m$ (Stokes spectral region of absorption) results in a red shift of the fluorescence spectrum with time, $\Delta v^f(t) = [v^f(t) - v^f(t = \infty)] > 0$. In this case, configurational relaxation results in *down-relaxation* of the fluorescence band, which is known as Time-Dependent Stokes Shift [29, 39, 40]. This type of relaxation can easily be explained in terms of molecular reorientation in solvates of similar configurations in the ground electronic state, and it was experimentally observed by Ware et al. and Mazurenko et al. long time ago [39, 48, 49]. As seen from the field diagram (Fig. 5) an initial spectral shift $\Delta v^f(t = 0)$ is a function of excitation frequency!

2. For excitation of solutes with 0–0 transitions $v_{00} > v_{II}^m$ (antiStokes spectral region of absorption), the situation is the opposite: at the initial instant of time, the spectra are red-shifted as compared to the steady state spectra, $\Delta v^f(t) < 0$. In this case, the return of the spectrum to its normal position during configurational relaxation will lead to a "blue" shift with time. From the physical point of view, this means that the intermolecular energy excess, which the solvates possess before excitation, is partially converted into emitted energy leading to an increase in the radiation frequency with time. That is why the process may be called the "up-relaxation" of the fluorescence spectra.

3. Finally, the third case corresponds to pumping of solutes with 0–0 transitions, $v_{00} > v_{II}^m$. In this case, the fluorescence spectrum, immediately after excitation, must be close to the steady state one and should not vary with time. From the physical point of view, this case corresponds to the situation, where solvates with a local field R_{II} (corresponding to the equilibrium configuration in the excited state) are excited.

All these features were observed experimentally for solutions of 3-amino-N-methylphthalimide, 4-amino-N-methylphthalimide, and for nonsubstituted rhodamine. The results were observed for cooled, polar solutions of phthalimides, in which the orientational relaxation is delayed. Exactly the same spectral behavior was observed [50] by picosecond spectroscopy for low viscosity liquid solutions at room temperature, in which the orientational relaxation rate is much higher. All experimental data indicate that correlation functions of spectral shifts $\Delta v^f(t)$, which are used frequently for describing the Time Dependent Stokes Shift, are essentially the functions of excitation frequency.

Using the Onsager model, the function $\Delta v^f(t)$ can be calculated for all time domains of dielectric relaxation of solvents measured experimentally for commonly used liquids (see, for example, [39]). Such simulations, for example, give for alcohols, at least, three different time components of spectral shift during relaxation, which are due to appropriate time domains of solvents relaxation.

It is necessary to distinguish the case of inhomogeneous broadening from the ground state heterogeneity of samples. In the case when we have in a sample two or more fluorophores with well overlapping spectra, simple registration of spectra at different wavelengths of excitation is not sufficient to distinguish these individual ground state forms from inhomogeneous broadening effects. In general case, the ground state heterogeneity can be analyzed and distinguished by studying the site-selective effects in excitation and in fluorescence [27].

2.3   Effect of Polarity

All inhomogeneous broadening effects are directly coupled with the polarity of all molecules in solutions. As seen from (11), the half-width of inhomogeneous broadening function $\Delta v_{0.5}$ depends essentially on function of dielectric reaction f, which, in turn, is determined (vide infra) in agreement with (19) by dielectric constant ε. Polarity plays a major role in solvatochromism as well as in many other physical, chemical, and biomedical phenomena. Solvent–solute interactions are commonly described in terms of van der Waals interactions and in terms of specific interactions like hydrogen bonding (H-bonding). All these interactions have electrostatic nature. The van der Waals interactions are also called as universal or nonspecific, as they take place always and in all types of systems.

What is interplay between these interactions and polarity?

Usually we call neutral molecule as polar one if it has considerable permanent electric dipole moment μ^0. The total dipole moment should include also an induced one, αR (α is a polarizability of the molecule, R is the intensity of electric field interacting with molecule), and may be presented as $\mu = \mu^0 + \alpha R$. Permanent part of dipole moment for nonsymmetrical organic molecules usually accepted to be essentially larger than induced one; that is why orientational forces or interactions of permanent electric dipoles are the most important in polar solutions [1, 2, 4, 12, 39].

After these preliminary notes, we can conclude that common term polarity expresses the complex interplay of all types of solute–solvent electrostatic interactions, i.e., nonspecific dielectric and specific such as H-bonding. Therefore, *polarity* cannot be characterized by a single parameter, although the "polarity" of a solvent (or other matrices) is often associated with the static macroscopic dielectric constant ε or with the dipoles of solvent molecules (microscopic characteristic). It is clear that polarity is the most significant factor contributing to the energy of solute solvation (or stabilization) (10) and, hence, to the position of energetic levels of the molecules in solution. For some molecules with the large difference of permanent dipole moments in the ground and excited states, $\Delta\mu$, solvatochromic dependences are very strong, and they may be used to determine the polarity.

There are single- and multiparameter approaches for determining the polarity and separation of contribution of different interactions to the total effect of polarity on spectroscopic characteristics. They are based on different theories of solvatochromic shifts of absorption and fluorescence bands.

In principle, we can suggest to apply expressions of the theory of inhomogeneous broadening (vide supra Sects. 2.1 and 2.2) for determination of polarity. So, (11) gives us the relation of halfwidth of broadening $\Delta v_{0.5}$ with characteristics of solute–solvent system. Then, if solute size, a, and a change of dipole moments while electronic transition properties, $\Delta \mu$, are known and $\Delta v_{0.5}$ is determined experimentally, we can with (11) calculate function $f(\varepsilon)$. Within the Onsager's model, the latter has simple analytic form that allows finding dielectric constant ε characterizing polarity of solvent.

2.3.1 Single-Parameter Approach

Empirical polarity scales are popular in chemistry. The Reinchard scale E_T (30) [51] is based on the application of the pyridinium-N-phenoxide betain dye revealing dramatic negative solvatochromism connected with the large $\Delta \mu$ difference (ground state dipole moment equals $\approx$15 D and practically zero in the excited state). The parameter E_T (30) is defined as the transition energy for the main absorption band of the dissolved dye in kilocalories per mole. The list of E_T (30) values is available [51], they range from 30.9 for nonpolar neutral n-heptane ($\varepsilon = 1.88$) to 63 in water ($\varepsilon = 81$). Such large differences in E_T (30) values evidence to very large solvation energy due to orientational interactions, which are close to the energy of the 0–0 transition in inert neutral solvent. In many investigations with other than betain dyes, the parameter E_T (30) has been used to characterize the solvent polarity and to check other dyes' reaction on polarity of solvent. The E_T (30) scale also provides possibility of selecting the H-bonding contribution to solvatochromism as protic solvents give different dependences than aprotic ones.

There are two important drawbacks of such an approach: (1) a polarity scale based on a particular class of probes, in principle, does not account, for example, sizes of probes, which should strongly effect the interactions; (2) betain dyes do not fluoresce, which restrict essentially the field of application of this approach, because in many cases, absorption spectrum could not be measured accurately (small volumes of samples, study of cells, and single molecules spectroscopy). Therefore, polarity-sensitive fluorescent dyes offer distinct advantage in many applications.

2.3.2 Multiparameter Approach

More advanced scale was proposed by Kamlet and Taft [52]. This phenomenological approach is very universal as may be successfully applied to the positions and intensities of maximal absorption in IR, NMR (nuclear magnetic resonance), ESR (electron spin resonance), and UV–VS absorption and fluorescence spectra, and to many other physical or chemical parameters (reaction rates, equilibrium constant, etc.). The scale is quite simple and may be presented as:

$$\tilde{v}^{f,a} = \tilde{v}_0^{f,a} + s\pi^* + a\alpha + b\beta \tag{12}$$

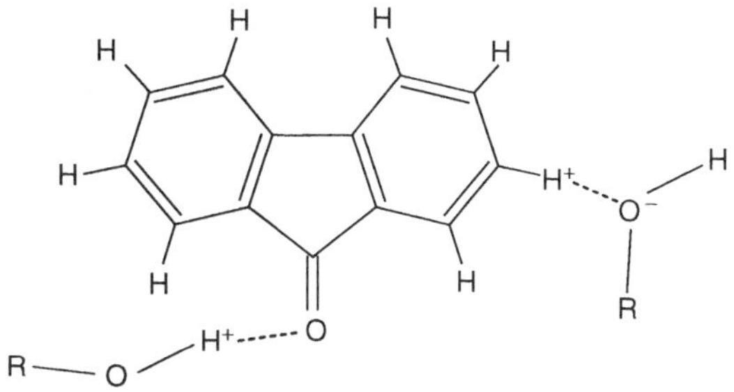

Fig. 6 Chemical structure of fluorenone (FL) with solvent molecules forming H-bond with carbonyl oxygen and hydrogen of FL

where $\tilde{v}^{f,a}$ and $\tilde{v}_0^{f,a}$ are the wavenumbers of the band maxima (*f*-fluorescence and *a*-absorption) in the solvent under consideration and in the reference solvent (generally neutral solvent CH), π^* is a measure of polarity effects of the solvent; the α scale is an index of solvent hydrogen bond donor acidity and β scale is an index of solvent hydrogen bond acceptor basicity. The coefficients s, a, and b give the sensitivity of a process to each of the individual contributions. The main advantage of the Kamlet–Taft treatment is to sort out the quantitative role of properties such as hydrogen bonding.

A good illustration is provided [53] by fluorenone (FL) in a group of $n = 19$ solvents, which possesses both donor and acceptor groups (see structural scheme of FL in Fig. 6), the multiple regression equations were obtained as:

$$\tilde{v}^a = (26423 \pm 275) - (3115 \pm 510)\pi^* - (2070 \pm 420)\alpha$$
$$- (161 \pm 462)\beta \ (n = 19, r = 0.85) \tag{13}$$

relative contributions of different factors are: π^*-58.3%, α-38.7%, β-3.0%.

$$\tilde{v}^f = (21681 \pm 145) - (1229 \pm 292)\pi^* - (2447 \pm 311)\alpha$$
$$- (650 \pm 336)\beta \ (n = 19, r = 0.93) \tag{14}$$

relative contributions are: π^*-28.4%, α-56.6%, β-15%.

Estimation of the standard error of the β term in the transitions of absorption $\tilde{v}^a$ and fluorescence $\tilde{v}^f$ showed that it was not a statistically significant variable in the regression equation. These results suggest that the influence of the β term on the $\tilde{v}^a$ may be considered as negligible. In this case, solvation is mainly determined by the dipole interaction (parameter π^*). The ratio of the α and π^* regression coefficients indicates the relative importance of dipole interaction over H-bond donation of solvent in the ground state (see (13), whereas the H-bond donation plays a significantly larger role in the excited fluorescent state (14). This H-bond creates complex of FL with solvent due to attraction of H atom of solvent to negatively charged carbonyl oxygen of the solute (see scheme in Fig. 6). In the excited state also is noticeable the contribution of another type of H-bond (β-15%), which is due to

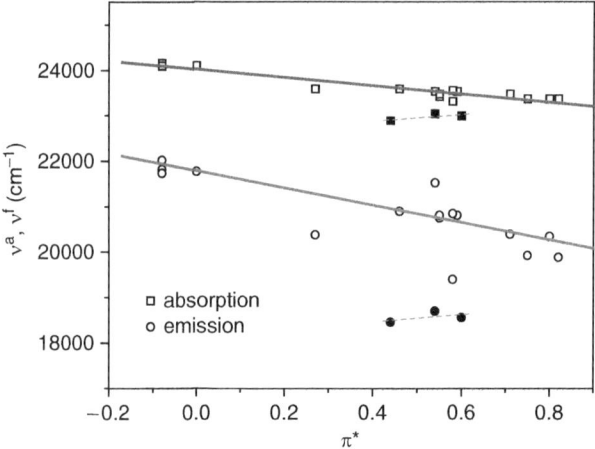

Fig. 7 Absorption and emission energies of fluorenone versus polarity parameter π^*. Solid marks denote the data for alcohols (courtesy of M. Józefowicz)

attraction of H atom of FL to negatively charged atoms of solvent. This situation is typical for carbonyls, for some solutes, like, for example, 4-amino-7-methylcoumarin [54] contribution of both types of H-bond may be comparable.

Figure 7 shows the positions of FL absorption and fluorescence spectra (in centimeter^{-1}) as a function of π^*. As it is seen, the graphs for different neutral solvents show linear dependence, whereas for protic solvents (alcohols), a different line can be drawn. Such graphs can be practically used for the determination of π^* in different samples and solvents. If the measured value will be placed on the line for alcohols, this means that the site of our probe incorporation possesses proton donor property.

2.3.3 Theory of Solvatochromic Shifts

The main equation, which allows understanding the physical sense of 0–0 frequency shifts in the spectra after inserting a molecule from gas phase into the solvent, is:

$$h\Delta v = hv_0 - hv = \Delta\mu R/h \qquad (15)$$

When $\Delta\mu = \mu_e - \mu_g$ is positive, we have positive solvatochromism resulting in the "red" shift of transitions; otherwise, when $\Delta\mu$ is negative, the spectra move to the "blue" and that is the case of negative solvatochromism. As follows from this expression, the change of dipole moments $\Delta\mu$ during electronic transition is a necessary condition for observed solvatochromic shift: the more this difference – the stronger is the solvatochromic shift. The second important parameter is the local reactive field R, which in accordance with (10), may be calculated, if the function of dielectric reaction of the solvent f is known. The most known and often used approach is the Onsager model, which gives an opportunity to account this function. There are few types of electromagnetic interactions, which contribute mainly to reaction function: orientational, inductive, and dispersive. As was shown by Bakhshiev [4, 55, 56], in the frames of the Onsager model with some natural approximations, the solvatochromic shifts due to orientational interactions for absorption Δv_{or}^a, fluorescence Δv_{or}^f, and their difference $v_{or}^{a,f}$, respectively, may be accounted as:

$$h\Delta v_{or}^a = \frac{2\mu_g(\mu_g - \mu_e \cos \gamma)}{a^3} f(\varepsilon_0, n_0) \tag{16}$$

$$h\Delta v_{or}^f = \frac{2\mu_e(\mu_g \cos \gamma - \mu_e)}{a^3} f(\varepsilon_0, n_0) \tag{17}$$

$$\Delta h v_{or}^{a,f} = h v_{or}^a - h v_{or}^f = \frac{2(\mu_g^2 - 2\mu_e\mu_g \cos \gamma + \mu_e^2)}{a^3} f(\varepsilon_0, n_0) \tag{18}$$

where

$$f(\varepsilon_0, n_0) = \left(\frac{\varepsilon_0 - 1}{\varepsilon_0 + 2} - \frac{n_0^2 - 1}{n_0^2 + 2} \right), \tag{19}$$

ε_0 and n_0 are dielectric constant for low frequencies and refractive index, respectively.

These are general equations that allow calculating solvatochromic shifts if we know electric characteristics of the solute (dipole moments in the ground and excited states, μ_g and μ_e, respectively, and the angle γ between them), its Onsager radius a, and the function of interactions $f(\varepsilon_0, n_0)$.

Quite often for known typical solutes, the angle γ is small or close to zero, and in this case, $\cos \gamma \approx 1$ (the last condition is valid for angles $\gamma \leqslant 25 - 30°$) and then (18) may be presented as:

$$\Delta v_{or}^{a,f} = h v_{or}^a - h v_{or}^f = \frac{2(\mu_g - \mu_e)^2}{a^3} f(\varepsilon_0, n_0) \tag{20}$$

The same expression was also deduced by Lippert [57] and Mataga [58], which neglected by a change of the angle γ; however, the function $f(\varepsilon_0, n_0)$ has another form than in (19) as the effect of polarizability of solvent molecules was not accounted. Mac Ray also accounted the solute polarization and obtained [59] the same equation as (20) with similar function $f(\varepsilon_0, n_0)$ as in (19).

Experimental plot $\Delta h v_{or}^{a,f}$ versus $f(\varepsilon_0, n_0)$ allows to calculate with (20) a change of dipole moments $\Delta\mu$ during electronic transition. Figure 8 presents such plot for

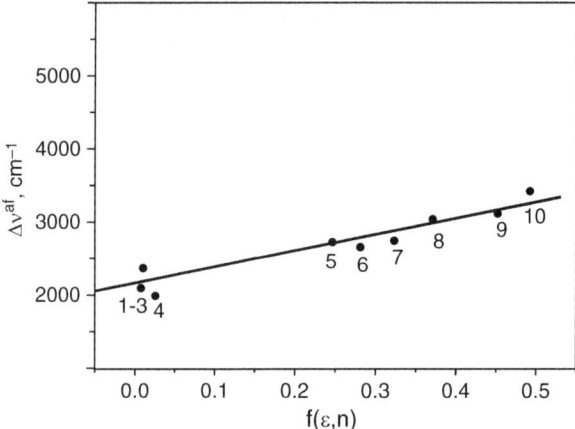

Fig. 8 Solvatochromic experimental plot $\Delta h \nu_{or}^{af}$ versus $f(\varepsilon_0, n_0)$ of fluorenone (courtesy of M. Józefowicz). The latter function was calculated for selected solvents 1-10 using (19)

FL, and the difference in dipole moments determined from the plot is 2.36 D if the Onsager radius is 0.33 nm [53]. The Onsager cavity radius was obtained from molecular models where the molar volumes were calculated by CAChe WS 5.0 computer program. The simplest method to estimate the cavity radius is to assume $a = (3V/4\pi)^{1/3}$, where V is the volume of the solute.

Of course, there are some uncertainties in this procedure, as the Onsager model describes the structures of solution and a solute only approximately. It can be noted that there is a good opportunity to calculate dipole moments, exactly, their ratio, in the simpler way using the relative shifts of absorption, and fluorescence spectra. As follows from (16) and (17), dividing them by proper parts we may obtain the following relation:

$$\frac{\Delta \nu_{or}^a}{\Delta \nu_{or}^f} = \frac{\mu_g(\mu_g - \mu_e \cos \gamma)}{\mu_e(\mu_g \cos \gamma - \mu_e)} \qquad (21)$$

This equation does not contain any information about sizes of the Onsager cavity and function of interaction and hence, this relation is quite general. If $\cos \gamma$ is close to unity, the expression can be simplified:

$$\frac{\Delta \nu_{or}^a}{\Delta \nu_{or}^f} = \frac{\mu_g}{\mu_e} \qquad (22)$$

The same equation was introduced by Suppan, where $\Delta \nu_{or}^{a,f}$ was treated as spectral shift of the probe spectra in two different solvents [60].

For probes with known structure and electric dipoles change $\Delta \mu = \mu_e - \mu_g$, the spectral shift measured experimentally $\Delta \nu_{or}^{af}$ enables to calculate the value of function $f(\varepsilon_0, n_0)$. Then it is possible to calculate dielectric constant ε_0 characterizing polarity of solvent if to use analytical form of this function, like, for example, (19). For polar surrounding the most important contribution to this function is provided by dielectric constant ε_0 as the other parameter n_0 varies weakly from solvent to solvent. Thus, the solvent sensitivity of a fluorophore to polarity of solvent or polar interactions in the solute–solvent system can be estimated by a solvatochromic (Lippert–Mataga, Bakhshiev, Kawski [61, 62]) plot. The most sensitive probes are those with the largest change in $(\mu_g - \mu_e)^2/a^3$ factor, i.e., having large change dipoles moments at excitation and small dimensions a. Thus, (16)–(22) provide a basis for the determination of important molecular parameters μ_g, μ_e, γ. These results can be compared with that obtained by quantum mechanical methods. Their more precise determination can be obtained by electrooptical methods [63].

When we perform experiment in such way that there is no interference of H-bonds or these bonds are stable and structure of solvent also does not varies essentially, solvatochromic plot demonstrates very good linearity as shown, for example, for some naphthylamine derivatives in ethanol–water mixtures. The linearity of solvatochromic plots is often regarded as an evidence for the dominant importance of nonspecific universal intermolecular interaction in the spectral shifts. Specific solvent effects lead to essential deviation of measured points from this linear plot.

2.4 Photoinduced Charge Transfer

All early theories of solvatochromism were developed for constant dipole moments in the excited singlet states. This is evidenced by (15)–(20), where μ_e is a constant dipole moment for definite probe. There are few interesting classes of organic probes with strong charge separation or ICT in the excited state, which can provide a considerable growth of the dipole moment. Well-known example is molecule of 4-(N,N'-dimethylamino)-benzonitrile (DMABN) and its derivatives. DMABN molecule has planar structure in the ground state. Simplified photophysics of CT probes may be explained by the four-level scheme similar to that presented in Fig. 9. The initially excited state is called the locally excited (LE) that is the Franck–Condon state for transition upon excitation of molecule. In solvents with low polarity, DMABN emits at the short wavelengths near 330 nm from the LE state. As the solvent polarity increases, a new fluorescence band appears, which is shifted to the long wavelength region and overlaps with the first band. This band is formed by ICT state, which is denoted commonly as TICT (twisted internal charge transfer) because dimethylamino group in this state is perpendicular to the rest of molecule. In DMABN, this state is due to a transfer of electron density from the amino N atom to the center of aromatic ring and it forms very rapidly. Emission from TICT state populates the Franck–Condon state, which is different from the

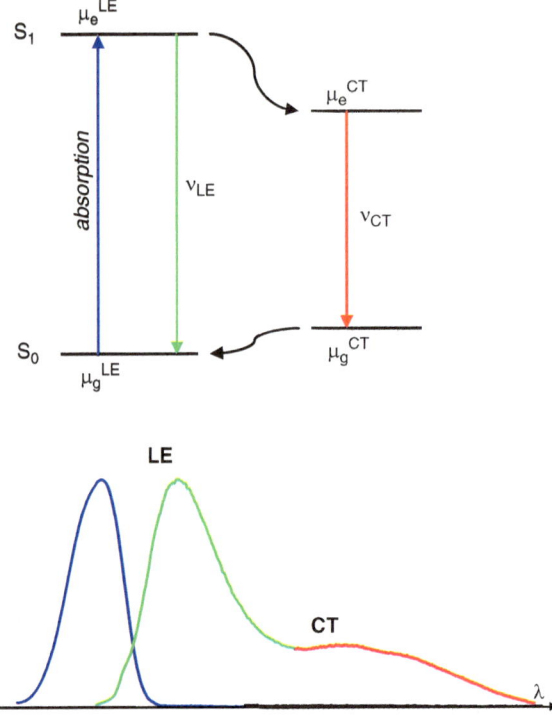

Fig. 9 Scheme of energy levels of absorption and fluorescence of solute with charge transfer

initial ground state. All four electronic states are very polar; their average values of dipoles moments are: $\mu_g^{LE} \sim 6$ D, $\mu_e^{LE} \sim 12$ D, $\mu_g^{CT} \sim 6$ D and $\mu_g^{CT} \sim 15$ D [64].

Due to different change of electric dipole moments during transitions in systems of LE and TICT states, the emission bands of such probes have characteristic solvatochromic plots [64]. Obviously, the TICT band demonstrates a higher sensitivity to solvatochromic function $f(\varepsilon_0, n_0)$ than LE band, as $\Delta\mu$ for this band reaches 9 D in comparison with $\Delta\mu = 6$ D for LE state. Further, in the polar solvents, long wavelength fluorescence grows in relative intensity, while the intensity of the first LE band decreases with polarity of the medium. All these features are seen in Fig. 10, where absorption and fluorescence spectra are shown for DMABN in different solvents. Thus, for ICT probes, not only solvatochromic shifts of different bands but also ratio of their intensities, I_{LE}/I_{ICT}, can be used for determination of polarity of the probe surrounding. As it is known [7, 22], ratiometric signals have a number of advantages in comparison with other methods of detection and probing.

There are a lot of probes with charge transfer, among which are Prodan, Laurdan, and derivatives, derivatives of stilbene, and Bodipy. For some of them,

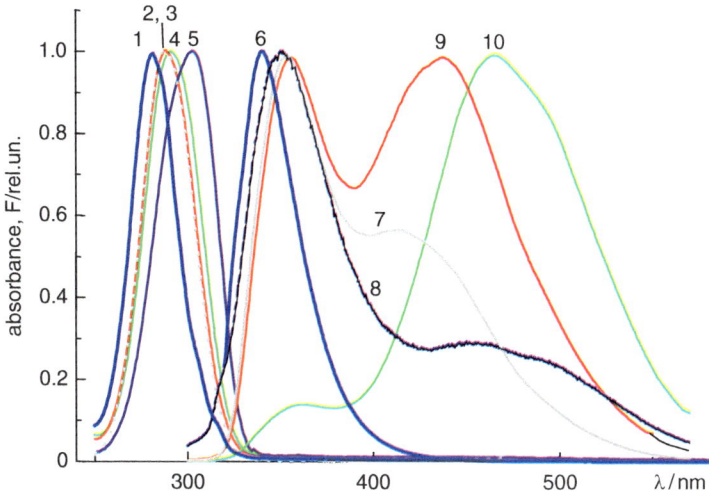

Fig. 10 Spectra of absorption (1–5) and fluorescence (6–10) for DMABN in different solvents: 1,6-cyclohexan; 2,7-dioxan; 3,9-tetrahydrofuran; 5,8-glycerol; 4,10-acetonitrile

for example, in DCM (4-dicyanomethylene-2-methyl-6-p-dimethylamino-styryl-4H-pyran), $\Delta\mu$ reaches almost 20 D and they demonstrate strongest response to variation of polarity.

A change in the ability of a solvent to form H-bonds can affect the nature of the lowest singlet states [1, 2, 65] of organic probes. Some aromatic carbonyl compounds often have low-lying, closely spaced n–π^* and π–π^* states. Inversion of these two states can be observed, when polarity and H-bonding ability of the solvent increase, because the n–π^* state shifts to higher energy whereas the π–π^* state moves to lower. This results in an increase in fluorescence quantum yield because radiative emission from n–π^* states is known to be less efficient than from π–π^* states. The other consequence is a red shift of the fluorescence spectrum. Inversion of n–π^* and π–π^* states was observed, for example, in DMABN and derivatives, and in 7-alkoxycoumarines.

The inhomogeneous broadening model offers additional possibilities in the study of CT reactions. At usual excitation at maximum of the main absorption band of CT solutes, we populate the LE states, and in polar solvents, there follows the relaxation to CT state. On the contrary, red-edge excitation of vitrified solutions of some solutes allows, in principle, to get directly CT states omitting the LE state. Such opportunities were demonstrated experimentally for 9,9′-Bianthryl molecules in propylene glycol [32, 66]. Similar results were obtained also with DMABN in glycerol [67] and Laurdan in glycerol [68]. Thus, we can conclude that selective excitation of different groups of solutes within their population displayed by inhomogeneously broadened spectra is an important tool for studying the mechanism of CT reactions. A change of wavelength of excitation at the red-edge of absorption allows studying the coupling of these reactions with polarity, dielectric relaxation, and other properties of solute environment. Hence, it is possible due to inhomogeneous broadening to modulate smoothly the CT character and rates and to observe involvement of inhomogeneous kinetics in these reactions.

We can resume that the dyes with high value of $\Delta\mu/a^3$ parameter demonstrate a high sensitivity to polar characteristics of environment. It is necessary to distinguish the dyes with rigid geometry having constant dipole in excited state μ_e, those with CT states (they have changeable in time dipole μ_e) and those with close lying n–π^* and π–π^* transitions. Formation of H-bonds provides the effects on spectra in the same direction as an increase of solvent polarity; hence, such contribution should be accounted while studying solute–solvent systems where specific interactions like H-bonding can occur. Detailed coverage of the topic, namely, solute–solvent systems appearing in TICT states can be found in this book [69].

2.5 *Effect of Specific Interactions and Viscosity*

The general or universal effects in intermolecular interactions are determined by the electronic polarizability of solvent (refraction index n_0) and the molecular polarity (which results from the reorientation of solvent dipoles in solution) described by dielectric constant ε. These parameters describe collective effects in solvate's shell. In contrast, specific interactions are produced by one or few neighboring molecules, and are determined by the specific chemical properties of both the solute and the solvent. Specific effects can be due to hydrogen bonding, preferential solvation, acid–base chemistry, or charge transfer interactions.

These bonds can occur between molecules (intermolecular) or within different atomic groups of a single molecule (intramolecular, like in molecules exhibiting ESIPT) [16–23, 70]. Two types of H-bond, which are due to solvent hydrogen bond donor acidity and solvent hydrogen bond acceptor basicity are depicted in Fig. 6 for FL. The H-bond can be placed somewhere between a covalent bond and an electrostatic intermolecular attraction and occurs in both inorganic molecules, such as water, and organic molecules, such as DNA. Usually, the contour of the broad spectra of organic probes as a result of H-bonding remains unchanged but demonstrates the red shift.

The method of revealing of H-bonds is very simple: an addition of low concentration, 1–3% of molar fraction, of alcohols (ethanol, methanol) to the solution in neutral solvent (CH, for example) results in a substantial spectral shift. Further addition of alcohols, up to 100%, gives much smaller shifts. A small percentage of alcohol may cause 50–80% of total spectral shift. Upon addition of the trace quantities of alcohol, one sees that the intensity of the initial spectrum is decreased, and new red-shifted spectrum appears. The appearance of new spectral component is a characteristic of specific solvent effects. Because the specific spectral shifts occur only at low concentration of alcohol, this effect is probably attributed to H-bonding to electronegative group in the molecule. The next experiment, which can support this conclusion, is an addition of aprotic solvent, for example,

Fig. 11 Spectral shifts
$\Delta v^{af} = v^a - v^f$ of fluorenone
versus molar fraction x_m of
THF (1) and EtOH (2)
(courtesy of M. Józefowicz)

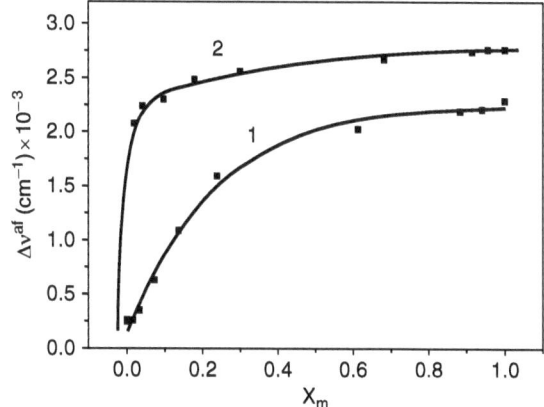

tetrahydrofuran (THF), to the same neutral solvent. In this case, we observe another
picture: a smooth red shift of the spectrum takes place gradually and a saturation of
the shift is seen at $\approx$60–80% of THF molar concentrations, for example, see [53].
All these peculiarities are seen in Fig. 11, where emission spectral shifts of FL in
CH are shown versus molar fractions of THF and ethanol (EtOH).

Interesting examples of specific solvent effects are provided by molecules with
structured emission spectra in nonpolar solvents, such as 2-acetylantracene (2-AA)
and some other derivatives of anthracene [71]. Addition of low concentrations of
EtOH results in a loss of the band structure, which is replaced by unstructured
emission shifted to longer-wavelengths. As the solvent polarity is increased further
with alcohol concentration, the fluorescence spectrum continues to shift gradually
to longer wavelengths. These spectra suggest that the fluorescence of 2-AA is
sensitive to both specific solvent effects and general solvent effects in more polar
solvents.

Specific solvent–solute interactions can occur either in the ground or in the
excited states. If the interaction only occurs in the excited state, then the polar
additive would not affect the absorption spectra. If the interaction occurs in the
ground state, some changes in the absorption spectrum are expected. In the case of
2-AA, the absorption spectra show loss of vibrational structure and a red shift upon
adding methanol. This suggests that 2-AA and alcohol are already hydrogen-
bonded in the ground state. An absence of changes in the absorption spectra
would indicate that there are no any specific interactions in the ground state.
Alternatively, weak H-bonding may occur in the ground state, and the strength of
this interaction may be increased after the excitation. If the probe and polar solvent
associate in the excited state only, then the appearance of specific solvent effects
will depend on the rates of diffusion of the probe and polar solvent.

Evidence for specific solvent–solute interactions can be seen in the Lippert or
others solvatochromic plots. One notices that the Stokes shift is generally larger in
H-bonding solvents (water, alcohols) than in solvents with less probability to form

H-bonds. Such behavior of the Stokes shifts in protic solvents is typical for specific solvent–solute interactions, and has been seen for many solutes: phthalimides, FLs, oxazines, and others [1, 2, 4, 50, 51, 72–74].

Situation with H-bonding also demands to take into account the fact that alcohols have ability to form various associates or even clusters at normal conditions. The most efficient method for determination of inhomogeneity in the excited states is fluorescence polarization measurements. These methods also frequently applied for studying of solvent viscosity, they may be provided in two variants: steady state and time-resolved. Relations for time-resolved and steady state fluorescence anisotropy may be given as [1, 2, 75]:

$$r(t) = r_0 \sum_i \exp(-t/\varphi_i) \tag{23}$$

$$\bar{r} = r_0 \sum_i \frac{\alpha_i}{1 + \tau/\varphi_i} \tag{24}$$

where φ_i is rotational correlation time, r_0 is the fundamental anisotropy, i.e., theoretical anisotropy in the absence of any fluorophore motions. Rotational motion can be hindered in the rigid sample. The experimental value, called the limiting anisotropy, is always slightly smaller than the theoretical value. When the absorption and fluorescence transition dipole moments are parallel, the theoretical value of r_0 is 0.4, but the experimental value usually ranges from 0.32 to 0.39. A totally asymmetric rotor has three different rotational diffusion coefficients, and in the case if the absorption and the emission transition moments are not directed along one of the principal diffusion axes, the decay of $r(t)$ is a sum of five exponentials. Steady state anisotropic measurements are then insufficient for fully characterizing of rotational motions, and time-resolved experiments are required.

Rotational correlation time is directly coupled to rotational diffusion coefficient:

$$\varphi = \frac{\eta V s}{kT} \tag{25}$$

where k is the Boltzmann constant, T is the absolute temperature, V is the volume of rotating unit, η is viscosity, and s is the coupling factor ($s = 1$ for "stick" and $s < 1$ for "slip" boundary conditions, which are introduced for accounting of solute–solvent interactions). Stick conditions determined from polarization measurements indicate directly to solute–solvent complexes in the excited state. From (25), one may obtain information about the values of viscosity η or about hydrodynamic volumes. The term *microviscosity* is often used, but no absolute values can be given, and the best we can do is to speak of an *equivalent viscosity*, i.e., the viscosity of a homogeneous medium, in which the response of the probe is the same. The main reason is that the size of probes is comparable to that of the surrounding molecules forming the probing microenvironment; therefore, the structure of solvate should be accounted and (25) is only the first approximation.

It should be expected that in inhomogeneous population of solutes those of them that are solvated stronger as a result of intermolecular H-bonding and, therefore, emitting in the long-wavelength part of the fluorescence spectrum would rotate slower than the molecules emitting at shorter wavelengths [72–74]. From experimental data obtained for oxazine 17 in different solvents [74], it follows that in aprotic solvents (dimethylformamide and dimethylsulfoxide), the values of normalized to viscosity rotation time, τ_c/η, are the same $\sim$180–190 ps cP^{-1} for the molecules emitting in short-wave (625 nm) and long-wave (690 nm) ranges of fluorescence spectrum. However, in proton donor solvents (alcohols), there are distributions of τ_c/η over the spectrum, for example, in ethanol molecules emitting in the short wavelength part of the band rotate faster $\sim$330 ps cP^{-1} than the molecules fluorescing in the red part$\sim$ 420 ps cP^{-1}. Deuteration of solvents leads to a sharp decrease of rotation time τ_c/η and, very importantly, to a smaller distribution of τ_c/η values within the short- and long-wavelength parts of the spectrum. The values τ_c/η determined for alcohols exceed that in aprotic solvents. The obtained data evidence for the existence of intermolecular H-bonding with different clusters of alcohols, which slows down the rotation of oxazine 17 and the existence of inhomogeneity due to the presence of complexes with different structures leading to spectral dependence of τ_c. Therefore, solvation of the solutes having carbonyl groups may be essentially determined by intramolecular H-bonds resulting in an appearance of multicentricity and inhomogeneous broadening of their spectra, which determines the spectral and energetic characteristics of fluorescence.

An understanding of specific and general solvent effects can provide a basis for interpreting the emission spectra of fluorophores that are bound to macromolecules or, in general, are located in different sites of structurally heterogeneous system.

In addition to the described above methods, there are computational QM–MM (quantum mechanics–classic mechanics) methods in progress of development. They allow prediction and understanding of solvatochromism and fluorescence characteristics of dyes that are situated in various molecular structures changing electrical properties on nanoscale. Their electronic transitions and according microscopic structures are calculated using QM coupled to the point charges with Coulombic potentials. It is very important that in typical QM–MM simulations, no dielectric constant is involved! Orientational dielectric effects come naturally from reorientation and translation of the elements of the system on the pathway of attaining the equilibrium. Dynamics of such complex systems as proteins embedded in natural environment may be revealed with femtosecond time resolution. In more detail, this topic is analyzed in this volume [76].

3   Concluding Remarks

A number of processes and excited state reactions in the S_1 singlet state of organic solutes is possible; they may compete with fluorescence and affect directly the quantum yield, the lifetime, and the spectrum of emission. We have reviewed briefly the most important, applied on a broad scale in fluorescence sensing, and among them are the electron transfer, formation of exciplexes and excimers, ESIPT, FRET, ICT coupled with changes of molecular geometry (TICT), solvatochromism, and hydrogen bonding.

An inherit property of polar solution of organic solutes is the fluctuations and heterogeneity of their environment structures that results in essential inhomogeneous broadening of their electronic spectra. Inhomogeneous broadening affects all spectral characteristics of organic molecules in solutions. The selective excitation of different groups of solutes contributing to inhomogeneously broadened spectra is an important methodology for studying the mechanism of excited state reactions such as CT, FRET, intermolecular relaxation in liquid solutions, and some others. A change of wavelength of excitation allows exciting the solutes from different sites and studying the coupling of excited state reactions at these sites with polarity, dielectric relaxation, and other properties of solute environment. The knowledge of physical mechanisms behind the fluorescence properties of organic dyes opens a door for determination of both qualitative and quantitative characteristics in their use as fluorescence reporters.

References

1. Lakowicz J (2006) Principles of fluorescence spectroscopy, 3rd edn. Springer-Verlag, New York
2. Valeur B (2007) Molecular fluorescence, 4th edn. Weinheim, Willey-VCH
3. Waluk J (2000) Conformational aspects of intra and intermolecular excited state proton transfer. In: Waluk J (ed) Conformational analysis of molecules in excited State. Willey-VCH, Weinheim, pp 57–112
4. Bachshiev NG (2005) Photophysics of dipole–dipole interactions. Solvation and complexation processes. S-Petersburg's University Publishing House, Sankt-Petersburg (in Rus)
5. Demtröder W (1988) Laser spectroscopy. Basic concepts and instrumentation, 3rd edn. Springer-Verlag, Berlin, Heidelberg, New York
6. Rettig W, Strehmel B, Schrader S, Seifert H (1999) Applied fluorescence in chemistry, biology and medicine. Springer, New York
7. Demchenko AP (2009) Introduction to fluorescence sensing. Springer-Verlag, Berlin, Heidelberg
8. Tichonov EA, Shpak MT (1979) Nonlinear optical phenomena in organic compounds. Naukova Dumka, Kijiv (in Rus)
9. Dobretsov GE (1989) Fluorescent probes in study of cells. Membranes and lipoproteins. Nauka, Moscow (in Rus)
10. Ratajczak H, Orville-Thomas WJ (1981) Molecular interactions.v.1 and 2. Wiley, New York
11. de Silva AP, Fox DB, Huxley AJM et al (2000) Combining luminescence, coordination and electron transfer for signaling purposes. Coord Chem Rev 205:41–57
12. Lakowicz JR (1999) Principles of fluorescence spectroscopy. Kluwer Academic, New York
13. Terenin AN (1967) Photonics of dye molecules. Nauka, Leningrad (in Rus)
14. Birks JB (1970) Photonics of aromatic molecules. Wiley-Inter-Science, London
15. Kapinus EI (1988) Photonics of molecular complexes. Naukova Dumka, Kijiv
16. Tomin VI, Oncul S, Smolarczyk G, Demchenko AP (2007) Dynamic quenching as a simple test for the mechanism of excited state reaction. Chem Phys 342:126–134

17. Tomin VI, Smolarczyk G (2008) Dynamic quenching of the multiband fluorescence of 3-hydroxyflavone. Opt Spectros 104:919–925
18. Tomin VI, Tomin VI, Jaworski R (2008) Investigation of reactions from the highest excited states of molecules by fluorescent spectroscopy methods. Opt Spectros 104:40–49
19. Tomin VI (2009) Effect of temperature and dynamic quenching on proton transfer in 3-hydroxyflavone. Opt Spectros 107:92–100
20. Formosinho SJ, Arnaut LG (1993) Excited state proton transfer reactions II. Intramolecular reactions. Photochem Photobiol A Chem 75:21–48
21. Weller A (1961) Fast reactions of excited molecules. Prog React Kinet 1:189–214
22. Demchenko AP (2005) Optimization of fluorescence response in the design of molecular biosensors. Anal Biochem 343:1–22
23. Sengupta PK, Kasha M (1979) Excited state proton transfer spectroscopy of 3-hydroxyflavone and quercetin. Chem Phys Lett 68:382–385
24. Hsieh CC, Ho ML, Chou PT (2010) Organic dyes with excited-state transformations (electron, charge and proton transfers), A.P. Demchenko (ed.), Advanced Fluorescence Reporters in Chemistry and Biology I, Springer Ser Fluoresc 8:225–266
25. Forster Th (1960) Transfer mechanisms of electronic excitation energy. Radiat Res Suppl 2:326–339
26. Ermolaev VL, Bodunov EN, Sveshnikova EB et al (1977) Nonradiative electronic energy transfer. Nauka, Leningrad
27. Demchenko AP (2008) Site-selective red-edge effects. Methods Enzymol 450:59–78
28. Weber G, Shinitzky M (1970) Failure of energy transfer between identical aromatic molecules on excitation at the red edge of the absorption spectrum. Proc Natl Acad Sci USA 65:823–830
29. Nemkovich NA, Rubinov AN, Tomin VI (1991) Inhomogeneous broadening of electronic spectra of dye molecules in solutions. In: Lakowicz JR (ed) Topics in fluorescence spectroscopy, principles, vol 2. Plenum, New York, pp 367–428
30. Zenkevich RAN, EI NNA, Tomin VI (1982) Directed energy transfer due to orientational broadening of energy levels in photosynthetic pigments solutions. J Lumin 26:367–376
31. Demchenko AP (1986) Ultraviolet spectroscope of proteins. Springer-Verlag, Heidelberg, New York
32. Demchenko AP, Sytnik AI (1991) Site selectivity in excited state reactions in solutions. J Chem Phys 95:10518–10524
33. Nemkovich NA, Gulis IM, TominV I (1982) Excitation-frequency dependence of the efficiency of directed nonradiative energy transfer in two-component solid solutions of organic compounds. Optic Spectros 53:140–143
34. Galley WC, Purkey RM (1970) Role of heterogeneity of the solvation site in electronic spectra in solution. Proc Natl Acad Sci U S A 67:1116–1121
35. Rubinov AN, Tomin VI (1970) Bathochromic luminescence in solutions of organic dyes at low temperatures. Opt Spectros 29:578–580
36. Personov RI, Al'shits LA, Bykovskaja LA (1972) The effect of fine structure appearance in laser-excited fluorescence spectra of organic compounds in solid solutions. Opt Commun 6:169–173
37. Marcus RA (1965) On the theory of shifts and broadening of electronic spectra of polar solutes in polar media. J Chem Phys 43:1261–1274
38. Mazurenko YT (1983) Statistics of solvation and solvatochromy. Opt Spectros 55:471–478
36. Mazurenko YT (1989) Spectroscopy of relaxational and statistical phenomena in solvate shells of molecules. In: Bachshiev N (ed) Solvatochromy. Problems and methods. Publishing House of Leningrad's State University, Leningrad, pp 122–190
40. Demchenko AP (2002) The red-edge effects: 30 years of exploration. J Lumin 17:19–42
41. Rubinov AN, Tomin VI, Zhivnov VA (1973) A shift of fluorescence spectrum of molecules in nonresonance light field. Opt Spectros 35:778–781
42. Tomin VI, Rubinov AN, Voronin VF (1973) An effect of quenchers on fluorescence spectra of polar dye solutions. Opt Spectros 34:1108–1111

43. Rubinov AN, Tomin VI (1971) Bathochromic luminescence of organic dyes in alcohol solutions and polymer matrices. Opt Spectros 32:424–428
44. Tomin VI, Rubinov AN, Kozma L (1973) Inhomogeneous broadening of α and β phosphorescence spectra of dyes. Acta Phys Chem Szeged 21:11–18
45. Nemkovich NA, Rubinov AN, Tomin VI (1980) Directed energy transfer in rigid single component solutions. JTP Lett 6:270–273
46. Gulis IM, Komiak AI, Tomin VI (1978) The electronic excitation energy transfer at conditions of dye spectra inhomogeneous broadening. Izv Akad Nauk SSSR, ser fiz 42:307–312
47. Nemkovich NA, Gulis IM, Tomin VI (1982) Excitation-frequency dependence of the efficiency of directed nonradiative energy transfer in two-component solid solutions of organic compounds. Opt Spectros 53:140–143
48. Ware WR, Lee SK, Brant GJ, Chow PP (1970) Nanosecond time-resolved emission spectroscopy: spectral shifts due to solvent–solute relaxation. J Chem Phys 54:4729–4737
49. Ware WR, Chow PP, Lee SK (1968) Time-resolved nanosecond emission spectroscopy: spectral shifts due to solvent–solutes relaxation. Chem Phys Lett 2(6):356–358
50. Rubinov AN, Tomin VI, Bushuk BA (1982) Kinetic spectroscopy of orientational states of solvated dye molecules in polar solutions. J Lumin 26:377–391
51. Reinchardt C (1988) Solvent effects in organic chemistry. Verlag Chemie, Weinheim
52. Kamlet MJ, Abboud J-L, Taft R (1977) The solvatochromic comparison method. 6. The π* scale of solvent polarities. JACS 99:6027–6038
53. Jozefowicz M (2007) Determination of reorganization energy of fluorenone and 4-hydroxyfluorenone in neat and binary solvent mixtures. Spectrochim Acta A 67:444–449
54. Kamlet MJ, Dickinson C, Taft RW (1981) Linear solvation energy relationship. Solvent effects on some fluorescent probes. Chem Phys Lett 77:69–72
55. Bakhshiev NG (1972) Spectroscopy of intermolecular interactions. Nauka, Leningrad
56. Bakhshiev NG (1989) Solvatochromy. Problems and methods. Publishing House of Leningrad's State University, Leningrad
57. von Lippert E (1957) Spektroskopische bistimmung des dipolmomentes aromatischer verbindungen im ersten angeregten singulettzustand. Z Electrochem 61:962–975
58. Mataga N, Kaifu Y, Koizumi M (1956) Solvent effects upon fluorescence spectra and the dipole moments of excited molecules. Bull Chem Soc Jpn 29:465–470
59. Mac Rae EG (1956) Theory of solvent effects of molecular electronic spectra: frequency shifts. J Chem Phys 61:562–572
60. Suppan P (1983) Excited-state dipole moments from absorption/fluorescence solvatochromic ratios. Chem Phys Lett 94:272–275
61. Billot L, Kawski A (1962) Zur Theorie des Einflusses von Lösungsmitteln auf die Elektronenspektren der moleküle. Z Naturforsch 17a:621–626
62. Kawski A (1992) Solvent-shift effect of electronic spectra and excitation state dipole moments. In: Rabek JR (ed) Progress in photochemistry and photophysics. CRC, New York, pp 1–47
63. Baumann W (1989) Determination of dipole moments in the ground and excited states. In: Rossiter BW, Hamilton J (eds) Physical methods of chemistry. 3B. Willey, New York, pp 45–131
64. Grabowski ZR, Rotkiewicz K, Rettig W (2003) Structural changes accompanying intramolecular electron transfer: focus on twisted intramolecular charge-transfer states and structures. Chem Rev 103(3899–4032):35
65. Seliskar C, Brand L (1971) Electronic spectra of 2-aminonaphthalene-6-sulfonate and related molecules. II. Effects of solvent medium on the absorption and fluorescence spectra. JACS 93:5414–5420
66. Demchenko AP, Sytnik AI (1991) Solvent reorganizational red-edge effect in intramolecular electron transfer. Proc Natl Acad Sci USA 88:9311–9314
67. Tomin VI, Hubisz K (2004) Band broadening in the electronic spectra of 4-dimethylaminobenzonitrile in polar solvents. Russ J Phys Chem 78:1114–1118

68. Tomin VI, Brozis M, Heldt J (2003) The red edge effects in laurdan solutions. Z Naturforsch 58a:109–117
69. Haidekker MA, Nipper M, Mustafic A, Lichlyter D, Dakanali M, Theodorakis EA (2010) Dyes with segmental mobility: molecular rotors, Ch. 10. In: Demchenko AP (ed) Advanced Fluorescence Reporters in Chemistry and Biology I. Springer Ser Fluoresc 8:267–307
70. Kasha M (1986) Proton-transfer spectroscopy. Perturbation of the tautomerization potential. J Chem Soc Faraday Trans 2 82:2379–2392
71. Timaki T (1982) The photodissociation of 1- and 2-acethylantracenes with methanol. Bull Chem Soc Jpn 55:1761–1767
72. Bushuk BA, Rubinov AN, Stupak AP (1987) Inhomogeneous broadening of spectra of dye solutions due to intermolecular hydrogen bonding. J Appl Spectros 47:1251–1254
73. Rubinov AN, Bushuk BA, Stupak AP (1983) Picosecond spectroscopy of intermolecular interactions in dye solutions. Appl Phys B 30:99–104
74. Bushuk BA, Rubinov AN, Stupak AP (1987) Inhomogeneous broadening of dye solutions spectra due to intermolecular hydrogen bond. Zh Prikl Spektr 47:934–938
75. Gajsenok VA, Sarzevski AM (1986) Anizotropy of absorption and emission of multiatomic molecules. Publishing House of Byelorussian State University, Minsk
76. Callis PR (2010) Electrochromism and solvatochromism in fluorescence response of organic dyes. A nanoscopic view, Ch. 10. In: Demchenko AP (ed) Advanced Fluorescence Reporters in Chemistry and Biology I. Springer Ser Fluoresc 8:309–330

Organic Dyes with Excited-State Transformations (Electron, Charge, and Proton Transfers)

Cheng-Chih Hsieh, Mei-Lin Ho, and Pi-Tai Chou

Abstract In this chapter, contemporary progress on organic dyes undergoing excited-state electron and/or proton transfers is reviewed via three aspects: (1) the fundamental view on the mechanistics of electron transfer reaction, i.e., adiabatic versus nonadiabatic processes, and their applications in, e.g., ion recognition; (2) excited-state intramolecular proton transfer (ESIPT) in view of hydrogen bonding configuration; and (3) the role of solvation and solvation dynamics in the proton transfer coupled charge transfer reactions. For (1), of particular emphasis are bipolar molecules, for which the electron donor (D) and acceptor (A) are linked by non-π-conjugated covalent bonds. As for altering the distance between D and A via rigid spacers, the electron transfer process can be fine-tuned from adiabatic to nonadiabatic. Relevant fundamentals and applications are briefly reviewed and discussed. For (2), prototypes of intramolecular ESIPT are classified into five-, six-, and seven-membered ring hydrogen-bonding systems. While five- and six-membered hydrogen-bonding systems are somewhat common, seven-membered one is rare. In this section, in addition to those of five- and six-membered classes, paradigms of seven-membered hydrogen-bonding systems are presented by using the analog of green fluorescent protein core chromophore. In (3), various mechanisms of protic solvent assisting excited-state proton transfer (ESPT) are reviewed, among which plausible mechanism is deduced and discussed. Particular attention is paid to reaction dynamics in alcohol and aqueous solutions, with an aim toward biological applications. Last but not least, the differences in how solvent diffusive reorganization and solvent relaxation affect the ESPT dynamics are discussed; the conclusions should provide more insight into the micro- and macrosolvation processes.

C.-C. Hsieh, M.-L. Ho, and P.-T. Chou (✉)
Department of Chemistry, National Taiwan University, No. 1, Sec. 4, Roosevelt Rd., Da-an District, Taipei 106, Taiwan
e-mail: chop@ntu.edu.tw

A.P. Demchenko (ed.), *Advanced Fluorescence Reporters in Chemistry and Biology I:*   225
Fundamentals and Molecular Design, Springer Ser Fluoresc (2010) 8: 225–266,
DOI 10.1007/978-3-642-04702-2_7, © Springer-Verlag Berlin Heidelberg 2010

Keywords Excited-state intramolecular proton transfer · Fluorescence dye · Photoinduced electron transfer · Proton coupled electron transfer · Relaxation dynamics

Contents

1 Photoinduced Electron Transfer (PET)

Excited-state electron transfer represents one of the most fundamental pathways in chemical and biological processes. This, together with its prospects for application, elaborated in later sections, has been attracting considerable attention. As a result, many review papers and books have been published to address this issue [1–6].

Among the various relevant research directions, one fundamental issue lies in the differentiation between adiabatic and nonadiabatic types of electron transfer. A rather strong coupling between electron donor (D) and acceptor (A) may result in a great mixing between two potential energy surfaces such that the electron transfer occurs essentially along the same potential energy surface in the excited state. Such a process is *adiabatic* and is commonly referred as an optical electron transfer. On the other hand, weak coupling leads to a small interaction of $\ll k_B T$ (298 K) between the potential surfaces of D and A, resulting in a nonnegligible barrier along the electron-transfer pathway. This *nonadiabatic* process is commonly defined as photo-induced electron transfer (PET). Optical electron transfer process responds concurrent with the optical excitation and the observed relaxation dynamics are usually manifested by solvent relaxation due to a large alternation in the dipolar vector, giving rise to emission solvatochromism [7–9]. Conversely, the relatively slow rate of PET may compete with the radiative pathway, i.e., the fluorescence, such that dual emission, consisting of the donor fluorescence (assuming only the donor is excited) and D^+A^- charge transfer emission, may be resolved. For the latter, the emission intensity is normally weak and, in most cases, is irresolvable due to its forbidden transition in character versus the neutral ground state. Numerous studies have focused on the D/A dyads linked by either a rigid or flexible framework to study the associated PET processes [10–15], among which key issues regarding through bond, through space, or structural tuning parameters have been thoroughly examined to gain detailed insights into the associated mechanism/theory.

1.1 Theoretical Background of Electron Transfer

For molecule that lacks excited-state electron transfer, upon excitation, the fluorescence quantum yield (Φ_{nor}) and the fluorescence lifetime (τ_{nor}) can be expressed as:

$$\Phi_{nor} = \frac{k_r}{k_r + k_{nr}} \quad \tau_{nor} = \frac{1}{k_r + k_{nr}} \tag{1}$$

where k_r, and k_{nr} denote the radiative and nonradiative decay rate constant (excluding electron transfer process), respectively. Considering the occurrence of an extra excited-state deactivation pathway, such as electron transfer induced by an increase of solvent polarity or a decrease of the effective distance between D and A, the above expression can be modified as:

$$\Phi = \frac{k_r}{k_r + k_{nr} + k_{et}} \quad \tau = \frac{1}{k_r + k_{nr} + k_{et}} \tag{2}$$

where k_{et} denotes the rate constant of electron transfer. Φ and τ are the observed yield and lifetime, respectively, of the normal fluorescence. By knowing the lifetime (τ_{nor}) and quantum yield (Φ_{nor}, see (1)) of a model compound in the absence of electron transfer process, one can thus estimate k_{et} with the following equations:

$$k_{et} = \frac{1}{\tau} - \frac{1}{\tau_{nor}} = \left(\frac{\Phi_{nor}}{\Phi} - 1 \right) / \tau_{nor} \tag{3}$$

k_{et} can thus be estimated by the steady state absorption and emission spectra, i.e., the associated quantum yield, in combination with the relaxation dynamics of the control molecule; or more precisely, it can be probed by time-resolved fluorescence and/or absorption spectroscopy [16, 17]. Electron spin resonance (EPR) [18], and theoretical approaches [19] can also provide valuable information regarding this process.

The facility of PET can be confirmed by a thermodynamic approach. For weakly interacting PET process, the associated free energy driving force of reaction (ΔG) can be estimated by the Rehm–Weller equation [20], expressed as

$$\Delta G = E_{ox}(D) - E_{red}(A) - E_{00} - \underbrace{\frac{(e^2/4\,\pi\varepsilon_s\varepsilon_0 r_c)}{}}_{\text{coulombic}}$$
$$- \underbrace{\frac{(e^2/8\pi\varepsilon_0)(1/r_{D^+} + 1/r_{A^-})(1/\varepsilon_{CV} - 1/\varepsilon_s)}{}}_{\text{solvation}} \tag{4}$$

where $E_{ox}(D)$ and $E_{red}(A)$ (in eV) are the oxidation and reduction potentials of donor and acceptor molecules, respectively, measured in the solvent with a dielectric constant ε_{CV}. E_{00} is the energy of the 0–0 transition of the chromophore where PET takes place. r_{D+} and r_{A-} are the effective ionic radii and r_c (in Å) is the center-to-center distance between donor cation and acceptor anion, respectively. ε_s denotes the dielectric constant of the solvent applied. Assuming a harmonic oscillator model for solvent–solute interaction so that the corresponding potential energy surface (PES) can be treated as a parabolic shape, a general relationship between driving force associated with the reaction (electron transfer) free energy (ΔG) and activation energy (ΔG^+) can be expressed as

$$\Delta G^+ = \frac{(\Delta G + \lambda)^2}{4\lambda} \quad \lambda = \lambda_i + \lambda_s \qquad (5)$$

where λ_i and λ_s are the intrinsic barriers corresponding to the internal and solvent reorganization energy, respectively (see Fig. 1).

The solvent reorganization term reflects the changes in solvent polarization during electron transfer. The polarization of the solvent molecule can be divided into two components: (1) the electron redistribution of the solvent molecules and (2) the solvent nuclear reorientation. The latter corresponds to a slow and rate-determining step involving the dipole moments of the solvent molecules that

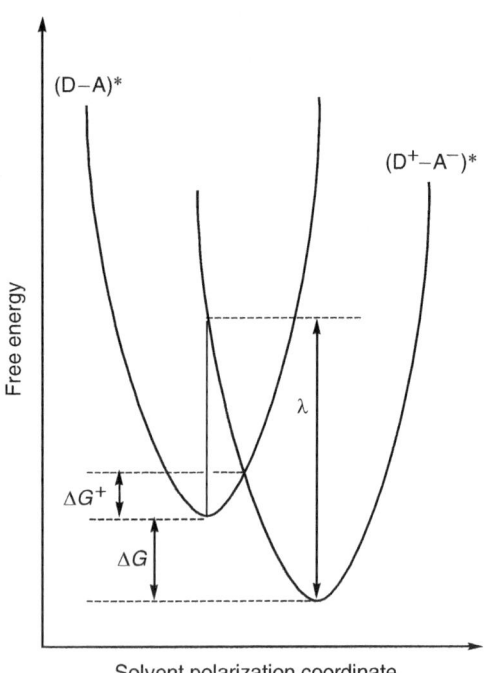

Fig. 1 The potential energy surface of nonadiabatic PET process

reorient themselves around the reactant molecules prior to electron transfer. Such an arrangement, which is most effective in stabilizing the transition state, results in the lowest free energy of activation for electron transfer. The solvent reorganization energy can be estimated by the Born-Hush approach [21],

$$\lambda_s = \frac{e^2}{4\pi\varepsilon_0} \left(\frac{1}{2r_{D^+}} + \frac{1}{2r_{A^-}} - \frac{1}{r_C} \right) \left(\frac{1}{n^2} - \frac{1}{\varepsilon_s} \right) \tag{6}$$

and the internal reorganization energy (λ_i) can be determined by measuring the charge transfer absorption and emission spectra in nonpolar solvent ($\lambda_s \sim 0$).

When the electron coupling between locally excited-state (LE) and charge transfer state (CT) is weak, the electron transfer rate k_{et} can be expressed as (7)

$$k_{et} = \frac{2\pi}{\hbar(4\pi\lambda k_B T)^{1/2}} |V|^2 \exp\left[\frac{-\Delta G^+}{k_B T} \right] \tag{7}$$

where $|V|$ involves an electron coupling matrix element between LE and CT, k_b is the Boltzman constant, $\hbar$ is the Planck's constant, and T denotes the temperature. Followed by the replacement of ΔG by ΔG^+ using (5) [a parabolic PES for solvent interaction (vide supra)], (7) can be rewritten to a well-known Marcus format [21–24] for the nonadiabatic PET reaction, expressed as

$$k_{et} = \frac{2\pi}{\hbar(4\pi\lambda k_B T)^{1/2}} |V|^2 \exp\left[\frac{-(\Delta G + \lambda)^2}{4\lambda k_B T} \right] \tag{8}$$

Experimentally, the plot for $ln(k_{et}T^{1/2})$ as a function of $1/T$ is linear. The energy barrier ΔG^+ and the coupling matrix can thus be extracted from the slope and intercept of the plot, respectively. Followed by (5) with λ deduced from λ_s plus λ_i, ΔG of the PET reaction can thus be resolved.

1.2 Adiabatic Versus Nonadiabatic Electron Transfer Across Linear Fused Oligo-Norbornyl Structures

One of the advances in the field of PET is the design of molecular devices, in which D and A pairs are ingeniously linked by covalent bridges (B) to form D–B–A dyads. Electron transfers between D and A across B in a controlled manner may thus display useful functionalities, such as molecular rectifiers [25], switches [26], bio-sensors [27], photovoltaic cells [28], and nonlinear optical materials [29]. Spacers that have been utilized are versatile, including small molecules, such as cyclohexane [30], adamantane [31], bicyclo[2.2.2]octane [32], steroids [33], and oligomers of various sizes, polynorbornanes [34] and ladderanes [35] being two examples. Among these numerous types of spacers, rigid linear rod-shaped structures, though not commonly seen [36, 37], are highly symmetrical, and may reduce the complexity of studies due to the constraint of geometrical and conformational variations.

Upon fixing the D/A chromophores, the variation of spacer length should fine-tune the electron-coupling matrix V (7) between LE and CT states. Such an approach should address a fundamental issue regarding the differentiation between adiabatic and nonadiabatic electron transfer tuned by subtle changes of the D/A distance. In this regard, one prototypical example is the study of a series of 1,4-dimethoxynaphthalene (D)-(bridge)-1,1-dicyanoethylene (A) systems, in which the bridge consists of norbornadienes, by Paddon-Row and coworkers [38]. Through pulse radiolysis, they were able to present the thermodynamic and optical data in a semiquantitative manner. Unfortunately, the corresponding ultraweak electron-transfer emission impedes further detailed investigation on spectroscopic and relaxation dynamics. The ultraweak emission might be expected due to the forbidden CT transition to the ground state and quenching via flexible skeleton motion. This obstacle could be circumvented via strategic design and synthesis of a new series of dyads **I–III** (Fig. 2) composed of 2,3-dimethoxynaphthalene as an

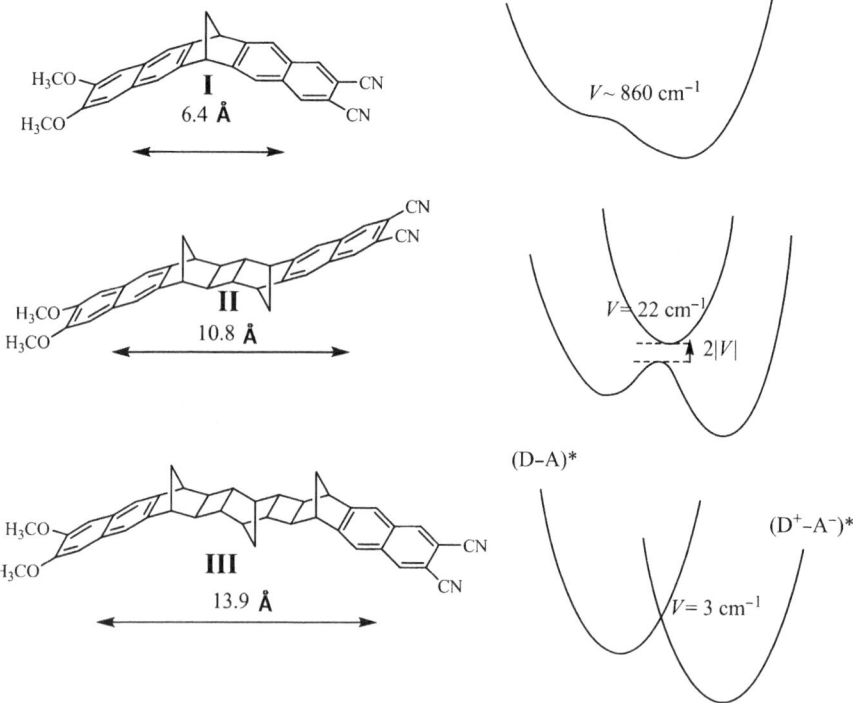

Fig. 2 The molecular structure of **I–III** and the representation of the potential energy surfaces for adiabatic to nonadiabatic electron transfer reactions

electron donor (D) and 2,3-dicyanonaphthalene as an acceptor (A), bridged by
n-norbornadiene ($n = 1$–3). These rigid dyads lead to the resolution of CT emissions and serve as an excellent paradigm to demonstrate the switch of excited-state
electron-transfer dynamics from an adiabatic to a nonadiabatic process [39]. Theoretical approaches have estimated the center-to-center distances between D and A
to be 6.4, 10.8, and 13.9 Å for **I**, **II**, and **III**, respectively. The tendency of the
coupling magnitude is consistent with the experimental observation, in which
system **I**, owing to its strong D/A coupling ($|V| \sim 860$ cm^{-1}), essentially undergoes
an adiabatic type of optical electron transfer, resulting in only charge-transfer
emission that is subject to remarkable solvatochromism, being shifted in peak
wavelength from 380 nm in cyclohexane to 510 nm in acetonitrile solvent. Con-
versely, due to the small electronic coupling of $|V|$ for systems **II** (22 cm^{-1}) and **III**
(3 cm^{-1}), a relatively much slower nonadiabatic electron transfer (i.e., PET) takes
place in systems **II** ($k_{et} \sim 2.5 \times 10^{10}$ s^{-1}) and **III** ($k_{et} \sim 10^9$ s^{-1}), giving rise to
dual emission in CH$_2$Cl$_2$ at 298 K. k_{et} in **II** is apparently much faster than that of **III**
in aprotic solvent, consistent with much bridge-distance-tuning PET formalism.

1.3   The Application of Electron Transfer Process Toward Visual Detection of Metal Ions

Applications based on the mechanism of excited-state electron transfer are ubiquitous in numerous fields. The corresponding electron transfer reactions, in most
cases, are referred to as *adiabatic* processes, for which the electron coupling
matrix is $\gg k_B T$. This is simply due to the fact that a D/A system is designed
such that D and A chromophores are cross talked via a π-conjugated spacer.
In other words, through the π-conjugation, the electron coupling matrix is large
such that the zeroth order PES for both LE and CT states have been strongly
perturbed. In this case, the Franck–Condon absorption profile is no longer equal to
the sum of individual D and A chromophores. On the other hand, for weak
coupling through either σ bond or space (with solvent perturbation), due to the
negligible interaction between LE and CT states, the Franck–Condon absorption
profile is thus more or less equal to the sum of the respective D and A absorption
profiles. As such, an empirical rule of thumb is that the lowest lying transition
band for a *nonadiabatic* type of electron transfer, referred to as PET, commonly
consists of the sum of D/A fragments, while significant perturbation of the
absorption profile is obvious (c.f. the sum of D and A) for the case of *adiabatic*
electron transfer (i.e., optical electron transfer). In fact, for the latter, fragments of
D and A cannot be clearly separated in most cases.

From the sensing point of view, the adiabatic design is superior to any nonadiabatic systems because of its strong mixing between LE and CT states, resulting in a
large transition moment for CT emission. Thus, the contrast of the on/off signal
based on the intensity ratio for LE versus CT is remarkable. Conversely, the CT

emission is commonly much weaker for nonadiabatic reactions, and cannot be exploited effectively as a signal transducer.

A prototypical example is electron-transfer-based cation sensors, which have been explored extensively in past decades [40–45]. Such a molecular device attached to, e.g., a polymer bead and capable of sensing cations can have practical applications for qualitative detection and quantitative determination of metal ions. Not surprisingly, the development of polymer-bead-attached sensors as reusable cation sensors is of current interest [46, 47]. For example, a molecule bearing π-conjugated A and D usually undergoes an adiabatic, intramolecular charge transfer (ICT) upon electronic excitation [40]. The ICT and hence elongation of the π electron conjugation occurring upon Franck–Condon excitation contributes considerably to the absorption profile. If the A–D molecule binds a metal ion at the acceptor site, a bathochromic shift occurs to account for the enhanced ICT. However, in most cases, a metal ion tends to bind at the electron-rich donor site, causing a hypsochromic shift, which often makes visual detection difficult [40–45].

To extend the A–D recognition concept, in the following section, we also focus on the optical properties and binding behavior of D–A–D molecules, which are of interest to investigate because they may be employed in multiple stage sensing of appropriate analytes over a large dynamic range of concentrations.

In general, several binding states may exist in equilibrium. The degree of π electron conjugation induced by ICT, in a qualitative sense, can be designated by the dipolar vector depicted in Fig. 3. For example, the binding of the A–D molecule with a metal ion at the acceptor site, forming MA–D, would enhance the acceptor strength and facilitate ICT, whereas the binding at the donor site, forming A–DM, would reduce ICT. When both the acceptor and donor sites are bound to metal ions, the original donor group D is changed to an electron withdrawing group DM, and the dipole is diminished. According to the preference of binding states in equilibrium, a bathochromic shift might be attributed to an increase in ICT, and a hypsochromic shift might be attributable to a decrease in ICT. The binding event of the D–A–D molecule with metal ions should be much more complicated than that for the A–D molecule. However, a simplified book-keeping mechanism can be drawn by focusing on the two major stages (the early and late stages), as shown in Fig. 3b. At the early stage, the D–A–D molecule may bind with a metal ion, either at the acceptor or at the donor sites, to form 1:1 complexes. An increase of ICT is obvious in the binding at the acceptor site, forming D–AM–D, due to the enhanced acceptor strength. When a metal ion binds at one of the donor sites, the resulting A–DM unit may also function as an enhanced electron-withdrawing group to facilitate ICT, a consequence that differs from the binding of the A–D molecule at the donor site. Therefore, one can predict that the binding of the D–A–D molecule with a metal ion at the early stage will cause a bathochromic spectral shift, regardless of the binding at the acceptor or donor site. When the D–A–D molecule is saturated with metal ions at the late stage, all the acceptor and donor sites are bound to metal ions to form a 1:3 complex (MD–AM–DM). A hypsochromic spectral shift thus occurs to account for the great decrease of dipole at this stage.

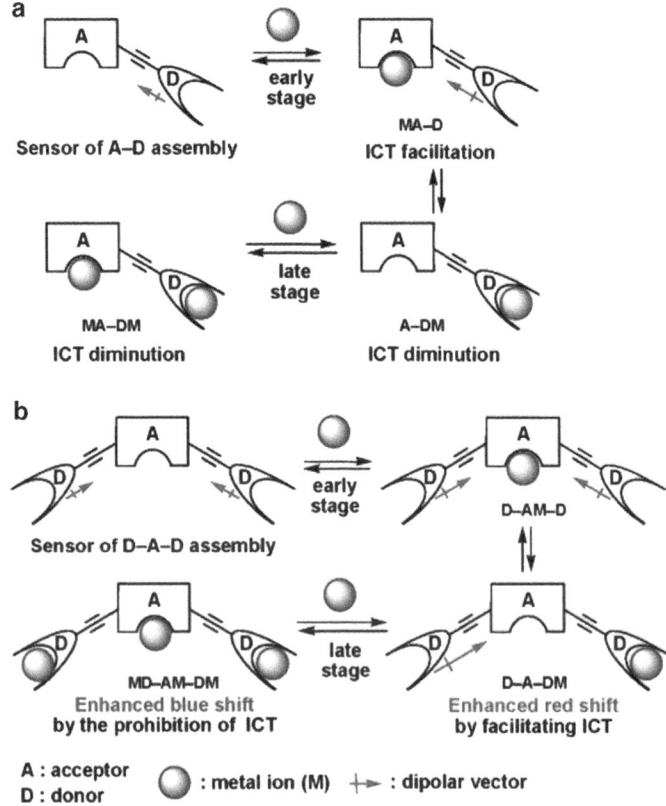

Fig. 3 Concept of metal ion sensor with conjugated A–D and D–A–D assemblies (reprint from ref. [58], Copyright © 2005 American Chemical Society)

1.3.1 The Conjugated D–A Bipolar System

A representative case demonstrated here is chenodeoxycholic acid-based ICT sensors for alkali metal ions, which have been immobilized on Merrifield resin and on Tentagel [48]. The fluorescence of the sensor beads is enhanced upon binding the cations. Recently, there has been intense interest in the application of near-IR (NIR) probes to detect metal cations and biological compounds [49–52]. Zhu et al. have designed and constructed a new class of NIR probe **MCy-1** with colorimetric assay to specifically detect the presence of Hg^{2+} over a wide range of other interfering cations (Fig. 4) [53]. They choose heptamethine cyanine as the fluorophore and di-thia-dioxa-monoaza crown ether moiety as the receptor to yield the probe. Blue-shifted absorption of **Mcy-1** relative to the parent dye, the acceptor model compound, occurs, and the absorption maximum has a blue-shift of about 88 nm. This blue-shift can be attributed to an efficient excited-state ICT process from the donor nitrogen atom on the di-thia-dioxa-monoaza macrocycles to the acceptor

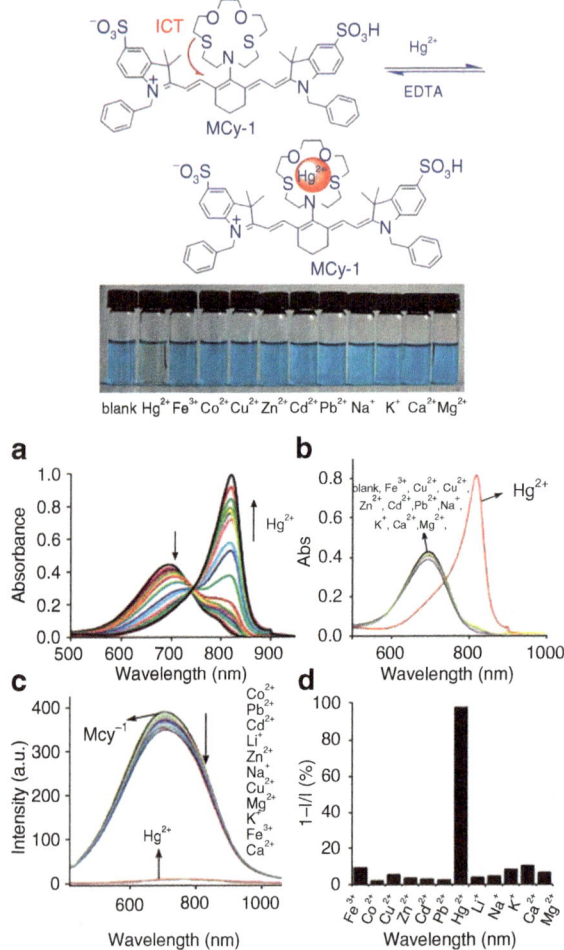

Fig. 4 (**a**) Changes in the absorption spectra of **Mcy-1** upon titration by Hg^{2+}; (**b**) absorption spectra changes of **Mcy-1** upon addition of different cations (10 equiv); (**c**) fluorescence responses of **Mcy-1** to various metal cations; (**d**) bars represent the percentage of fluorescence quenched (reprint from ref. [53], Copyright © 2008 American Chemical Society)

tricarbocyanine group [54]. In methanol solution, **Mcy-1** is characterized by an intense band centered at 695 nm ($\varepsilon = 86,000\ M^{-1} \cdot cm^{-1}$), which is responsible for the blue color of the solution [53]. The absorption at 695 nm decreases sharply with the gradual addition of Hg^{2+} to the solution of **Mcy-1**. At the same time, a new band at 817 nm ($\varepsilon = 190,000\ M^{-1} \cdot cm^{-1}$) increases prominently, with one isosbestic point at 740 nm (Fig. 4a). Such a large red-shift (122 nm) makes the color of the solution change from blue to almost colorless. According to the linear Benesi–Hildebrand expression [55], the 1:1 stoichiometry between the Hg^{2+} and **Mcy-1** was observed. A possible explanation for the red-shift is that the coordination of the Hg^{2+} to the ligand reduces the electron donating ability of the nitrogen atom at the macrocycle. Thus, the ICT process is not possible, and the blue-shift in absorption spectra is suppressed. In other words, a red-shift in absorption spectra is observed upon Hg^{2+} binding. ^1H NMR studies provide further information for the coordination process.

1.3.2 The Conjugated D–A–D Assembly

2,7-bis(1H-pyrrol-2-yl)ethynyl-1,8-naphthyridine, a push–pull conjugated molecule, exhibits a very large Stokes shift of fluorescence upon complexation with glucopyranoside, as reported previously [56]. As for further extension, 2,7-Bis{4-[di(2-hydroxyethyl)amino]phenylethynyl}-1,8-naphthyridine (BHPN, see Fig. 5), a conjugated molecule incorporating a central moiety of naphthyridine and two terminal moieties of di(hydroxyethyl)aniline connected by ethynyl bridges, is picked to represent a D–A–D assembly and to demonstrate the concept of the two-stage sensing with enhanced signal transduction [57, 58]. As shown by the schematic drawing (Fig. 5), the naphthyridine moiety acts as the acceptor site, whereas the di(hydroxyethyl)amino moiety functions as the donor site. The acceptor and donor moieties are connected by ethynyl bridges to form a conjugated scaffold. Instead of the commonly used macrocycles for metal ion detection [40–45], BHPN with the acyclic di(hydroxyethyl)aniline components renders a straightforward preparation and good solubility in aqueous media. BHPN shows two-stage color changes on binding with mercury(II) ion in Me_2SO/H_2O (1:1) solution with a bathochromic shift from 450 to 498 nm, and then an extraordinarily large hypsochromic shift to 378 nm (Fig. 6). The D–A–D constitution can serve as a protocol for the future design of a multiple-stage sensing system, which may eventually lead to practical application on the logic gates [57], based on its sensitivity and selectivity of metal ions and other possible analytes.

1.3.3 Ion Sensor in a Through-Space PET System

The modular nature of the sensor allows the design of different sensors based on this concept. Fluorogenic calix [4]arenes (1, Fig. 7) bearing a pendent ethyleneamine on

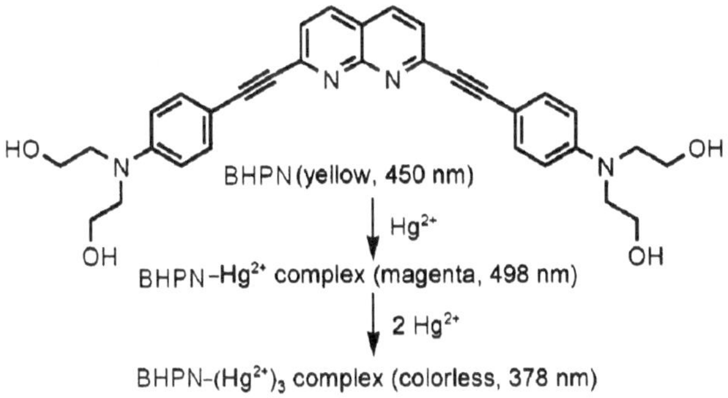

Fig. 5 Two-stage sensing property via a conjugated donor–acceptor–donor constitution of BHPN (reprint from ref. [58], Copyright © 2005 American Chemical Society)

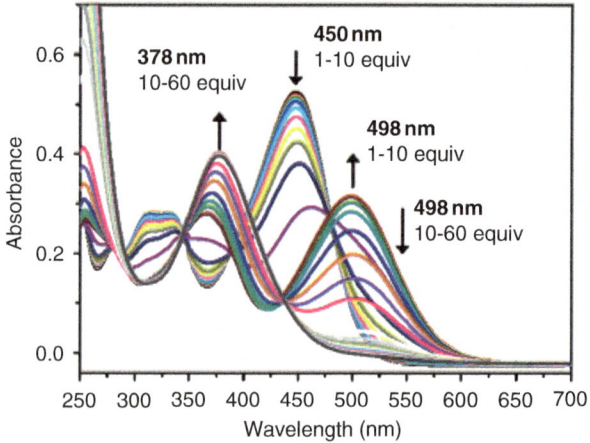

Fig. 6 UV-Vis titration of the D–A–D receptor BHPN (1×10^{-5} M in Me$_2$SO/H$_2$O, 1:1) with various amounts of Hg^{2+} ion (1×10^{-2} M in distilled water). The ratiometric visual detection of Hg^{2+} ion can thus be demonstrated by UV-Vis absorption spectra (reprint from ref. [58], Copyright © 2005 American Chemical Society)

their triazacrown rings was synthesized in the cone conformation [59]. **1** displays a relatively weak emission, reflecting that a PET process from the pendent amine group (–CH$_2$CH$_2$–NH$_2$) to the fluorogenic pyrenes mainly occurs. Addition of various metal ions or anions to the solution of **1** reduces the PET because the pendent alkylamine takes part in the complexation, causing their fluorescence spectra to be changed. When Pb^{2+}, a quenching metal ion, is added to **1**, their pyrene monomer emission is enhanced, and their excimer emission quenched, which is due to conformational changes of the facing carbonyl groups as well as to the participation of the ethyleneamine into the three-dimensional Pb^{2+} ion encapsulation. In contrast, upon addition of alkali metal ions to **1**, both monomer and excimer emissions are observed to increase, which is attributable to the chelating enhanced fluorescence effect and the retained conformations. For anion sensing, **1** shows a high selectivity for F$^-$ ions over other anions tested. When the F$^-$ ion is bound to **1** by hydrogen-bonding between the amide NH of the triazacrown ring and F$^-$, both their monomer and excimer emissions are weakened due to PET from the bound F$^-$ to the pyrene units.

2   Organic Dyes with Excited-State Proton Transfer Property

Proton transfer represents one of the most fundamental processes involved in chemical reactions as well as in living systems [60]. Vast numbers of scientists have been participating in relevant research, leading to thousands upon thousands

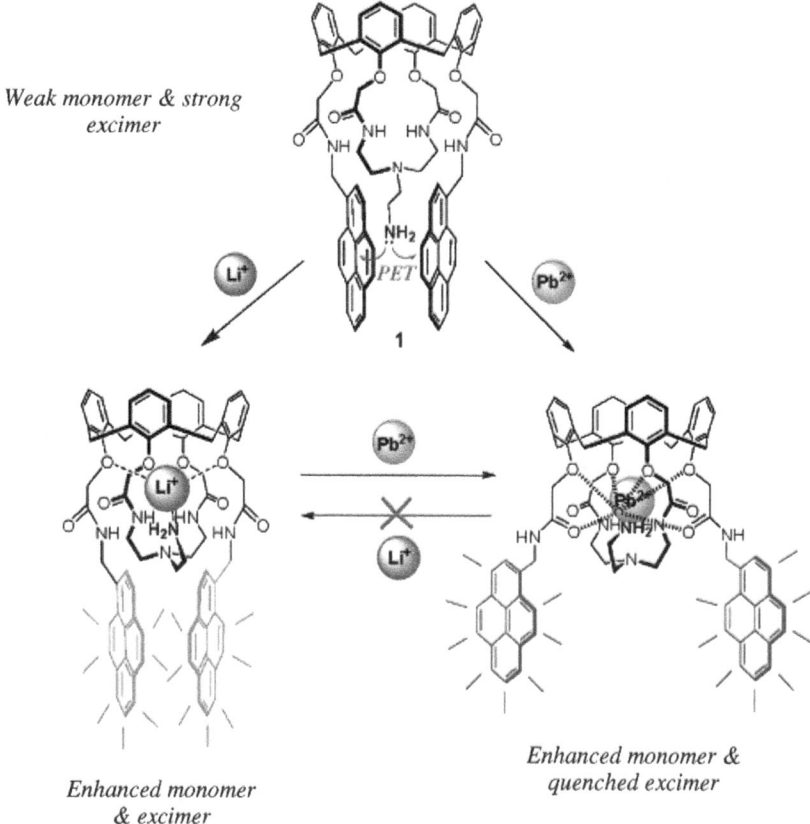

Fig. 7 Fluorescence ratiometry of monomer/excimer emissions in a through space PET system (reprint from ref. [59], Copyright © 2005 American Chemical Society)

of publications and numerous books on the topic. Various types of proton transfer reactions, depending on, e.g., the reaction in the ground or excited states, adiabatic versus nonadiabatic, strong versus weak hydrogen bonding, acidity (proton donor) and basicity (proton acceptor), have been identified [61–67]. At this time, proton transfer reaction seems to have the potential for unlimited extensions and aspects in view of fundamentals and applications. In 2008, Osamu Shimomura, Martin Chalfie, and Roger Tsien were awarded the Nobel Prize in Chemistry for the discovery and development of green fluorescent protein (GFP). GFP takes advantage of the presence of a core chromophore that undergoes excited-state proton transfer (ESPT) via the proton relay of water molecules, which results in highly effective and intense green fluorescence. The great consequence of GFP in biochemical and medical research provides a case in point to illustrate the importance of ESPT reactions in modern science.

2.1 Excited-State Intramolecular Proton Transfer

The fundamental approach to a proton transfer process, which is crucial to mimic many chemical and biological reactions, has relied deeply on studies of excited-state intramolecular proton transfer (ESIPT) reactions in the condensed phase.

The ESIPT reaction generally requires hydrogen-bonding formation between the vicinal proton donor (hydroxyl or amino proton) and acceptor groups (carbonyl oxygen or pyridyl nitrogen) that exist in the molecule (or complex) per se. As such, upon electronic excitation, proton transfer takes place along the excited-state PES, forming a proton-transfer tautomer, which, in theory, possesses significant differences from its corresponding normal species in structure and electronic configuration (see Fig. 8 for a prototype), resulting in a large Stokes shifted $S'_1 \rightarrow S'_0$ fluorescence (hereafter, the prime sign denotes the proton-transfer tautomer). Femtosecond time-resolved fluorescence spectroscopy has been employed since the early 1990s to investigate the early stage of the intrinsic ESIPT dynamics. Also, numerous ESIPT molecules have been discovered and investigated to shed light on their corresponding spectroscopy and dynamics. A partial list of the key ESIPT molecules includes salicylic acid and its associated various derivatives, [68] 2-hydroxybenzophenone, [69] 3-hydroxyflavone, [70] 5-hydroxyflavone, [71] 1,5-dihydroxyxanthraquinone, [72] 2-hydroxy-4,5-naphthotropone, [73] 2-(2'-hydroxyphenyl) benzoxazole, [74] and 2-(2'-hydroxyphenyl)benzothiazole [75]. An earlier summarization of ESIPT molecules, focusing on steady-state spectroscopy, the thermodynamics of ESIPT based on the pK_a of excited singlet and triplet states, as well as the nonradiative pathways of the proton-transfer tautomer, has been provided by Klöpffer [76]. A 1986 review written by Kasha [63] classified various types of ESIPT reaction in which solvent perturbation to manipulate the proton transfer

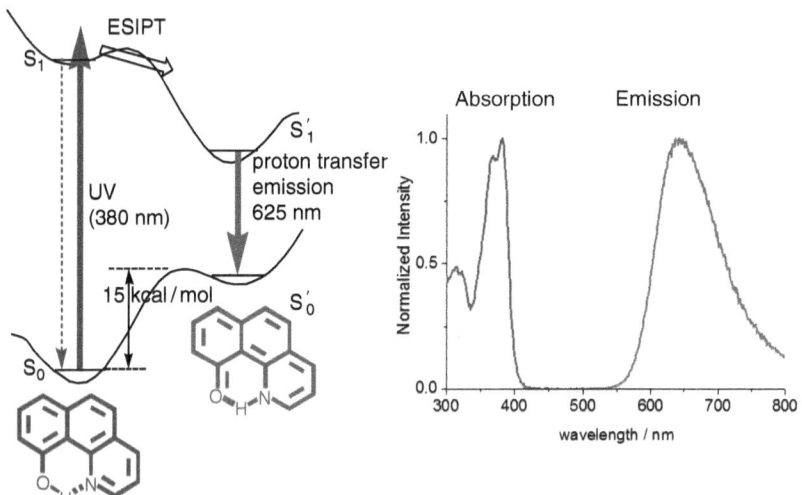

Fig. 8 A qualitative sketch of ESIPT cycle in HBQ

dynamics was especially emphasized. Subsequently, in 1989, a feature article reported by Barbara et al. [77] reviewed the dynamics of ESIPT on the picosecond timescale as well as the proton tunneling effect on a symmetrical double well potential. In 1993, Formosinho and Arnaut [78] reviewed several biorelated ESIPT molecules, with an emphasis on their application as fluorescence probes. Femtosecond time-resolved spectroscopy has been employed since the beginning of the 1990s to investigate the early stage of the intrinsic ESIPT dynamics. The results have indicated that for a highly unsymmetrical, exergonic type of intramolecular proton transfer such as that in 3-hydroxyflavone, [79] benzothiazole, [75] benzo-triazole [80], 10-hydroxy[*h*]benzoquinoline [81], and 5-hydroxyflavone, [71] the ESIPT timescale is less than several hundred femtoseconds, with a lack of the deuterium isotope effect. One can also refer to the article published by Zewail and coworkers [82] for a detailed overview of developments before 1996 regarding this relevant subject.

ESIPT with a highly unsymmetrical potential energy surface is generally conceived to be nearly barrierless (Fig. 8). The ultrafast reaction timescale may correspond to the period of certain low frequency, large-amplitude motions incorporating the change in nuclear distance associated with the hydrogen bond. This viewpoint has been recently confirmed by two approaches. On the one hand, based on an ultrashort ($\sim$20 fs) stimulated emission pumping experiment, vibrational coherence in the excited state has been observed for the ESIPT molecule 2-(2'-hydroxy-5'-methylphenyl)benzotria-zole [83]. The results indicate that anharmonic coupling between modes in the frequency range below 500 cm^{-1} and some high frequency motions play a key role in the proton transfer reaction. On the other hand, proton-transfer promoting modes incorporating low-frequency, large amplitude motion have been deduced from the anharmonicity effects of overtone and combination bands, as well as the anomalous intensity distribution in the resonance Raman spectra of several ESIPT molecules [84].

The application of ESIPT molecules has also been vastly developed. Prototypical examples are probes for solvation dynamics [85] and biological environments [86], the development of laser dyes [87], fluorescence recording [88], ultraviolet stabilizers [89], metal ion sensors [90], and recent applications in the field of organic light emitting devices (OLEDs, see Fig. 9) [91]. The following sections elaborate on a sequence of ESIPT molecules possessing five- to seven-membered ring hydrogen-bonding structures. Such a classification may be unprecedented and should be beneficial to those synthetic chemists interested in applications. In addition, readers with general backgrounds should gain much fundamental knowledge on the aspect of applications from the following content.

2.1.1 ESIPT via Five-Membered-Ring Hydrogen-Bonding System

3-Hydroxyflavone

Among those ESIPT molecules possessing five-membered-ring hydrogen bonds, 3-hydroxyflavone (3HF) has been considered a paradigm ever since its ESIPT

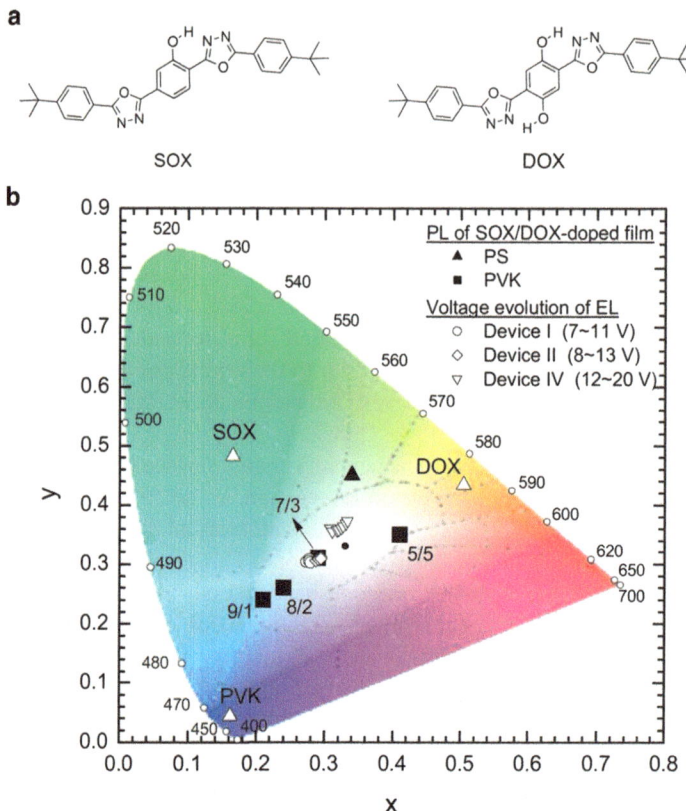

Fig. 9 (**a**) Molecular structures of novel ESIPT dyes, 2,5,-bis[5-(4-t-butylphenyl)-[1,3,4]oxadia-zol-2-yl]-phenol (SOX), and 2,5-bis[5-(4-t-butylphenyl)-[1,3,4]oxadiazol-2-yl]-benzene-1,4,-diol (DOX). (**b**) Emission colors in the Commission Internationale de L'Eclariage (CIE) chromaticity diagram. The *inner oval* and the *filled circle* at coordinate (*x,y*) of (0.33, 0.33) indicate the *white region* and the ideal color, respectively. Note that PS and PVK denote polystyrene and poly (N-vinylcarbazole) film (reprint from ref. [91], Copyright © 2005 Wiley-VCH)

property was discovered by Kasha and coworkers [70]. In nonpolar solvent such as cyclohexane, 3HF exhibits the $S_0 \rightarrow S_1$ ($\pi\pi^*$) transition maximized at 340 and 354 nm, while the fluorescence of 3HF shows an anomalously large Stokes shifted band maximized at ~526 nm (Φ = 0.36, τ_f = 3 ns) [92]. Upon methylation of 3HF to form 3-methoxyflavone, 3-methoxyflavone shows normal Stokes shifted fluorescence maximized at ~360 nm in cyclohexane [93]. It is thus clear that ESIPT takes place from the hydroxyl proton to the carbonyl oxygen, giving rise to the proton-transfer tautomer emission. Due to its relatively weak, five-membered ring hydrogen bond (c.f. six-membered ring hydrogen bond), 3HF was once adopted as a model compound to demonstrate the appreciable barrier during ESIPT. Subsequently, it was found that the retardation of ESIPT at low temperature is mainly due to the external hydrogen bond perturbation caused by traces of protic solvent impurity (e.g., water, alcohols) present in the nonpolar solvents [70]. In a dry, extensively purified nonpolar solvent such as cyclohexane, the time-resolved measurement renders an ultrafast time of ESIPT for 3HF (τ_{pt} <240 fs) [79, 94], and ESIPT is essentially barrierless, perhaps due to the low frequency skeletal motions associated with the hydrogen bond [83].

Solvent polarization effects on the ESPT reaction of 3HF derivatives in polar, aprotic solvents have been previously observed [95]. In the case of nonsubstituted 3HF, the change in dipole moment that accompanies ESPT is not large. As a result, the proton transfer event is fairly weakly coupled to the solvent polarization; i.e., there is a fairly small (<kT) solvent-induced barrier to proton transfer. However, the proton transfer dynamics could be dramatically altered by appropriate substitution on parent 3HF. Prototypical examples include N,N-dialkylamino-3-hydroxyflavone and 7-N,N-Diethylamino-3-hydroxyflavone. The N,N-dialkylamino group is very strongly electron-donating and, when conjugated with aromatic systems, is often associated with ICT character. The presence of an electron donating group will result in a large dipole moment difference between the normal and the tautomer forms in the excited-state. As a result, the proton transfer rate is significantly affected by the solvent polarization. Focus on this subject will be presented in Sect. 3.

2-Pyridyl Pyrazoles

From the fundamental viewpoint, searching for the ESIPT reaction with a finite, well-defined barrier is of great importance to gaining detailed insights into the reaction potential energy surface. Given similar proton donor/acceptor strengths, the intramolecular hydrogen-bonding strength is empirically in the order of 6-> 5-≫ 4-membered systems. In fact, the formation of an intramolecular four-membered ring hydrogen bond is prohibited because it is both steric and orientation hindrance. Alternatively, a five-membered-ring hydrogen-bonding system, associated with relatively weak intramolecular hydrogen bonds, may provide good opportunities. To achieve this goal, conjugated pyrrole–pyridine systems such as 5-(2-pyridyl)-1-H-pyrazoles [96], possessing an N–H...N type of five-membered hydrogen bond, have received much attention. Unlike the O–H site, which has certain rotational degrees of freedom, the orientation effect of the pyrrolic N–H site, being restricted toward a specific direction, is more critical for the hydrogen bond formation. Furthermore, the photoacidity of the pyrrolic hydrogen is, in general, weaker than that of the phenolic hydrogen. As a result, ESIPT in five-membered hydrogen-bonding pyrrole–pyridine systems may be associated with an appreciably large barrier.

The ESIPT time constants, having been determined to be 70 ~ 130 ps, are much slower than for typical ESIPT molecules (<150 fs) in nonpolar solvent. The temperature-dependent lifetime studies of analogs of 5-(2-pyridyl)-1-H-pyrazole (Fig. 10) systems in methylcyclohexane led to the deduction of a barrier of ~2 kcal/mol.

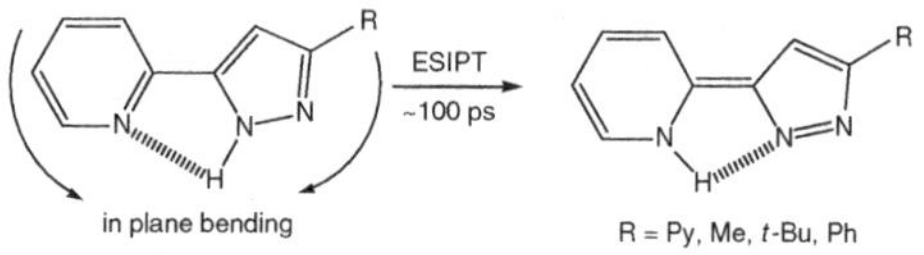

Fig. 10 Structures and the proposed ESIPT mechanism for the 5-(2-pyridyl)-1-*H*-pyrazole system

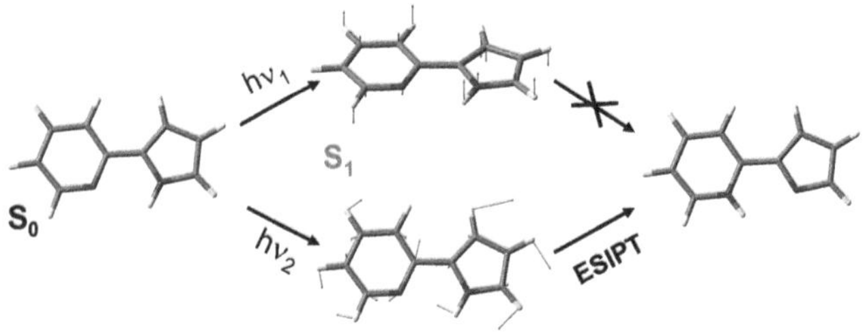

in plane bending

ESIPT
~100 ps

R = Py, Me, *t*-Bu, Ph

Fig. 11 Mode-selective ESPT in 2-(2′-Pyridyl)pyrrole indicates that the ESIPT reaction is associated with in-plane vibration modes (reprint from ref. [97], Copyright © 2005 American Chemical Society)

Theoretical calculations also indicated skeletal reorganization amid ESIPT processes. Recently, the vibrational modes associated with ESIPT were successfully demonstrated by the high resolution fluorescence excitation spectra of 2-(2′-pyridyl)pyrrole (PP) in supersonic jets [97]. The strengthening of intramolecular hydrogen bonds in PP is most pronounced for the lowest frequency vibration associated with the in-plane vibration modes, the 151 and 144 cm^{-1} in S_0 and S_1, respectively (Fig. 11). Conversely, excitation of out-of-plane modes weakens the hydrogen-bonding by increasing the N–N distance, effectively blocking the reaction. The results demonstrate a novel and unique system among ESIPT molecules wherein the intrinsic proton transfer is associated with a substantial energy barrier. This, in combination with the structural simplicity and diversity, makes the 2-pyridyl pyrazoles system an ideal model for probing the ESIPT dynamics, which are believed to bring up a broad spectrum of interests in the proton-transfer field.

2.1.2 ESIPT via Six-Membered-Ring Hydrogen-Bonding System

10-Hydroxybenzo[*h*]quinoline

Among ESIPT systems with six-membered-ring hydrogen bond, 10-hydroxybenzo[*h*] quinoline (HBQ, Fig. 8) perhaps possesses the strongest hydrogen bond. The hydrogen-bonding strength, measured by means of ^{1}H-NMR study, is estimated to be ~10 kcal/mol [98]. This may be due to the geometry restriction of phenanthrene, which spontaneously forces the optimization of hydrogen-bonding distance and orientation. Empirically, the stronger hydrogen-bonding leads to better coupling between normal and proton-transfer tautomer PES, possibly reducing the ESIPT barrier.

The dynamics of ESIPT in HBQ have also been investigated by steady-state and femtosecond time-resolved fluorescence spectroscopy [81]. In cyclohexane, the timescale for the proton transfer in the excited state was deduced to be <150 fs under the response limit. The initially prepared keto tautomer undergoes a ∼330 fs intramolecular vibrational energy redistribution (IVR) process [99]. Subsequently, a solvent-induced vibrational relaxation takes place in a timescale of 8–10 ps, followed by a relatively much longer, thermally cooled $S'_1 \rightarrow S'_0$ decay with a rate of ∼270 ps. Recently, the dynamics of the ESIPT of HBQ and the associated coherent nuclear motion were investigated in solution by femtosecond absorption spectroscopy. Especially, the lowest-frequency mode at 242 cm^{-1} is dephased significantly faster than the other three modes. This observation was regarded as a manifestation that the nuclear motion of the 242-cm^{-1} mode is correlated with the structural change of the molecule associated with the reaction coordinate (Fig. 12). In comparison with Raman measurements and DPT calculations, it was found that the nuclear motion of this lowest frequency mode involves a large displacement of the O–H group toward the nitrogen site, as well as in-plane skeletal deformation that assists the oxygen and nitrogen atoms to move closer to each other [99, 100].

Following ESIPT, the keto-tautomer emission of HBQ, with >10,000 cm^{-1} Stokes shift (emission absorption of 600 nm versus absorption peak of 390 nm in, e.g., cyclohexane, Fig. 8) and free of perturbation even in water due to the strong intramolecular hydrogen-bonding, has important applications in various fields. Sytnik et al. applied HBQ to probe enzyme kinetics and concluded that HBQ can distinguish static solvent-cage polarity from dynamical solvent dielectric relaxation and other solvent-cage effects (e.g., mechanical restriction of molecular conformation) [86, 101]. Robert et al. used HBQ as a fluorescence probe to examine the influence of organized media, especially the cyclodextrins in aqueous solution [102]. The excited-state proton-transfer dynamics of HBQ in aqueous solution have also been studied under various pH conditions [98]. The results led to the conclusion that the resonance charge transfer between hydroxyl oxygen and

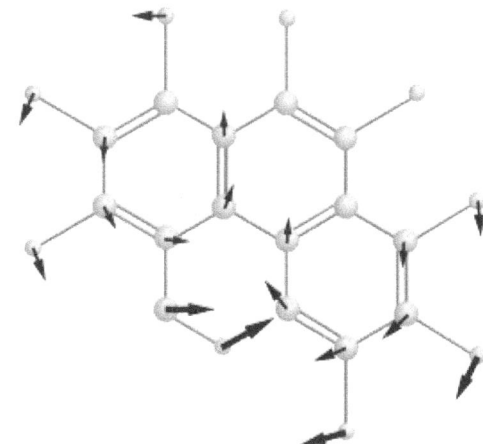

Fig. 12 Nuclear motion of the ground-state vibration at 243 cm^{-1} (calculated, 248 cm^{-1}) that corresponds to the lowest-frequency wavepacket motion observed in excited-state HBQ (reprint from ref. [99], Copyright © 2005 American Chemical Society)

benzoquinolinic nitrogen acts as a driving force for the proton-transfer reaction. The results also provide important insight for the solvent (i.e., H_2O)-assisted ESPT dynamics, which are limited by the proton donating and/or accepting rate associated with free water molecules.

2-(2'-Hydroxyphenyl)benzoxazole and 2-(2'-Hydroxyphenyl)benzo-thiazole

Other prototypes of the six-membered ring hydrogen-bonding system are 2-(2'-hydroxy)phenylbenzoxazole (HBO) and 2-(2'-hydroxyphenyl)benzothiazole (HBT) (Fig. 13). HBO and HBT undergo intramolecular proton transfer not only in the singlet excited state [74, 75, 103] but also in the triplet manifold state [104, 105]. Upon UV excitation ($\sim$350 nm), HBO and HBT undergo proton transfer in the singlet manifold, giving rise to a large Stokes shifted tautomer fluorescence maximized at 500 and 520 nm, respectively, in nonpolar solvents. For HBT, the absolute rate constant of ESIPT has been measured to be $\sim$150 fs, [75] indicating that the occurrence of the proton transfer reaction is not induced by a supposed much faster O–H vibrational timescale but rather by certain in-plane, low-frequency vibrational motions associated with changes of hydrogen-bonding distance. Time-resolved infrared vibration spectroscopy has resolved the transient absorption of the C=O stretching band ($\sim$1,530 cm^{-1}) marking the characteristic of the keto-tautomer structure in the S'_1 state. Monitoring of such a structural change revealed that the S'_1 state appears at a delay timescale of 30–50 fs after electronic excitation, reconfirming the ultrafast rate of ESIPT [106]. In sharp contrast, the rate of ground-state reverse proton transfer for HBT has been reported to be surprisingly slow (in a range of μs–ms). From the results of pump-probe, two-step laser-induced fluorescence, and the transient absorption spectra, the observed transient specie in the ground state was assigned to the *trans*-keto form, which was produced by *cis*-to-*trans* isomerization of the *cis*-keto form in the singlet excited state. Thus, the slow reverse proton transfer in the ground state, in fact, is ascribed to the rate of *trans*–*cis* isomerization prior to re-formation of the intramolecular hydrogen bond, which is a prerequisite to execution of proton transfer, while the intrinsic proton transfer in the ground state is believed to be ultrafast, i.e., with a small or even negligible barrier [107, 108]. The relative potential energy and dynamics of the enol, *cis*-keto, and *trans*-keto forms of HBO and HBT are summarized in Fig. 13.

Seminal studies on the dynamics of proton transfer in the triplet manifold have been performed on HBO [109]. It was found that in the triplet states of HBO, the proton transfer between the enol and keto tautomers is reversible because the two (enol and keto) triplet states are accidentally isoenergetic. In addition, the rate constant is as slow as milliseconds at 100 K. The results of much slower proton transfer dynamics in the triplet manifold are consistent with the earlier summarization of ESIPT molecules. Based on the steady-state absorption and emission spectroscopy, the changes of pK_a between the ground and excited states, and hence the thermodynamics of ESIPT, can be deduced by a Förster cycle [65]. Accordingly, compared to the pKa in the ground state, the decrease of pK_a in the

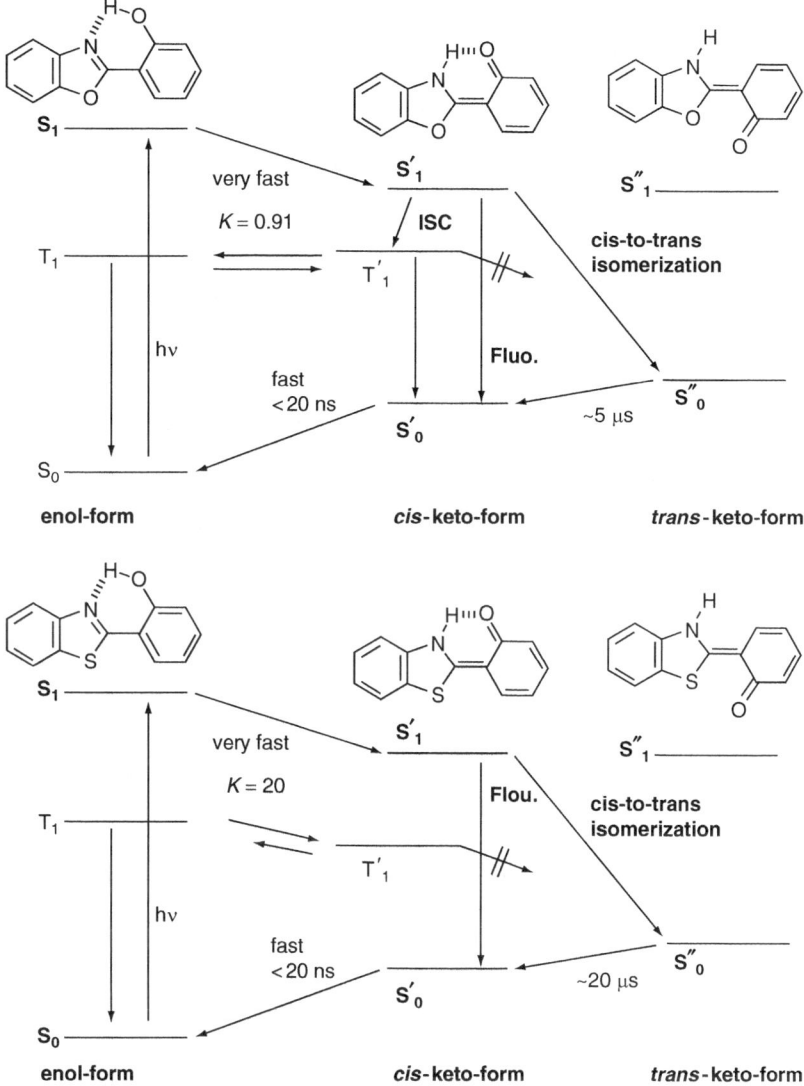

Fig. 13 Energy diagram of hydrogen atom transfer of HBO (*above*) and HBT (*below*) (reprint from ref. [108], Copyright © 2002 Royal Society of Chemistry)

triplet state is much smaller than that in the singlet excited state, qualitatively rationalizing the slow ESIPT dynamics in the triplet manifold.

As for the relevant application, recently, a specific photochromic compound, 1,2-bis(2′-methyl-5′-phenyl-3′-thienyl)perfluorocyclopentene (BP-BTE), and the analogs of HBO, 2,5-bis(5′-*tert*-butyl-benzooxazol-2′-yl)hydroquinone (DHBO), were employed in the high-contrast, reversible, photochromic switching of fluorescence emission and its perfect nondestructive readout (Fig. 14). Due to the large

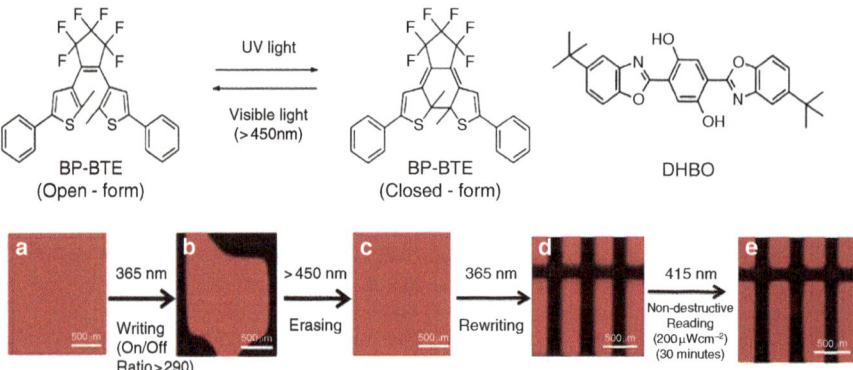

Fig. 14 The molecular structures of BP-BTE and DHBO (*above*). Microsized erasable ESIPT-fluorescence photoimaging on a spin-coated BP-BTE/DHBO-loaded PMMA film and its nondestructive readout capability; (**a**) initial open-form state; (**b**) writing; (**c**) erasing; (**d**) rewriting; and (**e**) continuous nondestructive reading under irradiation with relatively high-intensity 415 nm light (200 μW cm^{-2}) for 30 min. The dark region represents the area irradiated with the 365 nm UV light (*below*) (reprint from ref. [88], Copyright © 2006 American Chemical Society)

Stokes-shifted tautomer emission and high quantum yield in solid state (10%), DHBO was successfully used as an energy transfer donor in the photochromic switching system [88].

2.1.3 ESIPT via Seven-Membered-Ring Hydrogen-Bonding System

Seven-membered-ring hydrogen-bonding systems that undergo ESIPT are very rare. A case in point is a derivative of green fluorescence protein (GFP) core chromophore, 4-(4-hydroxybenzylidene)-1,2-dimethyl-1*H*-imidazol-5(4*H*)-one (*p*-HBDI), namely 4-(2-hydroxybenzylidene)-1,2-dimethyl-1*H*-imidazol-5(4*H*)-one (*o*-HBDI, see Fig. 15). GFP, which serves as an energy acceptor and emitter for bioluminescence in the sea pansy *Renilla reniformis* and the jellyfish *Aequorea victoria*, is the subject of much interest because of its applications in molecular biology and biochemistry [110]. GFP takes advantage of the presence of a chromophore that is anchored both covalently and via a hydrogen-bond network, 4-(4-hydroxybenzylidene)-1,2-dimethyl-1*H*-imidazol-5(4*H*)-one (*p*-HBDI, Fig. 15a), which undergoes ESPT [111] via the proton relay of water molecules and/or some residues to a remote residue such as E222 [112], resulting in a very effective and intense anion fluorescence. The properties of fluorophores of GFP and of other fluorescent proteins are described in the chapter of Merola et al in this book [148].

In view of chemistry, most of the research has been focusing on the chemical modification of *p*-HBDI [113] analogs at the C(1) position, such that the emission color can be tuned via the substituent effect [114]. Nevertheless, studies reveal a strong cutoff between the properties of wild-type GFP (or certain GFP mutants)

Fig. 15 (**a**) Structure isomers between *p*-HBDI and *o*-HBDI. (**b**) The absorption and emission spectra of *o*-HBDI in cyclohexane (*black solid line*) and solid film (*red solid line*, emission only) and *o*-MBDI (*blue solid line*) in cyclohexane (reprint from ref. [117], Copyright © 2007 American Chemical Society)

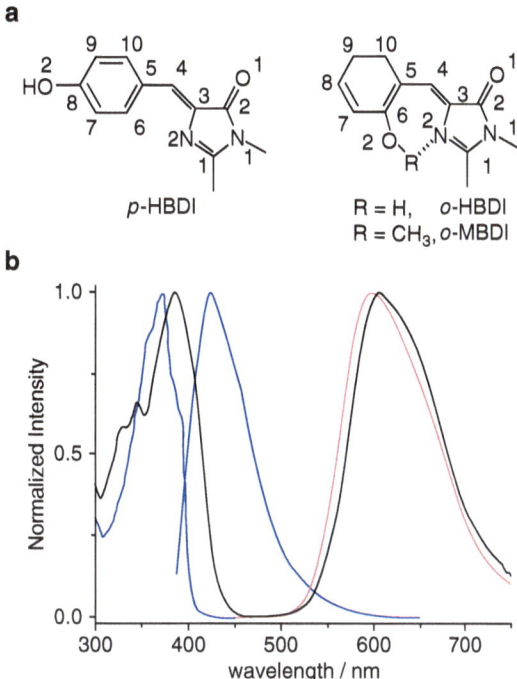

and the synthetic analog chromophores of *p*-HBDI. In view of photophysics, the fluorescence yield of the protein-free chromophore in fluid solvents is much weaker and strongly temperature dependent. The results suggest an efficient radiationless transition operating in *p*-HBDI, most probably induced by strong conformational relaxation along torsional deformation [115] of the two exocyclic C–C bonds, from the initially prepared Franck–Condon state to a nonfluorescent twisted intermediate [116]. More recently, it has been proposed that the shallow potential energy surface of the intermediates may conically intersect with that of the ground state, inducing the dominant radiationless deactivation [113]. Such a conformational relaxation is greatly suppressed in wild GFP by its proton relay, rigid environment.

Recently, a structural isomer of the core chromophore (*p*-HBDI), namely *o*-HBDI (Fig. 15a), has been strategically designed and synthesized [117]. *o*-HBDI possesses a seven-membered ring hydrogen bond, from which the ESIPT takes place, resulting in a remarkable proton transfer tautomer emission of ~605 nm in organic solvents such as cyclohexane (Fig. 15b). As compared to those generally observed weak emissions for *p*-HBDI ($\Phi_f \sim 10^{-4}$ and $\tau_f \sim 1.7$ ps in toluene) at room temperature [113], the tautomer emission yield of 3.1×10^{-3} with $\tau_f \sim 32$ ps in cyclohexane implies that the hydrogen-bonding strength may, in part, hinder the exocyclic C–C bond isomerization. ESIPT also takes place in the solid film, giving rise to a ~595 nm tautomer emission with a quantum yield as high as 0.4. The radiationless decay process of *p*-HBDI and *o*-HBDI and its analogs in solution should be associated with exocyclic C–C bond rotation. Although the

bioactivity of o-HBDI is pending further exploration, its future chemical derivation is versatile. It is believed that fine tuning the proton-transfer emission can be
achieved via the derivation at the C(1) position, while the radiationless quenching
process may be further reduced by anchoring bulky groups at the C(4) position,
generating a new series of isomers of p-HBDI with remarkable ESIPT properties.

As for the future prospects, from a fundamental viewpoint, specific modes
associated with motion in the seven-membered ring hydrogen-bonding system
that trigger the ESIPT reaction are of great interest. This may rely on the resolution
of the coherence characteristic. Due to the freezing of C(1)–C(2) rotation and hence
the strong proton-transfer emission for o-HBDI in solid film [117], synthesis of
various analogs via fine-tuning of the emission color is feasible. This may attract
great attention in the fields of, e.g., OLEDs or solar cells. For the latter, recent
development of o-HBDI analogous dyes suitable for dye-sensitized solar cells
(DSSC) has been reported [118].

3 Solvation Dynamics in the Proton Transfer Coupled Charge Transfer Reaction

3.1 Fundamental Background

In this section, we switch gears slightly to address another contemporary topic,
solvation dynamics coupled into the ESPT reaction. One relevant, important issue
of current interest is the ESPT coupled excited-state charge transfer (ESCT)
reaction. Seminal theoretical approaches applied by Hynes and coworkers revealed
the key features, with descriptions of dynamics and electronic structures of nonadiabatic [119, 120] and adiabatic [121–123] proton transfer reactions. The most
recent theoretical advancement has incorporated both solvent reorganization and
proton tunneling and made the framework similar to electron transfer reaction,
[119–126] such that the proton transfer rate k_{pt} can be categorized into two regimes:

(a) For nonadiabatic limit [120]:

$$k_{PT} = \frac{C^2}{\hbar} \sqrt{\frac{\pi}{E_S RT}} \exp\left(-\frac{\Delta G^{\neq}}{RT}\right) \tag{9}$$

$$\Delta G^{\neq} = \frac{(\Delta G_{RXN} + E_S)^2}{4E_S} \tag{10}$$

where ΔG_{RXN} and $\Delta G^{\neq}$ denote the reaction free energy and reaction barrier,
respectively; E_S is the solvent reorganization energy; C represents the proton
coupling's quantum average over the vibrational modes associated with the proton
motion; and $\hbar$ is the Planck constant.

(b) For adiabatic limit [122]:

$$k_{PT} \equiv \frac{\omega_s}{2\pi} \exp\left(\frac{-\Delta G_{ad}^{\neq}}{RT}\right) \qquad (11)$$

where ω_s stands for solvent fluctuation frequency in the reactant well, and the adiabatic reaction activation energy is expressed as

$$\Delta G_{ad}^{\neq} = \Delta G_0^{\neq} + \alpha_0 \Delta G_{RXN} + \alpha_0' \frac{(\Delta G_{RXN})^2}{2} \qquad (12)$$

where $\Delta G_0^{\neq}$ represents the intrinsic barrier at $\Delta G_{RXN} = 0$; α_0, the Brønsted coefficient, is the derivative of $\Delta G^{\#}$ with respect to ΔG_{RXN} evaluated at $\Delta G_{RXN} = 0$; and α_0' is the derivative of α_0 with respect to ΔG_{RXN} evaluated at $\Delta G_{RXN} = 0$.

To which category the proton-transfer reaction is ascribed strongly depends on coupling between reactant and product electronic states, which has significant dependence on the distance between proton donor and acceptor, i.e., hydrogen-bonding strength, or in other words, hydrogen bonding length. The larger electronic coupling not only reduces the barrier height but also narrows the width. It is interesting to note that compared to adiabatic electron transfer, the tunneling probability in nonadiabatic proton transfer should be much more sensitive to interatomic separation simply because a proton is much heavier ($\sim 2,000$ times) than an electron. Moreover, in reality, the separation between proton donor and acceptor would not be fixed, but fluctuating, manipulated by any vibrational motion associated with changes of hydrogen bonding. As a result, proton tunneling probability varies with respect to types of vibrations. In this approach, Hynes and coworkers have extended the nonadiabatic proton transfer model [119, 120] by incorporating a low frequency vibrational motion between proton donor and acceptor, i.e., the Q modes. The results indicate that the rate constant for nonadiabatic proton transfer is the sum of $k_{n \to m}$, the rate constant from the nth vibrational state of Q mode in reactant.

$$k_{PT} = \sum_n \sum_m P_n k_{n \to m} = \sum_n \sum_m P_n \frac{C_{nm}^2}{\hbar} \sqrt{\frac{\pi}{E_S RT}} \exp\left(-\frac{\Delta G_{n \to m}^{\neq}}{RT}\right) \qquad (13)$$

where P is the thermal distribution of the nth excited vibrational state of the Q mode, C_{nm} is the quantum average of proton coupling, and $\Delta G_{n \to m}^{\neq}$ is the activation barrier for $n \to m$ transition. With explicit computation of C_{nm}, and considering reaction symmetry within the Q mode, a new term, $E_\alpha = \hbar^2 \alpha^2 / 2m_H$, whose physical interpretation is the coupling term between Q mode and solvent polarity, is introduced, accompanied by removal of the summation term in (13):

$$k_{PT} = \frac{\langle C^2 \rangle}{\hbar} \sqrt{\frac{\pi}{(E_S + \tilde{E}_\alpha)RT}} \exp\left\{ -\frac{(\Delta G_{RXN} + E_S + \tilde{E}_\alpha)^2}{4(E_S + \tilde{E}_\alpha)RT} \right\} \qquad (14)$$

$$\tilde{E}_\alpha = E_\alpha (1/2)\beta\hbar\omega_Q \coth((1/2)\beta\hbar\omega_Q) \qquad (15)$$

where E_α is a quantum energy term associated with the tunneling probability's variation with the Q vibration, $\tilde{E}_a$ is an isotope-dependent parameter via E_α with $\beta = 1/RT$, and ω_Q denotes the vibrational frequency for the Q vibration. Note that the rate constant expression (14) resembles (9); the only differences are that the coupling probability is thermally averaged and that the Q mode vibrational reorganization energy has a certain contribution. Other intramolecular modes could also be incorporated by similar treatment.

One aspect in this modern model is the quantum character of the proton in a proton transfer reaction. In other words, the proton motion is treated as quantum rather than classical. Accordingly, the motion is commonly faster than the rearrangement of solvent molecules, and thus Born–Oppenheimer approximation can be analogously made for the correlation between proton and solvent molecules in this reaction. As a consequence, the equilibrium between the moving proton and the surrounded solvent molecules is established at each instant, and the reaction activation free energy is thus essentially dominated by the solvent reorganization, rather than being given by the height of proton migration barrier, since charge redistribution is involved. Furthermore, the solvent coordinate, rather than the proton coordinate, serves as the reaction coordinate within the proton transfer process. If we approximate the free energy along the solvent coordinate by a simple parabola, the amplitude of reaction activation energy can be analytically determined by three factors: (1) *reaction energy (vertical displacement)*, (2) *dipole moment difference between normal and tautomer forms (horizontal displacement)*, and (3) *solvent polarity (curvature)*.

In the following sections, we will review several ESIPT systems that have been strategically designed to probe solvent polarization coupled reaction dynamics. The experimental results render firm support of the modern theoretical model.

3.2 Excited-State Charge Transfer Coupled Proton Transfer Reaction

To probe as well as to signify the solvent polarity effect, it is important to fully comprehend the occurrence of ESPT accompanied by large changes in dipole moment, such that a proton transfer event could be greatly subject to the solvent polarity effect [127–129]. To avoid bimolecular complexity, a unimolecular system, i.e., a system invoking the ESIPT, has received particular attention. To anticipate a great change in dipole moment suited for probing solvation dynamics

Fig. 16 A generalized ESCT/
ESIPT system and its possible
reaction patterns. (* denotes
the electronically excited
state, PT and CT symbolize
proton transfer and charge
transfer species, respectively)
(reprint from ref. [145],
Copyright © 2008 American
Chemical Society)

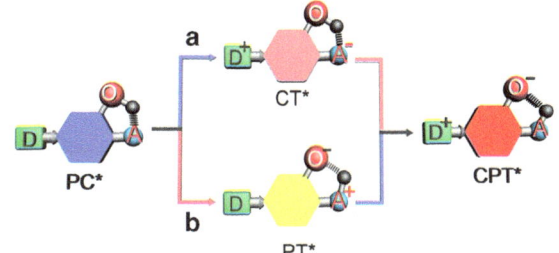

amid ESIPT, prototypical molecules are strategically designed to incorporate both
ESIPT and excited-state intramolecular charge transfer (ESICT) properties. A
conceptual ESIPT/ESICT coupled system is generalized and depicted in Fig. 16,
in which D and A denote the electron donor and acceptor, respectively. In most
cases, D and A are separated by a chromophore (represented by a benzene-like
structure), and their relative positions are suited for charge transfer reactions via, for
example, π electron delocalization in the excited state. In most designed systems,
a hydroxyl (or amino) hydrogen forms an intramolecular hydrogen bond with A
[130–136]. On the one hand, electronically exciting ground state normal form PC to
its excited state PC^* may cause charge transfer, forming a charge transfer species
CT^*. On the other hand, the hydrogen-bonded H atom may act as a strong photo-
acid, such that proton transfer takes place upon excitation, resulting in a proton
transfer tautomer denoted by PT^* (Fig. 16). Thus, depending on the reaction time
domain, studies of ESICT vs. ESIPT can be classified into two categories: (A)
When the rate of ESICT is faster than that of ESIPT, following $PC^* \rightarrow CT^*$ charge
transfer, the $CT^* \rightarrow CPT^*$ proton transfer reaction then takes place. (B) When
ESIPT takes place prior to ESICT, the overall reaction may be described as a
$PC^* \rightarrow PT^*$ proton transfer, followed by a $PT^* \rightarrow CPT^*$ charge transfer process.

Based on the above concept, a number of potential ESICT/ESIPT coupled
systems have been ingeniously designed and investigated. Prototypical examples
include *N,N*-dialkylamino-3-hydroxyflavones [128, 131–136], 2-hydroxy-4-(di-
p-tolyl-amino)benzaldehyde [127], and 2-(2′-hydroxy-4′-dietheylaminophenyl)
benzothiazole, [129] in which the *N,N*-dialkyl group is strategically designed to
act as an electron donor, while the carbonyl oxygen or the nitrogen group within the
parent ESPT moiety serves as an electron acceptor.

Up to this stage, the results have indicated that most systems, so far, can be
ascribed to case (A), namely an ultrafast ($\ll 150$ fs) ESCT prior to ESIPT (Fig. 17).
Such an adiabatic type of ESCT, i.e., an optical electron transfer process, can be
rationalized by a strong π-electron overlapping between donor and acceptor moieties, such that the electronic coupling matrix is much larger than that of the Marcus
type of weak-coupling electron transfer process [137–139]. For this case, following
ultrafast ESICT, which is even assumed to occur during the Franck–Condon
excitation (optical electron transfer), it is found that ESIPT always takes place
and that its rate is relatively complicated by the competitive solvation relaxation

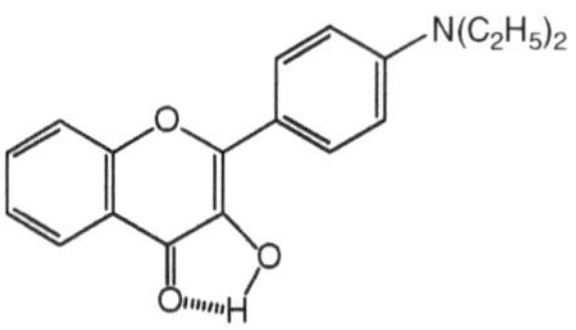

4'-N, N-diethylamino-3-hydroxyflavone

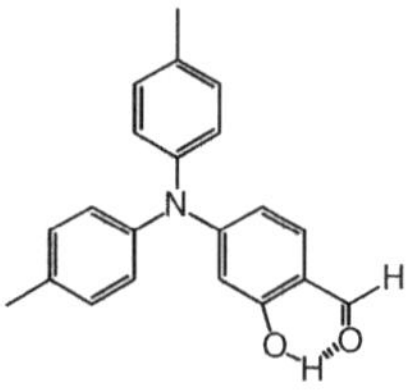

2-hydroxy-4-(di-p-tolyl-amino)benzaldehyde

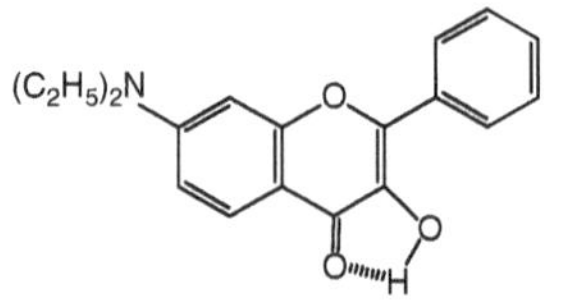

7-N, N-diethylamino-3-hydroxyflavone

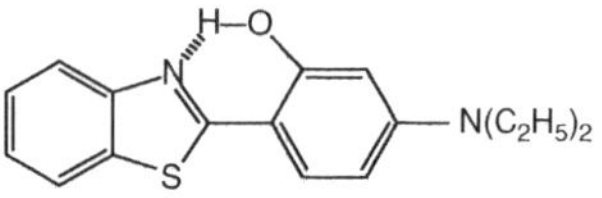

2-(2'-hydroxy-4'-dietheylaminophenyl)benzothiazole

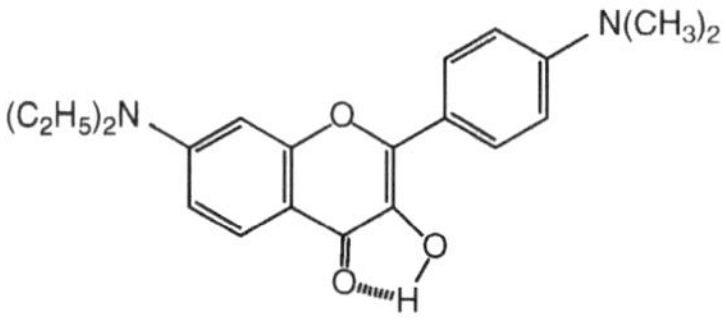

4'-N, N-dimethylamino-7-N, N-diethylamino-3-hydroxyflavone

Fig. 17 Molecular structures of some representative ESICT/ESIPT compounds

process, i.e., the $CT^* \rightarrow CT_{eq}^*$ process (the subscript "eq" denotes the solvent equilibrated state). After reaching the solvent equilibration, due to the difference in equilibrium polarization between CT_{eq}^* and CPT_{eq}^*, $CT_{eq}^* \rightarrow CPT_{eq}^*$ proton transfer reaction is associated with a solvent-induced barrier. The results, reflected from the steady-state approach, commonly show dual emission, consisting of CT_{eq}^* and CPT_{eq}^* emission (see Fig. 18 for the case of 4'-N,N-diethylamino-3-hydroxy-flavone, while the relaxation pathways are qualitatively depicted in Fig. 19).

From the chemistry point of view, one intriguing issue of the above approach lies in that the degree of charge transfer in the excited state can be fine-tuned via chemical modification, such that proof of concept can be developed systematically. As depicted in Fig. 17, three analogs of 3HF, 4'-N,N-diethylamino-3-hydroxyflavone ***(i), 7-N,N-diethylamino-3-hydroxyflavone (ii), and 4'-N,N-dimethylamino-7-N, N-diethylamino-3-hydroxyflavone (iii) have been employed as an ingenious approach to fine-tuning the ESICT-coupled ESIPT reaction via the dipolar functionality of the molecular framework [135]. Both i and ii exhibit remarkable dual emission due to the differences in solvent-polarity environment between ESICT and ESIPT states, while the interplay of two charge-transfer entities in iii leads to ESIPT decoupling from the solvent-polarity effect, resulting in a unique proton-transfer tautomer emission. The results make further rational design of the ESICT/ ESIPT coupled systems feasible simply by tuning the net dipolar effect. Accordingly, systematic investigation of the correlation in regards to the difference in dipolar vectors between ESICT and ESIPT versus solvent-polarity-induced barriers becomes possible.

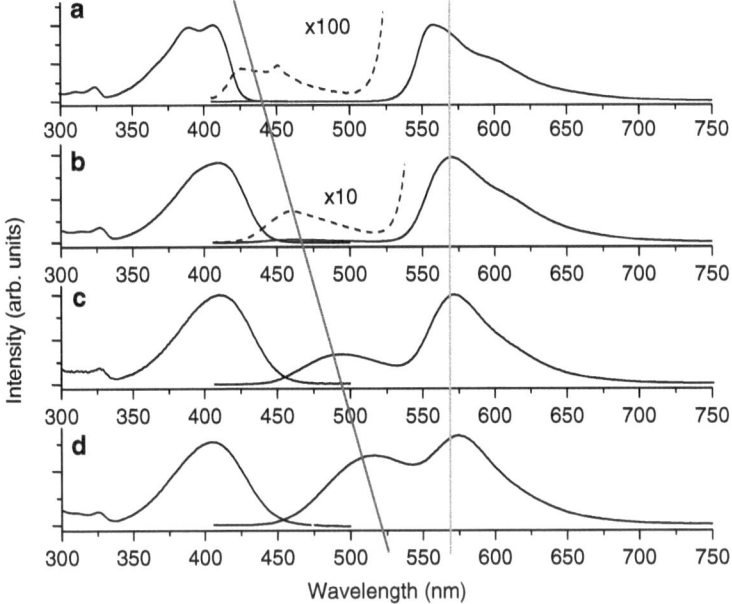

Fig. 18 Static absorption and fluorescence spectra of 4′-*N,N*-diethylamino-3-hydroxyflavone in (**a**) cyclohexane, (**b**) benzene, (**c**) dichloromethane, and (**d**) acetonitrile at 298 K (reprint from ref. [136], Copyright © 2005 American Chemical Society)

The above approach is mainly applied in polar, aprotic solvents. As for the protic solvent, which serves as an ideal model to mimic the biosystem, the solvent hydrogen bonding perturbation leads to the rupture of solute intramolecular hydrogen bonds and hence makes the study extremely complicated. For approaching ESCT/ESPT coupled reaction in protic solvents, strategic design lies in molecules lacking intramolecular hydrogen bonds, such that ESPT takes place via the assistance of protic solvent molecules [63]. Under this criterion, a case in point is the 7-azaindole (7AI) type of ESIPT systems. Strategically, the electron donor/acceptor functionality has been incorporated into 7AI, such that charge transfer influencing ESIPT can be examined in protic solvents. Prototypes of those designated 7AI analogs consist of 5-cyano-7-azaindole (5CNAI), 3-cyano-7-azaindole (3CNAI) [140], 3,5-dicyano-7-azaindole (3,5CNAI), and dicyanoethenyl-7-azaindole (DiC-NAI) (Fig. 20) [141]. In this approach, on the one hand, the cyano moiety serves as an electron withdrawing group, such that ESICT may take place from pyrrolic nitrogen to the cyano substituent. On the other hand, the pyrrolic hydrogen and the pyridinyl nitrogen act as proton donating and accepting groups, respectively, with a lack of intramolecular hydrogen bonds, such that proton transfer takes place under the assistance of the solvent relay (Fig. 21).

As a result, similar to 7AI, 3CNAI, and 3,5CNAI undergo methanol-catalyzed excited-state double proton transfer (ESDPT), revealing dual (normal and proton transfer) emission. However, proton transfer is prohibited for 5CNAI and DiCNAI

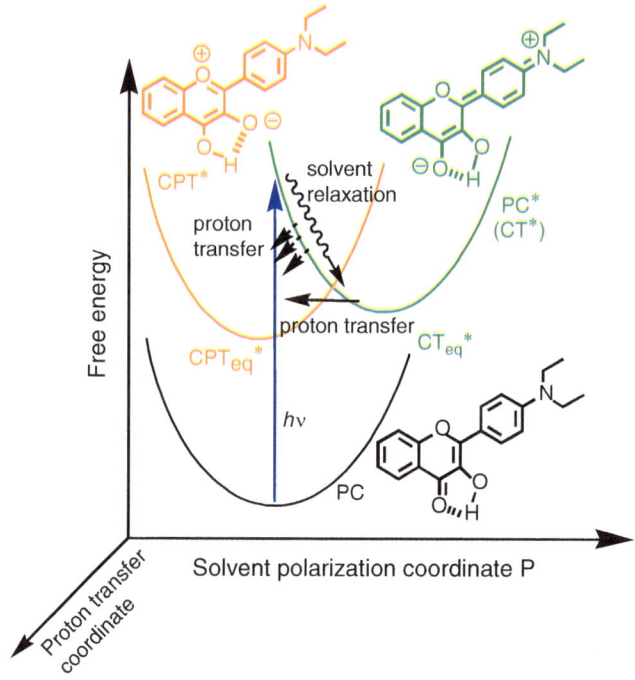

Fig. 19 Relaxation processes for case (A) ESICT/ESIPT system using 4′-*N,N*-diethylamino-3-hydroxyflavone as an example

in methanol, which is supported by the observation of a unique, normal (nonproton transfer) emission. The normal emission of each derivative reveals a significant charge transfer property by its prominent solvatochromism, though to different extents. The ESDPT rate, which is determined by the rise/decay of tautomer/normal form emission through dynamical experiments, is shown to be quite different through this series of 7AI derivatives (Table 1).

The widely accepted 7AI ESDPT mechanism, which involves equilibrium between free 7AI and 1:1 solvent-complex [143], cannot explain the above mentioned dynamical data adequately, for the main chromophore in these compounds remains almost unchanged. The difference in the solvent complex equilibrium constant cannot account for the large proton rate difference, either. Thus, it is proposed that the solvent-induced barrier plays a significant role in this system. According to the theoretical approach, the dipole moment differences between normal form and tautomer in the first excited state have been calculated to follow the trend 5CNAI > 3,5, CNAI > 3CNAI ∼ 7AI. The trend of the observed rate of ESPT process is in good correlation with the proposed solvent-polarity-induced barrier resulting from the difference in the changes of dipole moments between the equilibrium polarization of normal (N_{eq}^{*}) and tautomer species (T_{eq}^{*}) in the solvent coordinate (Fig. 19) [141]. Moreover, as shown in Fig. 20, it is obvious that the barrier is increased upon increasing the difference in dipole moment (either magnitude or direction) between

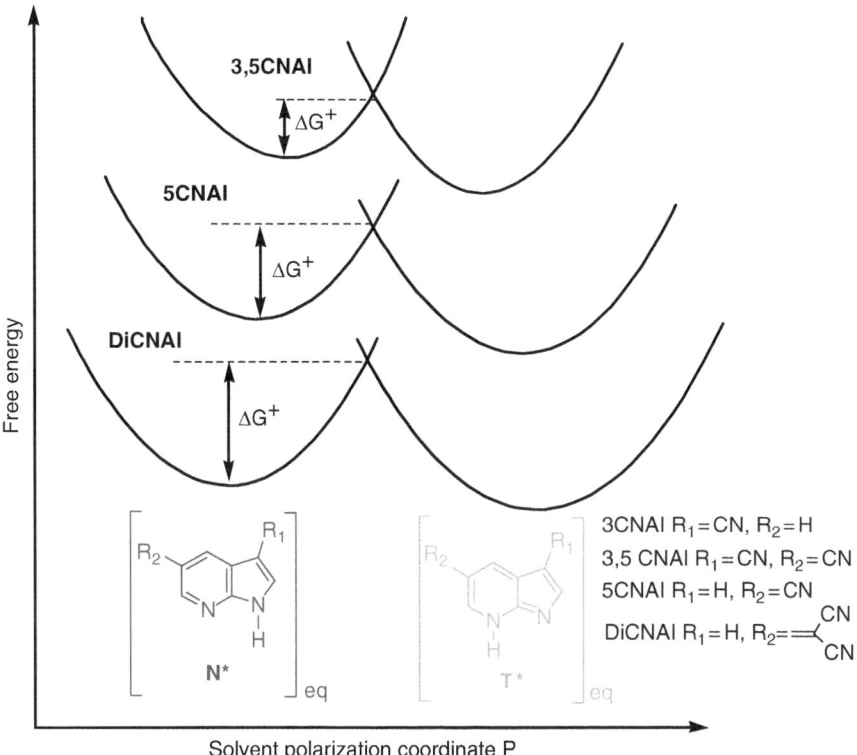

Fig. 20 The proposed mechanism of ESPT incorporating a solvent-polarity-induced barrier in protic solvents following optical charge transfer and solvent relaxation. See full name of each compound in text (reprint from ref. [141], Copyright © 2008 Wiley-VCH)

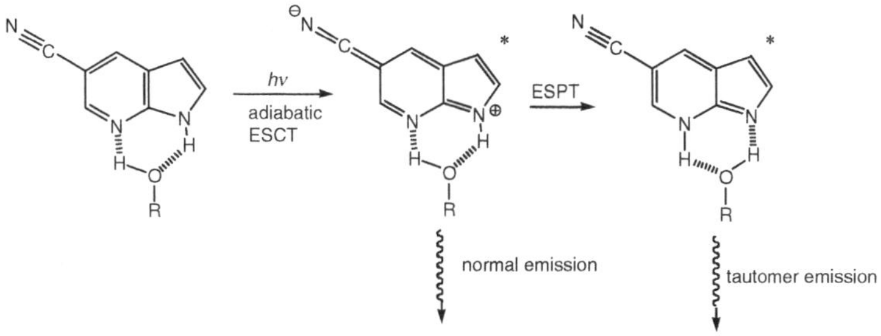

Fig. 21 The proposed ESPT/ESCT coupled mechanism of 5CNAI in protic solvents

normal and tautomer forms, consistent with the theoretical prediction regarding horizontal displacement of dipole separation described in Sect. 3.1.

The above 7AI analogs serve as one of the few experimental proofs for the solvent-induced barrier in proton transfer reaction. It is thus believed that through

Table 1 Photophysical properties of 7AI and its correlated cyano analogs in methanol

	λ_{abs}/nm	λ_{em}/nm $(\Phi)^a$	τ/ns[b]
7AI[c]	288	N: 374	τ: 0.146
		T: 503 (0.07)	τ_1: 0.134 (−0.44)
			τ_2: 0.654 (0.56)
3CNAI	285	N: 343	τ: 0.23
		T: 480 (0.02)	τ_1: 0.24 (−0.49)
			τ_2: 5.88 (0.51)
3,5CNAI	294	N: 377	τ: 0.69
		T: 515 (0.03)	τ_1: 0.71 (−0.52)
			τ_2: 1.13 (0.48)
5CNAI	297	N: 395 (0.20)	τ: 4.8
DiCNAI	351	N: 600 (0.0014)	τ: 0.20

[a]The reported Φ is the sum of the normal (N) and tautomer (T) emission bands
[b]Data in parentheses are the fitted preexponential factors
[c]Photophysical properties of 7AI are taken from ref. [142]

other ingenious design, the systematic investigation of the ESCT/ESDPT coupled reaction is possible in protic solvents, which may be crucial to gaining fundamental insights into the current research fields regarding, for example, proton-coupled electron transfer in a living system.

3.3 Excited-State Proton Transfer Coupled Charge Transfer Reaction

The theoretical and experimental progress elaborated in Sect. 3.1 has fundamental importance, in that it clearly addresses the role of solvent polarity channeling into the proton transfer dynamics. Unfortunately, up to this stage, most experimental model systems applied the ESCT/ESIPT dynamics are ascribed to case (A), in which ESCT takes place prior to ESIPT. Thus, upon Franck–Condon excitation, the associated reaction dynamics have been complicated by the competitive solvent relaxation to reach the equilibrium polarization. As such, the study of early ESCT/ESIPT reaction dynamics is commonly limited by the rate of solvent relaxation. To overcome this hurdle, it is of great fundamental interest, as well as urgent, to seek an ideal system to probe ESCT/ESIPT coupling reactions free from early solvent relaxation processes. In theory, an ideal case in point stems from case (B), for which a molecule is designed such that it undergoes ESIPT prior to the ESCT reaction. Recently, via the strategic design and synthesis of molecule 2-((2-(2-hydroxyphenyl) benzo[d]oxazol-6-yl)methylene)-malononitrile (diCN–HBO) [144], the study of EISPT coupled ESCT, i.e., process (B), has become feasible.

From a molecular structure point of view, for diCN–HBO, the lone pair electrons of the benzo-nitrogen atom are intrinsically involved in the π-electron resonance to establish the aromaticity, such that its electron donating strength, compared with those of alkyl and aryl amines, is negligibly weak. Thus, upon Franck–Condon

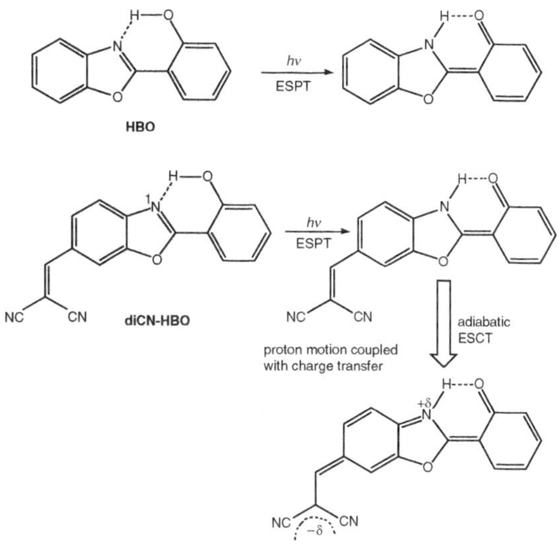

Fig. 22 Proposed ESPT reaction for HBO and ESIPT/ESCT-coupled reaction for diCN–HBO (reprint from ref. [145], Copyright © 2008 American Chemical Society)

excitation of diCN–HBO, the degree of charge transfer should be negligible. On the other hand, similar to its parent molecule 2-(2′-hydroxyphenyl)benzoxazole (HBO) [74, 103], ESPT is expected to take place from the hydroxyl proton to the N1 nitrogen, resulting in a proton transfer tautomer, i.e., a keto form. Once the proton-transfer tautomer is formed, the N1 nitrogen atom becomes the secondary alkyl amino nitrogen and thus should act as a good electron donor (Fig. 22).

Unlike most of the ESCT/ESIPT systems, in which ESCT takes place prior to ESIPT, diCN–HBO undergoes ESIPT, concomitantly accompanied by the charge transfer process, such that the ESIPT reaction dynamics are directly coupled with solvent polarization effects. The long-range solvent polarization interactions result in a solvent-induced barrier that affects the overall proton transfer reaction rate. Dual emission has thus been observed; the proton transfer tautomer emission peak moves drastically in solvents bearing different polarities. In cyclohexane, the rate constant of ESPT of diCN–HBO was determined to be 1.1 ps, which is apparently slower than that of 150 fs for the parent molecule HBO. Upon increasing solvent polarity, the ESPT rate constants were also determined to be 1.00 ± 0.13 ps in benzene, 0.60 ± 0.05 ps in CH_2Cl_2, and 0.31 ± 0.03 ps in CH_3CN, respectively, values which reveal that increasing the solvent polarity tends to render an increase in the rate of ESPT (Fig. 23) [145]. The overall reaction dynamics can be described by a mechanism incorporating both solvent polarization and proton-transfer reaction coordinates (Fig. 24). The proton transfer tautomer possessing large degrees of charge transfer character is obviously stabilized upon increasing solvent polarity, while the PC^* state is not influenced, and thus the corresponding solvent-induced barrier is reduced. It is worthy of note that unlike the parent ESIPT molecule HBO, which executes ultrafast ($\sim$150 fs) ESIPT in nonpolar solvent such as cyclohexane [74], the ESIPT/ESICT coupled system diCN–HBO undergoes a finite time

Fig. 23 Time-resolved
fluorescence decay of
diCN–HBO in various
solvents monitored at the
normal emission (PC*)

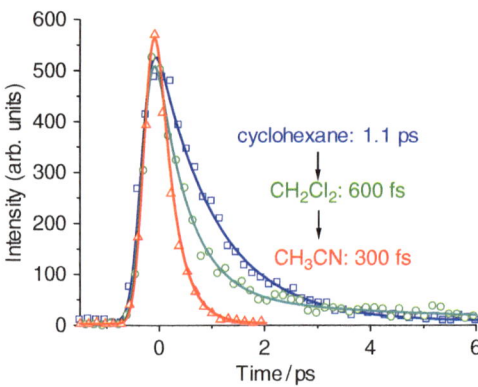

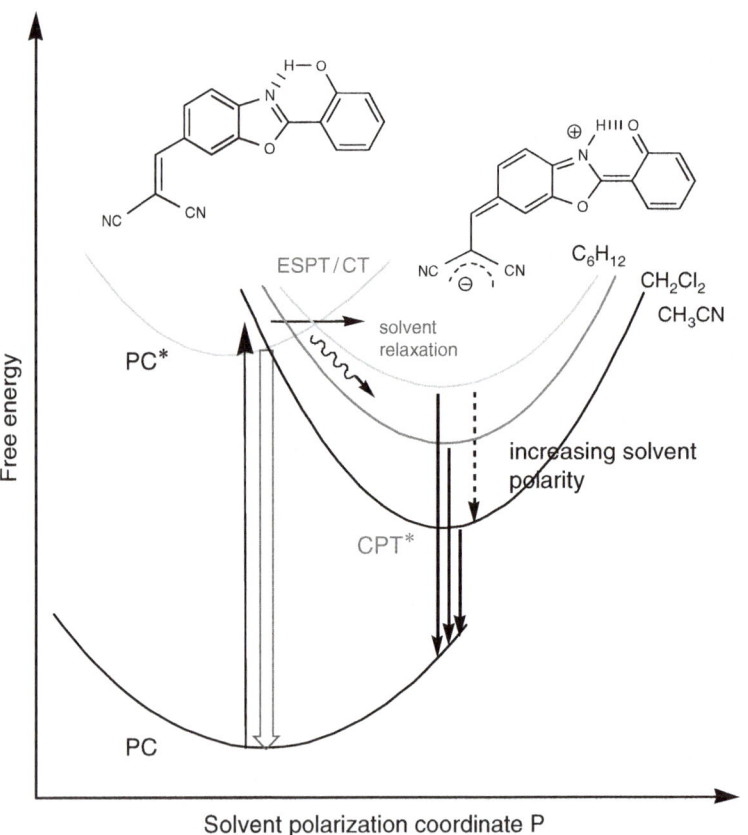

Fig. 24 Proposed ESPT/ESCT reaction/relaxation dynamics using diCN–HBO as a model (reprint
from ref. [145], Copyright © 2008 American Chemical Society)

constant (1.1 ps) proton transfer in similar cyclohexane. The result, on the one hand, can be rationalized by the nonnegligible quadruple moment effect from cyclohexane [146], which induces the small solvent-perturbed barrier. On the other hand, a dynamic polarization model in nonpolar solvents has been recently proposed by Hamaguchi and coworkers [147]. Because of the large dipolar change between PC* and CPT* in the case of diCN–HBO, the induced dipole/dipole interaction is considered to be nonnegligible, inducing an appreciable barrier in the ESPT/ESCT coupled reaction. Nonetheless, this diCN–HBO system serves as the first ESIPT/ESCT example in which ESIPT takes place prior to ESCT process. And again, the existence and importance of the solvent-induced barrier to the overall proton transfer dynamics are revealed.

As a closing remark for this chapter, we have succinctly reviewed three pivotal topics relevant to excited-state electron and proton transfer reactions. In the Sect. 1, on bipolar D–A dyads linked by covalent bonds, studies of excited state electron transfer via theoretical and experimental approach as well as via application viewpoints in, e.g., ion sensor have been reviewed. In Sect. 2, prototypes of ESIPT molecules classified by different membered-ring hydrogen-bonding in sequence have been elaborated. Depending on orientation and strength of the hydrogen bond, active modes inducing ESIPT have been discussed. Last but not least, once in polar solvents, due to the common changes in dipole moment during ESIPT, solvent polarity is expected to play a crucial role, which then channels into the reaction dynamics. In this chapter, theory on solvent-polarity-induced proton transfer reaction dynamics is first discussed, followed by its verification based on two types of designed experiments, namely the charge transfer-induced proton transfer reaction (case A) and proton transfer-induced charge transfer reaction (case B). Both prove to be prototypical models for proof of the concept. We thus hope that through reading this chapter, the reader has gained more fundamental knowledge as well as perspectives on application to organic dyes with excited-state transformations such as electron, charge, and proton transfers.

References

1. Kuznetsov AM, Ulstrup J (1999) Electron transfer in chemistry and biology: an introduction to the theory. Wiley, New York
2. May V, Kühn O (2004) Charge and energy transfer dynamics in molecular systems. Wiley-VCH, Weinheim, New York
3. Wagenknecht HA (2008) Charge transfer in DNA: from mechanism to application. Wiley-VCH, Weinheim, New York
4. Kavarnos GJ, Turro NJ (1986) Photosensitization by reversible electron transfer: theories, experimental evidence, and examples. Chem Rev 86:401–449
5. Babara PF, Mayer TJ, Ratner MA (1996) Contemporary issues in electron transfer research. J Phys Chem 100:13148–13168
6. D'Alessandro DM, Keene FR (2006) Current trends and future challenges in the experimental, theoretical and computational analysis of intervalence charge transfer (IVCT) transitions. Chem Soc Rev 35:424–440

7. Maroncelli M, Macinnis J, Flaming FR (1989) Polar solvent dynamics and electron-transfer reactions. Science 4899:1674–1681

8. Heitele H (1993) Dynamic solvent effects on electron-transfer reactions. Angew Chem Int Ed Engl 32:359–377

9. Horng ML, Gardecki JA, Papazyan A et al (1995) Subpicosecond measurements of polar salvation dynamics: Coumarin 153 revisited. J Phys Chem 99:17311–17337

10. Paddon-Row MN (1994) Investigating long-range electron-transfer processes with rigid, covalently linked donor-(norbornylogous bridge)-acceptor systems. Acc Chem Res 27:18–25

11. Jortner J, Bixon M, Langenbacher T (1998) Charge transfer and transport in DNA. Proc Natl Acad Sci USA 95:12759–12765

12. Davis WB, Svec WA, Ratner MA et al (1998) Molecular-wire behavior in p-phenylenevinylene oligomers. Nature 396:60–63

13. Adams DM, Brus L, Chidsey CED et al (2003) Charge transfer on the nanoscale: current status. J Phys Chem B 107:6668–6697

14. Guldi DM, Aminur Rahman GM, Sgobba V et al (2006) Multifunctional molecular carbon materials-from fullerenes to carbon nanotubes. Chem Soc Rev 35:471–487

15. Wiberg J, Guo L, Pettersson K et al (2007) Charge recombination versus charge separation in donor-bridge-acceptor systems. J Am Chem Soc 129:155–163

16. Wan C, Fiebig T, Kelley SO et al (1999) Femtosecond dynamics of DNA-mediated electron transfer. Proc Natl Acad Sci USA 96:6014

17. Kamat PV (2002) Photophysical, photochemical and photocatalytic aspects of metal nanoparticles. J Phys Chem B 106:7729–7744

18. Levanon H, Möbius K (1997) Advanced EPR spectroscopy on electron transfer processes in photosynthesis and biomimetic model systems. Annu Rev Biophys Biomol Struct 26:495–540

19. Hsu CP (2009) The electronic couplings in electron transfer and excitation energy transfer. Acc Chem Res 42:509–518

20. Rehm D, Weller A (1970) Kinetics of fluorescence quenching by electron and H-atom transfer. Isr J Chem 8:259

21. Marcus RA (1963) On the theory of oxidation–reduction reactions involving electron transfer. V. Comparison and properties of electrochemical and chemical rate constants. J Phys Chem 67:853–857

22. Marcus RA (1964) Chemical and electrochemical electron-transfer theory. Annl Rev Phys Chem 15:155–196

23. Marcus RA (1965) On the theory of electron-transfer reactions. VI. Unified treatment for homogeneous and electrode reactions. J Chem Phys 43:679

24. Marcus RA, Sutin N (1985) Electron transfers in chemistry and biology. Biochim Biophys Acta 811:265–322

25. McCreery RL (2004) Molecular electronic junctions. Chem Mater 16:4477–4496

26. Hayes RT, Wasielewski MR, Gosztola D (2000) Ultrafast photoswitched charge transmission through the bridge molecule in a donor-bridge-acceptor system. J Am Chem Soc 122:5563–5567

27. Fan C, Plaxco KW, Heeger AJ (2005) Biosensor based on binding-modulated donor–acceptor distances. Trends Biotechnol 23:186–192

28. Galoppini E (2004) Linkers for anchoring sensitizers to semiconductor nanoparticles. Coordin Chem Rev 248:1283–1297

29. Bella SD (2001) Second-order nonlinear optical properties of transition metal complexes. Chem Soc Rev 30:355–366

30. Chattoraj M, Chung DD, Paulson B et al (1994) Mediated electronic energy transfer: effect of a second acceptor state. J Phys Chem 98:3361–3368

31. Balzani V, Juris A, Venturi M (1996) Luminescent and redox-active polynuclear transition metal complexes. Chem Rev 96:759–833

32. Zimmerman HE, Goldman TD, Hirzel TK et al (1980) Rod-like organic molecules. energy-transfer studies using sinelo-photon counting. J Org Chem 45:3933–3951
33. Tung CH, Zhang LP, Li Y et al (1997) Intramolecular long-distance electron transfer and triplet energy transfer. Photophysical and photochemical studies on a norbornadiene-steroid-benzidine system. J Am Chem Soc 119:5348–5354
34. Warrener RN (2000) New adventures in the synthesis of hetero-bridged *syn*-facially fused nornornadines ("[*n*]polynorbornadienes") and their topological diversity. Eur J Org Chem 2000:3363–3380
35. Warrener RN, Pitt IG, Butler DN (1983) The synthesis of new linear and angular systems useful as rigid rods and spacers in the design of molecules. J Chem Soc Chem Commun 1340–1341
36. Warrener RN, Abbenante G, Kennard CHL (1994) A tandem cycloaddition protocol for the controlled synthesis of [n]ladderanes: new rods and spacers. J Am Chem Soc 116:3645–3646
37. Chow TJ, Chiu NR, Chen HC et al (2003) Photoinduced electron transfer reaction tuned by donor-acceptor pairs via rigid linear spacer heptacyclo[6.6.0.02, 6.03, 13.04, 11.05, 9.010, 14]tetradecane. Tetrahedron 59:5719–5730
38. Oevering H, Paddon-Row MN, Heppener M et al (1987) Long-range photoinduced through-bond electron transfer and radiatice recombination n via nonconjugated bridges: distance and solvent dependence. J Am Chem Soc 109:3258–3269
39. Chen KY, Hsieh CC, Cheng YM et al (2006) Tuning excited state electron transfer from an adiabatic to nonadiabatic type in donor-bridge-acceptor systems and the associated energy-transfer process. J Phys Chem A 110:12136–12144
40. de Silva AP, Gunaratne HQN, Gunnlaugsson T et al (1997) Signaling recognition events with fluorescent sensors and switches. Chem Rev 97:1515–1566
41. Xu H, Xu X, Dabestani R et al (2002) Supramolecular fluorescent probes for the detection of mixed alkali metal ions that mimic the function of integrated logic gates. J Chem Soc Perkin Trans 2:636
42. Koskela SJM, Fyles TM, James TD (2005) A ditopic fluorescent sensor for potassium fluoride. Chem Commun 7:945–947
43. Uchiyama S, McClean GD, Iwai K et al (2005) Membrane media create small nanospaces for molecular computation. J Am Chem Soc 127:8920–8921
44. Farruggia G, Iotti S, Prodi L et al (2005) 8-hydroxyquinoline derivatives as fluorescent sensors for magnesium in living cells. J Am Chem soc 128:344–350
45. Kele P, Nagy K, Kotschy A (2006) The development of conformational-dynamics-based sensor. Angew Chem Int Ed 45:2565–2567
46. Arimori S, Bell ML, Oh CS et al (2001) Molecular fluorescence sensors for saccharides. Chem Commun 18:1836–1837
47. Bronson RT, Michaelis DJ, Lamb RD et al (2005) Efficient immobilization of a cadmium chromosensor in a thin film: feneration of a cadmium sensor prototype. Org Lett 7:1105–1108
48. Nath S, Maitra U (2006) A simple and general strategy for the design of fluorescent cation sensor beads. Org Lett 8:3239–3242
49. Citterio D, Sasaki S, Suzuki K (2001) A new type of cation responsive chromoionophore with spectral sensitivity in the near-infrared spectral range. Chem Lett 30:552–553
50. Coskun A, Yilmaz MD, Akkaya EU (2007) Bis(2-pyridyl)-substituted boratriazaindacene as an NIR-emitting chemosensor for Hg(II). Org Lett 9:607–609
51. Killoran J, McDonnell SO, Gallagher JF et al (2008) A substituted BF$_2$-chelated tetraaryla-zadipyrromethene as an intrinsic dual chemosensor in the 650–850 nm spectral range. New J Chem 32:483–489
52. Kiyose K, Kojima H, Urano Y et al (2006) Development of a ratiometric fluorescent zinc ion probe in near-infrared region, based on tricarbo-cyanine chromophore. J Am Chem Soc 128:6548–6549

53. Zhu M, Yuan M, Liu X et al (2008) Visible near-infrared chemosensor for mercury ion. Org Lett 10:1481–1484
54. Peng X, Song F, Lu E et al (2005) Heptamethine cyanine dyes with a large Stokes shift and strong fluorescence: a paradigm for excited-state intramolecular charge transfer. J Am Chem Soc 128:6548–6549
55. Benesi HA, Hildebrand JH (1949) A Spectrophotometric investigation of the interaction of iodine with aromatic hydrocarbons. J Am Chem Soc 71:2703–2707
56. Fang JM, Selvi S, Liao JH et al (2004) Fluorescent and circular dichroic detection of monosaccharides by molecular sensors: bis[(pyrrolyl)ethynyl]naphthyridine and bis [(indoili)ethynyl]naphthyridine. J Am Chem Soc 126:3559–3566
57. Rurack K, Koval'chuck A, Bricks JL et al (2001) A Simple bifunctional fluoroionophore signaling different metal ions either independently or cooperatively. J Am Chem Soc 123:6205–6206
58. Huang JH, Wen WH, Sun YY et al (2005) Two-stage sensing property via a conjugated donor–acceptor–donor constitution: application to the visual detection of mercuric ion. J Org Chem 70:5827–5832
59. Lee SH, Kim SH, Kim SK et al (2005) Fluorescence ratiometry of monomer/excimer emissions in a space-through PET system. J Org Chem 70:9288–9295
60. Müller A, Ratajack H, Junge W et al (1992) Studies in physical and theoretical chemistry; electron and proton transfer in chemistry and biology, vol 78. Elsevier, Amsterdam, The Netherlands
61. Waluk J (2000) Conformational analysis of molecules in excited states. Wiley-VCH, New York
62. Elsaesser TH, Bakker HJ (2002) Ultrafast hydrogen bonding dynamics and proton transfer processes in the condensed phase. Springer, Heidelberg
63. Kasha M (1986) Proton-transfer spectroscopy: perturbation of the tautomerization potential. J Chem Soc Faraday Trans 2(82):2379–2392
64. Chou PT (2001) The host/guest type of excited-state proton transfer; a general review. J Chin Chem Soc 48:651–682
65. Tolbert LM, Solntsev KM (2002) Excited-state proton transfer: from constrained systems to "super" photoacids to superfast proton transfer. Acc Chem Res 35:19–27
66. Waluk J (2003) Hydrogen-bonding-induced phenomena in bifunctional heteroazaaromatics. Acc Chem Res 36:832–838
67. Dermota TE, Zhong Q, Castleman AW (2004) Ultrafast dynamics in cluster systems. Chem Rev 104:1861–1886
68. Rodríguez-Santiago L, Sodupe M, Oliva A et al (1999) Hydrogen atom or proton transfer in neutral and single positive ions of salicylic acid and related compounds. J Am Chem Soc 121:8882–8890
69. Lamola AA, Sharp LJ (1966) Environmental effects on the excited states of o-hydroxy aromatic carbonyl compounds. J Phys Chem 70:2634–2638
70. McMorrow D, Kasha M (1984) Intramolecular excited-state proton transfer in 3-hydroxy-flavone. Hydrogen-bonding solvent perturbations. J Phys Chem 88:2235–2243
71. Chou PT, Chen YC, Yu WS et al (2001) Spectroscopy and dynamics of excited-state intramolecular proton-transfer reaction in 5-hydroxyflavone. Chem Phys Lett 340:89–97
72. Van Benthem MH, Gillispie GD (1984) Intramolecular hydrogen bonding. 4. Dual fluorescence and excited-state proton transfer in 1, 5-dlhydroxyanthraqulnone. J Phys Chem 88:2954–2960
73. Jang DJ, Kelley DF (1985) Time-resolved and steady-state fluorescence studies of the excited-state intramolecular proton transfer and relaxation of 2-hydroxy-4, 5-naphthotropone. J Phys Chem 89:209–211
74. Wang H, Zhang H, Abou-Zied OK et al (2003) Femtosecond fluorescence upconversion studies of excited-state proton-transfer dynamics in 2-(20-hydroxyphenyl)benzoxazole (HBO) in liquid solution and DNA. Chem Phys Lett 367:599–608

75. Frey W, Laermer F, Elsaesser T (1991) Femtosecond studies of excited-state proton and deuterium transfer in benzothiazole compounds. J Phys Chem 95:10391–10395
76. Klöpffer W (1977) Intramolecular proton transfer in electronically excited molecules. In: Pitts JN Jr, Hammond GS, Gollnick K (eds) Advances in photochemistry, vol 10. Wiley, New York
77. Barbara PF, Walsh PK, Brus LE (1989) Picosecond kinetic and vibrationally resolved spectroscopic studies of intramolecular excited-state hydrogen atom transfer. J Phys Chem 93:29–34
78. Arnaut LG, Formosinho SJ (1993) Excited-state proton transfer reactions I. Fundamentals and intermolecular reactions. J Photochem Photobiol A 75:1–20
79. Schwartz BJ, Peteanu LA, Harris CB (1992) Direct observation of fast proton transfer: femtosecond photophysics of 3-hydroxyflavone. J Phys Chem 96:3591–3598
80. Frey W, Elsaesser T (1992) Femtosecond intramolecular proton transfer of vibrationally hot molecules in the electronic ground state. Chem Phys Lett 189:565–570
81. Chou PT, Chen YC, Yu WS et al (2001) Excited-state intramolecular proton transfer in 10-hydroxybenzo[h]quinoline. J Phys Chem A 105:1731–1740
82. Douhal A, Lahmani F, Zewail AH (1996) Proton-transfer reaction dynamics. Chem Phys 207:477–498
83. Chudoba C, Riedle E, Pfeiffer M et al (1996) Vibrational coherence in ultrafast excited state proton transfer. Chem Phys Lett 263:622–628
84. Pfeiffer M, Lenz K, Lau A et al (1997) Analysis of the vibrational spectra of heterocyclic aromatic molecules showing internal proton and deuterium transfer. J Raman Spectrosc 28:61–72
85. Parsapour F, Kelley DF (1996) Torsional and proton transfer dynamics in substituted 3-hydroxyflavones. J Phys Chem 100:2791–2798
86. Sytnik A, Kasha M (1994) Excited-state intramolecular proton transfer as a fluorescence probe for protein binding-site static polarity. Proc Natl Acad Sci USA 91:8627–8630
87. Sakai K, Tsuzuki T, Itoh Y et al (2005) Using proton-transfer laser dyes for organic laser diodes. Appl Phys Lett 86:081103
88. Lim SJ, Seo J, Park SY (2006) Photochromic switching of excited-state intramolecular proton-transfer (ESIPT) fluorescence: a unique route to high-contrast memory switching and nondestructive readout. J Am Chem Soc 128:14542–14547
89. Catalán J, del Valle JC, Claramuntb RM (1996) Photophysics of the 2-(2'-hydroxyphenyl) perimidine: on the fluorescence of the enol form. J Lumin 68:165–170
90. Roshal AD, Grigorovich AV, Doroshenko AO et al (1998) Flavonols and crown-flavonols as metal cation chelators. The different nature of Ba^{2+} and Mg^{2+} complexes. J Phys Chem A 102:5907–5914
91. Kim S, Seo J, Jung HK et al (2005) White luminescence from polymer thin films containing excited-state intramolecular proton-transfer dyes. Adv Mater 17:2077–2082
92. Chou P, McMorrow D, Aartsma TJ et al (1984) The proton–transfer laser. Gain spectrum and amplification of spontaneous emission of 3-hydroxyflavone. J Phys Chem 88:4596–4599
93. Etter MC, Urbańczyk-Lipkowska Z, Baer S et al (1986) The crystal structures and hydrogen-bond properties of three 3-hydroxyflavone derivatives. J Mol Struct 144:155–167
94. Ameer-Beg S, Ormson SM, Brown RG et al (2001) Ultrafast measurements of excited state intramolecular proton transfer (ESIPT) in room temperature solutions of 3-hydroxyflavone and derivatives. J Phys Chem A 105:3709–3718
95. Swinney TC, Kelley DF (1993) Proton transfer dynamics in substituted 3-hydroxyflavones: solvent polarization effects. J Chem Phys 99:211–221
96. Yu WS, Cheng CC, Cheng YM et al (2003) Excited-state intramolecular proton transfer in five-membered hydrogen-bonding systems: 2-pyridyl pyrazoles. J Am Chem Soc 125:10800–10801
97. Kijak M, Nosenko Y, Singh A et al (2007) Mode-selective excited-state proton transfer in 2-(2'-pyridyl)pyrazole isolated in a supersonic jet. J Am Chem Soc 129:2738–2739

98. Chou PT, Wei CY (1996) Photophysics of 10-hydeoxybenzo[h]quinoline in aqueous solution. J Phys Chem 100:17059–17066
99. Takeuchi T, Tahara T (2005) Coherent nuclear wavepacket motions in ultrafast excited-state intramolecular proton transfer: sub-30-fs resolved pump-probe absorption spectroscopy of 10-hydroxybenzo[h]quinoline in solution. J Phys Chem A 109:10199–10207
100. Kim CH, Joo T (2010) Coherent excited state intramolecular proton transfer probed by time-resolved fluorescence. Phys Chem Chem Phys. doi:10.1039/b915768a
101. Sytnik A, Del Valle JC (1995) Steady-state and time-resolved study of the proton-transfer fluorescence of 4-hydroxy-5-azaphenanthrene in model solvents and in complexes with human serum albumin. J Phys Chem 99:13028–13032
102. Roberts EL, Chou PT, Alexander TA et al (1995) Effects of organized media on the excited-state intramolecular proton transfer of 10-hydroxybenzo[h]quinoline. J Phys Chem 99:5431–5437
103. Abou-Zied OK, Jimenez R, Thompson EHZ et al (2002) Solvent-dependent photoinduced tautomerization of 2-(2′-hydroxyphenyl)benzoxazole. J Phys Chem A 106:3665–3672
104. Chou PT, Martinez ML, Studer SL (1992) The role of the *cis*-keto triplet state in the proton transfer cycle of 2-(2′-hydroxyphenyl)benzothiazole. Chem Phys Lett 195:586–590
105. Ikegami M, Arai T (2000) Laser flash photolysis study on hydrogen atom transfer of 2-(2-hydroxyphenyl)benzoxazole and 2-(2-hydroxyphenyl)benzothiazole in the triplet excited state. Chem Lett 9:996–997
106. Rini M, Dreyer J, Nibbering ETJ et al (2003) Ultrafast vibrational relaxation processes induced by intramolecular excited state hydrogen transfer. Chem Phys Lett 374:13–19
107. Brewer WE, Martinez ML, Chou PT (1990) Mechanism of the ground-state reverse proton transfer of 2-(2-hydroxyphenyl)benzothiazole. J Phys Chem 94:1915–1918
108. Ikegami M, Arai T (2002) Photoinduced intramolecular hydrogen atom transfer in 2-(2-hydroxyphenyl)benzoxazole and 2-(2-hydroxyphenyl)-benzothiazole studied by laser flash photolysis. J Chem Soc Perkin Trans 2:1296–1301
109. Al-Soufi W, Grellmann KH, Nickel B (1991) Keto-enol tautomerization of 2-(2′-hydroxyphenyl)benzoxazole and 2-(2′-hydroxy-4′-methylphenyl) benzoxazole in the triplet state: hydrogen tunneling and isotope effects. 1. Transient absorption kinetics. J Phys Chem 95:10503–10509
110. Tsien RJ (1998) The green fluorescence protein. Annu Rev Biochem 67:509–544
111. Agmon N (2005) Proton pathways in green fluorescence protein. Biophys J 88:2452–2461
112. Stoner-Ma D, Melief EH, Nappa J et al (2006) Proton relay reaction in green fluorescent protein (GFP): polarization-resolved ultrafast vibrational spectroscopy of isotopically edited GFP. J Phys Chem B 110:22009–22018
113. Mandal D, Tahara T, Meech SR (2004) Excited-state dynamics in the green fluorescence protein chromophore. J Phys Chem B 108:1102–1108
114. He X, Bell AF, Tonge PJ (2002) Synthesis and spectroscopic studies of model red fluorescent protein chromophores. Org Lett 4:1523–1526
115. Schaefer T (1975) A relationship between hydroxyl proton chemical shifts and torsional frequencies in some ortho-substituted phenol derivatives. J Phys Chem 79:1888–1890
116. Gepshtein R, Huppert D, Agmon N (2006) Deactivation mechanism of the green fluorescent chromophore. J Phys Chem B 110:4434–4442
117. Chen KY, Cheng YM, Lai CH et al (2007) Ortho green fluorescence protein synthetic chromophore; Excited-state intramolecular proton transfer via a seven-membered-ring hydrogen-bonding system. J Am Chem Soc 129:4534–4535
118. Chung WT, Chen BS, Chen KY et al (2009) Fluorescent protein red Kaede chromophore; one-step, high-yield synthesis and potential application for solar cells. Chem Comm 45:6982–6984
119. Borgis D, Hynes JT (1991) Molecular-dynamics simulation for a model nonadiabatic proton transfer reaction in solution. J Chem Phys 94:3619–3628

120. Borgis D, Hynes JT (1996) Curve crossing formulation for proton transfer reactions in solution. J Phys Chem 100:1118–1128
121. Kiefer PM, Hynes JT (2002) Nonlinear free energy relations for adiabatic proton transfer reactions in a polar environment. I. Fixed proton donor–acceptor separation. J Phys Chem A 106:1834–1849
122. Kiefer PM, Hynes JT (2002) Nonlinear free energy relations for adiabatic proton transfer reactions in a polar environment. II. Inclusion of the hydrogen bond vibration. J Phys Chem A 106:1850–1861
123. Hynes JT, Tran-Thi TH, Grunucci G (2002) Intermolecular photochemical proton transfer in solution: new insights and perspectives. J Photochem Photobiol A Chem 154:3–11
124. German ED, Kuznetsov AM (1981) Dependence of the hydrogen kinetic isotope effect on the reaction free energy. J Chem Soc, Faraday Trans 1 77:397–412
125. German ED, Kuznetsov AM, Dogonadze RR (1980) Theory of the kinetic isotope effect in proton transfer reactions in a polar medium. J Chem Soc, Faraday Trans 2 76:1128–1146
126. Morillo M, Cukier RI (1990) On the effects of solvent and intermolecular fluctuations in proton transfer reactions. J Chem Phys 92:4833–4838
127. Chou PT, Yu WS, Cheng YM et al (2004) Solvent-polarity tuning excited-state charge coupled proton-transfer reaction in p-N, N-ditolylaminosalicylaldehydes. J Phys Chem A 108:6487–6498
128. Cheng YM, Pu SC, Yu YC et al (2005) Spectroscopy and femtosecond dynamics of 7-N, N-diethylamino-3-hydroxyflavone. The correlation of dipole moments among various states to rationalize the excited-state proton transfer reaction. J Phys Chem A 109:11696–11706
129. Cheng YM, Pu SC, Hsu CJ et al (2006) Femtosecond dynamics on 2-(2'-hydroxy-4'-diethylaminophenyl)benzothiazole: solvent polarity in the excited-state proton transfer. ChemPhysChem 7:1372–1381
130. Gormin D, Kasha M (1988) Triple fluorescence in aminosalicylates. Modulation of normal, proton-transfer, and twisted intramolecular charge-transfer (TICT) fluorescence by physical and chemical perturbations. Chem Phys Lett 153:574–576
131. Chou PT, Martinez ML, Clements JH (1993) The observation of solvent-dependent proton-transfer/charge-transfer lasers from 4'-diethylamino-3-hydroxyflavone. Chem Phys Lett 204:395–399
132. Parsapour F, Kelley DF (1996) Torsional and proton transfer dynamics in substituted 3-hydroxyflavones. J Phys Chem 100:2791–2798
133. Shynkar VV, Mély Y, Duportail G et al (2003) Picosecond time-resolved fluorescence studies are consistent with reversible excited-state intramolecular proton transfer in 4'-(dialkylamino)-3-hydroxyflavones. J Phys Chem A 107:9522–9529
134. Ameer-Beg S, Ormson SM, Poteau X et al (2004) Ultrafast measurements of charge and excited-state intramolecular proton transfer in solutions of 4'-(N, N-dimethylamino) derivatives of 3-hydroxyflavone. J Phys Chem A 108:6938–6943
135. Chou PT, Huang CH, Pu SC et al (2004) Tuning excited-state charge/proton transfer coupled reaction via the dipolar functionality. J Phys Chem A 108:6452–6454
136. Chou PT, Pu SC, Cheng YM et al (2005) Femtosecond dynamics on excited-state proton/charge-transfer reaction in 4'-N, N-diethylamino-3-hydroxyflavone. The role of dipolar vectors in constructing a rational mechanism. J Phys Chem A 109:3777–3787
137. Shephard MJ, Paddon-Row MN, Jordan KD (1993) Electronic coupling through saturated hydrocarbon bridges. Chem Phys 176:289–304
138. Paddon-Row MN, Shephard MJ (1997) Through-bond orbital coupling, the parity rule, and the design of "superbridges" which exhibit greatly enhanced electronic coupling: a natural bond orbital analysis. J Am Chem Soc 119:5355–5365
139. Napper AM, Head NJ, Oliver AM et al (2002) Use of U-shaped donor-bridge-acceptor molecules to study electron tunneling through nonbonded contacts. J Am Chem Soc 124:10171–10181

140. Chou PT, Yu WS, Wei CY et al (2001) Water-catalyzed excited-state double proton transfer in 3-cyano-7-azaindole: the resolution of the proton-transfer mechanism for 7-azaindoles in pure water. J Am Chem Soc 123:3599–3600
141. Hsieh CC, Chen KY, Hsieh WT et al (2008) Cyano analogues of 7-azaindole: probing excited-state charge-coupled proton transfer reactions in protic solvents. ChemPhyChem 9:2221–2229
142. Négrerie M, Gai F, Bellefuille SM et al (1991) Photophysics of a novel optical probe: 7-Azaindole. J Phys Chem 95:8663–8670
143. Mentus S, Maroncelli M (1998) Solvation and the excited-state tautomerization of 7-azaindole and 1-azacarbazole: computer simulations in water and alcohol solvents. J Phys Chem A 102:3860–3876
144. Seo J, Kim S, Park SY (2004) Strong solvatochromic fluorescence from the intramolecular charge-transfer state created by excited-state intramolecular proton transfer. J Am Chem Soc 126:11154–11155
145. Hsieh CC, Cheng YM, Hsu CJ et al (2008) Spectroscopy and femtosecond dynamics of excited-state proton transfer induced charge transfer reaction. J Phys Chem A 112:8323–8332
146. Craven IE, Hesling MR, Laver DR et al (1989) Polarizability anisotropy, magnetic anisotropy, and quadrupole moment of cyclohexane. J Phys Chem 93:627–631
147. Iwata K, Ozawa R, Hamaguchi H (2002) Analysis of the solvent- and temperature-dependent Raman spectral changes of S1 trans-stilbene and the mechanism of the trans to cis isomerization: dynamic polarization model of vibrational dephasing and the C=C double-bond rotation. J Phys Chem A 106:3614–3620
148. Merola F, Levy B, Demachy I, Pasquier H (2010) Photophysics and Spectroscopy of Fluorophores in the Green Fluorescent Protein Family. In: Demchenko AP (ed) Advanced Fluorescence Reporters in Chemistry and Biology I. Springer Ser Fluoresc 8:347–383

Dyes with Segmental Mobility: Molecular Rotors

Mark A. Haidekker, Matthew Nipper, Adnan Mustafic, Darcy Lichlyter, Marianna Dakanali, and Emmanuel A. Theodorakis

Abstract Molecular rotors are fluorescent molecules that are characterized by the ability to form twisted states through the rotation of one segment of the structure with respect to the rest of the molecule. Intramolecular rotation changes the ground-state and excited-state energies, and molecular rotors deexcite from the twisted state either without photon emission or with a different wavelength than from the LE state. Intramolecular rotation is strongly dependent on the solvent. Solvent polarity, hydrogen bond formation, isomerization, excimer formation, and steric hindrance are predominant forms of solvent–fluorophore interaction. Of highest importance is steric hindrance, because it links the solvent's microviscosity to the formation rate of TICT states, which, in turn, determines the spectral emission. For this reason, molecular rotors have found a wide range of applications as fluorescent sensors of microviscosity and solvent free volume. Application examples include bulk viscosity measurement, probing dynamics of polymer formation, protein sensing and probing of protein aggregation, and microviscosity probing in living cells.

Keywords Biofluids · Chemosensors · Emission spectroscopy · Mechanosensors · Optical properties · Polarity · Rheology · Twisted intramolecular charge transfer · Viscosity

Contents

M.A. Haidekker (✉), M. Nipper, A. Mustafic, and D. Lichlyter
University of Georgia, Athens, GA 30602, USA
e-mail: mhaidekk@uga.edu

M. Dakanali and E.A. Theodorakis
Department of Chemistry and Biochemistry, University of California, San Diego, La Jolla, CA 92093-0358, USA

A.P. Demchenko (ed.), *Advanced Fluorescence Reporters in Chemistry and Biology I:*   267
Fundamentals and Molecular Design, Springer Ser Fluoresc (2010) 8: 267–308,
DOI 10.1007/978-3-642-04702-2_8, © Springer-Verlag Berlin Heidelberg 2010

1   Introduction

The term *molecular rotor* refers to a group of fluorescent molecules with an intramolecular charge transfer (ICT) mechanism, which undergo a twisting motion in the excited state. This family of fluorophores is often known as twisted intramolecular charge transfer (TICT) complexes. After photon absorption, a molecular rotor can return to the ground state either from the locally excited (LE) state or from the twisted state. The energy gaps between the LE and twisted states to the ground state are very different, and the deexcitation from the twisted state has either a red-shifted emission wavelength or no emission at all. One of the main reasons why molecular rotors are intensively investigated is the dependency of the twisted-state deexcitation rate on the local environment, namely on the solvent's microviscosity and polarity.

The anomalous dual fluorescence emission of p-N,N-dimethylamino benzonitrile (DMABN) in polar solvents was first reported by Ernst Lippert in 1962. Emission spectra of DMABN in solvents of different polarity show a dual emission, where the red-shifted emission is stronger relative to the primary emission when the solvent polarity increases. Furthermore, it can be observed that overall emission intensity is reduced in more polar solvents, but higher solvent viscosity increases the emission intensity. Spectra of DMABN in different solvents are shown in the chapter of Tomin in this book [1].

Molecular rotors have in common that fluorescent excitation leads to an ICT, in the case of DMABN from the nitrogen in the dimethylamino electron donor group to the nitrile electron acceptor group (Fig. 1). The ICT leads to a highly polar

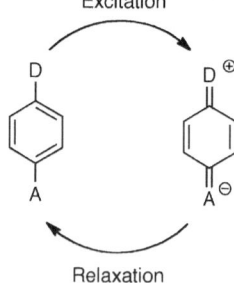

Fig. 1 Mesomeric structures of a *para*-substituted benzene intramolecular charge transfer (ICT) complex in the ground state and in the dipolar excited state

Fig. 2 Intramolecular twisting in DMABN. In the excited state, the molecule has a high propensity to rotate the dimethylamino group out of the planar ground-state configuration [3]. This process changes the dipole energy, and relaxation from the twisted state causes the red-shifted emission band

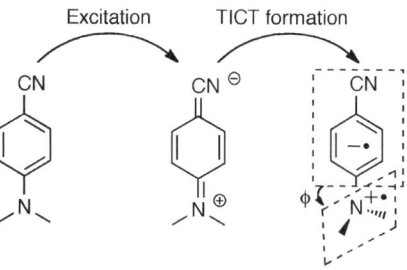

structure in the excited state. The TICT hypothesis was first formulated by Rotkie-wicz et al. [2] whereby the molecule can undergo an intramolecular twisting motion around a single bond. Twisted-state formation of DMABN is sketched in Fig. 2. Deexcitation of the molecule can occur either from the planar LE state or from the twisted state, and both cases lead to different deexcitation energies. In the case of DMABN, the red-shifted emission band can be explained by a deexcitation from the lower-energy twisted state.

Twisted-state formation is strongly influenced by the surrounding media. Steric hindrance can reduce the formation of twisted states, and polar solvents show reorientation around the charged dipole and may even form hydrogen bonds with the fluorophore. The ability to respond to multiple solvent effects makes customization of these fluorophores possible. Different TICT-forming fluorophores can be used for monitoring photochemical effects, quenching, and competitively coupled and consecutively coupled product species [4]. A common example of a competitive scheme can be found in the stilbene group. Stilbenes can form nonradiative secondary products in direct competition with the formation of radiative products. If the reaction dynamics are favorable for the formation of nonradiative products, intramolecular fluorescence quenching takes place [5], evidenced by a lower quantum yield. These examples demonstrate how the twisted-state mechanism can form the basis for the unique sensitivity of molecular rotors toward the solvent. There are several mechanisms through which the solvent influences the fluorescent properties of a TICT molecule. Of highest relevance are excimer formation, isomerization, polar reorientation of the solvent molecules, and steric hindrance of intramolecular rotation by the solvent. This chapter focuses on solvent interaction through steric hindrance. An overview of the interaction mechanisms of TICT fluorophores with the solvent, an overview of commonly used families of TICT-forming molecules, and practical applications of TICT complexes for sensing solvent properties are given in this chapter.

2 Photophysical Properties of Molecular Rotors

Two phenomena need to be examined to understand the interaction of a molecular rotor with its environment. First, changes of the ground-state and excited-state energies between LE and twisted states need to be examined, and second, the interaction of the fluorophore in its excited state with the solvent needs to be considered.

In most fluorescent processes, the lowest unoccupied molecular orbital (LUMO) and the highest occupied molecular orbital (HOMO) are the two relevant levels considered in electron energy change. HOMO–LUMO photoexcitation of a fluorophore promotes one of the two present electrons to an antibonding state that is usually denoted by $*$. In addition to the σ and π orbitals (a typical fluorescence-related transition is π–$\pi*$), electrons can also be promoted from nonbonding electrons such as those that surround oxygen or nitrogen. These n-electrons can move to either the $\sigma*$ or $\pi*$ antibonding state. ICT complexes are characterized by the presence of electron donor and acceptor groups that are connected via a π-conjugation system. The donor group usually provides n-electrons. The π-conjugation allows interaction between the donor and acceptor groups so that the nonbonding electrons of the donor can be delocalized to the unoccupied orbitals of the acceptor. A key component is the spatial separation of the donor and acceptor groups with the π-system so that the orbitals of the donor and acceptor groups have negligible spatial overlap.

The promotion to a higher energy level can cause the electron to relax to a number of vibrational states before the molecule's return to the ground state and consequential photon release. This relaxation is the energy loss that is predominantly responsible for the wavelength shift between absorption and fluorescence (Stokes shift). In TICT complexes, the formation of the twisted state can be interpreted as a dominant vibrational state that leads to major energy loss. The twisting of the molecule is a result of an unbalanced dipole moment upon photon absorption and requires the ability of the nitrogen atom in the donor group to undergo a change from a pyramidal configuration in the ground state to a planar configuration in the charge transfer state [6]. A qualitative sketch of the energy levels of a TICT molecule can be seen in Fig. 3 [7–9]. The ground-state energy level

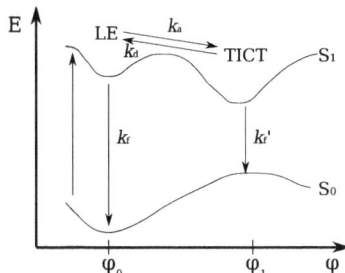

Fig. 3 Ground-state and excited-state energies in the planar and twisted configurations of a TICT molecule. Photon absorption usually elevates the molecule to the LE state in the planar configuration, where the angle of intramolecular rotation, φ_0, is close to zero. From the LE state, the molecule can return to the ground state with a rate k_f, or it undergoes intramolecular rotation with a rate k_a. The twisted state is characterized by a larger intramolecular rotation angle φ_1, usually 90°. From the twisted state, the molecule can return to the ground state with a rate k_f' or it returns to the LE state. The return rate to the LE state, k_d, is usually very small, but can be increased in nonpolar solvents

is higher in the twisted configuration, and the molecule has a natural tendency to return to the planar state. The excited-level TICT state has a lower energy than the LE state, but the states are separated by a small energy maximum that needs to be overcome when the molecule enters the TICT state from the LE state. It has also been found that any effects that delay the planar formation, i.e., placing the amino group within a heterocyclic ring [6], lead to an overall increase in the energy barrier between LE and TICT states. The energy barrier between the LE and TICT states is solvent-dependent. For DMABN, a polar solvent lowers the barrier, and viscous solvents generally elevate the barrier [9]. Consequently, the conversion rates are solvent-dependent. The fluorescent lifetimes $\tau_f = k_f^{-1}$ and $\tau_f' = k_f'^{-1}$ have been determined to be near 2.5–3 ns. The TICT formation rate k_a is in the order of 10^{10} s^{-1}, and the return rate k_d is strongly solvent-dependent and has been found to vary in magnitude from 10^8 to 10^{10} s^{-1} [10, 11]. Since these rates are highly dependent on the dye and the solvent, the values are given for orientation only. The actual rotational motion takes place within less than 20 ps.

From the excited-state conversion rates (k_a and k_d), it is possible to calculate the ratio of the quantum yields between emission from the twisted and LE states. The ratio of the steady-state quantum yields from the locally excited state ϕ_{LE} and the TICT state ϕ_{TICT} for DMABN and other aminobenzonitrile derivatives is described in (1) [12],

$$\frac{\phi_{TICT}}{\phi_{LE}} = \frac{k_f'}{k_f} \cdot \frac{k_a}{k_d + 1/\tau_{CT}'} \tag{1}$$

where τ_{CT}' is the TICT lifetime, k_f' is the deexcitation rate from the twisted state, k_f is the deexcitation rate from the LE state, k_a is the rate of TICT formation from the LE state, and k_d is the reverse rate, i.e., the return rate to the LE state from the TICT state. Generally, $k_a \gg k_d$ so that relaxation from the TICT state dominates. Molecular rotors that have radiationless deexcitation from the TICT state exhibit a low quantum yield from a single (LE) emission band.

Two examples for the energy levels of the S_0 and S_1 states in dependency of the intramolecular rotation angle were obtained through computational methods by Stsiapura et al. [13] and Allen et al. [14] for thioflavin T and 9-(dicyanovinyl)-julolidine (DCVJ), two molecules known for their ability to form TICT states. The energy levels are shown in Fig. 4 to provide quantitative data that relate to Fig. 3. In the ground state, the molecule tries to find the energy-minimizing conformation. In the case of DCVJ, the minimum energy exists when the molecule is fully planar. Thioflavin T prefers a 37° twist angle as the energy-minimizing configuration. In a similar manner, the molecules rapidly assume their minimum-energy configuration in the excited state by performing an intramolecular rotation. In both thioflavin T and DCVJ, the minimum excited-state energy is reached when the donor and acceptor planes are rotated 90° relative to each other. Excited-state intramolecular rotation takes place very rapidly, typically in the low picosecond range, when the rotation is not hindered by the solvent. When the energy gaps between S_1 and S_0

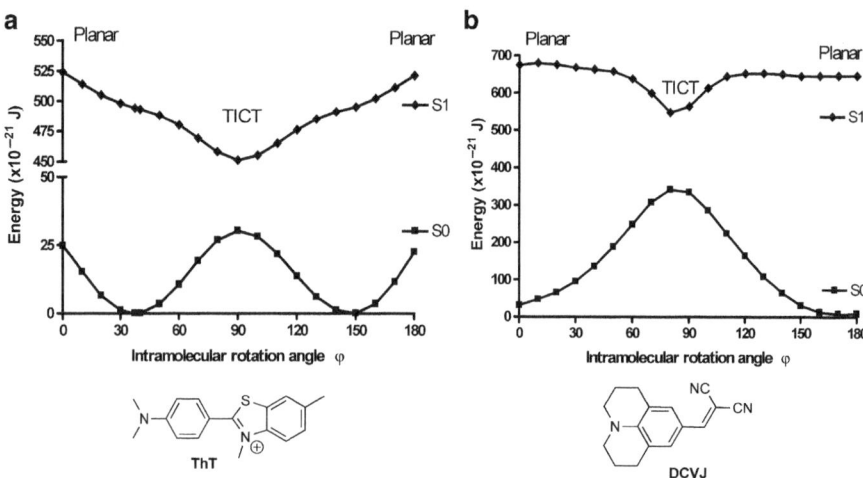

Fig. 4 Ground-state and excited-state energies of the TICT complexes thioflavin T (a) and 9-(dicyanovinyl)-julolidine (DCVJ) (b) as a function of the intramolecular rotation angle (data from Stsiapura et al. [13] and Allen et al. [14]). In both cases, energy levels were determined by quantum mechanical simulations. For thioflavin T, the energy difference between S_1 and S_0 corresponds to approximately 400 nm in the planar state and 470 nm in the twisted state. In the case of DCVJ, the energy differences correspond to 310 and 960 nm, respectively. The DCVJ energy levels reflect a rotation around the vinyl double bond

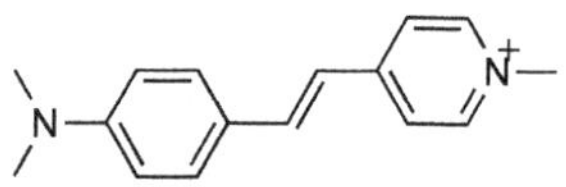

Fig. 5 Structure of
p-(dimethylamino)
stilbazolium (*p*-DASPMI)

states between the planar and TICT configurations are relatively small, dual-band emission occurs. An example is DMABN where the energy gap between the S_1 and S_0 states is about 30% smaller in the TICT state than in the planar state. With a larger donor–acceptor distance, the energy gap between S_1 and S_0 states becomes dramatically smaller in the TICT configuration than in the planar configuration. When returning to the ground state from the TICT state, many molecular rotors cross an energy gap that is so low that photon emission cannot be detected [15]. The TICT-forming dye *p*-(dimethylamino)stilbazolium (*p*-DASPMI, Fig. 5) was found to have an energy gap of 3 eV (corresponding to 412 nm) in the planar configuration and approximately 1.2 eV (corresponding to 1,030 nm) after a 90° twist around the central double bond, and relaxation from this twisted state is considered radiation-less [16]. The molecular rotor DCVJ (Fig. 4b) was found to have an energy gap more than three times larger in the planar configuration than in the TICT configuration [14]. In the extreme case, the energy gap can actually become negative [17]. The computational models that were used to compute the energy gaps did not consider solvent influences, and the actual spectra show a peak emission at higher wavelengths than the wavelengths that correspond to the energy gaps. DCVJ, for example, has a peak emission near 500 nm, whereas the planar-state energy gap (Fig. 4b) corresponds to 310 nm. A molecular rotor with radiationless relaxation from the twisted state exhibits only a single emission band, but its quantum yield is highly dependent on the environment. It is therefore important to distinguish between molecular rotors that have solvent-dependent dual emission bands and those that have a single emission band with a solvent-dependent quantum yield. The latter group is particularly interesting, because some representatives show a good separation between the influence of polarity and viscosity [14, 18, 19]. Popular examples are the julolidine-based dyes, among them DCVJ (Fig. 4). Radiationless relaxation from the TICT state occurs because of the low TICT $S_1 - S_0$ energy gap. The LE state has a relatively low dipole moment and shows only a minimal solvatochromic shift. Solvent microviscosity strongly influences TICT formation rate and therefore the quantum yield of LE emission. Consequently, the emission wavelength is weakly polarity-dependent, whereas the quantum yield is almost exclusively determined by solvent viscosity [14, 18, 19].

The predominant mechanisms of the interaction between TICT molecules and the solvent deserve further explanation. TICT fluorophores are known to be sensitive to solvent polarity. DMABN with its red-shifted second emission band is a particularly well-researched example. Any ICT complex has an increased dipole moment in the excited state compared to the ground state because of the photoinduced charge separation. In the example of DMABN, the ground-state dipole moment is approximately 6 D (1 D (Debye) $\approx 3.33564 \times 10^{-23}$ C m), and the excited-state dipole moments are approximately 12 D in the LE state and 15 D in the TICT state. Polar solvent molecules orient themselves along the fluorophore dipole by aligning their electric fields. Upon relaxation, the solvent molecules return to the ground-state orientation. As a consequence, the fluorophore exhibits a bathochromic shift that reflects the energy expended for the reorientation of the solvent molecules. The magnitude of this effect depends on the solvent polarity, that is, its dielectric constant ε. In addition, solvent polarity directly influences the TICT formation rate k_a and the reverse rate k_d in a manner that emission from the TICT states increases in more polar solvents [12]. It is possible to design the chemical structure of TICT molecules such that the ratio between intrinsic dual-band fluorescence and outside polar effects can be optimized [4].

The polarity of a molecule is often closely correlated with the ability to form hydrogen bonds with solvent molecules that are in close proximity to the fluorophore. Hydrogen bond formation can lead to spectral shifts as well as changes in the ordering of states. State change ordering occurs in heterocycles and carbonyl compounds, which results in a lower energy difference between the lowest $(n \rightarrow \pi^*)$ and $(\pi \rightarrow \pi*)$ singlet states as a consequence of ground-state hydrogen bonding [20–22]. Hydrogen bonds also have a propensity to alter intermolecular charge transfer rates of TICT molecules and, in turn, increase the TICT formation rate of the fluorophore [21]. Therefore, hydrogen bond formation has similar effects on the fluorescent emission as polar–polar interaction.

One of the most popular applications of molecular rotors is the quantitative determination of solvent viscosity (for some examples, see references [18, 23–27] and Sect. 5). Viscosity refers to a bulk property, but molecular rotors change their behavior under the influence of the solvent on the molecular scale. Most commonly, the diffusivity of a fluorophore is related to bulk viscosity through the Debye–Stokes–Einstein relationship where the diffusion constant D is inversely proportional to bulk viscosity η. Established techniques such as fluorescent recovery after photobleaching (FRAP) and fluorescence anisotropy build on the diffusivity of a fluorophore. However, the relationship between diffusivity on a molecular scale and bulk viscosity is always an approximation, because it does not consider molecular-scale effects such as size differences between fluorophore and solvent, electrostatic interactions, hydrogen bond formation, or a possible anisotropy of the environment. Nonetheless, approaches exist to resolve this conflict between bulk viscosity and apparent microviscosity at the molecular scale. Förster and Hoffmann examined some triphenylamine dyes with TICT characteristics. These dyes are characterized by radiationless relaxation from the TICT state. Förster and Hoffmann found a power–law relationship between quantum yield and solvent viscosity both analytically and experimentally [28]. For a quantitative derivation of the power–law relationship, Förster and Hoffmann define the solvent's micro-friction κ by applying the Debye–Stokes–Einstein diffusion model (2)

$$\kappa = 8\pi r^3 \eta \tag{2}$$

where η is the dynamic bulk viscosity of the solvent and r is the effective radius of one phenyl group of the triphenylamine dye. Förster and Hoffmann interpret the three arms of the dye as rotational masses that obey the classical differential equation of rotation, where the constant for the first-order term is the electrostatic force that returns the phenyl groups to their initial angle. Three special cases emerge from this approach. First, when the solvent viscosity is very low, the fluorescence quantum yield reaches a minimum that is solvent-independent. Second, in the range of extremely high viscosities, radiative deexcitation dominates with a negligible rotational deexcitation rate, and the quantum yield follows (3),

$$\phi_F \approx \frac{\tau_s}{\tau_0}\left(1 - \frac{6\sigma^2}{\eta^2}\right) \tag{3}$$

where σ is a dye-dependent constant that contains all dye-specific and viscosity-independent variables, τ_0 is the natural lifetime (i.e., the lifetime in the absence of any nonradiative relaxation events), and τ_s is the lifetime in the absence of all deactivation processes through conformational changes. The dye-dependent constant σ has units of viscosity. Under the assumption of the size of a phenyl group of $r \approx 2 \times 10^{-10}$ m and the energy resulting from the potential differences $\alpha \approx 10^{-9}$ J, a typical value for this constant is $\sigma = 100$ Pa s [28]. It can be seen that the limiting case of (3) for $\eta \gg \sigma$ yields the maximum achievable quantum yield of $\phi_F \approx \tau_s/\tau_0$. The third and most important case is obtained for intermediate viscosity values when $\eta \ll \sigma$ and the quantum yield obeys (4):

$$\phi_F \approx 0.893 \cdot \frac{\tau_s}{\tau_0} \left(\frac{\eta}{\sigma}\right)^{2/3} \tag{4}$$

Most of the constants in (4) are dye-dependent constants that can be determined experimentally. It is therefore straightforward to combine those constants into a single dye-dependent constant C and rewrite (4) as (5),

$$\phi_F = C \cdot \eta^x \tag{5}$$

which is the empirical form of the relationship between viscosity and quantum yield most often referred to as the Förster–Hoffmann equation. In (5), all variables are assumed as unitless (in fact, C has units of viscosity to the power of $-x$ as evident from (4)). The special case of $x = 2/3$ emerges from an integration step that leads to (4). It can be interpreted as a maximum value when $\eta \ll \sigma$, and experimental results may yield lower values of x.

Loutfy and coworkers [29, 30] assumed a different mechanism of interaction between the molecular rotor molecule and the surrounding solvent. The basic assumption was a proportionality of the diffusion constant D of the rotor in a solvent and the rotational reorientation rate k_{or}. Deviations from the Debye–Stokes–Einstein hydrodynamic model were observed, and Loutfy and Arnold [29] found that the reorientation rate followed a behavior analogous to the Gierer–Wirtz model [31]. The Gierer–Wirtz model considers molecular free volume and leads to a power–law relationship between the reorientation rate and viscosity. The molecular free volume can be envisioned as the void space between the packed solvent molecules, and Doolittle found an empirical relationship between free volume and viscosity [32] (6),

$$\eta = A \cdot \exp\left(B\frac{V_0}{V_f}\right) \tag{6}$$

where A and B are solvent-specific empirical constants, V_f is the fluid free volume, and V_0 is the van der Waals volume of the fluorescent dye. Loutfy and Arnold [29] showed experimentally that the nonradiative decay rate (i.e., the intramolecular rotation rate) behaves in an analogous manner, and that the nonradiative decay rate can be described in terms of free volume through (7) [30],

$$k_{nr} = k_{nr}^0 \exp\left(-x'\frac{V_0}{V_f}\right) \tag{7}$$

where k_{nr}^0 is an intrinsic, solvent-independent relaxation rate, and x' is a solvent-dependent constant.

Substitution of the exponential term from (6) into (7) replaces the free-volume term by the solvent viscosity and provides an equation for the viscosity-dependent nonradiative decay rate, from which the quantum yield emerges (8):

$$\phi_F \approx \frac{k_r}{k_{nr}^0} \cdot \left(\frac{\eta}{A}\right)^x \tag{8}$$

In (8), the solvent-independent constants k_r, k_{nr}^0, and A^x can be combined into a common dye-dependent constant C, which leads directly to (5). The radiative decay rate τ_r can be determined when rotational reorientation is almost completely inhibited, that is, by embedding the molecular rotor molecules in a glass-like polymer and performing time-resolved spectroscopy measurements at 77 K. In one study [33], the radiative decay rate was found to be $k_r = 2.78 \times 10^8 \text{ s}^{-1}$, which leads to the natural lifetime $\tau_0 = 3.6$ ns. Two related studies where similar fluorophores were examined yielded values of $\tau_0 = 3.3$ ns [25] and $\tau_0 = 3.6$ ns [29]. It is likely that values between 3 and 4 ns for τ_0 are typical for molecular rotors.

Contrary to the derivation proposed by Förster and Hoffmann where the value of the exponent x was specified to be rigidly 2/3 as the result of an integration, Loutfy and coworkers allow for a variable value of x that depends both on the dye and the solvent. Since the parameter A that originates from (6) is also solvent-dependent, the power–law relationship between viscosity and quantum yield needs calibration for any dye–solvent combination before the measurement of bulk viscosity with molecular rotors becomes possible. The principle of increasing quantum yield in solvents of higher viscosity is demonstrated in Fig. 6. Viscosity was modulated with mixtures of ethylene glycol and glycerol, which have different viscosities but somewhat similar polarity. Emission intensity follows a power–law relationship as predicted in (5). Linear regression yields exponents $x = 0.54$ in both cases and intercept values C_0 (the value of C for a viscosity of unity) of $C_0 = 144 \times 10^3$ for CCVJ and $C_0 = 71 \times 10^3$ for DCVJ.

Neither the rigorous derivation of the viscosity sensitivity by Förster and Hoffmann nor the more empirical derivation by Loutfy et al. account for all experimentally observed phenomena. More recently, molecular simulations have been employed to model the interaction between molecular rotors and the solvent. Parusel examined the energy states and dipole moments of DMABN derivatives to examine whether a planar or twisted configuration in the ground state accounts for the dual emission [34] and ruled out a mechanism where the molecule is twisted in the ground state and planar in the excited state (known as PICT mechanism), leaving the TICT mechanism as acceptable model. In addition, Parusel and Köhler found that the chain length of the donor group plays only a minimal role in the photophysical properties of DMABN derivatives [35]. Quiñones et al. [36] performed electronic structure modeling of DMABN in the gas phase and in the presence of polar solvents. In the gas phase, they found that DMABN assumes a nonplanar conformation where the dimethylamino group bends – rather than rotates – out of the molecule plane ("wagging angle"), whereas DMABN assumes a fully

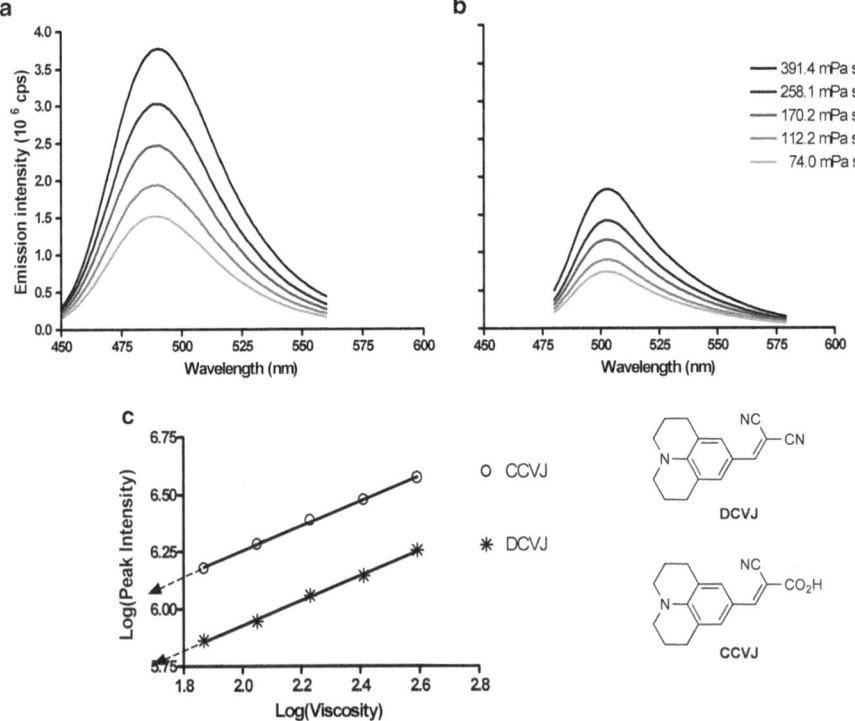

Fig. 6 Emission spectra of the molecular rotors CCVJ (**a**) and DCVJ (**b**) in a viscosity gradient with different mixture ratios of ethylene glycol and glycerol. As predicted by (5), a higher emission intensity, caused by a higher fluorescent quantum yield, is seen in higher-viscosity solvents. Plotting the peak intensities as a function of viscosity in a double-logarithmic scale reveals the power–law relationship described in (5), and the constants x and C can be determined from the diagram: x is the slope of the regression line, and C is the extrapolated intercept (*arrows*) of the regression lines with log(viscosity) = 0 or $\eta = 1$ m Pa s

planar conformation in the presence of polar solvents. In a related study, Zerner et al. [37] investigated ICT complexes with a malononitrile ($NC-CH_2-CN$) acceptor group, and compared the measured absorption spectra of an extensive number of structurally related compounds with computationally determined behavior, particularly in the presence of solvents. In a similar manner, Cao et al. [38] examined a hemicyanine dye that belongs to the popular group of stilbene-derived molecular rotors, and Allen et al. [14] studied the photophysical properties of DCVJ, one of the most widely applied molecular rotors. Structural changes and fluorescent response were simulated by Sudholt et al. [39] in a simulation of 255 molecules (one molecule being DMABN, surrounded by 254 solvent molecules). Structural changes, energy levels, spectral shifts, and lifetime dynamics were reported in vacuum and under influence of cyclopentane and acetonitrile as solvents. An overview of quantum chemical calculations performed with DMABN and related compounds can be found in a comprehensive review paper by Grabowski et al. [3]

together with an in-depth overview of the physical-chemistry foundations of molecular rotors. Whereas quantum chemical computations and molecular simulations have become more prominent in recent years, additional high-level models that reflect temperature, multiple solvents, and solvents in motion would further advance the understanding of TICT molecules with their environment. In addition, a shift of the focus from DMABN to the class of single-emission molecular rotors would reflect the latter's dominant use in practical applications.

3 Chemical Classes of Molecular Rotors

A large number of fluorophores builds on the ICT mechanism. It is therefore not surprising that the formation of TICT states has been observed in a large number of structurally different compounds. Common to the structure of all molecular rotors is a motif that consists of an electron donor unit in π-conjugation with an electron acceptor group. This motif is shown in Fig. 7. The universal donor group (electron donating group, EDG) is an alkylated nitrogen atom. The acceptor groups (electron accepting group, EAG) vary significantly and include functionalities such as nitriles, carboxylic esters, or aromatic rings. The two groups are connected via a π-conjugation system that facilitates electron transfer from donor to acceptor upon photoexcitation. DMABN (**1**) is a representative example of such design. Modifications that do not alter the extent of π-conjugation between the donor and acceptor groups will be summarized in the class of DMABN-related structures. Extension of the π-conjugation system increases the distance between the donor and acceptor units and thus significantly changes the fluorescent profile of such probes. An example is the structure of (*p*-(dialkylamino)-benzylidene)malononitrile, DBMN (**2**) [33]. DBMN is one example of fluorophore that exhibits nonradiative deactivation from the TICT state, which occurs when the S_1–S_0 energy gap in the twisted state becomes much smaller than in the planar state (cf. Fig. 3). Within the class of DBMN-related structures are included molecules with longer π-conjugation motifs such as stilbenes and 2-dicyanomethylene-3-cyano-2,5-dihydrofuran (DCDHF) flurophores. A third class of molecular rotors includes triphenylmethane-based fluorophores, and one representative example is crystal violet

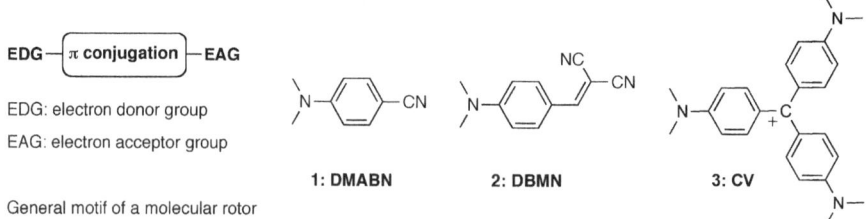

Fig. 7 General motif and representative structures of molecular rotors

[28]. In a separate class are included compounds such as porphyrins and BODIPY that have not yet been thoroughly explored as molecular rotors. A comprehensive review on porphyrin-based structures has been recently published [40]. The photophysical behavior of BODIPY dyes has been interpreted through the TICT mechanism [41, 42]. However, chemical computations by Kee et al. [41] present a very atypical picture where the planar excited-state configuration (0° or 180°) is energetically preferred with a lowest-energy angle of 55° in the ground state (only 35° from perpendicularity). A low degree of rotation would correspond to a low sensitivity toward the environment. Because the involvement of TICT states in the photophysical behavior of porphyrins and BODIPY has not conclusively been proven, these dyes are not covered in this chapter.

3.1 DMABN-Related Fluorescent Probes

3.1.1 Modifications of the Electron Donor Group

The general motif of DMABN (1) can be optimized for individual applications by modifying the electron donor or electron acceptor substituents. Often, such modifications have an effect on the sensitivity and overall fluorescent profile of the probes. The dimethyl amino donor of DMABN can be replaced by aliphatic or alicyclic amino groups. If there are no steric restrictions to the internal rotation, these probes exhibit the main characteristic features of DMABN with only some quantitative differences in sensitivity, radiative rates, and energy levels. Compounds 4 and 5 (Fig. 8) serve as an example of this approach: similar to DMABN, they both exhibit dual fluorescence, but the charge transfer band is more intense for 4 due to the larger ring structure of the donor group [43]. Related studies with dialkyl homologs of DMABN, such as 6 and 7, have shown a preference for higher emission from the planar LE state as the length of the alkyl groups is increasing [12]. Based on this observation, it was proposed that increase of the alkyl group length increases the dipole moment of the molecule. This stabilizes the excited state and therefore leads to an increase of fluorescence intensity.

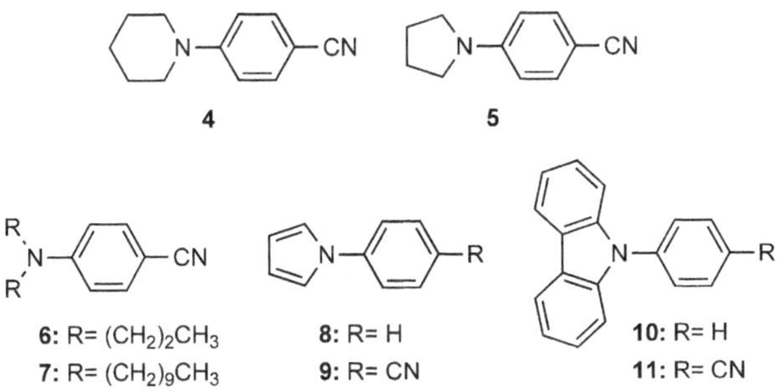

Fig. 8 Representative structures of DMABN-based rotors with different electron donor groups

Nitrogen-containing heteroaromatic rings have also been used as the electron donor groups (Fig. 8). In the case of pyrrole, the relatively low ionization potential of this ring produces a measurable fluorescence band from the LE state even when unsubstituted benzene is used as the acceptor (e.g., compound **8**). The absorption spectra of **8** and **9** show no significant changes in solvents of increasing polarity. However, the fluorescence emission of **9** is more red-shifted in polar solvents, presumably due to the electron accepting CN group. Carbazoles have a similar oxidation potential to pyrrole and can be also used as electron donors. Interestingly, **10** exhibits a solvent-independent fluorescence emission, whereas **11** behaves predictably as a molecular rotor.

3.1.2 Modifications of the π-Conjugation System

Substitution at the periphery of the phenyl ring plays a minor role in the fluorescent profile of the probe unless they affect the rotation of the donor group. For example, both compounds **12** and **13** (Fig. 9) have dual fluorescence that is similar to DMABN [44]. However, the steric congestion that is due to substitution at the periphery of the phenyl ring can destabilize the planar structure of the excited state and increase the rate of crossing to a twisted state. For instance, compound **13** has a more pronounced charge transfer band than compound **12**. Similarly, substitution of the phenyl ring with electron donating groups, such as in compound **14**, results in only small differences in the fluorescent profile of this probe as compared to DMABN [45]. In the cases of polysubstituted phenyl rings, the steric effects dominate over the electronic effects [46].

Replacing the phenyl group by a naphthalene reduces the energy gap between the LE and the charge transfer state. Compound **15** emits a dual fluorescence that was interpreted in terms of the TICT model [47]. The presence of the donor and acceptor groups along the long axis of naphthalene of rotor **16** increases significantly both Stokes shift and intensity over **15**. This can be explained by considering that substituents along the long axis of naphthalene decrease significantly the lowest singlet excited state [6]. This results in an increase of the energy "hump"

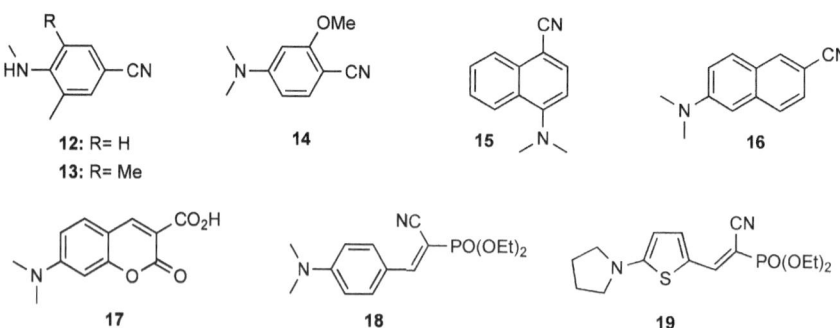

12: R= H
13: R= Me
14
15
16
17
18
19

Fig. 9 Representative structures of DMABN-based rotors with different π-conjugation systems

between that excited state and the TICT state, thus rendering the compound brighter, but less sensitive to environmental changes. On the other hand, substituents along the short axis of naphthalene, such as in rotor **15**, decrease the second lowest excited state and thus increase the coupling of this state with the TICT states.

Replacing the phenyl group by a coumarin motif produces an important subclass of molecular rotors, represented by **17**, which have found several applications in viscosity and polarity measurements [48]. On the other hand, compound **19** has a red-shifted emission as compared to **18** due to the effective delocalization of the π-electrons in the presence of the thiophene chromophore [49].

3.1.3 Modifications of the Electron Acceptor Group

Replacing the nitrile by a carbonyl group does not significantly affect the fluorescence profile of the probe. For instance, dual fluorescence is observed for compounds **20** and **21** in polar solvents (Fig. 10). However, the fluorescence profile of the carboxylic acid-containing compound **20** is further complicated by the protolytic dissociation in the same solvents [50]. Moreover, esters with long carboxylic side chains, such as **22**, have been reported to form excimers in organic solvents [51]. In the case of amides, the substituents on the amide nitrogen can attenuate the acceptor effect. Thus, the nonsubstituted amide **23** is more red-shifted as compared to the fully substituted amide **24** [52]. Both amides have dual fluorescence in polar solvents, although with low quantum yields.

Replacing the nitrile group by a benzothiazole produces an important subclass of fluorescent compounds represented by thioflavin T (**25**, Fig. 10). It is not clear if this compound undergoes deactivation via intramolecular rotation that would meet the criterion for a molecular rotor. The steady-state absorption and emission properties of thioflavin T has been attributed to micelle formation [53, 54], dimer and excimer formation [55, 56], and deactivation through intramolecular rotation [57].

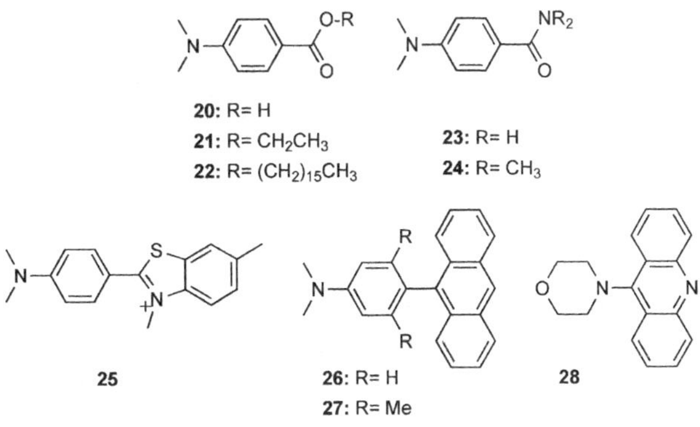

20: R= H
21: R= CH₂CH₃
22: R= (CH₂)₁₅CH₃

23: R= H
24: R= CH₃

25

26: R= H
27: R= Me

28

Fig. 10 Representative structures of DMABN-based rotors with different electron acceptor group

Anthracene has also been used as an acceptor (Fig. 10). In solution, **26** emits a single fluorescence band that is somewhat structured in nonpolar solvents and becomes broad and structureless with increasing polarity [58]. The strongly hindered molecule **27** also exhibits a similar behavior, but its absorption spectrum is better structured [59]. The rate of formation of a charge transfer state is higher for **27** than for **26**. Based on this observation, it appears that the twist around the anthryl–phenyl C–C bond plays a significant role in the fluorescence profile of the probes [60]. Acridines, such as **28**, behave similarly to anthracene except that acridine is a better electron acceptor [61].

3.2 DBMN-Related Structures

3.2.1 Julolidines

Incorporation of the nitrogen donor within a fused ring system, such as the tricyclic julolidine motif, prohibits rotation across the C–N bond. For this reason, compound **29** has a single fluorescence band that is not sensitive to the environment (Fig. 11). Compound **30** has an extended π-conjugation system, due to the additional double bond, which participates in the fluorescence excitation and emission of this molecule. This compound, widely known as DCVJ, exhibits a single-band fluorescence emission that is red-shifted compared to **29**. Photoexcitation of DCVJ takes place by electron transfer from the julolidine nitrogen to one of the nitrile groups. Relaxation takes place either via fluorescence emission or via nonradiative deexcitation from the TICT state that involves intramolecular rotation around the vinyl double bond. If such rotation is hindered by reduced molecular free volume (corresponding to high viscosity of the environment), the relaxation occurs via an increased fluorescence emission. In contrast, in solvents of low viscosity, the relaxation proceeds mainly via the nonradiative TICT pathway. Overall, this results in a fluorescence emission where the quantum yield depends on the viscosity of the environment in a power–law fashion as described in (5) [62].

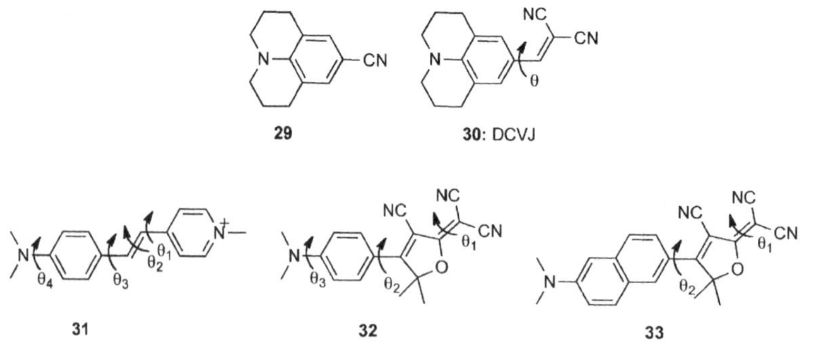

Fig. 11 Representative structures of DBMN-related structures with extended π-conjugation system

3.2.2 Stilbenes

Extending the conjugation between the donor and acceptor groups influences the
fluorescence profile of the probes. A molecular rotor based on the stilbene motif is
compound **31** (Fig. 11) [16]. One important difference to the aniline- and julolidine-
based molecular rotors is the larger number of bonds around which the molecule
can twist and form TICT states of different energy levels. Strehmel et al. [16]
proposed that rotation around the three carbon–carbon bonds of the stilbene motif
$(\theta_1, \theta_2,$ and $\theta_3)$ is possible. In addition, rotation around the nitrogen–carbon bond of
the dimethylamino donor is possible (θ_4), but it has a very low rotation rate due to a
high energy barrier [16]. According to Strehmel et al. [16], rotation around the
stilbene double bond is responsible for the creation of a lowered $S_1 - S_0$ energy gap
that leads to radiationless relaxation from the twisted state, and this rotation mode is
the preferred TICT mode, because there is no energy barrier between the planar and
double-bond TICT state. Conversely, Cao et al. [38] and Pillai et al. [63] found that
rotation around the stilbene double bond is not possible and attribute the radiationless relaxation from a TICT rotation around the bond marked with θ_3. Because of
the larger number of bonds that allow twisted states, stilbene-based dyes show a
more complex solvent interaction than aniline- or julolidine-based dyes. Similar
observations have been reported for pyridinium- and diazene-based stilbenes [64].

3.2.3 2-dicyanomethylene-3-cyano-2,5-dihydrofuran (DCDHF) Fluorophores

The conjugation between the donor and acceptor groups can be extended by the
incorporation of other ring structures, such as the dihydrofuran motif of compound
32 (Fig. 11) [65]. This has led to the design of the DCDHF fluorophores that exhibit
two types of sensitivity to local environment: solvatochromism as a result of the
charge transfer character of the excitation and viscosity dependence due to the
suppression of the twisting motion that permits nonradiative deexcitation pathways.
Extending the π-conjugation by replacing the phenyl ring with a naphthalene group
(**33**) and other polyaromatic rings or modifying the substituents of the nitrogen
donor can fine-tune the fluorescence profile of these molecules.

3.3 Ionic Dyes

The pioneering work Förster and Hoffmann [28] on the viscosity dependence of the
fluorescence quantum yield of triphenylmethane dyes (TPM) has set the foundation
for several reports in these dyes (Fig. 12). It was found that both an ability to twist
around the carbocationic center and the donor–acceptor properties are important
[66]. Specifically, a strong intramolecular quenching is observed for **34** that is
virtually absent (two orders of magnitude slower quenching rate) in the bridged

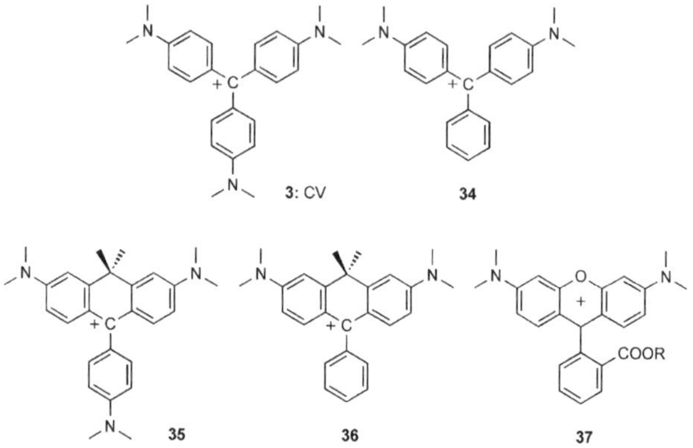

Fig. 12 Representative structures of triphenylmethane dyes

derivative **36**. Comparison of **3** with **35** shows that the quenching rate is only about ten times slower. This suggests that the flexibility of the third ring is not relevant to the fluorescence unless it contains a donor group. These observations also apply to related rhodamine dyes (compound **37**, Fig. 12) [67].

4 The Measurement of Molecular Rotor Fluorescence

Molecular rotors are useful as reporters of their microenvironment, because their fluorescence emission allows to probe TICT formation and solvent interaction. Measurements are possible through steady-state spectroscopy and time-resolved spectroscopy. Three primary effects were identified in Sect. 2, namely, the solvent-dependent reorientation rate, the solvent-dependent quantum yield (which directly links to the reorientation rate), and the solvatochromic shift. Most commonly, molecular rotors exhibit a change in quantum yield as a consequence of nonradiative relaxation. Therefore, the fluorophore's quantum yield needs to be determined as accurately as possible. In steady-state spectroscopy, emission intensity can be calibrated with quantum yield standards. Alternatively, relative changes in emission intensity can be used, because the ratio of two intensities is identical to the ratio of the corresponding quantum yields if the fluid optical properties remain constant. For molecular rotors with nonradiative relaxation, the calibrated measurement of the quantum yield allows to approximately compute the rotational relaxation rate k_{or} from the measured quantum yield ϕ_F through (9) [29],

$$\frac{k_{or}}{k_r} = \frac{1}{\phi_F} - \frac{1}{\phi_0} \tag{9}$$

where k_r is the radiative relaxation rate and ϕ_0 is the quantum yield in the absence of rotational relaxation. The dye-dependent constants k_r and ϕ_F need to be determined by other means. Often, ϕ_0 is set to unity for a rough approximation, and $k_r = \tau_0^{-1}$ is obtained from fluorescence in glass at 77 K. When the fluid is measured under conditions that alter the viscosity, but leave all other factors unchanged, the newly measured intensity F_2 relates to the altered viscosity η_2 through (5) and gives rise to (10),

$$\frac{F_2}{F_1} = \left(\frac{\eta_2}{\eta_1}\right)^x \tag{10}$$

because all dye-dependent constants in (5) as well as the instrument factors cancel out. Equation (10) can be solved for η_2. This approach is feasible, for example, to measure temperature-induced changes or when measuring cells exposed to viscosity-modulating treatments [68].

Molecular rotors with a dual emission band, such as DMABN or N,N-dimethyl-[4-(2-pyrimidin-4-yl-vinyl)-phenyl]-amine (DMA-2,4; **38**, Fig. 13) [64], allow to use the ratio between LE and TICT emission to eliminate instrument- and experiment-dependent factors analogous to (10). One example is the measurement of pH with the TICT probe p-N,N-dimethylaminobenzoic acid **39** [69]. The use of such an intensity ratio requires calibration with solvent gradients, and influences of solvent polarity may cause solvatochromic shifts and adversely influence the calibration. Probes with dual emission bands often have points in their emission spectra that are independent from the solvent properties, analogous to isosbestic points in absorption spectra. Emission at these wavelengths can be used as an internal calibration reference.

A different approach to design a self-calibrating dye was proposed [70], in which a viscosity-sensitive molecular rotor (2-cyano-3-(4-dimethylaminophenyl) prop-2-enoic acid) was covalently linked to a reference dye, 7-methoxycoumarin-3-carboxylic acid, which exhibited no viscosity sensitivity (**40**, Fig. 13). A ratiometric measurement, that is, rotor emission relative to reference emission, was shown to be widely independent of dye concentration [70]. However, the design of such a ratiometric dye poses some challenges because of resonance energy transfer from

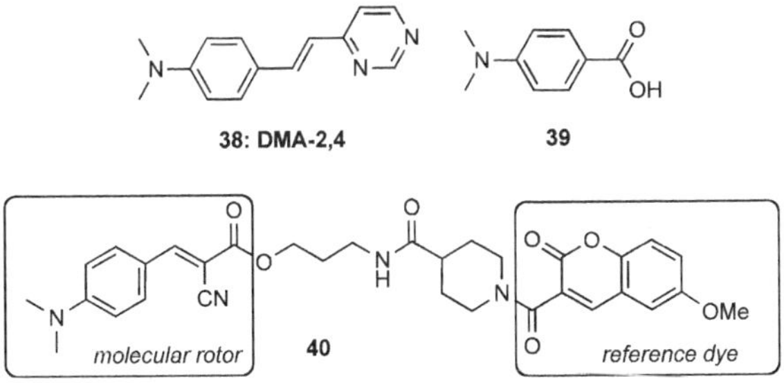

38: DMA-2,4   **39**

molecular rotor   **40**   reference dye

Fig. 13 Self-calibrating dyes DMA-2,4 and p-N,N-dimethylaminobenzoic acid. Compound **40** is an engineered ratiometric dye composed of a viscosity-sensitive molecular rotor and a nonviscosity-sensitive reference dye [70]

the shorter-wavelength dye to the longer-wavelength dye. With suitable filters, both emission bands can be acquired simultaneously. For this reason, such a dye is particularly attractive in microscopy experiments where local concentration changes would influence purely intensity-based results.

Another approach to reduce some of the influence of the instrument factors is by using specialized spectroscopic instruments with the capability to acquire additional information, namely, fluid absorption and scattering. Spectrofluorometers exist that can simultaneously acquire fluorescence emission and spectral absorption. In practice, it is difficult to distinguish dye absorption from fluid absorption, but dye concentration and its absorption coefficient are usually known. Therefore, a corrected emission intensity can be obtained that corresponds to the idealized emission intensity in the absence of solvent absorption. From the corrected intensity, the quantum yield can be deduced. Correction for fluid absorption becomes even more complex when the dye concentration is high and when the solvent absorption is strongly wavelength-dependent. The presence of scatterers further complicates the correction. In two relatively simple cases of forward-scattering microspheres and of starch solutions, the average excitation path length was found to be increased, and the presence of the scatterer increased fluorescence intensity. By measuring the scattering intensity, a corrected fluorescence emission was found that almost completely eliminated the influence of the scattering agent [71]. Higher scatterer content, however, would again reduce the measured intensity, and additional studies need to be performed to obtain correction formulas or algorithms for different types and concentrations of scatterers, and for combinations of scatterers and absorbers.

The measurement of fluorescent decay dynamics, i.e., fluorescence lifetime measurements, promise to overcome several of the challenges discussed above. Most importantly, lifetime and quantum yield are directly related through (11),

$$\phi_F = \frac{k_r}{k_r + k_{nr}} = \frac{\tau}{\tau_0} \tag{11}$$

where τ is the lifetime of the fluorophore and τ_0 is the natural lifetime, i.e., the lifetime in the absence of nonradiative processes (see (9)). Under consideration of (5), it can be seen that the lifetime τ is proportional to η^x, but without the instrument- and concentration-dependent factors that influence steady-state intensity. Calibration is still necessary to compute viscosity from lifetime, but fewer constants factor into the calibration for lifetime measurements than when steady-state intensity is used. However, molecular rotors typically exhibit extremely short lifetimes. Loutfy and Law [33] found lifetimes of p-N,N-dialkylaminobenzylidenemalononitriles between 3.2 and 11.1 ps in ethyl acetate. Ethyl acetate and many other solvents have a very low viscosity, and these lifetimes are typical for the types of molecular rotors with nonradiative relaxation from the TICT state. Many lifetime instruments are limited to the nanosecond to high picosecond range, and instruments that provide accurate results below 100 ps are usually specialized and expensive devices. However, specialized microscopes exist that allow spatially resolved lifetime measurements, termed fluorescence lifetime imaging (FLIM). A FLIM microscope uses pulsed excitation and gated acquisition [72]. For this purpose, ultrafast lasers and a suitable high-speed, high-efficiency camera are required. Although this equipment is expensive and not readily available, the advantage of a straightforward calibration (11) and lifetime measurement widely independent of local variations of dye concentration makes this method attractive for cell studies.

Steady-state behavior and lifetime dynamics can be expected to be different because molecular rotors normally exhibit multiexponential decay dynamics, and the quantum yield that determines steady-state intensity reflects the average decay. Vogel and Rettig [73] found decay dynamics of triphenylamine molecular rotors that fitted a double-exponential model and explained the two different decay times by contributions from Stokes diffusion and free volume diffusion where the orientational relaxation rate k_{or} is determined by two Arrhenius-type terms:

$$k_{or} = A_{\text{Stokes}} \exp\left(-\frac{E_{\text{Stokes}}}{kT}\right) + A_{\text{FV}} \exp\left(-\frac{E_{\text{FV}}}{kT}\right) \qquad (12)$$

Here, A_{Stokes} and A_{FV} are the contribution weights from Stokes- and free-volume diffusion, respectively, and E_{Stokes} and E_{FV} are the apparent activation energies for Stokes- and free-volume diffusion. Moreover, functional aspects of the solvent play an important role as well. It has been observed that polyfunctional alcohols, such as propanediol and glycerol, increase the orientational deactivation rate [73, 74], whereby single alcohols appear to have stick boundary conditions in the hydrodynamic model, and polyfunctional alcohols appear to have slip conditions, thus increasing diffusivity. Multiexponential relaxation dynamics that were dominated by solvent relaxation constants were also found by Dutta and Bhattacharyya [75], who reported lifetime constants in the low picosecond range, in the low nanosecond range, and higher, whereby the low nanosecond range carried significant information about the type of solvent, such as cyclodextrins, micelles, or polymers. Law [25] related relaxation times to the diffusional rotation of the solvent molecules, whereby rotational motion of the solvent molecules increased the rotational decay rate of the molecular rotors. Hara et al. [76] found triple-exponential decay functions when they applied pressure to the solvents to cause a pressure-induced viscosity increase. These few examples demonstrate to what extent lifetime experiments have the capability to reveal the complexity of the interaction of molecular rotors with their solvent. The analysis of lifetime experiments is complex, because many levels of solvent–rotor interaction, such as diffusion, electrostatic and polar interaction, and hydrogen bonding influence the lifetime dynamics and lead to complex decay patterns. This level of complexity cannot be seen in steady-state experiments, and it can be expected that lifetime experiments with instruments capable of capturing ultrashort decay rates provide a deeper insight into TICT dynamics and TICT-formation under the influence of different solvents.

5 Applications of Molecular Rotors

Much of this chapter covered the interaction of molecular rotors with the surrounding solvent. Elucidating this interaction is fundamental to the understanding of the TICT mechanism and, as the behavior of TICT fluorophores becomes better understood, leads to better understanding of the behavior of fluids. Of particular interest is the analysis of diffusion behavior, because it was found that molecular rotor fluorescence agrees more with a free-volume model [32, 77] than with the Debye–Stokes–Einstein model. The understanding of molecular rotors as free-volume probes leads to numerous practical applications where polymerization processes, covalent and electrostatic binding, and changes in microviscosity can be monitored in real-time. Examples where molecular rotors enjoy high popularity are polymer formation processes, sensing of conformational changes in proteins and other macromolecules, and physiological processes in the cell.

5.1 Measurement of Bulk Fluid Viscosity

Fluorescence-based bulk viscosity measurement is one application that advertises itself almost immediately, particularly with julolidine-derived molecular rotors where the quantum yield is widely independent from solvent polarity. Solving (5) for η and assuming proportionality of quantum yield and emission intensity leads to (13),

$$\eta = \xi \cdot F_{em}^{1/x} \tag{13}$$

where F_{em} is the peak emission intensity, ξ combines instrument- and solvent-dependent factors with the constant C in (5), and x is the exponent introduced in (5). Bulk viscosity measurement with conventional rheometers is a tedious process that requires several minutes of measurement time, followed by meticulous cleaning of the equipment. Rheometers are generally sensitive and prone to errors. In contrast, fluorescent intensity measurements could be performed in seconds, and by using disposable cuvettes, no cleaning would be necessary. Experiments have shown that fluorescence-based fluid measurements have the potential to surpass mechanical measurements in precision [78]. Interestingly, molecular rotors react to bulk viscosity changes in the manner described by (13) even if the viscosity is modulated by macromolecules that are several orders of magnitude larger than the rotor molecule itself [78, 79]. Two challenges need to be overcome. First, the constants C, ξ, and x are dependent on the type of solvent (e.g., alcohols, protein solutions, colloid solutions) and need to be calibrated for each type of fluid. Moreover, hydrophobic protein binding of some molecular rotors needs to be considered, although hydrophilic derivatives can be found that minimize protein binding [80]. Second, absorption and turbidity of the fluid influence the intensity measurement and need to be accounted for (see Sect. 4). Although the basic principle of bulk fluid viscosity measurement with molecular rotors has been established, more research is necessary before this method can supplement or replace conventional rheometry.

5.2 Probing Polymerization Dynamics with Molecular Rotors

The interdependence of free volume and quantum yield in TICT-forming fluorophores allows their use in probing static and dynamic changes of free volume for polymers and therefore as probes of polymerization dynamics [29]. In polymers, viscosity variations due to changes in temperature are often described with the Williams–Landel–Ferry (WLF) model [81]. Viscosity can be interpreted as a function of free volume (6) [32]. In (7), molecular rotors with nonradiative relaxation from the TICT state were introduced as free-volume probes. During polymerization, the mobility of molecular rotors is progressively hindered as more rigid polymeric structures are formed from monomers. Concurrently, the solvent viscosity increases and the free volume decreases, while the viscosity-dependent quantum yield of molecular rotors increases. In addition, the free volume V_f of polymer– diluent mixtures of glass-forming polymers changes with temperature according to (14) [82],

$$V_f = \frac{\alpha_p V_p T_{g,P} + \alpha_d (1 - V_p) T_{g,d}}{\alpha_p V_p + \alpha_d (1 - V_p)} \tag{14}$$

where V_p is the polymer volume fraction, $T_{g,P}$ is the polymer glass temperature, $T_{g,d}$ is the diluent glass temperature, α_p is the thermal expansion coefficient of the polymer, and α_d is the expansion coefficient of the diluent. Consequently, the reorientation rate of molecular rotors is influenced both by the chemical process and by temperature.

Mechanical and chemical methods for qualitative and quantitative measurement of polymer structure, properties, and their respective processes during interrelation with their environment on a microscopic scale exist. Bosch et al. [83] briefly discuss these techniques and point out that most conventional techniques are destructive because they require sampling, may lack accuracy, and are generally not suited for *in situ* testing. However, the process of polymerization, that is, the creation of a rigid structure from the initial viscous fluid, is associated with changes in the microenvironment on a molecular scale and can be observed with free-volume probes [83, 84].

Some TICT-forming fluorescent probes containing the p-N,N-dialkylamino benzylidene malononitrile motif (usually related to **30** in Fig. 11) have been applied to monitor and quantify polymerization reaction, crosslinking, chain relaxation, thermal transitions and relaxations, degradations (thermal, photochemical, physical), transport of small molecules, microphase separation, crystallization, gelation, and gel swelling. It has been experimentally confirmed that absorption and fluorescence maximum of these dyes move to longer wavelengths with an increase of the environment's dielectric constant. For dye **30**, DCVJ, the quantum yield increased markedly as the medium underwent a transition from fluid to glass, a phenomenon attributed to a decrease in polymer free volume during polymerization [30]. In some specific examples, an increasing quantum yield accompanied by a decrease in nonradiative relaxation was observed during polymerization processes of water-soluble copolymers derived from 1-methyl-3-vinylpyridinium salts [85], methacrylate monomers [64], and polysaccharide polymers [48]. In another example, the polymerization process of polyacrylamide, type-I collagen and a tetramethoxysilane (TMOS) sol-gel was monitored with CCVJ (**41**, Fig. 14). Fluorescence steady-state intensity was related to the oscillatory behavior of magnetoelastic amorphous metal–glass. The emission intensity of CCVJ was increased in type-I collagen and TMOS sol-gel, but CCVJ was degraded by the ammonium persulfate in the polyacrylamide gel. On the other hand, CCVJ fluorescence provided distinctly different dynamics in the hydrolysis and crosslinking phases [86]. Benjelloun et al. [85] used a derivative of DMABN to monitor the formation of hydrophobic microdomains in amphiphilic polymers in water. The fluorescent probe exhibited enhanced emission from the LE band at higher concentrations of the polymers and with higher chain lengths, exposing the increased formation of hydrophobic microdomains. Zhu et al. [87] applied the hydrophobic molecular rotor FCVJ (CCVJ farnesyl ester **42**, Fig. 14) to obtain measurements of the relationship between viscosity and molecular weight (M) of polypropyleneoxide polymer melts. The hydrophobic nature of FCVJ allows it to integrate in a hydrocarbon-based polymer. For the polymer melts with low molecular weight (425–2,000 g mol^{-1}), viscosity followed the Rouse model that describes single-chain dynamics, and where viscosity is proportional to the molecular weight. Conversely, in polymer melts with high molecular weight (2.7–4 kg mol^{-1}), viscosity followed a power–law relationship with the molecular weight, $\eta \propto M^{3.4}$, which agrees with the reptation model, where a more complex chain interaction restricts the chain motion. Aggregation

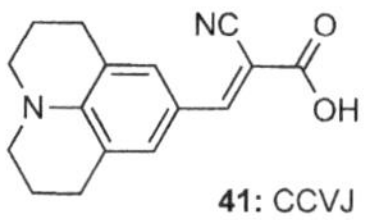

41: CCVJ

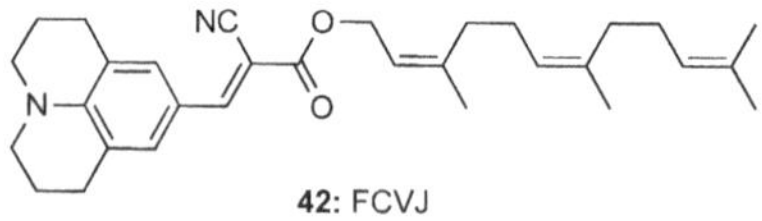

42: FCVJ 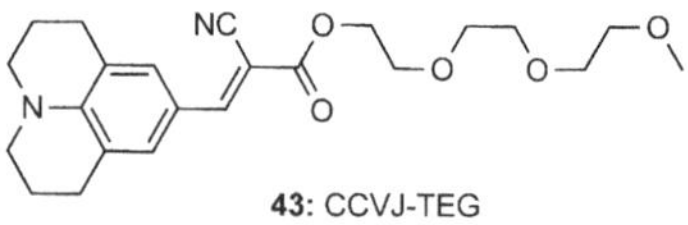   **43: CCVJ-TEG**

Fig. 14 Water-soluble CCVJ, hydrophobic CCVJ farnesyl ester (FCVJ) for cell membrane applications [88], and hydrophilic CCVJ triethyleneglycol ester (CCVJ-TEG) [89]

of proteins and saccharides are special cases of biopolymer formation and are discussed in detail in the next section.

5.3 Applications of Molecular Rotors in Protein Sensing and Sensing of Other Macromolecules

Most of the molecules introduced in this chapter are hydrophobic. Even those molecules that have been functionalized to improve water-solubility (for example, CCVJ and CCVJ triethyleneglycol ester **43**, Fig. 14) contain large hydrophobic structures. In aqueous solutions that contain proteins or other macromolecules with hydrophobic regions, molecular rotors are attracted to these pockets and bind to the proteins. Noncovalent attraction to hydrophobic pockets is associated with restricted intramolecular rotation and consequently increased quantum yield. In this respect, molecular rotors are superior protein probes, because they do not only indicate the presence of proteins (similar to antibody-conjugated fluorescent markers), but they also report a constricted environment and can therefore be used to probe protein structure and assembly.

Nile Red (**44**, Fig. 15), a highly polarity-sensitive fluorophore, is nearly quenched in aqueous solutions and emits at 570 nm in nonpolar solvents such as toluene, and peak emission ranges to 636 nm in polar liquids. Originally, this dye was used as a lipid and lipoprotein stain. Although some researchers describe Nile Red as a TICT molecule with electron transfer from the diethylamino group to the aromatic ring system, controversial explanations of its mechanism exist. According to the TICT interpretation, nonradiative relaxation form the TICT state dominates in polar solvents, but in nonpolar solvents, the TICT process is considered unfavorable and consequently the lifetime and quantum yield increase dramatically [90–92]. Other researchers suggest, in contrast to the TICT interpretation, that hydrogen bonding associated with solvent–dye interaction (i.e., alcohols) lowers the fluorescent lifetime via vibrational dissipation [93]. Whereas the actual mechanism is still under investigation, Nile Red sensitivity to environmental polarity is valuable in a variety of protein conformation investigations. For example, GRP94, the endoplasmic reticulum Hsp90 paralog (Heat shock protein 90) binds a variety of peptides. Using a known immunodominant peptide epitope of the vesicular stomatitis virus, acrylodan and Nile Red, the activation of peptide binding was found to be accompanied by enhanced peptide and solvent accessibility to hydrophobic binding sites [94]. This is just one of many examples of Nile Red's

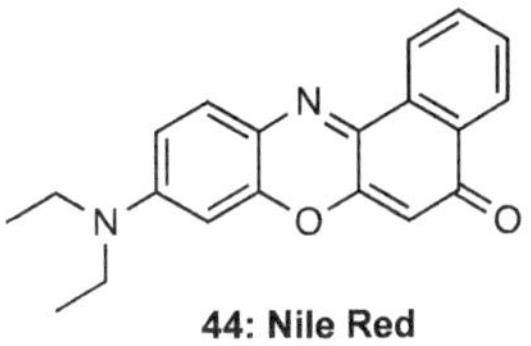

Fig. 15 Structure of Nile red   **44: Nile Red**

ability to determine aggregation, denaturation, and the conformation state of lipid/ protein interfaces.

Thioflavin T (**25** in Fig. 10) was originally used to stain amyloid deposits in tissues [95] and later for the quantification of amyloid fibrils in the presence of precursor proteins [96, 97]. Thioflavin T interacts with the β-sheet structure of the amyloid protein along the shallow groove formed by adjacent Tyr and Leu residues on the surface [98, 99]. Applications with Thioflavin T have advanced from simple histological stains for connective tissue to more complex dual staining techniques that compensate for buffer conditions and species-specific backgrounds [100, 101]. Each experimental system should be carefully analyzed for possible misinterpretations. One example is 4,5-dianilinophthalimide (DAPH), which disintegrates amyloid fibers involved in Alzheimer's disease [102–104]. When DAPH was studied with β-lactoglobulin, a whey protein, Thioflavin T reported disintegration with decreased fluorescence, but birefringence and TEM tests did not. It was discovered that DAPH likely binds in the β sheet channels, thus hindering Thioflavin T binding [105] and disguising the low rate of β-lactoglobulin disintegration. Other applications of Thioflavin T include the direct observation of amyloid growth with total internal reflection microscopy [106] and ligand reactions at the acylation site of acetylcholinesterase [107]. Thioflavin T can also be used with other types of helical structures. It has been demonstrated that Thioflavin T can bind to Type I collagen prepared from the atelocollagen of yellowfin tuna skin [108]. Transthyretin, a thyroid hormone carrier, is a protein linked to amyloid diseases such as senile systemic amyloidosis and familial amyloidotic polyneuropathy and was investigated with several TICT dyes, 1-anilinonaphthalene-8-sulfonate (ANS), 4-4-*bis*-1-phenylamino-8-naphthalene sulfonate (*Bis*-ANS), 4-(dicyanovinyl)-julolidine (DCVJ), and Thioflavin T (ThT). Both steady-state and time-fluorescence assays, static and kinetic, characterized the protofibrillar states exposing conformational changes, aggregation, and misfolding [109].

Conformation and aggregation of proteins are not the only macromolecule applications for TICT fluorophores. For example, Bosch et al. [110, 111] have developed a saccharide sensing molecule containing a boron–nitrogen bond that displays fluorescence in both the LE state and the twisted internal charge transfer state. Without the presence of saccharides (in this case, either D-fructose, D-glucose, D-galactose, or D-mannose), the molecule emits at 404 nm from its TICT state when excited at 274 nm. With the addition of saccharides, the B–N bond is broken and a boronate species forms. The fluorescent emission blue-shifts to 362 nm (the LE state emission) since the nitrogen lone pair is free to conjugate with the π-system on the aniline component. Tan et al. [112] continued to modify this base structure to create another saccharide sensor, 4-*N*-methyl-*N'*-(2-dihydroxyborylbenzyl)-benzonitrile, a close derivative of the DMABN structure, and at the same time a boronate analog, which is a selective sensor for fluoride ions. Twisted intramolecular charge state fluorophores are also a valuable tool for nucleic acid–protein interactions. In Fig. 16, the fluorophore component **45** for nucleic acid monitoring is compared to DCVJ. The core fluorophore is conjugated to 5'-modified DNA and then mixed with bovine serum albumin (BSA) and single-stranded DNA binding protein

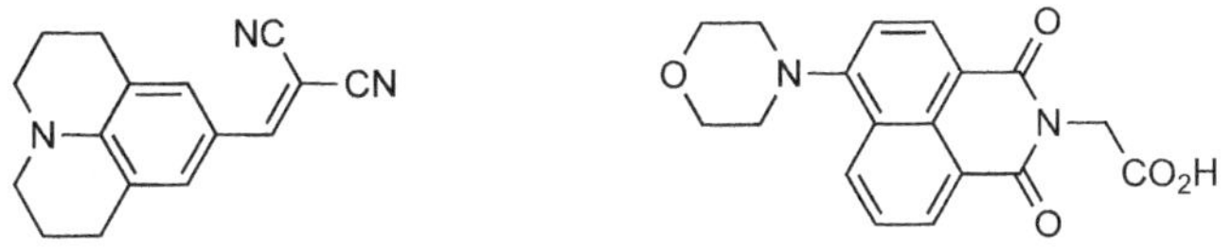

30: DCVJ   **45**

Fig. 16 DCVJ compared to the core fluorophore for the DNA–Dye–BSA/SSP assay (**45**). The 5′ DNA strand is conjugated to the carboxylic acid group, leaving the morpholine group to access the hydrophobic pockets

(SSP). In the presence of BSA, the fluorescence intensity increased (expressed as a ratio against dye–DNA alone) indicating preference for the hydrophobic pocket where a potential TICT mechanism is associated with inhibited intramolecular rotation. When SSP was added, the fluorescent intensity decreased, indicating that the fluorophore was removed from the hydrophobic pockets [113]. As a probe, this assay could be tuned for a variety of protein–DNA monitoring in biological mixtures. Further study into the photophysical energy states during each phase of the assay could give interesting insight into complex biological processes.

A popular group of macromolecules are cyclodextrins, artificial polysaccharides that form a nanoscale truncated hollow cone with hydrophilic ends and a hydrophobic core. Cyclodextrins can be used to deliver hydrophobic substances to aqueous environments (e.g., drug delivery [114]), or they can withdraw hydrophobic or organometallic compounds from solutions. The exact nature of the hydrophobic core, particularly with respect to the formation of interlocked molecular systems (rotaxanes) and with respect to cyclodextrin dimer formation are still under investigation, and molecular rotors advertise themselves as probes for the restricted microenvironment of the cyclodextrin hydrophobic core. The principle of the interaction of cyclodextrins with a fluorophore can be demonstrated with the polarity-sensitive probe 6-propionyl-2-(dimethylamino)naphthalene (PRODAN). Although TICT formation of PRODAN is debated, the strong red-shift of the probe in polar media is well documented. In a study by Al-Hassan and Khanfer [115], PRODAN binding to cyclodextrin is demonstrated. Aqueous solutions of PRODAN and α-cyclodextrin, the smallest cyclodextrin with six glucopyranoside units, did not show a difference to aqueous PRODAN solution in the absence of cyclodextrin, but a blue-shift and increase of emission intensity was observed in solutions of β-cyclodextrin (seven glucopyranoside units) and γ-cyclodextrin (eight glucopyranoside units). The blue-shift indicates that the dye migrates into the hydrophobic cyclodextrin core. The migration process can be observed in time-resolved spectroscopy [115], where equilibration takes place with a time constant of several hours. Similar observations were made by Nakamura et al. with DMABN and glucosyl-conjugated β-cyclodextrin. Emission intensity of both DMABN emission bands increased upon association with cyclodextrin, and from the spectra, Nakamura et al. determined the association constants. Furthermore, a higher concentration of DMABN caused multiple DMABN molecules to be captured in the hydrophobic cavity, and their mutual influence increased the polar (red-shifted or

TICT) emission band. Similarly, the presence of other solvents increased the polar emission band, indicating that solvent molecules accumulated inside the cyclodextrin cavity and caused DMABN to react to the more polar environment. With a carboxylic-acid analog of DMABN, p-(N,N-diethylamino)benzoic acid (DEABA), Kim et al. [116] examined hydrogen-bonding effects in α- and β-cyclodextrins. Analysis of the fluorescent spectra and lifetimes allowed to conclude that DEABA associates differently with α- and β-cyclodextrins. The size of DEABA was estimated to be 9.3 Å long and 6.2 Å wide. Consequently, the diethylamino group would not fit into the core of α-cyclodextrin (diameter of 5 Å), forcing the carboxy group into the core. Conversely, β-cyclodextrin with an inner diameter of 6.5 Å allows the diethylamino group to fit; this is the preferred orientation due to the hydrophobic nature of the core. In this case, however, the carboxylic group is exposed to the solvent. In β-cyclodextrin, the red-shifted TICT emission increases, whereas in α-cyclodextrin, the TICT band decreases relative to the LE band. Similar results were reported by Panja and Chakravorti [117] who associated α-cyclodextrin with 4-N,N-dimethylamino cinnamaldehyde (DMACA) and found that two different configurations existed: first, where the dimethylamino group was inside the cavity and second, where the but-2-enal group was inside the cavity. Wang et al. [118] demonstrated that a DMABN-cyclodextrin combination can act as a biosensor. Wang et al. covalently bound DMABN and biotin to β-cyclodextrin and found that the presence of avidin and cholic acid analogs increased the LE emission of DMABN, indicating that the proximity of avidin and some cholic acid analogs exerted a nonpolar influence on the DMABN fluorophore.

5.4 Molecular Rotors as Microviscosity Probes in the Cell

Cellular biomechanics are primarily determined by the cell membrane, the cytoskeleton, and the cytoplasm. Whereas the cytoskeleton can be imagined as a relatively rigid framework, both the cytoplasm and the cell membrane have viscoelastic properties that change in various states of disease. The cell membrane plays a particularly important role since its viscosity influences the activity of membrane-bound proteins [119]. Consequently, changes in membrane viscosity have been linked with alterations in various physiological processes in the cell, particularly in conjunction with some disease states. Increased viscosity of red blood cell and platelet membranes has been observed in diabetic patients, and the viscosity change has been proposed to contribute to the reduced ability of the insulin receptor (a membrane-bound protein) to undergo aggregation [120, 121]. On the other hand, decreased membrane viscosity in leukocytes of patients with Alzheimer's disease has been postulated to facilitate aggregation of the Amyloid Precursor Protein (a transmembrane protein), a fragment of which is deposited in the brain as insoluble plaque [122]. Patients with liver disorders, including alcoholism-based diseases, showed higher erythrocyte membrane viscosity, which correlates highly with the severity of liver dysfunction [123]. Moreover, increased membrane viscosity in leukocytes has been connected to the aging process [124]. These examples constitute only a few examples of an enormous literature body on the relationship between cell membrane viscosity and disease.

The effects of cytoplasmic viscosity are widely unexplored. This may be attributed to the relatively difficult measurement methods available. In the cytoplasm, magnetic microparticles were predominantly used to obtain information on viscoelastic properties [125–127]. Clearly, the use of magnetic microparticles demands the use of very expensive equipment; the observation of the particles is time-consuming, thus limiting temporal resolution, and the interaction between the particles and the cellular environment may also cause measurement artifacts.

It becomes clear that the investigations related to the viscosity of cellular components strongly depends on the availability of viscosity probing methods that allow detection of viscosity changes on a microscopic scale and with very short response times. Fluorescence-based methods advertise themselves, because they meet the two demands of high spatial and temporal resolution. Two widely used methods to obtain viscosity information, predominantly employed in the cell membrane, are fluorescence recovery after photobleaching (FRAP) and fluorescence anisotropy. Both methods are very different from each other, but both use the diffusivity of specific fluorophores to obtain information about the viscosity of the environment. More recently, molecular rotors have been employed as microviscosity probes in the cell. Each of the three fluorescent methods has its own advantages and disadvantages. FRAP has been established as the gold-standard of membrane viscosity measurement, but a single recovery experiment may take several minutes, and the bleached spot size must be chosen to balance high spatial resolution (small spot) with low noise and low measurement errors (large spot). The bleaching pulse introduces energies high enough to cause protein crosslinking and free radical formation, which may alter the environment, particularly in live cells. Moreover, the necessary confocal equipment is costly. Anisotropy experiments can be performed with conventional epifluorescent microscopes, and anisotropy measurements have a dramatically better spatial and temporal resolution than FRAP. On the other hand, local viscosity can only approximately be computed from the polarization anisotropy ratio, and any minor misalignment or imperfection of the polarizers – including any tendency of lasers or monochromators to polarize light – causes major measurement errors. By using molecular rotors, the simplest equipment is sufficient, and emission intensity from molecular rotors can be observed in real time and with a spatial resolution limited only by the optical system. On the other hand, intensity-based measurements can be confounded by local concentration changes, illumination inhomogeneities, and optical properties (scattering, absorption) of the sample. As introduced in Sect. 4, relative measurements, ratiometric dyes, and lifetime imaging can eliminate some of the confounding factors.

Early studies focused on the behavior of molecular rotors in vesicles [128] and lipid bilayers [18, 26]. Humphrey-Baker et al. [128] found that an indocyanine dye associates with micellar systems in aqueous suspension. The dye migrates into the micelles and shows an increased quantum yield and a bathochromic shift of emission. Although Humphrey-Baker et al. identify modulation of the quantum yield by rigidization of the dye in the hydrophobic environment of the micelles, a detailed explanation of TICT formation is not given. Kung and Reed [18] performed spectroscopic studies of DCVJ in several solvents and liposome preparations. Kung and Reed found a weak linear relationship between the solvent's dielectric constant and peak emission wavelength where the full range from nonpolar (benzene) to polar (alcohols) caused less than 30 nm of bathochromic shift. A viscosity gradient created with mixtures of alcohols demonstrated the power–law relationship between viscosity and quantum yield (5) with exponent $x = 0.6$. In DPPC liposomes, a sharp change of the exponent was seen when the temperature was increased over the transition temperature. This experiment demonstrates how a molecular rotor reflects the sudden change in free-volume behavior in phospholipids between the gel and the liquid-crystal phase. This effect was later exploited by our own group to validate the viscosity-sensitivity of molecular rotors bound to the hydrophobic tails of phospholipids [129]. Lukac [26] found a very similar behavior when examining a derivative of (p-(dialkyl amino)-benzylidene)malononitrile in DPPC and DSPC vesicles. Interestingly, Lukac found a more pronounced temperature-dependent bathochromic shift in DSPC vesicles than in DPPC vesicles, which was attributed to a different localization of the molecular rotor in DSPC vesicles with their longer tail chains. The study by Kung and Reed demonstrated a key advantage of julolidine-based molecular rotors, namely, the separation of the effects of polarity and viscosity, which only influence peak emission and quantum yield, respectively [18]. This effect was later confirmed in similar studies [14, 19]. More recently, Nipper et al. [130] related apparent viscosity reported by FCVJ (**42**) and apparent viscosity reported by FRAP and found a linear relationship with good correlation.

Molecular rotors were also used in several studies to probe live cells, namely, the cell membrane, the cytoplasm, and the cytoskeleton. Viriot et al. [27] present a review of applications of several conjugated molecular rotors to probe microdomains in polymers, investigate the thermotropic behavior of liposomes, and stain the membranes of endothelial cells. Our own group has performed a number of studies related to changes in membrane viscosity in endothelial cells as a consequence of fluid shear stress. A study with DCVJ-stained cells showed a reversible decrease of membrane viscosity under fluid shear stress [68], allowing to conclude that the cell membrane is the most likely mechanoreceptor of the cell. DCVJ has a high affinity for hydrophobic parts of the cell, including the cell membrane and the cytoskeleton. We found that interior parts of endothelial cells were stained with DCVJ, which diminishes the relative fluorescent response from the membrane. For this reason, we sought DCVJ derivatives that were more membrane-specific. Two notable derivatives were FCVJ and CCVJ-conjugated phospholipids [129]. The phospholipid-bound molecular rotors were shown to be highly sensitive viscosity probes in liposomes and cultured endothelial cells [129], but they also showed a certain cytotoxicity, likely because the introduction of the phospholipids changes the natural phospholipid composition of the membrane. Hydrocarbon chains like geranyl and farnesyl are known to improve membrane localization [131, 132]. Consequently, membrane localization of the CCVJ farnesyl ester (FCVJ) was improved over DCVJ, and intensity responses were dramatically stronger [88]. A particularly interesting application is the examination of lipid rafts with molecular rotors. Pure intensity-based microscopy, for example with DCVJ, will be confounded by locally elevated concentration in the caveolin-rich lipid rafts [133], but ratiometric dyes and lifetime methods may overcome this challenge.

Fewer studies have focused on the cell cytoplasm. One challenge is the geometric inhomogeneity of the cell cytoplasm, where thickness and local concentration gradients influence the measured intensity. Luby-Phelps et al. [134] proposed to use the indocyanine dyes Cy3 and Cy5 as a dye pair where Cy3 acts as a molecular rotor with viscosity-dependent nonradiative deactivation channel, and the relatively rigid Cy5 acts as reference fluorophore. Fluorescence excitation and emission differ by about 100 nm between those dyes. Luby-Phelps et al. performed microinjection of the dye mixture into cells followed by ratiometric microscopic imaging and found that cytoplasmic viscosity was not significantly higher than water and did not significantly vary over the projected cell area. Conversely, Kuimova et al. [135] performed lifetime imaging on cells stained with an indacene-derived dye. The lifetime τ of the dye depended on the viscosity η of its environment with $\tau = z\,\eta^{\alpha}$ where z and α are constants that were calibrated with alcohol mixtures. In lifetime imaging, the lifetime τ is spatially resolved. Kuimova et al. found viscosity values of 140 ± 40 mPa s in contrast to the values found by Luby-Phelps. However, the microscopy images in the study by Kuimova et al. [135] exhibited a grainy distribution of the dye, which may be an indication that the dye was either preferentially bound to proteins or underwent micelle formation. Whereas Cy3 and Cy5 are water-soluble by merit of a charged nitrogen atom, the indacene dye with its weak dipole between the N^{+}–B^{-} pair and its long aliphatic chain appears to be widely hydrophobic. Protein binding of the indacene dye could explain the discrepancy.

The cellular cytoskeleton, primarily composed of microfilaments, microtubules, and intermediate filaments, provides structural support and enables cell motility. The cytoskeleton is composed of biological polymers and is not static. Rather, it is capable of dynamic reassembly in less than a minute [136]. The cytoskeleton is built from three key components, the actin filaments, the intermediate filaments, and the microtubules. The filaments are primarily responsible for maintaining cell shape, whereas the microtubules can be seen as the load-bearing elements that prevent a cell from collapsing [136]. The cytoskeleton protects cellular structures and connects mechanotransductive pathways. Along with mechanical support, the cytoskeleton plays a critical role in many biological processes.

Two of the cytoskeletal components, the actin filaments and the microtubules have been studied with molecular rotors. The main component of the actin filaments is the actin protein, a 44 kD molecule found in two forms within the cell: the monomeric globulin form (G-actin) and the filament form (F-actin). Actin binds with ATP to form the microfilaments that are responsible for cell shape and motility. The rate of polymerization from the monomeric form plays a vital role in cell movement and signaling. Actin filaments form the cortical mesh that is the basis of the cytoskeleton. The cytoskeleton has an active relationship with the plasma membrane. Functional proteins found in both structures give actin the ability to form regions in the cell with distinct physical features. Such compartmentalization is often referred to as cell polarity. The microtubule structure is a tubular assembly of heterodimeric tubulin proteins. Tubulin is composed of α- and β- subunits that weigh around 55 kD each. Microtubules have distinct regions promoting growth from only one end. Any attempt to monitor tubule formation must focus on the advancing side of the chain. Microtubules participate in exocytosis, intracellular transport, spindle formation, and organism motility [137].

Traditional fluorophore–antibody staining has been extensively used to locate local concentrations of microfilaments and microtubules, and to track their growth. These techniques are limited by their ability to only detect a single target. When used to monitor actin aggregation, traditional fluorescent methods are limited by their ability to only measure concentration. It is the ability to track both monomeric and polymeric substrates that makes molecular rotors a powerful tool for studying protein reaction kinetics and protein assembly. The molecular rotor DCVJ has been established as a probe for monitoring bulk polymerization kinetics [83] and as a probe for various biological processes [27, 138]. The molecular rotor DCQ (1-(2-Hydroxyethyl)-6-[(2,2-dicyano)vinyl]-2,3,4-trihydroquinoline) has exhibited sensitivity to the polymerization of G-Actin to F-Actin [138, 139]. DCQ readily binds to actin, a process that is accompanied by an increase of DCQ emission intensity [138]. When actin is exposed to metallic ions, such as Mg^{2+}, a transformation from G-actin to F-actin takes place. This transformation is accompanied by a further increase of DCQ emission, and DCQ can be seen as a reporter for the conformational changes that actin undergoes during assembly. The molecular rotor DCVJ has been shown to preferentially bind to tubulin [140]. Kung and Reed [140] used DCVJ to observe DCVJ binding to tubulin and tubulin aggregation. DCVJ binding to tubulin proteins is accompanied with an increase of emission intensity that indicates hydrophobic, noncovalent binding with an accompanying restriction of intramolecular rotation. Assembly of DCVJ-carrying tubulin to higher-order microstructures was found to induce a conformational change in the region of the DCVJ binding site so as to further restrict the rotational mobility of the dye, and tubulin assembly can be monitored by increased DCVJ emission intensity. Moreover, the increase of emission intensity differs between tubulin sheets and microtubules, an observation that indicates differences in the morphology of sheet and tubular aggregates at the molecular level. With the ability to measure dynamic changes in both actin and tubulin, cytoskeletal changes can be tracked in real time and correlated with cellular function.

A more specialized application of molecular rotors in the cell is the measurement of intracellular potentials. Conventional techniques for measurements of potential in cells, organelles, and neurons have used microelectrodes as the main instrument for detection of changes in the cell potential. These techniques are technically challenging and unsatisfactory in terms of spatial resolution and signal-to-noise ratio. A number of styryl-related dyes emerged [141] where voltage-sensitivity was attributed to TICT formation [142]. Key to understanding the sensitivity of styryl dyes, such as aminostilbazolium (Fig. 17) is to recognize that there are two ground

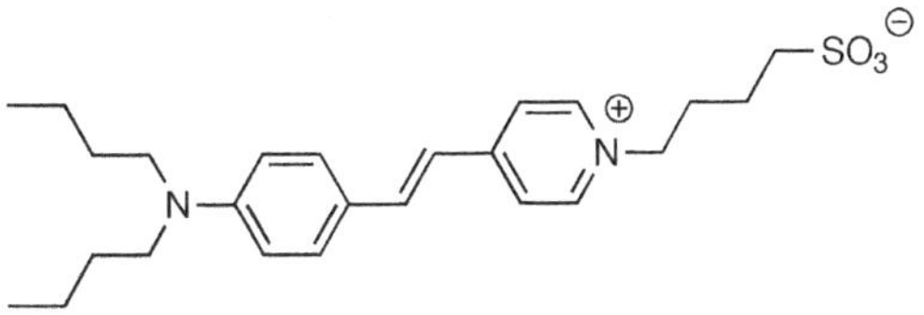

Fig. 17 Chemical structure of
the styryl dye: (dibutylamino)
stilbazolium butylsulfonate

states (the *cis*- and *trans*-isomers, separated by an energy barrier) and three relaxation modes, that is, regular fluorescence emission and nonradiative conversion into
the *trans*-isomer, and a nonradiative deexcitation from a twisted state into either the
cis-or the *trans*-isomer. Twisting takes place around the central double bond (θ_2 in
Fig. 11). Isomerization of the aminostilbazolium dye from the *cis*- to the *trans*-state
takes place predominantly in amphiphilic molecules and organic solvents. Ephard
and Fromherz [142] suggest two possible mechanisms for the voltage sensitivity of
aminostilbazolium in phospholipid membranes. First, an electrostatic field parallel
to the molecule's main axis would interact with the excited-state redistribution of
the charges in the molecule. Such a field would hinder a charge redistribution
toward the anilino group, thus increasing the energy barrier that leads to the twisted
state. Consequently, the horizontal transition rate into the twisted state is slowed
down and fluorescence emission increases. Second, the electrostatic field could
dislocate the dye inside the heterogeneous membrane/water interface, for example,
by tilting the anilino group toward the aqueous phase. Consequently, the dye would
experience a more polar environment with lower viscosity, and decreased fluorescence intensity would be observed. A later study by Jones and Bohn [143] showed
that the presence of an electrostatic potential caused an increase in fluorescence
intensity, which would indicate that the voltage-sensitivity of the dye is controlled
by a charge redistribution that elevates the barrier for the TICT state, but Jones and
Bohn interpret their experiments such that phospholipid protonation and dye
dislocation are the predominant factors.

6   Future Directions

The examples in the previous section give a comprehensive overview of application areas where molecular rotors have become important fluorescent reporters.
Current work on the further development of molecular rotors can broadly be
divided into three areas: photophysical description, structural modification, and
application development. Although a number of theories exist that describe the
interaction between a TICT fluorophore and its environment, the detailed mechanism of interaction that includes effects such as polarity, hydrogen bonding, or
size and geometry of a hydrophobic pocket are not fully understood. Molecular
simulations have recently added considerable knowledge, particularly with

respect to the change of energy levels when the molecule undergoes conformational changes. However, few studies exist that provide detailed models of rotor–solvent interaction with multiple solvent molecules. Of particular interest are models that predict the interaction of molecular rotors with proteins, polymers that undergo polymerization, solvent mixtures, and solvent molecules with diffusional motion.

Chemical modifications of the structure are possible to optimize molecular rotors for specific applications. Examples are the functionalization of molecular rotors with recognition groups that allow specific protein binding, integration in a polymer matrix, or hydrophilic modifications that increase solubility and decrease the likelihood of binding to hydrophobic regions. Moreover, modifications of the actual molecular rotor dipole are possible to tune the photophysical properties [144]. All three elements of the EDG-π-EAG motif (Fig. 7) can be modified. It can readily be seen that an increased donor–acceptor distance causes a bathochromic shift and a tendency toward single emission from the LE state. Substitution of the π-system, for example, with a methoxy group, can increase the overall quantum yield. Additional points of rotation can be introduced (e.g., **38** in Fig. 13).

With further understanding how molecular rotors interact with their environment and with application-specific chemical modifications, a more widespread use of molecular rotors in biological and chemical studies can be expected. Ratiometric dyes and lifetime imaging will enable accurate viscosity measurements in cells where concentration gradients exist. The examination of polymerization dynamics benefits from the use of molecular rotors because of their real-time response rates. Presently, the reaction may force the reporters into specific areas of the polymer matrix, for example, into water pockets, but targeted molecular rotors that integrate with the matrix could prevent this behavior. With their relationship to free volume, the field of fluid dynamics can benefit from molecular rotors, because the applicability of viscosity models (DSE, Gierer–Wirtz, free volume, and WLF models) can be elucidated. Lastly, an important field of development is the surface-immobilization of molecular rotors, which promises new solid-state sensors for microviscosity [145].

7 Conclusion

A number of fluorescent dyes with internal charge transfer mechanism allow the molecule to twist (rotate) between the electron donor and electron acceptor moieties of the fluorescent dipole. In most cases, the twisted conformation is energetically preferred in the excited S_1 state, whereas the molecule prefers a planar or near-planar conformation in the ground state. For this reason, photoexcitation induces a twisting motion, whereas relaxation to the ground state returns the molecule to the planar conformation. Moreover, the $S_1 - S_0$ energy gap is generally smaller in the twisted conformation, and relaxation from the twisted state causes either a red-shifted second emission band or – if the energy gap is small enough – non-radiative relaxation. These fluorescent molecular rotors are attractive as reporters in chemistry and biology, because the rate of twisted-state formation is influenced by the environment. Polarity plays a role in stabilizing the twisted state of some molecular rotors, and steric hindrance reduces TICT formation. The rate of TICT formation can therefore be determined with fluorescence steady-state or lifetime measurements. Very frequently, molecular rotors are employed as reporters for microviscosity, but the relationship between apparent viscosity and molecular-scale processes that influence TICT formation needs to be considered. Free-volume theory seems to be best suited to explain experimental results, predominantly the power–law relationship between bulk viscosity and quantum yield seen in some molecular rotors with nonradiative relaxation pathways. As such, molecular rotors have successfully been applied to probe polymerization and aggregation processes and to obtain spatially resolved local viscosity information. Whereas the molecular-level mechanisms of some TICT molecules, primarily aniline- and benzylidene malononitrile-related compounds are fairly well understood, the hypothesized involvement of twisted states in the fluorescent behavior of other families (e.g., porphyrins and BODIPY derivatives) is not fully understood and deserves more research.

References

1. Tomin V (2010) Physical principles behind spectroscopic response of organic fluorophores to intermolecular interactions. In: Demchenko AP (ed) Advanced Fluorescence Reporters in Chemistry and Biology I. Springer Ser Fluoresc 8:189–223
2. Rotkiewicz K, Grellmann KH, Grabowski ZR (1973) Reinterpretation of the anomalous fluorescence of p-N, N-dimethylaminobenzonitrile. Chem Phys Lett 19:315–318
3. Grabowski ZR, Rotkiewicz K, Rettig W (2003) Structural changes accompanying intramolecular electron transfer: focus on twisted intramolecular charge-transfer states and structures. Chem Rev 103(10):3899–4032
4. Rettig W, Lapouyade R (1994) Fluorescence probes based on twisted intramolecular charge transfer (TICT) states and other adiabatic photoreactions. Topics in fluorescence spectroscopy 4:109–149
5. Lapouyade R, Czeschka K, Majenz W, Rettig W, Gilabert E, Rulliere C (1992) Photophysics of donor–acceptor substituted stilbenes. A time-resolved fluorescence study using selectively bridged dimethylamino cyano model compounds. J Phys Chem 96(24):9643–9650
6. Zachariasse KA, Grobys M, von der Haar T, Hebecker A, Il'ichev YV, Jiang YB, Morawski O, Knhnle W (1996) Intramolecular charge transfer in the excited state. Kinetics and configurational changes. J Photochem Photobiol Chem 102(1S1):59–70
7. Grabowski ZG, Dobkowski J (1983) Twisted intramolecular charge transfer (TICT) excited states: energy and molecular structure. Pure Appl Chem 55(2):245–252
8. Gregoire G, Dimicoli I, Mons M, Donder-Lardeux C, Jouvet C, Martrenchard S, Solgadi D (1998) Femtosecond dynamics of "TICT" state formation in small clusters: the dimethylaminobenzomethyl ester acetonitrile system. J Phys Chem A 102(41):7896–7902
9. Rulliere C, Grabowski ZG, Dobkowski J (1987) Picosecond absorption spectra of carbonyl derivatives of dimethylaniline: the nature of the TICT excited states. Chem Phys Lett 137(5): 408–413

10. Bulgarevich DS, Kajimoto O, Hara K (1995) High-pressure studies of the viscosity effects on the formation of the twisted intramolecular charge-transfer (TICT) state in 4,4'-diaminodiphenyl sulfone (DAPS). J Phys Chem 99(36):13356–13361

11. Il'ichev YV, Kuhnle W, Zachariasse KA (1998) Intramolecular charge transfer in dual fluorescent 4-(dialkylamino) benzonitriles. Reaction efficiency enhancement by increasing the size of the amino and benzonitrile subunits by alkyl substituents. J Phys Chem A 102(28): 5670–5680

12. Schuddeboom W, Jonker SA, Warman JM, Leinhos U, Kühnle W, Zachariasse KA (1992) Excited-state dipole moments of dual fluorescent 4-(dialkylamino) benzonitriles. Influence of alkyl chain length and effective solvent polarity. J Phys Chem 96:10809–10819

13. Stsiapura VI, Maskevich AA, Kuzmitsky VA, Turoverov KK, Kuznetsova IM (2007) Computational study of thioflavin T torsional relaxation in the excited state. J Phys Chem A 111(22):4829–4835

14. Allen BD, Benniston AC, Harriman A, Rostron SA, Yu C (2005) The photophysical properties of a julolidine-based molecular rotor. Phys Chem Chem Phys 7(16):3035–3040

15. Rettig W, Strehmel B, Majenz W (1993) The excited states of stilbene and stilbenoid donor–acceptor dye systems. A theoretical study. Chem Phys 173(3):525–537

16. Strehmel B, Seifert H, Rettig W (1997) Photophysical properties of fluorescence probes. 2. A model of multiple fluorescence for stilbazolium dyes studied by global analysis and quantum chemical calculations. J Phys Chem B 101(12):2232–2243

17. Rettig W, Klock A (1985) Intramolecular fluorescence quenching in aminocoumarines. Identification of an excited state with full charge separation. Can J Chem 63(7):1649–1653

18. Kung CE, Reed JK (1986) Microviscosity measurements of phospholipid bilayers using fluorescent dyes that undergo torsional relaxation. Biochemistry 25:6114–6121

19. Haidekker MA, Brady TP, Lichlyter D, Theodorakis EA (2005) Effects of solvent polarity and solvent viscosity on the fluorescent properties of molecular rotors and related probes. Bioorg Chem 33(6):415–425

20. Diverdi LA, Topp MR (1984) Subnanosecond time-resolved fluorescence of acridine in solution. J Phys Chem 88(16):3447–3451

21. Guilbault GG (1990) Practical fluorescence. CRC, Boca Raton, FL

22. El-Sayed MA, Kasha M (1959) Interchange of orbital excitation types of the lowest electronic states of 2 ring N-heterocyclics by solvation. Spectrochim Acta 15:758–759

23. Haidekker MA, Akers W, Lichlyter D, Brady TP, Theodorakis EA (2005) Sensing of flow and shear stress using fluorescent molecular rotors. Sensor Lett 3:42–48

24. Kuimova MK, Botchway SW, Parker AW, Balaz M, Collins HA, Anderson HL, Suhling K, Ogilby PR (2009) Imaging intracellular viscosity of a single cell during photoinduced cell death. Nat Chem 1(1):69–73

25. Law KY (1980) Fluorescence probe for microenvironments: Anomalous viscosity dependence of the fluorescence quantum yield of p-N, N-dialkylaminobenzylidenmalononitrile in 1-alkanols. Chem Phys Lett 75(3):545–549

26. Lukac S (1984) Thermally induced variations in polarity and microviscosity of phospholipid and surfactant vesicles monitored with a probe forming an intramolecular charge-transfer complex. J Am Chem Soc 106:4386–4392

27. Viriot ML, Carré MC, Geoffroy-Chapotot C, Brembilla A, Muller S, Stoltz J-F (1998) Molecular rotors as fluorescent probes for biological studies. Clin Hemorheol Microcirc 19:151–160

28. Förster Th, Hoffmann G (1971) Die Viskositätsabhängigkeit der Fluoreszenzquantenausbeuten einiger Farbstoffsysteme [effect of viscosity on the fluorescence quantum yield of some dye systems]. Z Phys Chem 75:6376

29. Loutfy RO, Arnold BA (1982) Effect of viscosity and temperature on torsional relaxation of molecular rotors. J Phys Chem 86:4205–4211

30. Loutfy RO (1986) Fluorescence probes for polymer free-volume. Pure Appl Chem 58 (9):1239–1248

31. von Gierer A, Wirtz K (1953) Molekulare Theorie der Mikroreibung [Molecular theory of microfriction]. Z Naturforschung 8a:523–538

32. Doolittle AK (1952) Studies in Newtonian flow III. The dependence of the viscosity of liquids on molecule weight and free space (in homologous series). J Appl Phys 23(2): 236–239

33. Loutfy RO, Law KY (1980) Electrochemistry and spectroscopy of intramolecular charge-transfer complexes. p-N, N-dialkylaminobenzylidenemanononitriles. J Phys Chem 84: 2803–2808

34. Parusel ABJ (2001) Excited state intramolecular charge transfer in N, N-heterocyclic-4-aminobenzonitriles: a DFT study. Chem Phys Lett 340(5–6):531–537

35. Parusel ABJ, Köhler G (2001) Influence of the alkyl chain length on the excited-state properties of 4-dialkyl-benzonitriles. A theoretical DFT/MRCI study. Int J Quantum Chem 84(2):149–156

36. Quiñones E, Ishikawa Y, Leszczynski J (2000) Conformational properties of dimethylaminobenzonitrile in gas phase and polar solvents: ab initio HF/6-31G (d, p) and MP2/6-31G (d, p) investigations. J Mol Struct: THEOCHEM 529(1–3):127–134

37. Zerner MC, Reidlinger C, Fabian WMF, Junek H (2001) Push–dyes containing malononitrile dimer as acceptor: synthesis, spectroscopy and quantum chemical calculations. J Mol Struct: THEOCHEM 543(1–3):129–146

38. Cao X, Tolbert RW, McHale JL, Edwards WD (1998) Theoretical study of solvent effects on the intramolecular charge transfer of a hemicyanine dye. J Phys Chem A 102(17):2739–2748

39. Sudholt W, Staib A, Sobolewski AL, Domcke W (2000) Molecular-dynamics simulations of solvent effects in the intramolecular charge transfer of 4-(N, N-dimethylamino) benzonitrile. Phys Chem Chem Phys 2(19):4341–4353

40. Kottas GS, Clarke LI, Horinek D, Michl J (2005) Artificial molecular rotors. Chem Rev 105 (4):1281–1376

41. Kee HL, Kirmaier C, Yu L, Thamyongkit P, Youngblood WJ, Calder ME, Ramos L, Noll BC, Bocian DF, Scheidt WR (2005) Structural control of the photodynamics of boron–dipyrrin complexes. J Phys Chem B 109(43):20433

42. Kollmannsberger M, Rurack K, Resch-Genger U, Daub J (1998) Ultrafast charge transfer in amino-substituted boron dipyrromethene dyes and its inhibition by cation complexation: a new design concept for highly sensitive fluorescent probes. J Phys Chem A 102:10211–10220

43. Rettig W (1980) External and internal parameters affecting the dual fluorescence of p-cyano-dialkylanilines. J Lumin 26:21–46

44. Zachariasse KA, von der Haar T, Hebecker A, Leinhos U, Kuhnle W (1993) Intramolecular charge transfer in aminobenzonitriles: requirements for dual fluorescence. Pure Appl Chem 65(8):1745–1750

45. Rotkiewicz K, Rettig W, Detzerd N, Rothe A (2003) Substituent-induced coupling of the two lowest excited singlet states of 2-methoxy-derivatives of 4-(N, N-dimethylamino)- and 4-(N-methylamino)benzonitrile. Phys Chem Chem Phys 5:998–1002

46. Shinohara Y, Arai T (2008) Effect of methoxy substituents on the excited state properties of stilbene. Bull Chem Soc Jpn 81(11):1500–1504

47. Lippert E, Ayuk AA, Rettig W, Wermuth G (1981) Adiabatic photoreactions in dilute solutions of p-substituted N, N′-dialkylanilines and related donor–acceptor compounds. J Photochem 17:237–241

48. Even P, Chaubet F, Letourneur D, Viriot ML, Carre MC (2003) Coumarin-like fluorescent molecular rotors for bioactive polymers probing. Biorheology 40(1):261–263

49. Yang X, Jiang X, Zhao C, Chen R, Qin P, Sun L (2006) Donor–acceptor molecules containing thiophene chromophore: synthesis, spectroscopic study and electrogenerated chemiluminescence. Tetrahedron Lett 47:4961–4964

50. Cowley DJ, Peoples AH (1977) Rotational isomerism and dual luminescence in dipolar dialkylamino-compounds. J Chem Soc Chem Commun:352–353

51. Zhen Z, Tug C-H (1991) Hydrophobic effects on photophysical and photochemical processes: excimer fluorescence and aggregate formation of long-chain alkyl 4-(N, N-dimethylamino) benzoate in water–organic binary mixtures. Chem Phys Lett 180(3):211–215
52. Braun D, Rettig W, Delmond S, Letard J-F, Lapouyade R (1997) Amide derivatives of DMABN: a new class of dual fluorescent compounds. J Phys Chem A 101:6836–6841
53. Kumar S, Singh AK, Krishnamoorthy G, Swaminathan R (2008) Thioflavin T displays enhanced fluorescence selectively inside anionic micelles and mammalian cells. J Fluoresc 18(6):1199–1205
54. Khurana R, Coleman C, Ionescu-Zanetti C, Carter SA, Krishna V, Grover RK, Roy R, Singh S (2005) Mechanism of thioflavin T binding to amyloid fibrils. J Struct Biol 151(3): 229–238
55. Naik LR, Naik AB, Pal H (2009) Steady-state and time-resolved emission studies of thioflavin-T. J Photochem Photobiol A Chem 204:161–167
56. Retna Raj C, Ramaraj R (1997) Cyclodextrin induced intermolecular excimer formation of thioflavin T. Chem Phys Lett 273(3–4):285–290
57. Stsiapura VI, Maskevich AA, Kuzmitsky VA, Uversky VN, Kuznetsova IM, Turoverov KK (2008) Thioflavin T as a molecular rotor: fluorescent properties of thioflavin T in solvents with different viscosity. J Phys Chem B 112(49):15893–15902
58. Herbich J, Kapturkiewicz A (1991) Radiative and radiationless depopulation of the excited intramolecular charge transfer states: aryl derivatives of aromatic amines. Chem Phys 158:143–153
59. Siemiarczuk A, Grabowski ZR, Krówczynski A, Asher M, Ottolenghi M (1977) Two emitting states of excited p-(9-Anthryl)-N, N-dimethylaniline derivatives in polar solvents. Chem Phys Lett 51:315–320
60. Siemiarczuk A, Ware WR (1987) Complex excited-state relaxation in p-(9-Anthryl)-N, N-dimethylaniline derivatives evidenced by fluorescence lifetime distributions. J Phys Chem 91:3677–3682
61. Herbich J, Dobkowski J, Rulliére C, Nowacki J (1989) Low-temperature dual fluorescence in 9-morpholinoacridine picosecond TICT state formation? J Lumin 44:87–95
62. Haidekker MA, Theodorakis EA (2007) Molecular rotors-fluorescent biosensors for viscosity and flow. Org Biomol Chem 5(11):1669–1678
63. Pillai ZS, Sudeep PK, George Thomas K (2003) Effect of viscosity on the singlet-excited state dynamics of some hemicyanine dyes. Res Chem Intermed 29(3):293–305
64. Bosch P, Peinado C, Martin V, Catalina F, Corrales T (2006) Fluorescence monitoring of photoinitiated polymerization reactions synthesis, photochemical study and behaviour as fluorescent probes of new derivatives of 4-dimethylaminostyryldiazines. J Photochem Photobiol A Chem 180(1–2):118–129
65. Lord SJ, Conley NR, Lee HD, Nishimura SY, Pomerantz AK, Willets KA, Lu Z, Wang H, Liu N, Samuel R, Weber R, Semyonov A, He M, Twieg RJ, Moerner WE (2009) DCDHF fluorophores for single-molecule imaging in cells. ChemPhysChem 10:55–65
66. Rettig W, Vogel M, Lippert E, Otto H (1986) The dynamics of adiabatic photoreactions as studied by means of the time structure of synchrotron radiation. Chem Phys 103:381–390
67. Vogel M, Rettig W (1985) Efficient intramolecular fluorescence quenching in triphenylmethane dyes involving excited states with charge separation and twisted conformations. Ber Bunsenges 89(9):962–968
68. Haidekker MA, L'Heureux N, Frangos JA (2000) Fluid shear stress increases membrane fluidity in endothelial cells: a study with DCVJ fluorescence. Am J Physiol Heart Circ Physiol 278(4):H1401–H1406
69. Jiang Y (1994) pH Dependence of the twisted intramolecular charge transfer(TICT) of p-N, N-dimethylaminobenzoic acid in aqueous solution. J Photochem Photobiol A Chem 78 (3):205–208
70. Haidekker MA, Brady TP, Lichlyter D, Theodorakis EA (2006) A ratiometric fluorescent viscosity sensor. J Am Chem Soc 128:398–399

71. Milich KN, Akers W, Haidekker MA (2005) A ratiometric fluorophotometer for fluorescence-based viscosity measurement with molecular rotors. Sensor Lett 3:237–243
72. Suhling K, French PMW, Phillips D (2005) Time-resolved fluorescence microscopy. Photochem Photobiol Sci 4(1):13–22
73. Vogel M, Rettig W (1987) Excited state dynamics of triphenylmethane-dyes used for investigation of microviscosity effects. Ber Bunsenges Phys Chem 91:1241–1247
74. Moog RS, Ediger MD, Boxer SG, Fayer MD (1982) Viscosity dependence of the rotational reorientation of rhodamine B in mono- and polyalcohols. Picosecond transient grating experiments. J Phys Chem 86:4694–4700
75. Dutta P, Bhattacharyya K (2004) Ultrafast chemistry in complex and confined systems. J Chem Sci 116:5–16
76. Hara K, Bulgarevich DS, Kajimoto O (1996) Pressure tuning of solvent viscosity for the formation of twisted intramolecular charge-transfer state in 4, 4′-diaminophenyl sulfone in alcohol solution. J Chem Phys 104(23):9431–9436
77. Gierer VA, Wirtz K (1953) Molekulare Theorie der Mikroreibung. Z Naturforsch 8(Part A):532–538
78. Akers W, Haidekker MA (2005) Precision assessment of biofluid viscosity measurements using molecular rotors. J Biomech Eng 127(3):450–454
79. Akers W, Haidekker MA (2004) A molecular rotor as viscosity sensor in aqueous colloid solutions. J Biomech Eng 126(3):340–345
80. Akers WJ, Cupps JM, Haidekker MA (2005) Interaction of fluorescent molecular rotors with blood plasma proteins. Biorheology 42(5):335–344
81. Williams ML, Landel RF, Ferry JD (1955) The temperature dependence of relaxation mechanisms in amorphous polymers and other glass-forming liquids. J Am Chem Soc 84:2803–2808
82. Kelley FN, Bueche F (1961) Viscosity and glass temperature relations for polymer-diluent systems. J Polym Sci 50(154):549–556
83. Bosch P, Catalina F, Corrales T, Peinado C (2005) Fluorescent probes for sensing processes in polymers. Chem Eur J 11(15):4314
84. Paczkowski J, Neckers DC (1991) Twisted intramolecular charge-transfer phenomenon as a quantitative probe of polymerization kinetics. Macromolecules 24(10):3013–3016
85. Benjelloun A, Brembilla A, Lochon P, Adibnejad M, Viriot ML, Carré MC (1996) Detection of hydrophobic microdomains in aqueous solutions of amphiphilic polymers using fluorescent molecular rotors. Polymer (Guildford) 37(5):879–883
86. Haidekker MA, Lichlyter D, Ben Johny M, Grimes CA (2006) Probing polymerization dynamics with fluorescent molecular rotors and magnetoelastic sensors. Sensor Lett 4:257–261
87. Zhu D, Haidekker MA, Lee J-S, Won Y-Y, Lee JC (2007) Application of molecular rotors to the determination of the molecular weight dependence of viscosity in polymer melts. Macromolecules 40:7730–7732
88. Haidekker MA, Ling T, Anglo M, Stevens HY, Frangos JA, Theodorakis EA (2001) New fluorescent probes for the measurement of cell membrane viscosity. Chem Biol 8(2): 123–131
89. Haidekker MA, Brady TP, Chalian SH, Akers W, Lichlyter D, Theodorakis EA (2004) Hydrophilic molecular rotor derivatives-synthesis and characterization. Bioorg Chem 32(4): 274–289
90. Sarkar N, Das K, Nath DN, Bhattacharyya K (1994) Twisted charge transfer processes of Nile red in homogeneous solutions and in faujasite zeolite. Langmuir 10(1):326–329
91. Hazra P, Chakrabarty D, Chakraborty A, Sarkar N (2004) Intramolecular charge transfer and solvation dynamics of Nile red in the nanocavity of cyclodextrins. Chem Phys Lett 388(1–3): 150–157
92. Jee AY, Park S, Kwon H, Lee M (2009) Excited state dynamics of Nile red in polymers. Chem Phys Lett 477(1–3):112–115

93. Cser A, Nagy K, Biczók L (2002) Fluorescence lifetime of Nile red as a probe for the hydrogen bonding strength with its microenvironment. Chem Phys Lett 360(5–6):473–478

94. Wearsch PA, Voglino L, Nicchitta CV (1998) Structural transitions accompanying the activation of peptide binding to the endoplasmic reticulum Hsp90 chaperone GRP94. Biochemistry 37(16):5709–5719

95. Vassar PS, Culling CF (1959) Fluorescent stains, with special reference to amyloid and connective tissues. Arch Pathol 68:487

96. Voropai ES, Samtsov MP, Kaplevskii KN, Maskevich AA, Stepuro VI, Povarova OI, Kuznetsova IM, Turoverov KK, Fink AL, Uverskii VN (2003) Spectral properties of thioflavin T and its complexes with amyloid fibrils. J Appl Spectrosc 70(6):868–874

97. Wood SJ, Maleeff B, Hart T, Wetzel R (1996) Physical, morphological and functional differences between pH 5.8 and 7.4 aggregates of the Alzheimer's amyloid peptide A. J Mol Biol 256(5):870–877

98. Biancalana M, Makabe K, Koide A, Koide S (2009) Molecular mechanism of thioflavin-T binding to the surface of b-rich peptide self-assemblies. J Mol Biol 385(4):1052–1063

99. Wu C, Biancalana M, Koide S, Shea JE (2009) Binding modes of thioflavin-T to the single-layer beta-sheet of the peptide self-assembly mimics. J Mol Biol 394(4):627–633

100. Sen P, Fatima S, Ahmad B, Khan RH (2009) Interactions of thioflavin T with serum albumins: spectroscopic analyses. Spectrochim Acta Part A: Mol Biomol Spectrosc 74(1): 94–99

101. Eisert R, Felau L, Brown LR (2006) Methods for enhancing the accuracy and reproducibility of Congo red and thioflavin T assays. Anal Biochem 353(1):144–146

102. Blanchard BJ, Chen A, Rozeboom LM, Stafford KA, Weigele P, Ingram VM (2004) Efficient reversal of Alzheimer's disease fibril formation and elimination of neurotoxicity by a small molecule. Proc Natl Acad Sci U S A 101(40):14326

103. Feng BY, Toyama BH, Wille H, Colby DW, Collins SR, May BCH, Prusiner SB, Weissman J, Shoichet BK (2008) Small-molecule aggregates inhibit amyloid polymerization. Nat Chem Biol 4(3):197

104. Wang H, Duennwald ML, Roberts BE, Rozeboom LM, Zhang YL, Steele AD, Krishnan R, Su LJ, Griffin D, Mukhopadhyay S (2008) Direct and selective elimination of specific prions and amyloids by 4, 5-dianilinophthalimide and analogs. Proc Natl Acad Sci 105 (20):7159

105. Kroes-Nijboer A, Lubbersen YS, Venema P, van der Linden E (2009) Thioflavin T fluorescence assay for [beta]-lactoglobulin fibrils hindered by DAPH. J Struct Biol 165(3):140

106. Ban T, Hamada D, Hasegawa K, Naiki H, Goto Y (2003) Direct observation of amyloid fibril growth monitored by thioflavin T fluorescence. J Biol Chem 278(19):16462–16465

107. De Ferrari GV, Mallender WD, Inestrosa NC, Rosenberry TL (2001) Thioflavin T is a fluorescent probe of the acetylcholinesterase peripheral site that reveals conformational interactions between the peripheral and acylation sites. J Biol Chem 276(26):23282

108. Morimoto K, Kawabata K, Kunii S, Hamano K, Saito T, Tonomura B (2009) Characterization of type I collagen fibril formation using thioflavin T fluorescent dye. J Biochem 145(5): 677

109. Lindgren M, Sörgjerd K, Hammarström P (2005) Detection and characterization of aggregates, prefibrillar amyloidogenic oligomers, and protofibrils using fluorescence spectroscopy. Biophys J 88(6):4200–4212

110. Bosch LI, Mahon MF, James TD (2004) The B–N bond controls the balance between locally excited (LE) and twisted internal charge transfer (TICT) states observed for aniline based fluorescent saccharide sensors. Tetrahedron Lett 45(13):2859–2862

111. Arimori S, Bosch LI, Ward CJ, James TD (2001) Fluorescent internal charge transfer (ICT) saccharide sensor. Tetrahedron Lett 42(27):4553–4555

112. Tan W, Zhang D, Zhu D (2007) 4-N-Methyl-N-(2-dihydroxyboryl-benzyl) amino benzonitrile and its boronate analogue sensing saccharides and fluoride ion. Bioorg Med Chem Lett 17(9):2629–2633

113. Fülöp A, Arian D, Lysenko A, Mokhir A (2009) A simple method for monitoring protein–DNA interactions. Bioorg Med Chem Lett 19(11):3104–3107
114. Albers E, Muller BW (1995) Cyclodextrin derivatives in pharmaceutics. Crit Rev Ther Drug Carrier Syst 12(4):311–337
115. Al-Hassan KA, Khanfer MF (1998) Fluorescence probes for cyclodextrin interiors. J Fluoresc 8(2):139–152
116. Kim YH, Cho DW, Yoon M, Kim D (1996) Observation of hydrogen-bonding effects on twisted intramolecular charge transfer of p-(N, N-diethylamino) benzoic acid in aqueous cyclodextrin solutions. J Phys Chem 100(39):15670–15676
117. Panja S, Chakravorti S (2002) Photophysics of 4-(N, N-dimethylamino)cinnamaldehyde/ alpha-cyclodextrin inclusion complex. Spectrochim Acta A Mol Biomol Spectrosc 58(1): 113–122
118. Wang J, Nakamura A, Hamasaki K, Ikeda H, Ikeda T, Ueno A (1996) A fluorescent molecule-recognition sensor with a protein as an environmental factor. Chem Lett 4:303–304
119. Shinitzky M (1984) Membrane fluidity and cellular functions. In: Shinitzky M (ed) Physiology of membrane fluidity. CRC, Boca Raton, FL, pp 1–51
120. Nadiv O, Shinitzky M, Manu H, Hecht D, Roberts CT Jr, LeRoith D, Zick Y (1994) Elevated protein tyrosine phosphatase activity and increased membrane viscosity are associated with impaired activation of the insulin receptor kinase in old rats. Biochem J 298(Pt 2):443–450
121. Osterode W, Holler C, Ulberth F (1996) Nutritional antioxidants, red cell membrane fluidity and blood viscosity in type 1 (insulin dependent) diabetes mellitus. Diabet Med 13(12): 1044–1050
122. Zubenko GS, Kopp U, Seto T, Firestone LL (1999) Platelet membrane fluidity individuals at risk for Alzheimer's disease: a comparison of results from fluorescence spectroscopy and electron spin resonance spectroscopy. Psychopharmacology (Berl) 145(2):175–180
123. Shiraishi K, Matsuzaki S, Ishida H, Nakazawa H (1993) Impaired erythrocyte deformability and membrane fluidity in alcoholic liver disease: participation in disturbed hepatic microcirculation. Alcohol Alcohol Suppl 1A:59–64
124. Maczek C, Bock G, Jurgens G, Schonitzer D, Dietrich H, Wick G (1998) Environmental influence on age-related changes of human lymphocyte membrane viscosity using severe combined immunodeficiency mice as an in vivo model. Exp Gerontol 33(5):485–498
125. Möller W, Takenaka S, Rust M, Stahlhofen W, Heyer J (1997) Probing mechanical properties of living cells by magnetopneumography. J Aerosol Med 10(3):171–186
126. Butler JP, Kelly SM (1998) A model for cytoplasmic rheology consistent with magnetic twisting cytometry. Biorheology 35(3):193–209
127. Valberg PA, Albertini DF (1985) Cytoplasmic motions, rheology, and structure probed by a novel magnetic particle method. J Cell Biol 101(1):130–140
128. Humphry-Baker R, Grätzel M, Steiger R (1980) Drastic fluorescence enhancement and photochemical stabilization of cyanine dyes through micellar systems. J Am Chem Soc 102(2):847–848
129. Haidekker M, Brady T, Wen K, Okada C, Stevens H, Snell J, Frangos J, Theodorakis E (2002) Phospholipid-bound molecular rotors: synthesis and characterization. Bioorg Med Chem 10(11):3627–3636
130. Nipper ME, Majd S, Mayer M, Lee JC, Theodorakis EA, Haidekker MA (2008) Characterization of changes in the viscosity of lipid membranes with the molecular rotor FCVJ. Biochim Biophys Acta 1778(4):1148–1153
131. Barbu VD (1991) Isoprenylation of proteins: what is its role? C R Seances Soc Biol Fil 185(5):278–289
132. Kohl NE, Conner MW, Gibbs JB, Graham SL, Hartman GD, Oliff A (1995) Development of inhibitors of protein farnesylation as potential chemotherapeutic agents. J Cell Biochem Suppl 22:145–150
133. Härtel S, Tykhonova S, Haas M, Diehl HA (2002) The susceptibility of non-UV fluorescent membrane dyes to dynamical properties of lipid membranes. J Fluoresc 12(3):465–479

134. Luby-Phelps K, Mujumdar S, Mujumdar RB, Ernst LA, Galbraith W, Waggoner AS (1993) A novel fluorescence ratiometric method confirms the low solvent viscosity of the cytoplasm. Biophys J 65(1):236–242
135. Kuimova MK, Yahioglu G, Levitt JA, Suhling K (2008) Molecular rotor measures viscosity of live cells via fluorescence lifetime imaging. J Am Chem Soc 130(21):6672–6673
136. Lodish HF (2008) Molecular cell biology, 6th edn. W.H. Freeman, New York
137. Dustin P (1984) Structure and Chemistry of microtubules. In: Microtubules, Springer-Verlag, New York, 19–94
138. Iio T, Takahashi S, Sawada S (1993) Fluorescent molecular rotors binding to actin. J Biochem 113:196–199
139. Sawada S, Iio T, Hayashi Y, Takahashi S (1992) Fluorescent rotors and their applications to the study of G–F transformation of actin. Anal Biochem 204:110–117
140. Kung CE, Reed JK (1989) Fluorescent molecular rotors: a new class of probes for tubulin structure and assembly. Biochemistry 28:6678–6686
141. Grinvald A, Fine A, Farber IC, Hildesheim R (1983) Fluorescence monitoring of electrical responses from small neurons and their processes. Biophys J 42(2):195–198
142. Ephardt H, Fromherz P (1989) Fluorescence and photoisomerization of an amphiphilic aminostilbazolium dye as controlled by the sensitivity of radiationless deactivation to polarity and viscosity. J Phys Chem 93(22):7717–7725
143. Jones MA, Bohn PW (2001) Total internal reflection fluorescence and electrocapillary investigations of adsorption at the water–dichloroethane electrochemical interface. 2. Fluorescence-detected linear dichroism investigation of adsorption-driven reorientation of di-N-butylaminonaphthylethenylpyridiniumpropylsulfonate. J Phys Chem B 105(11):2197–2204
144. Sutharsan J, Lichlyter D, Wright NE, Dakanali M, Haidekker MA, Theodorakis EA (2010) Molecular rotors: synthesis and evaluation as viscosity sensors. Tetrahedron 66:2582–2588
145. Lichlyter D, Haidekker MA (2009) Immobilization techniques for molecular rotors – towards a solid-state viscosity sensor platform. Sens Actuators B Chem 139:648–656

Electrochromism and Solvatochromism in Fluorescence Response of Organic Dyes: A Nanoscopic View

Patrik R. Callis

Abstract Methods are described that allow prediction and understanding of solvatochromism and fluorescence quenching of dyes embedded in a nanometer-scale medium, e.g., solvent, protein, and membranes. Spectra of the dye are calculated at the microscopic level using quantum mechanics coupled to the point charges representing the medium by Coulombic potentials, while the whole system propagates by classical molecular mechanics. This view avoids the necessity to define macroscopic parameters such as dielectric constant, viscosity, and effective molecular radius, while providing an accuracy of 5–10 nm in UV-VIS wavelength. A major advantage is the ability to expose large fluctuations, which translate to heterogeneity on timescales longer than the excited state lifetime, which are now routinely accessible for large assemblies. Extensive applications to the fluorescence of tryptophan in proteins are reviewed, and the delightful intricacies surrounding the mechanisms of voltage-sensitive dyes in membranes are briefly reviewed and discussed from the view of an outsider. A brief summary of two molecular dynamics studies involving voltage-sensitive dyes is given, pointing to promise for this type of investigation.

Keywords Electrochromism · Fluorescence · Proteins · QM–MM · Solvatochromism · Tryptophan · Voltage-sensitive dyes

Contents

P.R. Callis
Department of Chemistry and Biochemistry, Montana State University, Bozeman, MT 59717, USA
e-mail: pcallis@montana.edu

A.P. Demchenko (ed.), *Advanced Fluorescence Reporters in Chemistry and Biology I:*
Fundamentals and Molecular Design, Springer Ser Fluoresc (2010) 8: 309–330,
DOI 10.1007/978-3-642-04702-2_9, © Springer-Verlag Berlin Heidelberg 2010

1 Introduction

This chapter deals exclusively with underlying physical principles of *optical* electrochromism and solvatochromism of organic dyes. Both phenomena refer to changes in absorption and/or fluorescence spectra caused by a net local electric field change acting on an electron density shift accompanying optical excitation to an excited state of a probe molecule. As used in this chapter, electrochromism will refer to the response to applied fields external to the molecular system, e.g., the relatively small fields due to a membrane potential, on the order of 10^5 V cm^{-1}, resulting in wavelength shifts on the order of $\sim$1 nm (50 cm^{-1}). Solvatochromism will refer to the much larger response caused by polar solvents, for which the average fields are $>10^7$ V cm^{-1}, resulting in shifts up to $\sim$100 nm (5,000 cm^{-1}).

The reader may be aware that the terms electrochromism and electrochromic are also used in a large segment of literature to mean dyes whose absorption spectra are changed through redox reactions induced by electrochemical means. This subject is not included in this chapter.

A major advantage of fluorescence as a sensing property stems from the sensitivity to the precise local environment of the intensity, i.e., quantum yield (Φ_f), excited state lifetime (τ_f), and peak wavelength (λ_{max}). In particular, it is the *local* electric field strength and *direction* that determine whether the fluorescence will be red or blue shifted and whether an electron acceptor will or will not quench the fluorescence. An equivalent statement, but more practical, is that these quantities depend primarily on *the change in average electrostatic potential (volts) experienced by the electrons during an electronic transition* (See Appendix for a brief tutorial on electric fields and potentials as pertains to electrochromism). The reason this is more practical is that even at the molecular scale, the instantaneous electric

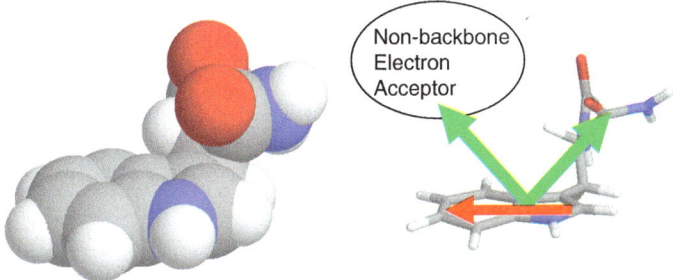

Fig. 1 *N*-formyltryptophanamide used for QM calculations on a tryptophan trimmed from a protein structure, also showing the two kinds of electron density shifts that control Trp fluorescence wavelength (*red*) and intensity/lifetime (*green*)

field produced by a dynamic medium of point charges varies enormously from atom to atom, and computing the molecular energy is quite awkward.

Figure 1 provides the essence of what is important using the tryptophan amino acid as an example. For λ_{max}, it is the stabilization of the electron density shift from the pyrrole to the benzene part of the indole ring during the ground to excited state ($S_0 \rightarrow S_1$) that determines the wavelength shift. For fluorescence quenching by electron transfer, it is stabilization of the electron (or hole) density shift from the chromophore as a whole to the electron acceptor (or donor) during intramolecular or intermolecular photoinduced electron transfer.

The aim of this Chapter is to review a method by which fluorescence properties of organic dyes can, in general, be predicted and understood at a microscopic (nm scale) by interfacing quantum methods with classical molecular dynamics (MD) methods. Some review of our extensive applications [1] of this method to the widely exploited intrinsic fluorescence probe in proteins, the amino acid tryptophan (Trp) will be followed by a discussion of electrochromic membrane voltage-sensing dyes.

2   Quantum Mechanical–Classical Mechanical (QM–MM) Methods for Solvatochromism and Electrochromism Predictions

2.1   Introduction and Background

Our method has evolved during many studies over the last two decades. These include studies on the effect of strong internal electric fields in crystals on optical transition dipole directions of nucleic acid bases [2, 3], QM–MM predictions of time-dependent solvatochromism on 3-methylindole (3MI) in water [4], and on tryptophan in several proteins [5–8]. More recently, the same techniques have been extended to the successful prediction and understanding of the perplexing variability of Trp fluorescence quantum yields and lifetimes in proteins, apparently caused by electron transfer to the backbone amides. The latter may be understood as "solvatochromism" that affects the energy gap between the fluorescing state (S_1) and the lowest charge transfer (CT) state [9–18]. This effect will be seen to play a role in the response of certain voltage-sensitive dyes. Exploratory studies of this nature have also been done on flavins in proteins [19].

2.2   General Methods and Principles

The methods and procedures we have used were influenced to various extents by earlier work [20–23]. In some of those, and in our studies, quantum mechanics is applied only to the chromophore of interest, using Michael Zerner's spectroscopically calibrated INDO/S-CIS method (Zindo) [24, 25], incorporating a modification that allows for input of electrostatic potentials and fields generated from the partial charges of every atom in the environment, e.g., protein and solvent. The response of Zindo to electric fields [5] is quite accurate, as indicated by a more fundamental quantum chemical study on indole [26]. Our approach to computing fluorescence wavelength, quantum yields, and lifetimes is a hybrid quantum mechanical-molecular mechanics (QM/MM) technique in which quantum mechanics is involved only in determining charges to the atoms of the chromophore and for interrogating the electronic transition energy as a function of the electric potentials produced by the atoms in the environment. The environment is modeled as a collection of point charges as given by a MM/MD force field. We have used CHARMM [27] for most of our applications. The chromophore dynamics and atomic coordinates are governed entirely by the MD with QM-modified charges. The transition energy calculation, with the potentials and fields added, is performed on the chromophore only. The effect of the environment surrounding the chromophore atoms is incorporated directly into the QM calculation through a straightforward modification of the matrix elements of the Fock operator in vacuum involving atomic orbitals μ and v:

$$F_{\mu\mu} = F^0_{\mu\mu} - eV_\alpha$$
$$F_{\mu v} = F^0_{\mu v} + e\vec{E} \cdot \vec{r}_{\mu v}$$

in which $-e$ is the electron charge, V_α is the electrostatic potential at quantum mechanical atom α created by *all* nonQM protein and solvent atoms, k; E_a is the associated electrostatic field. The potential and field are evaluated at the QM atoms, α, via straightforward Coulomb sums:

$$V_\alpha = \sum_k q_k / r_{\alpha k}$$
$$\vec{E}_\alpha = \sum_k (q_k / r_{\alpha k}^3) \vec{r}_{\alpha k}$$

where the summations extend over all nonQM atoms. In other words, the quantum system sees the protein and solvent simply as a collection of point charges, whose values are taken from the CHARMM 22 [27] forcefield. No dielectric constant is assumed, and electronic polarizability is not included.

Electronic excitation of the chromophore is simulated by instantly switching the charges (and in some studies also the geometry) on the QM system to excited state (S_1) values. Most of the internal Stark effect (ISE) is expressed implicitly by the difference of the potentials at different atoms.

The chromophore atoms are assigned charges given by the density matrix in the Löwdin basis from the INDO/S calculation:

$$q_\alpha = -e \left(\sum_\mu P_{\mu\mu} - Z_\alpha \right)$$

(where $P_{\mu\mu}$ are diagonal elements of the density matrix for all valence atomic orbitals (μ) centered on atom α, Z_α is the atomic core charge, and $-e$ is the electron charge). The charge distribution of the excited state is sufficiently sensitive to the local field that these charges should be updated by a QM calculation at least every 10 fs of simulated time to capture the fastest relaxation times [28, 29]. The interface for informing the MD of the QM charges is accomplished by the USERSB subroutine that is native to CHARMM.

From the starting structures (PDB file), the full complement of hydrogens is added using a utility within CHARMM. The entire protein is then solvated within a sphere of TIP3P model waters, with radius such that all parts of the protein were solvated to a depth of at least 5 Å. A quartic confining potential localized on the surface of the spherical droplet prevented "evaporation" of any of the waters during the course of the trajectory. The fully solvated protein structure is energy minimized and equilibrated before the production simulation.

A wavelength to compare with experiment comes from the average over the trajectory of calculated transition energies following equilibration. Examination of transition energy time correlation functions and direct-response results do not reveal a relaxation component beyond ~5 ps of the $S_0 \rightarrow S_1$ excitation that is significant compared to fluctuations during a single trajectory. Averaging several hundred trajectories of ns length are now possible [30, 31], and are able to capture slower relaxations and presumably could result in more precisely predicted wavelength maxima, and ns relaxation times [32, 33].

2.3 Analysis Tools

We have endeavored to decompose the shifts due to protein and solvent environment into contributions from individual amino acid residues, solvent molecules, and, in some cases, individual atoms. This can be done effectively at a given point in a trajectory from the summation over atom electron density changes ($\Delta\rho$) accompanying excitation weighted by the potentials at those atoms:

$$\sum_{\alpha} V_{\alpha}\Delta\rho_{\alpha}$$

where V_{α} is computed as before, and the changes in density are taken from the INDO/S-CIS calculation at that configuration. The contribution to the V_{α} values due to charges from any chosen groupings of atoms in the environment is therefore straightforward. The transition energy shift estimated this way correlates very well with the transition energy shifts coming directly from the program because they use the same Coulomb sum.

3 Indoles and Tryptophan in Proteins

3.1 Overview of Results

Following an exploratory study [34], Vivian and Callis [6] have predicted the fluorescence wavelengths of 19 tryptophans in 16 proteins, starting with crystal structures and using a hybrid quantum mechanical–classical molecular dynamics method, with the assumption that only electrostatic interactions of the tryptophan ring electron density with the surrounding protein and solvent affect the transition energy. With only one adjustable parameter, the scaling of the quantum mechanical atomic charges as seen by the protein/solvent environment, the mean absolute deviation between predicted and observed fluorescence maximum wavelength is 6 nm, as shown in Fig. 2. The modeling of electrostatic interactions, including hydration, in proteins is vital to understanding function and structure, and this study helps assess the effectiveness of current electrostatic models.

The combination of the CHARMM22 force field, explicit water and INDO/ S-CIS Löwdin charges on Trp (scaled by 0.80), and ab initio geometry difference (CIS-HF/3–21G) gives a good quantitative prediction for the fluorescence wavelength maximum, with the only fitting parameter being the charge-scaling factor for the Trp ring. No protein dielectric constant was imposed.

The numerous quantum chemical computations dating back many years and summarized previously [35], consistently showed that the increased dipole in the ^1L_a state of indole and 3MI have – without exception – predicted that electron density is shifted from the pyrrole ring to the benzene ring upon excitation to the ^1L_a state. This means that positively charged residues near the benzene end or negative charges near the pyrrole end of the Trp ring will shift λ_{max} to longer wavelengths (produce a red shift), with the opposite configuration producing a blue shift. Because these shifts are due to the electric field imposed by the protein and solvent, they may be termed an ISE, by analogy to the familiar shifting of energy levels via

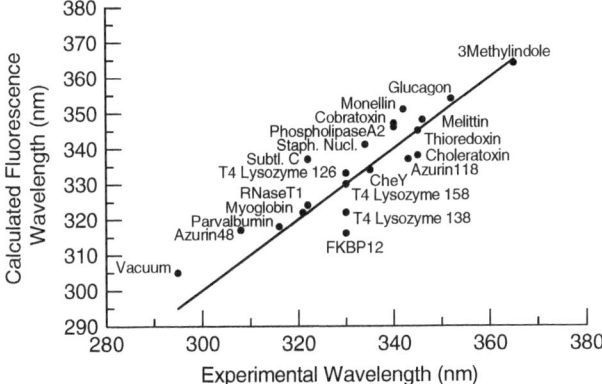

Fig. 2 Plots of QM–MM calculated versus experimental fluorescence maximum wavelengths for 19 Trps in 16 proteins and for 3-methylindole in water. Charges on the Trp ring are multiplied by 0.80 and the calculated values are averages over the 2,400 values calculated during the last 24 ps of 30-ps QM–MM trajectories

an applied (external) field. Pierce and Boxer have verified the magnitude and direction of the dipole moment change for Trp in solution by external Stark experiments [36]. The ISE had emerged as a useful concept to understand spectral shifts for a wide range of chromophores embedded in a "host" medium, including polyenes [37, 38], porphryrins [39] and other probes [40–42], but had seen little attention for the explanation of Trp fluorescence. This was at least, in part, because of uncertainties surrounding the nature of the fluorescing state, which we feel is no longer an issue. A primary contention in the extension of the ISE hypothesis to proteins is that the source of the field is irrelevant; only the magnitude and sign of its projection on the long axis of the indole ring matters in determining the spectral shift.

This study has solidified the notion that λ_{max} is determined almost entirely by electrostatics. For indole, it is the electric potential difference across the long axis of the indole ring, because that is the direction of electron density shift upon excitation, a general concept that must apply to any fluorescent probe molecule. This means that the relative direction of a charge from the chromophore is crucial: a red shift (shift to longer wavelength) results if density is shifted toward a positive charge or away from a negative charge.

In general, both water and protein contribute to the shifts in widely varying ratios for different Trps, and the ratio is, in most cases, difficult to anticipate [6]. When charged groups are nearly touching Trp, they usually dominate the mechanism of the shift, and waters may even create a blue shift in such environments. Water exposure, per se, is not sufficient for a large red shift. If the exposure is only along one edge of the Trp, only modest red shifts from water are possible. However, if one or both faces of the benzene ring are water-exposed, the wavelength of the fluorescence peak is usually near 350 nm. In such cases, a few waters make particularly large red shifts ($\sim$20 nm) due to exciplex-like H-bonds with negatively charged C atoms of the benzene ring in the ^1L_a excited state.

Looking more closely, for those Trps that are buried with no exposure to water, only Trp48 of azurin is in a truly hydrocarbon environment. Other completely buried Trps are in ribonuclease T1, myoglobin, che-y, T4 lysozyme (Trp138), and parvalbumin. These are red shifted relative to azurin typically because of nearby negative charges from protein near the pyrrole ring. Those with one edge exposed to water are subtilisin C., T4 lysozyme (Trps126 and 158), and Staph nuclease, which fall in the intermediate range. FKBP12 is unique in lying in a deep hydrophobic pocket and having one face accessible to water. In our simulations, the water was unable to spend much time red shifting the spectrum. Those with one face and an edge exposed are phospholipase A2, cobra toxin, cholera toxin, monellin, azurin Trp118. These are all well red shifted due to interaction with water. For the small peptides glucagon and melittin, the Trp has both faces exposed to water, which is typical for the most red shifted Trps in proteins. For the bare chromophore, 3-methylindole, there is no shielding from water by a side chain and backbone, and the red shift is even greater than ever found in proteins.

A surprising finding is that for 16 of the 19 Trps, protein contributes a red shift to the steady state Stokes shift, a result that is statistically very improbable. The extreme bias toward red shifts for the protein contributions suggests that protein electric fields relative to the modest ground-state dipole of the Trp residue may be important in the evolution of the protein folds.

3.2   Orientational Dielectric Compensation

In typical QM–MM simulations, no dielectric constant is included. Orientational dielectric effects come naturally from reorienting and translation of the elements of the system, providing the system comes to equilibrium. What is left out of the model is electronic polarization of molecules, which makes a minor contribution.

One striking example of how such microscopic dielectric effects affect results is the finding that water can cause significant (10–20 nm) red shifts for Trps that are essentially buried [6]. This appears to be due to the collective action of regions of water up to 25 nm distant that are probably oriented by the charges and/or shape of the protein. This was particularly evident in the case of S. nuclease, for which Trp140 is sandwiched between two positively charged lysines. Protein was found to make a $\sim$ 90 nm red shift, primarily due to these two close lysines, K133 and K110, but this was largely reversed by a large blue shift contribution from water, leading to a reasonable prediction of the steady state λ_{max} [6]. This intriguing result inspired a revealing study by Qiu et al. [43], in which the ultrafast time-resolved Stokes shift (TRSS) was determined for the WT and four mutants in which one of the charged side chains was replaced by a neutral one. They found that, remarkably, all five proteins gave a steady state λ_{max} of $\sim$332.5 $\pm$ 0.5 nm. Subsequent QM–MM simulations for these mutants showed that deleting a nearby charged residue has two large, compensating effects (dielectric compensation): there is a large direct change in Coulombic stabilization/destabilization of the electron density shift in the indole ring, and there is the collective effect due to change in orientational polarization of dozens of water molecules in response to the deleted charge. [unpublished work by the author]

3.3   Indole Ring Electronic Polarization

In addition to the dielectric compensation just described, protein and solvent are also coupled through the considerable polarization of the ^1L_a excited state caused by the large relaxation of the solvent/protein reaction field. This is not included in recent simulations that aim to capture properties requiring long simulations [30, 31]. In our QM–MM simulations, the computed dipole changes from $\sim$5 D to $\sim$8 D while in the excited state, during the time interval of $\sim$1 ps after excitation waters orient in response to the large excited dipole of the excited state. The formation of this Onsager reaction field therefore creates a shift due to the field from protein charges as well as from water – even in the complete absence of protein motion. This is essentially a quadratic ISE.

The method described here can be applied to gain insight about environmental effects on the absorption and fluorescence of other chromophores and environments, including dyes used to probe the membrane potentials.

3.4   Quantitative Quantum Yield/Lifetime Predictions from Electron Transfer Quenching

The QM–MM method was extended and applied [17, 18] to the long-standing question as to the origin of the quantum yield variation in different proteins – a puzzle because the quantum yield is barely affected by solvent when the indole ring is not in a protein. That work provided the first quantitative evidence that the full 30-fold range of Trp fluorescence quantum yields (and lifetimes) observed in proteins is due primarily to different rates of electron transfer (ET) from the excited indole ring to the empty antibonding MO of one of two nearest backbone amides. The dependence on protein environment arises mainly from the average local electric potential difference between the Trp ring and acceptor amide and from the amplitude of potential difference fluctuation caused by protein and solvent motions. This can be strongly influenced by the *location of nearby charges* relative to the transfer direction, including a very pronounced effect from the hydrogen bonding of a single water to the carbonyl group of the electron accepting amide. In other words, a negative charge near the chromophore ring and/or positive charge near the electron acceptor will increase the electron transfer rate (decrease the Φ_f), because these arrangements stabilize the charge transfer (CT) state. In the opposite case, there will be minimal electron transfer, and the Trp will be highly fluorescent.

The method has been applied to a number of interesting proteins [10, 11, 13, 14, 16].

3.5   Fluctuations, Relaxation, Heterogeneity, and Nonexponential Fluorescence Decay

The four quantities in the subtitle are intimately connected. Figure 3 shows graphically one of the most important benefits from MD results: the surprising amplitude and frequency of fluctuations. This figure shows the energies of S_1 (red) and of the lowest charge transfer (CT) state due to electron transfer from the indole ring to the nearby amide (black), both relative to the ground state for Trp68, a buried – and particularly weakly fluorescing – Trp in the human eye lens protein, γD-crystallin [10, 11, 13].

Emission. The red curve shows that the absorption and fluorescence rapidly fluctuate in time over a range of $\sim$2,000 cm^{-1}, even for this buried Trp, whose λ_{max} is at 330 nm. Such spectral broadening is even greater for dyes in moderately polar fluid solvents. The magnitude of these fluctuations is directly related to the Stokes shift magnitude through statistical mechanical principles [44]. In more rigid (glass-like) media, especially at lower temperature, the fluctuations are mostly frozen out on the timescale of excited states, but the inhomogeneity largely persists on a spatial scale. The well-studied *red edge effect* [45–47] (see also the chapter of Tomin in this volume [84]) arises from the ability to selectively excite a sub

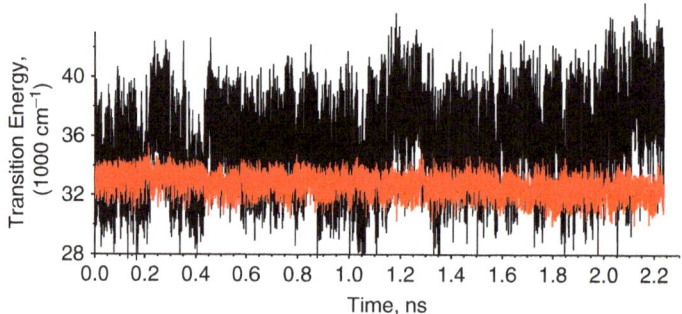

Fig. 3 Transition energy for $S_0 \rightarrow S_1$ (*red*) and $S_0 \rightarrow$ CT(black) for a tryptophan during a 2 ns QM–MM trajectory of the human eye lens protein γD-crystallin showing typical fluctuations due to rapid changes in local electrostatic potentials at the atoms of the chromophore. This Trp has a low quantum yield because the CT state is near the S_1 state much of the time. Heterogeneity in lifetime and wavelength are evident in both states because regions of 100 ps are seen having distinctly different average energies

population of solute molecules corresponding to the low energy wing of such a distribution. In the absence of major redistribution of solvent in the excited state prior to emission, one finds increasing emission λ_{max} as λ_{exc} is increased.

Quenching. Fig. 3 also shows the tremendous fluctuations associated with the CT state energy, again modulated by relative changes in potential difference between the indole ring and the amide on the order of 1 V. In accordance with Marcus theory, electron transfer can occur when the vibronic states of S_1 and CT are in resonance. The weak fluorescence ($\Phi_f = \sim 0.01$) of Trp68 in this protein stems from the low energy of the CT state, and especially the large down-going fluctuations. For both the S_1 and CT states, regions of ~ 100 ps can be seen to have distinctly different average values of energy, pointing to heterogeneity in both λ_{max}, τ_f, and Φ_f. The reason for most of this heterogeneity is largely due to a narrow channel connecting Trp68 to bulk water that allows water to be trapped in a position to donate an H-bond to the carbonyl oxygen of the amide for times as long as 100 ps, leading to a low CT state energy. For periods of time when this H-bond is broken, the CT energy is higher, and the τ_f is longer. Similarly, λ_{max} is made longer when a water resides with its oxygen accepting an H-bond from the Trp ring nitrogen [10, 11].

Nonexponential decay. Tryptophan fluorescence decay curves typically exhibit nonexponential behavior, and are often satisfactorily fit to a sum of two or three exponential terms. This is likely an unavoidable consequence of the combined sensitivity of λ_{max} and Φ_f to the electrostatic environment in proteins. There is no consensus regarding the primary cause(s) of the nonexponential decay. Opinions – and therefore interpretations – remain divided between the view that discrete subpopulations exhibit different *population* decay times, and at the other extreme, the view that the excited population is homogeneous, but has a time-dependent fluorescence spectrum that shifts to longer wavelengths. The development and increased use of ultrafast methods now consistently reveals the fast solvent shifts from solvent relaxation that were always known to exist for proteins at ambient temperature, but which were not accessible to the widely used time-correlated single photon counting (TCSPC) method [10, 14, 48–51]. In these studies, divergence of opinion has expanded to include the intriguing arena of 10–100 ps, wherein TRSSs are more readily embraced than in the ns region [32, 33]. At the same time, a better understanding of what controls electron transfer-based quenching (including experimental observation) has raised expectations of the existence of fast-decaying blue shifted subpopulations, which also create red shifting of the fluorescence spectrum in time. Two recent observations have greatly strengthened the evidence for such subpopulations: (1) The discovery that 5-fluorotryptophan (5FTrp), when incorporated into proteins, practically always shows single exponential decay of excited state population [52]. The finding that 5FTrp has a 0.2 eV higher ionization potential (making it a much poorer electron donor) strongly implicates heterogeneity in the electrostatic environment as a cause for nonexponential decay of Trp fluorescence [15]; (2) The nonnatural amino acid, Aladan, has a large change in dipole upon excitation, and like Trp, it can be used as a probe for water exposure in proteins. The free probe in solution exhibits nonexponential fluorescence decay, and an average fluorescence lifetime that *decreases* with increasing solvent polarity. In a recent study of the protein GB1 [53], the time-resolved fluorescence spectrum of Aladan, when in all but the most buried protein sites, exhibited the remarkable behavior of *blue shifting* at times longer than 0.5 ns. This is almost certainly proof of heterogeneity in environment, with the bluer fluorescence of those molecules in less solvent-exposed environments persisting longer than those in the more polar environments. It is just the reverse of what is seen for Trp, for which it is apparently the blue shifted species that usually disappear first, giving the appearance of a red-shifting solvent relaxation [8], thereby creating an ambiguous interpretation. The certain knowledge of this type of heterogeneity for Aladan gives credibility to its existence for Trp, where it is much harder to detect.

4 Voltage-Sensing Dyes

4.1 Introduction and Background

Early observations of voltage sensitivity of the optical absorbance and emission of dyes attached to neurons revealed remarkable linearity: the shape of transient changes in the optical signals almost exactly reproduced those of the voltage signals with relatively little noise [54–56] (see also the chapter of Clarke in this volume [85]). Initially, the general mechanism by which these dyes sensed the small voltages changes appeared variable, depending on dye and membrane type. Loew et al. emphasized desirability of the nonspecific electrochromism interpretation based on the Stark effect [56, 57] and have been active in synthesizing and studying many dyes that appear to have pure electrochromic responses to small membrane voltage transients [58, 59]. Other sensitive dyes, however, gave wavelength-dependent voltage sensitivity that was at odds with the pure electrochromic effect. In particular, high sensitivity (2–10%) changes in signal were observed when exciting at the absorption maximum, where the pure Stark effect should show vanishing response [60]. RH-421 is an example of class of dyes broadly known as styryl dyes designed to fit unobtrusively into lipid bilayers while exhibiting as large as possible electrochromic behavior. Figure 4 shows an example of the basic design with the shortest conjugation length. Surveys of the fluorescence properties of these dyes reveal that the fluorescence quantum yield is extremely sensitive to environment, changing over two orders of magnitude depending on polarity and viscosity of solvent. Recent MD simulation studies [61, 62] suggest that indeed such dyes randomly sample a wider range of environment than might have been imagined, meaning that there is much possibility to observe heterogeneity effects, which may be dynamically coupled to voltage changes [63, 64].

Mechanisms of voltage sensitivity in the styryl-type dyes depend on their unusual trends in solvent and lipid-induced shifts of absorption and fluorescence

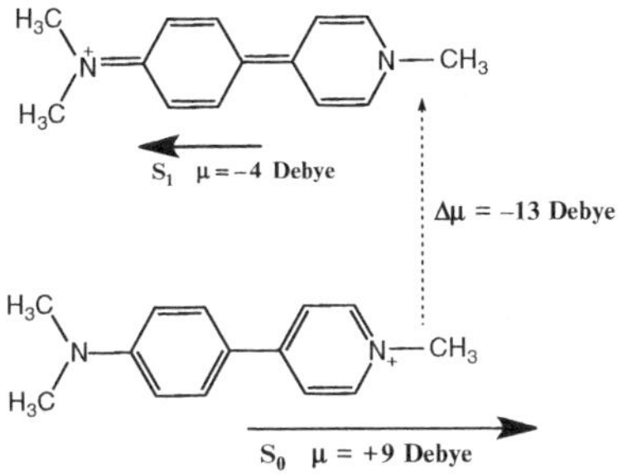

Fig. 4 The shortest member of the family of styryl-type voltage sensing dyes. This figure shows the dipole moments computed by INDO/S-CIS [unpublished work by the author] for the ground state and for the excited state following vertical absorption of a photon while remaining planar. The sign of the dipole is seen to reverse

S_1 $\mu = -4$ **Debye**

$\Delta\mu = -13$ **Debye**

S_0 $\mu = +9$ **Debye**

maxima, and in quantum yield variation [57, 58]. In particular, water causes a large blue shift for absorption and a large red shift for fluorescence relative to nonpolar solvents. Perhaps surprisingly, lipid vesicles typically create a 10–20 nm *larger blue shift* of absorption than water, even though lipids often are assumed to provide a nonpolar environment. Fluorescence in the lipid, however, *does* behave as if the chromophore is in nonpolar solvent. These observations are understood by considering the unusual solvation environment for the dye, which mimics a lipid molecule, having a polar head including the positively charged pyridinium, which is part of the chromophore. While the positively charged pyridinium lies at the membrane water interface and is stabilized by water and negative counter ions, the other end of the chromophore lies deep within the membrane in a largely nonpolar environment. Upon excitation, there is a large electron density shift from the aniline end (deep in the membrane) to the pyridinium; the dipole moment actually changes sign in a short time compared to the time for the solvent to reorganize (Franck–Condon concept). (See Fig. 4) the new dipole is therefore *destabilized* by the negative potential near the pyridinium, thus creating the blue shift as in water [58]. It is unclear, however, why the blue shift of absorption is greater than in water. That the fluorescence is similar to that in nonpolar solvent is qualitatively understood by the relaxation of water after excitation near the pyridinium to what is appropriate for a nonpolar solute, and of course, there is no polar solvent to stabilize the increase in positive charge near the aniline in the membrane. Finally, the fluorescence quantum yield increases by ~100-fold upon binding to lipid, taking on values typical of nonpolar solvents. However, it is not clear to what extent an increase in effective viscosity plays a role in this increase of quantum yield.

4.2   Mechanisms of Voltage Sensitivity

A particularly systematic and fundamental series of studies by Fromherz and associates may be traced during that last two decades, building upon the early work from the laboratories of Grinvald, Waggoner, Loew, and coworkers [54–57, 65].

Owing to the small changes induced by typical membrane potentials, the voltage sensitivity may be accurately expressed as a sum of independent terms. A fractional change in signal (absorbance or fluorescence) per 100 mV of applied membrane potential change was given by Fromherz et al. [66] as:

$$\frac{\delta I}{I} = \frac{\Delta\varphi_F}{\varphi_F} + \frac{f'(\bar{\nu}_F)}{\bar{\nu}_F}\Delta M_F + \frac{a'(\bar{\nu}_A)}{(\bar{\nu}_A)}\Delta M_A$$

wherein $\Delta\varphi_F$, ΔM_F, ΔM_A are the changes in quantum yield, shift of fluorescence maximum, and shift in absorption maximum, respectively. Each term is normalized by its corresponding value, where f and a are the magnitudes of the fluorescence and absorbance spectra; f' and a' are their slopes at the wavenumber of observation. The term with a' corresponds to the pure electrochromic response, and carries the signature of maximum response in the steep wings of absorption bands and will vanish at the peak maxima; in the absence of quantum yield variation and solvato-chromic shift following excitation, the f' term would also be a pure electrochromic response. Using this equation, the voltage response as a function of excitation and emission wavelengths was fit by finding appropriate and reasonable parameters [66]. The $\Delta\varphi_F$ term was suspected to be important because measurements of φ_F as a function of solvent and in lipid bilayers had shown a spectacular dependence on dielectric constant and viscosity [58, 59, 66–69], with φ_F decreasing by two orders of magnitude or more as solvent is varied from chloroform to water and as viscosity was decreased by a factor of 10. Fluorescence lifetime was found to follow the quantum yield linearly. The extreme quenching of fluorescence in polar solvents is strongly linked to an intramolecular CT state in which an electron is transferred from the amino group to the pyridinium group involving twisting about a bond involved in conjugation [66, 67, 69]. Therefore, there is definite reason to expect even small voltage-induced changes in local environment could change the quantum yield by a few percent, either by slight changes in "polarity" or "viscosity".

A related class of dyes showing similar dielectric and viscosity behavior are the dicyano–dihydrofurans (DCDHFs), which contain an amine donor and a DCDHF acceptor linked by a conjugated unit (benzene, thiophene, alkene, styrene, etc.). These dyes are sufficiently bright and stable that they may be used in single molecule studies. Willets et al. [70] have presented experimental and theoretical studies that explore the mechanism leading to this environmental sensitivity, both through experimental and theoretical work. This work provided insight into how certain twists within the DCDHF molecule affect radiative and nonradiative processes, and should help in the design of new reporter molecules for biological and materials science applications.

Purely electrochromic dyes. Transient membrane potentials have typical amplitudes on the order of 100 mV across a typical 4 nm thick bilayer. This means an average electric field within the membrane of 2.5×10^7 Vm^{-1} or 2.5×10^5 V cm^{-1}. While this may seem like a very large field, in fact it produces a small wavelength shift of only about 0.4 nm (20 cm^{-1}) in λ_{max} for typical dyes absorbing maximally near 500 nm (see Appendix). Despite the rather small expected shift in λ_{max}, fractional changes in signal of 2–10% are readily obtained from many dyes because of a large fractional change in extinction coefficient per nm of shift on the steep sides of absorption bands, typically 5–10% per nm shift. In comparison, the solvatochromic shift for the same dye induced by water (dielectric constant, $\varepsilon = 80$) is $\sim 3,000$ cm^{-1}, relative to a nonpolar solvent. This corresponds to an average Onsager reaction electric field produced by the oriented solvent dipoles of $\sim 4 \times 10^7$ V cm^{-1}.

For some purposes, it is desirable to rid the voltage response of quantum yield effects. Through an evolution of synthesizing dyes with ever fewer bonds about which fluorescence-dissipating twists might occur, Fromherz and coworkers produced a series of ANellated hemicyaNINE (ANNINE) dyes [71, 72]. Without CC single and double bonds – and therefore no photorotamerism or photoisomerism – this series of dyes uniformly reaches the ideal of pure electrochromic response. In addition, they have greater voltage sensitivity than for homologous styryl-type hemicyanines, and may be used as simple probes of polarity and local electric fields in colloids and polymers. Strong nonlinear optical effects are also expected.

In an even more spectacular development, Fromherz et al. [73] have exploited the strong, pure Stark-shift voltage sensitivity of ANNINE-6 for excitation at the extreme red edges of the excitation spectrum where the absorptivity is as much as 100 times smaller than at the peak; they have achieved unprecedented large fractional fluorescence changes with membrane voltage on cultured HEK293 cells for both one- and two-photon excitation. For 100 mV hyper- and depolarization voltage swings, the fluorescence changes reached 50 and -28% respectively for one-photon excitation at 514 nm, and 70 and -40% at 1,040 nm for two-photon. Such fractional sensitivities are ~ 5 times larger than what is commonly found with other voltage-sensing dyes, approaching the theoretical limit given by the spectral Boltzmann tail [73].

4.3   Heterogeneity and Lifetime Studies

In a medium such as a lipid bilayer membrane, the prospect for finding a probe molecule in a variety of environments in which motion may be restricted during the fluorescence lifetime is large. Indeed, there is considerable evidence from red edge excitation studies that this is the case [47, 64, 74–77]. These studies indicate that water or other polar entities penetrate to some degree well into the nonpolar tail portion of membranes.

Another powerful tool for examining this issue is the use of time-resolved fluorescence spectra, especially when combined with the technique of Time-Resolved Area Normalized Emission Spectra (TRANES) developed by Periasamy and coworkers [78–80]. In this method, separate decay curves are collected over a wide range of emission wavelengths and reconstructed into time-resolved spectra, which are then normalized to constant area. In this model-free approach, it is possible to deduce the nature of heterogeneity of the fluorescent species from the resulting pattern. In a study of several commonly used fluorescent membrane probes, the venerable voltage-sensitive styryl-type dye, RH-421, was found to exhibit TRANES that shifted dramatically with time in eggPC vesicles, but with no isoemissive point. The shift in maximum was from 600 to 660 nm during a 10 ns time window [80]. That such shifting could be interpreted as slow solvent relaxation was ruled out by the absence of negative amplitude decay associated spectra caused by the growing in of intensity on the red tail of the emission. This conclusion was further reinforced by the observation of a red edge effect for the same system [64, 80]. For the spectrum to shift to longer wavelengths in time requires that the short wavelength-emitting species also have shorter lifetimes, such that they decay away to leave only the long wavelength-emitting species. This seems paradoxical relative to the bulk behavior, in which the absorption and fluorescence both blue shift and the quantum yield greatly increases (lifetime becomes much longer) upon binding to vesicles – suggesting that the short wavelength component would have the long lifetime. If that were the case, the wavelength would shift to the *blue* with increasing time, as in the case of Aladan discussed above [53].

Experiment [80], however, requires that the red-shifted fluorescence comes from molecules that are in an environment quite different from the average, behaving as if in a nonpolar environment (long wavelength, long lifetime). The red-shifted component, therefore, appears to be from chromophores that are embedded more deeply into the membrane. In the case of Aladan, the absorption and fluorescence wavelengths are both longer in the more water-exposed protein sites, but the lifetime is shorter in water [53].

Apparently, a variety of probes sensitive to different environmental aspects are required to obtain an experimental picture of membrane structure and dynamics.

Hydroxychromone dyes. A relatively new class of dyes that are promising for probing membrane potential are the hydroxychromone dyes. Their main mode of response is alteration of a tautomeric proton transfer equilibrium, which provides convenient ratiometric recording measurements of membrane potential, independent of those based on styryl dyes [81]. Use of these probes for investigating membranes has recently been reviewed [82]. In addition, the ratio of the tautomer bands is also very sensitive to local environment, being sensitive to lipid composition, and showing red edge excitation shifts in lipid bilayers [47]. Heterogeneity of location was reported for these dyes. It was related to partial formation of H-bonds with water molecules. This type of heterogeneity can be eliminated by proper probe design [83].

4.4 Prospect for Molecular Dynamics and QM–MM Studies

The above discussion sets the stage for the type of QM–MM studies we have performed for tryptophan in proteins. The use of MD simulations to study membranes is now a mature field, and has recently been reviewed in the context of the present book by Demchenko and Yesylevskyy [82]. A number of questions might be answered regarding details of the mechanism of voltage-sensitive dyes by application of such QM–MM procedures. Two MD studies have appeared recently that involve styryl-type voltage-sensitive dyes in membranes: one in which much attention is paid to the location and dynamics of the dye [61], and another in which QM–MM using semiempirical quantum mechanics is used to predict and understand second harmonic generation (SHG) generated at the lipid–water interface by the dye [62].

In the former, Hinner et al. [61] made 50-ns simulations on a POPC lipid bilayer in water with a voltage-sensitive dye di-4-ASPBS (dibutyl-amino-styryl-pyridinium-butyl-sulfonate) and several of its variations with a different alkyl chain length or tethered counter ion. In addition to examining the dynamics of the dye, they also studied how the free energy of binding was affected by the structural variations of the different dyes. The chromophore and the hydrophobic tail of di-4-ASPBS were found to be inserted into the hydrophobic interior of the membrane with the charged pyridinium ring of the chromophore residing within the charged headgroup region of the membrane, and the tethered sulfonate headgroup of the amphiphile is located in the membrane/water interface. These findings are in accord with the interpretation of experimental data on similar styryl-type probes.

Interesting findings from this study are: (1) the chromophore tilt that has been determined in optical dichroism experiments is the average over a much broader distribution of angles than had been previously imagined; (2) the average tilt of the chromophore is decreased with increasing lipophilic chain length, which should lead to an increased voltage sensitivity and could explain trends that have been observed experimentally; (3) attaching polar groups to the long lipophilic chains of an amphiphilic probe affects the chromophore orientation only weakly, because the hydrophobic chains still anchor the chromophore as far as possible in the membrane interior, even though the polar groups reside at the membrane/water interface; and (4) the linear increase in binding free energy with increasing lipophilic chain length that has been observed experimentally for chain lengths of up to six methylene groups continues at least up to dodecyl chains according to the simulations; this is of interest because very large binding constants, as would be exhibited by very long-tailed amphiphiles, are difficult to access by experiment.

The authors will provide an all-atom GROMOS topology file, which will greatly facilitate laboratories interested in pursuing QM–MM simulations of voltage-dependent optical responses.

In a slightly earlier study by Rusu et al. [62], the amphiphilic potential-sensitive styryl dye of the aminonaphthyl-ethenylpyridinium class, di-8-ANEPPS (a molecule with nonlinear optical properties), was embedded in a dipalmitoylphosphatidylcholine (DPPC) bilayer, replacing one of the phospholipid molecules. External electrical fields with different strengths were applied across the membrane to simulate the action potential of a heart-muscle cell. QM–MM was applied during 10 ns simulations to compute wavelengths and oscillator strengths of electronic transitions for the embedded chromophore. The results were then used in a sum-over-states treatment to calculate the second-order hyperpolarizability to obtain the intensity of the second harmonic as a function of applied membrane potential changes. Their results agreed well with experimental measurements. Extensions of this study would seem possible, in which analysis could address open questions with regard to details of heterogeneity, possible red edge effects, effective dielectric constants, and rigidity on the optical properties for almost any probe molecule.

5 Conclusions

Much is known now about voltage-sensing dyes. It appears that from the point of view of pure electrochromic response, further improvements will be more incremental. Much of what is not known has to do with structural changes within the membrane caused by the passage of nerve impulses. Much more information can be obtained on this subject by dyes that are sensitive to parameters such as order, rigidity, and polarity of the local microenvironment. From progress that has been made by studying tryptophan in proteins, and recent molecular dynamics simulation involving dyes in membranes, the prospect is bright for obtaining QM–MM predictions of a variety of optical properties for dyes in membranes. This would include getting good leads for molecules that have not yet been synthesized.

Appendix: Notes on Electrostatics

Potential has the units of volts, i.e., joules/coulomb. An electron has lower energy in a more positive potential (closer to positive charge or farther from negative charge). Using Coulombs law, at a point 1 Å distant from a proton the potential in SI units is:

$$\frac{9 \times 10^9 \times 1.6 \times 10^{-19}}{1 \times 10^{-10}} = 14.4 \, \text{volts}$$

Therefore the energy of an electron 1 Å from a proton is -14.4 eV or -331 kcal mol^{-1} or $-1{,}385$ kJ mol^{-1}. The electric fields in proteins are on the order of a few times 10^7 V cm^{-1}, i.e., a few tenths of V Å^{-1} [26, 34].

The energy of a dipole in a constant electric field in SI units is given by:

$$E\Delta\mu\cos\theta$$

where E is the field in V m^{-1}, $\Delta\mu$ is the change in dipole in Coulomb m, and θ is the angle between the dipole vector and the field vector.

A typical dipole change upon exciting a sensing dye is at least 10 Debye, corresponding to a shift of charge e by about 2 Å. For a field of 10^7 V cm^{-1}, this gives an energy change of:

$$\Delta U = 10^9 \text{Vm}^{-1} \times 1.6 \times 10^{-19}\text{C} \times 2 \times 10^{-10}\text{m}$$
$$= 32 \times 10^{-20}\text{J} = 0.2\text{eV} = 1{,}600 \, \text{cm}^{-1}$$

A field of 2.5×10^7 V cm^{-1} therefore gives a commonly found shift of 4,000 cm^{-1}. This corresponds to about a 40 nm shift at 310 nm and about a 100 nm shift at 500 nm. For a 100 mV change over a 4 nm thick membrane, the field is 2.5×10^5 V cm^{-1}, which translates to about a 1 nm shift at 500 nm.

References

1. Callis PR (2009) In: Geddes CD (ed) Reviews in fluorescence 2007. Springer, NY, pp 199–248
2. Sreerama N, Woody RW, Callis PR (1994) Theoretical study of the crystal field effects on the transition dipole moments in methylated adenines. J Phys Chem 98:10397–10407
3. Theiste D, Callis PR, Woody RW (1991) Effects of the crystal field on transition moments in 9-ethylguanine. J Am Chem Soc 113:3260–3267
4. Muino PL, Callis PR (1994) Hybrid simulations of solvation effects on electronic-spectra – indoles in water. J Chem Phys 100(6):4093–4109
5. Callis PR, Burgess BK (1997) Tryptophan fluorescence shifts in proteins from hybrid simulations: an electrostatic approach. J Phys Chem 101:9429–9432
6. Vivian JT, Callis PR (2001) Mechanisms of tryptophan fluorescence shifts in proteins. Biophys J 80:2093–2109
7. Broos J, Tveen-Jensen K, de Waal E, Hesp BH, Jackson JB, Canters GW, Callis PR (2007) The emitting state of tryptophan in proteins with highly blue-shifted fluorescence. Angew Chem Int Ed Engl 46(27):5137–5139
8. Pan CP, Callis PR, Barkley MD (2006) Dependence of tryptophan emission wavelength on conformation in cyclic hexapeptides. J Phys Chem B 110(13):7009–7016
9. Muiño PL, Callis PR (2009) Solvent effects on the fluorescence quenching of tryptophan by amides via electron transfer. Experimental and computational studies. J Phys Chem B 113:2572–2577
10. Xu J, Chen J, Toptygin D, Tcherkasskaya O, Callis PR, King J, Brand L, Knutson J (2009) Femtosecond fluorescence spectra of tryptophan in human gamma-crystallin mutants: site-dependent ultrafast quenching. J Am Chem Soc 131(46):16751–16757
11. Chen J, Callis PR, King J (2009) Mechanism of the very efficient quenching of tryptophan fluorescence in human gamma D- and gamma S-crystallins: the gamma-crystallin fold may have evolved to protect tryptophan residues from ultraviolet photodamage. Biochemistry 48(17):3708–3716
12. Callis PR, Petrenko A, Muino PL, Tusell JR (2007) Ab initio prediction of tryptophan fluorescence quenching by protein electric field enabled electron transfer. J Phys Chem B 111(35):10335–10339
13. Chen JJ, Flaugh SL, Callis PR, King J (2006) Mechanism of the highly efficient quenching of tryptophan fluorescence in human gamma D-crystallin. Biochemistry 45(38):11552–11563
14. Xu JH, Toptygin D, Graver KJ, Albertini RA, Savtchenko RS, Meadow ND, Roseman S, Callis PR, Brand L, Knutson JR (2006) Ultrafast fluorescence dynamics of tryptophan in the proteins monellin and IIA(Glc). J Am Chem Soc 128(4):1214–1221
15. Liu T, Callis PR, Hesp BH, de Groot M, Buma WJ, Broos J (2005) Ionization potentials of fluoroindoles and the origin of nonexponential tryptophan fluorescence decay in proteins. J Am Chem Soc 127(11):4104–4113
16. Kurz LC, Fite B, Jean J, Park J, Erpelding T, Callis P (2005) Photophysics of tryptophan fluorescence: link with the catalytic strategy of the citrate synthase from Thermoplasma acidophilum. Biochemistry 44(5):1394–1413
17. Callis PR, Liu T (2004) Quantitative prediction of fluorescence quantum yields for tryptophan in proteins. J Phys Chem B 108:4248–4259

18. Callis PR, Vivian JT (2003) Understanding the variable fluorescence quantum yield of tryptophan in proteins using QM-MM simulations. Quenching by charge transfer to the peptide backbone. Chem Phys Lett 369:409–414
19. Callis PR, Liu T (2006) Short range photoinduced electron transfer in proteins: QM-MM simulations of tryptophan and flavin fluorescence quenching in proteins. Chem Phys 326 (1):230–239
20. Warshel A (1982) Dynamics of reactions in polar solvents. Semiclassical trajectory studies of electron transfer and proton-transfer reactions. J Phys Chem 86:2218–2224
21. Warshel A (ed) (1991) Computer modeling of chemical reactions in enzymes and solutions. Wiley-Interscience, New York, NY
22. Marchi M, Gehlen JN, Chandler D, Newton M (1993) Diabatic surfaces and the pathway for primary electron transfer in a photosynthetic reaction center. J Am Chem Soc 115:4178–4190
23. Gehlen JN, Marchi M, Chandler D (1994) Dynamics affecting the primary charge transfer in photosynthesis. Science 263:499–502
24. Ridley J, Zerner M (1973) Intermediate neglect of differential overlap (INDO) technique for spectroscopy: pyrrole and the azines. Theor Chim Acta(Berl) 32:111–134
25. Thompson MA, Zerner MC (1991) A theoretical examination of the electronic structure and spectroscopy of the photosynthetic reaction center from *Rhodopseudomonas viridis*. J Am Chem Soc 113:8210–8215
26. donder-Lardeux C, Jouvet C, Perun S, Sobolewski AL (2003) External electric field effect on the lowest excited states of indole: ab initio and molecular dynamics study. Phys Chem Chem Phys 5(22):5118–5126
27. MacKerell AD Jr, Bashford D, Bellott M, Dunbrack RL, Evanseck JD, Field MJ, Fischer S, Gao J, Ha S, Joseph-McCarthy D, Kuchnir L, Kuczera K, Lau FTK, Mattos C, Michnick S, Ngo T, Nguyen DT, Prodhom B, Reiher WE III, Roux B, Schlenkrich M, Smith JC, Stote R, Straub J, Watanabe M, Wiorkiewicz-Kuczera J, Yin D, Karplus M (1998) All atom empirical potential for molecular modeling and dynamics studies of proteins. J Phys Chem B 102:3586–3616
28. Maroncelli M (1993) The dynamics of solvation in polar liquids. J Mol Liq 57:1–37
29. Jimenez R, Fleming GR, Kumar PV, Maroncelli M (1994) Femtosecond solvation dynamics of water. Nature 369:471–473
30. Li TP, Hassanali AAP, Kao YT, Zhong DP, Singer SJ (2007) Hydration dynamics and time scales of coupled water-protein fluctuations. J Am Chem Soc 129(11):3376–3382
31. Li TP, Hassanali AA, Singer SJ (2008) Origin of slow relaxation following photoexcitation of W7 in myoglobin and the dynamics of its hydration layer. J Phys Chem B 112(50):16121–16134
32. Toptygin D, Savtchenko RS, Meadow ND, Brand L (2001) Homogeneous spectrally- and time-resolved fluorescence emission from single-tryptophan mutants of IIA^{Glc}. J Phys Chem B 105:2043–2055
33. Toptygin D, Gronenborn AM, Brand L (2006) Nanosecond relaxation dynamics of protein GB1 identified by the time-dependent red shift in the fluorescence of tryptophan and 5-fluorotryptophan. J Phys Chem B 110(51):26292–26302
34. Callis PR, Burgess BK (1997) Tryptophan fluorescence shifts in proteins from hybrid simulations: an electrostatic approach. J Phys Chem B 101(46):9429–9432
35. Callis PR (1997) ^1L_a and ^1L_b transitions of tryptophan: applications of theory and experimental observations to fluorescence of proteins. Meth Enzymol 278:113–150
36. Pierce DW, Boxer SG (1995) Stark effect spectroscopy of tryptophan. Biophys J 68: 1583–1591
37. Honig B, Dinur U, Nakanishi K, Balogh-Nair V, Gawinowicz MA, Arnaboldi M, Motto MG (1979) An external point-charge model for wavelength regulation in visual pigments. J Am Chem Soc 101:7084–7086
38. Kohler BE, Woehl JC (1995) Measuring internal fields with atomic resolution. J Chem Phys 102:7773–7781
39. Varadarajan R, Lambright DG, Boxer SG (1989) Electrostatic interactions in wild-type mutant recombinant human myoglobins. Biochemistry 28:3771–3781
40. Sitkoff D, Lockhart DJ, Sharp KA, Honig B (1994) Calculation of electrostatic effects at the amino terminus of an alpha helix. Biophys J 67:2251–2260

41. Lockhart DJ, Kim PS (1993) Electrostatic screening of charge and dipole interactions with the helix backbone. Science 260:198–202
42. Lockhart DJ, Kim PS (1992) Internal Stark effect measurement of the electric field at the amino terminus of an alpha helix. Science 257:947–951
43. Qiu WH, Kao YT, Zhang LY, Yang Y, Wang LJ, Stites WE, Zhong DP, Zewail AH (2006) Protein surface hydration mapped by site-specific mutations. Proc Natl Acad Sci USA 103 (38):13979–13984
44. Loring RA (1990) Statistical mechanical calculation of inhomogeneously broadened absorption line shapes in solution. J Phys Chem 94:513–515
45. Demchenko AP (2002) The red-edge effects: 30 years of exploration. Luminescence 17(1):19–42
46. Demchenko AP (2008) Site-selective red-edge effects. Fluoresc Spectrosc 450:59–78
47. Duportail G, Klymchenko A, Mely Y, Demchenko AP (2002) On the coupling between surface charge and hydration in biomembranes: experiments with 3-hydroxyflavone probes. J Fluoresc 12(2):181–185
48. Shen XH, Knutson JR (2001) Subpicosecond fluorescence spectra of tryptophan in water. J Phys Chem B 105(26):6260–6265
49. Pal SK, Peon J, Bagchi B, Zewail AH (2002) Biological water: femtosecond dynamics of macromolecular hydration. J Phys Chem B 106(48):12376–12395
50. Qiu W, Li T, Zhang L, Yang Y, Kao Y-T, Wang L, Zhong D (2008) Ultrafast quenching of tryptophan fluorescence in proteins: interresidue and intrahelical electron transfer. Chem Phys 350:154–164
51. Zang C, Stevens JA, Link JJ, Guo LJ, Wang LJ, Zhong DP (2009) Ultrafast proteinquake dynamics in cytochrome c. J Am Chem Soc 131(8):2846–2852
52. Broos J, Maddalena F, Hesp BH (2004) In vivo synthesized proteins with monoexponential fluorescence decay kinetics. J Am Chem Soc 126:22–23
53. Abbyad P, Shi XH, Childs W, McAnaney TB, Cohen BE, Boxer SG (2007) Measurement of solvation responses at multiple sites in a globular protein. J Phys Chem B 111(28): 8269–8276
54. Waggoner AS, Grinvald A (1977) Mechanisms of rapid optical changes of potential sensitive dyes. Ann N Y Acad Sci 303:217–241
55. Grinvald A, Hildesheim R, Farber IC, Anglister L (1982) Improved fluorescent-probes for the measurement of rapid changes in membrane-potential. Biophys J 39(3):301–308
56. Loew LM, Bonneville GW, Surow J (1978) Charge shift optical probes of membrane-potential – theory. Biochemistry 17(19):4065–4071
57. Loew LM, Scully S, Simpson L, Waggoner AS (1979) Evidence for a charge-shift electrochromic mechanism in a probe of membrane-potential. Nature 281(5731):497–499
58. Fluhler E, Burnham VG, Loew LM (1985) Spectra, membrane-binding, and potentiometric responses of new charge shift probes. Biochemistry 24(21):5749–5755
59. Wuskell JP, Boudreau D, Wei MD, Jin L, Engl R, Chebolu R, Bullen A, Hoffacker KD, Kerimo J, Cohen LB, Zochowski MR, Loew LM (2006) Synthesis, spectra, delivery and potentiometric responses of new styryl dyes with extended spectral ranges. J Neurosci Meth 151(2):200–215
60. Grinvald A, Fine A, Farber IC, Hildesheim R (1983) Fluorescence monitoring of electrical responses from small neurons and their processes. Biophys J 42(2):195–198
61. Hinner MJ, Marrink SJ, de Vries AH (2009) Location, tilt, and binding: a molecular dynamics study of voltage-sensitive dyes in biomembranes. J Phys Chem B 113(48):15807–15819
62. Rusu CF, Lanig H, Othersen OG, Kryschi C, Clark T (2008) Monitoring biological membrane-potential changes: a CI QM/MM study. J Phys Chem B 112(8):2445–2455
63. Clarke RJ (1997) Effect of lipid structure on the dipole potential of phosphatidylcholine bilayers. Biochim Biophys Acta-Biomembr 1327(2):269–278
64. Vitha MF, Clarke RJ (2007) Comparison of excitation and emission ratiometric fluorescence methods for quantifying the membrane dipole potential. Biochim Biophys Acta-Biomembr 1768(1):107–114

65. Waggoner AS (1979) Dye indicators of membrane-potential. Annu Rev Biophys Bioeng 8:47–68
66. Fromherz P, Dambacher KH, Ephardt H, Lambacher A, Muller CO, Neigl R, Schaden H, Schenk O, Vetter T (1991) Fluorescent dyes as probes of voltage transients in neuron membranes – progress report. Ber Bunsen-Ges Phys Chem 95(11):1333–1345
67. Fromherz P, Heilemann A (1992) Twisted internal charge-transfer in (aminophenyl)pyridinium. J Phys Chem 96(17):6864–6866
68. Ephardt H, Fromherz P (1993) Fluorescence of amphiphilic hemicyanine dyes without free double-bonds. J Phys Chem 97(17):4540–4547
69. Ephardt H, Fromherz P (1989) Fluorescence and photoisomerization of an amphiphilic aminostilbazolium dye as controlled by the sensitivity of radiationless deactivation to polarity and viscosity. J Phys Chem 93(22):7717–7725
70. Willets KA, Callis PR, Moerner WE (2004) Experimental and theoretical investigations of environmentally sensitive single-molecule fluorophores. J Phys Chem B 108(29): 10465–10473
71. Hubener G, Lambacher A, Fromherz P (2003) Anellated hemicyanine dyes with large symmetrical solvatochromism of absorption and fluorescence. J Phys Chem B 107(31): 7896–7902
72. Kuhn B, Fromherz P (2003) Anellated hemicyanine dyes in a neuron membrane: Molecular Stark effect and optical voltage recording. J Phys Chem B 107(31):7903–7913
73. Kuhn B, Fromherz P, Denk W (2004) High sensitivity of stark-shift voltage-sensing dyes by one- or two-photon excitation near the red spectral edge. Biophys J 87(1):631–639
74. Chattopadhyay A, Mukherjee S (1999) Red edge excitation shift of a deeply embedded membrane probe: implications in water penetration in the bilayer. J Phys Chem B 103(38): 8180–8185
75. Shrivastava S, Haldar S, Gimpl G, Chattopadhyay A (2009) Orientation and dynamics of a novel fluorescent cholesterol analogue in membranes of varying phase. J Phys Chem B 113(13):4475–4481
76. Sen P, Satoh T, Bhattacharyya K, Tominaga K (2005) Excitation wavelength dependence of solvation dynamics of coumarin 480 in a lipid vesicle. Chem Phys Lett 411(4–6):339–344
77. Adhikari A, Dey S, Das DK, Mandal U, Ghosh S, Bhattacharyya K (2008) Solvation dynamics in ionic liquid swollen P123 triblock copolymer micelle: a femtosecond excitation wavelength dependence study. J Phys Chem B 112(20):6350–6357
78. Koti ASR, Periasamy N (2001) Application of time resolved area normalized emission spectroscopy to multicomponent systems. J Chem Phys 115(15):7094–7099
79. Koti ASR, Krishna MMG, Periasamy N (2001) Time-resolved area-normalized emission spectroscopy (TRANES): a novel method for confirming emission from two excited states. J Phys Chem A 105(10):1767–1771
80. Ira ASR, Koti GK, Periasamy N (2003) TRANES spectra of fluorescence probes in lipid bilayer membranes: an assessment of population heterogeneity and dynamics. J Fluoresc 13(1):95–103
81. Clarke RJ, Kane DJ (1997) Optical detection of membrane dipole potential: avoidance of fluidity and dye-induced effects. Biochim Biophys Acta Biomembr 1323(2):223–239
82. Demchenko AP, Yesylevskyy SO (2009) Nanoscopic description of biomembrane electrostatics: results of molecular dynamics simulations and fluorescence probing. Chem Phys Lipids 160(2):63–84
83. Demchenko AP, Mely Y, Duportail G, Klymchenko AS (2009) Monitoring biophysical properties of lipid membranes by environment-sensitive fluorescent probes. Biophys J 96:3461–3470
84. Tomin VI (2010) Physical principles behind spectroscopic response of organic fluorophores to intermolecular interactions. In: Demchenko AP (ed) Advanced Fluorescence Reporters in Chemistry and Biology I. Springer Ser Fluoresc 8:189–223
85. Clarke R (2010) Electric field sensitive dyes. In: Demchenko AP (ed) Advanced Fluorescence Reporters in Chemistry and Biology I. Springer Ser Fluoresc 8:331–344

Electric Field Sensitive Dyes

Ronald J. Clarke

Abstract Electric field sensitive dyes allow electrical events in biological membranes to be detected optically by converting changes in electric field strength into a fluorescence or UV absorbance response. Their response mechanisms to a change in electric field can involve movement of the dye as a whole (either across or within the membrane) or the movement of the dye's electrons, with the mechanism followed by a particular dye depending on its molecular structure. The response times can vary from nanoseconds (for electron movement) to seconds (for dye movement across the entire membrane). Applications of the dyes include the quantification of plasma membrane potential, the surface potential, and the intra-membrane dipole potential, as well as following the kinetic activity of electrogenic ion pumps, such as the Na^+,K^+-ATPase.

Keywords Dual-wavelength ratiometry · Electrochromism · Ion-transporting membrane proteins · Membrane dipole potential · Phototoxicity

Contents

R.J. Clarke
School of Chemistry, University of Sydney, Sydney, NSW 2006, Australia
e-mail: r.clarke@chem.usyd.edu.au

A.P. Demchenko (ed.), *Advanced Fluorescence Reporters in Chemistry and Biology I: Fundamentals and Molecular Design*, Springer Ser Fluoresc (2010) 8: 331–344,
DOI 10.1007/978-3-642-04702-2_10, © Springer-Verlag Berlin Heidelberg 2010

Electric field sensitive dyes have been designed to interact with the membranes of biological cells or cell organelles and produce an optical response (e.g., a change in UV/visible absorbance or fluorescence intensity) to changes in the electrical potential difference across the membrane or the electric field strength within the membrane. They provide a valuable alternative or complementary approach to direct electrical methods, i.e., the use of electrodes. They are particularly useful in the case of small cells, cell organelles, vesicles, or, in some cases, even open membrane fragments, into which electrodes are difficult or even impossible to insert. A further advantage of electric field sensitive dyes over electrophysiological methods is that they allow a much better spatial resolution of electrical field intensity and electric field transients via fluorescence microscopy. They are, therefore, invaluable tools in researching the mechanisms of electrical processes in biology, for example, the activity of ion channels, transporters, and pumps within, for example, the membranes of excitable cells of muscle and nerve, which are responsible for healthy heart and brain function.

Electric field sensitive dyes respond to changes in electrical membrane potential by a variety of different mechanisms with widely varying response times depending on their chemical structure and their interaction with the membrane. An understanding of the mechanisms of dye response and their response mechanisms is important for an appropriate choice of a probe for a particular application. The purpose of this chapter is, therefore, to provide an overview of the dyes presently available, how they respond to voltage changes, and give some examples of how they have been applied. Finally, because there is still scope for the development of new dyes with improved properties, some directions for future research will be discussed.

Up to now, the development of electric field sensitive dyes has been concentrated in a relatively small number of laboratories, notably in those of Cohen [1, 2] (Yale University), Waggoner [3–5] (Amherst College), Grinvald and Hildesheim [6] (Weizmann Institute, Israel), Loew [7, 8] (University of Connecticut), and more recently in the laboratories of Fromherz [9, 10] (Max-Planck-Institute for Biochemistry, Germany) and Demchenko, Klymchenko, and Mély [11–15] (Palladin Institute of Biochemistry, Kiev, Ukraine, and the University of Strasbourg, France). Many of the dyes synthesized have been screened for the magnitude of their fluorescence response to voltage pulses on squid axons. From this screening process and from theoretical calculations, several classes of dyes have been identified, which give measurable responses. Normally, the dyes are grouped into two categories based on their response times:

1. Fast-response probes (response times less than milliseconds): styrylpyridinium and annellated hemicyanine dyes, merocyanine dyes, and 3-hydroxychromone dyes
2. Slow-response probes (response times greater than milliseconds): cationic carbocyanine and rhodamine dyes and anionic oxonol dyes

The basis for the different response times of these probes is their response mechanism. In order to produce a change in fluorescence, a change in electric field must induce some movement either of the dye molecule as a whole or of its electrons. The degree of movement determines the speed of the fluorescence response.

1   Slow Dyes

Slow-response probes, that is, carbocyanine, rhodamine, and oxonol dyes (see Fig. 1) respond via a movement of the dye across the entire membrane. These dyes have a net charge (positive in the case of the carbocyanine and rhodamine dyes; negative in the case of the oxonol dyes). When the membrane potential changes, this, therefore, perturbs the equilibrium distribution of the dyes across the membrane. For example, if the electrical potential of the cytoplasm of a cell becomes more negative relative to the extracellular fluid, this would cause a positively charged probe to accumulate within the cell, whereas a negatively charged probe would favor the extracellular fluid. Because the total volumes of the cytoplasm and the extracellular fluids are generally very different, such a perturbation of the distribution of probe across the membrane would also perturb the partition (or binding) equilibrium between the membrane and aqueous phases on each side of the membrane. For example, a redistribution of dye into the cytoplasm would be expected to increase the proportion of membrane-bound dye because the cytoplasmic volume is generally much smaller than that of the extracellular fluid. The quantum yields of voltage-sensitive probes are normally much higher when they are membrane bound. Therefore, an increase in the proportion of membrane-bound dye

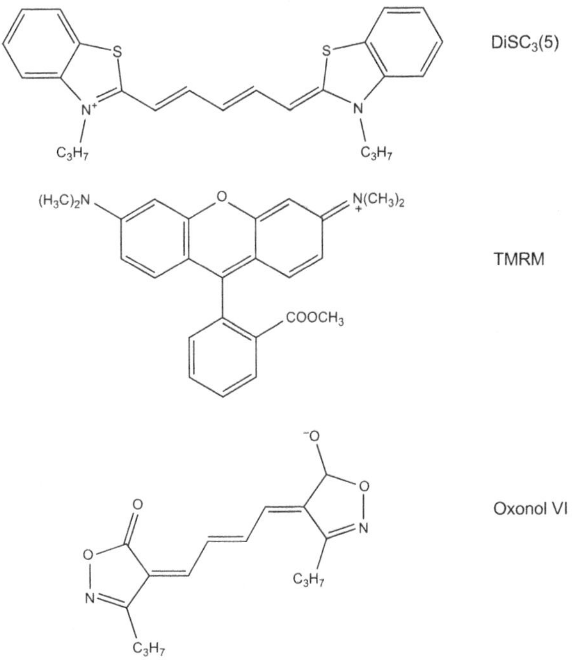

Fig. 1 Examples of the structures of a few slow-response electric field sensitive dyes: 3, 3'-dipropylthiadicarbocyanine (DiSC$_3$(5)), tetramethylrhodamine methyl ester (TMRM), and bis-(3-propyl-5-oxoisooxazol-4-yl)pentamine oxonol (oxonol VI)

would be expected to cause an increase in fluorescence at low dye concentrations. At high dye concentrations, however, increased dye accumulation within the membrane can cause a decrease in fluorescence because of the proximity of dye molecules to one another within the membrane and fluorescence quenching via fluorophore–fluorophore interactions (either via an inner filter effect or energy transfer). However, regardless of the direction of the fluorescence response, the rate-determining step for such slow-response probes is the movement of dye across the membrane from one leaflet to the other. This is a relatively slow process and dye response times are generally in the millisecond-second timescale.

Although dyes responding via transmembrane movement are slow, if kinetic resolution is not an issue, they do have the advantage that the magnitude of the change in fluorescence intensity observed is generally much greater than in the case of fast-response dyes (see later). This is because the change in environment produced by the membrane potential is much greater. For slow-response probes, there is a net movement of probe between the aqueous and the membrane environments, whereas the fast-response dyes remain in the membrane phase. Based on measurements on squid axons, fast-response dyes typically show a 2–10% fluorescence change per 100 mV change in membrane potential, whereas for slow-response probes, the fluorescence response is typically 100% per 100 mV. For equilibrium monitoring of the membrane potential, therefore, slow-response probes are preferred.

2 Fast Dyes

In the case of the fast-response probes, a major voltage-sensing mechanism appears to be electrochromism (see the chapter of Callis in this volume [31]). The styryl-pyridinium and the annellated hemicyanine dyes (see Fig. 2) belong to this category. These dyes undergo a significant charge shift on excitation, i.e., the electron distribution is very different in the excited state compared to the ground state (see Fig. 3). Any local electric field within the membrane hence causes a different degree of stabilization or destabilization of the ground and excited states of membrane-bound probes. This causes a shift in the dye's fluorescence excitation spectrum, which is then evident as either an increase or decrease in fluorescence intensity depending on the wavelength of fluorescence excitation. Because this mechanism only involves a redistribution of the probe's electrons, its response mechanism is expected to be very fast, i.e., femtoseconds.

It is possible, however, that the electrochromic response of some styrylpyridinium probes, for example, RH421 (see Fig. 2), is enhanced by a reorientation of the dye molecule as a whole within the membrane. There is a steep gradient in polarity on going from the aqueous environment across the lipid headgroup region and into the hydrocarbon interior of a lipid membrane. Therefore, any small reorientation of a probe within the membrane is likely to lead to a change in its local polarity and hence a solvatochromic shift of its fluorescence excitation spectrum. Such a

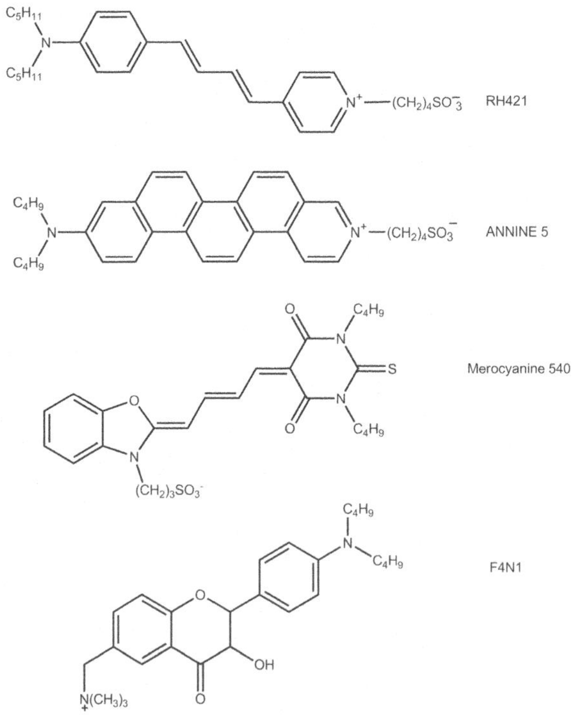

Fig. 2 Examples of the structures of a few fast-response electric filed sensitive dyes: N-(4-sulpho-butyl)-4-(4-(4-(dipentylamino)phenyl)butadienyl)pyridinium inner salt (RH421, a styrylpyridinium dye), ANNINE 5 (an annellated hemicyanine dye), merocyanine 540, and N-[(4′-dimethylamino)-3-hydroxy-6-flavonyl]methyl-N,N-trimethyl ammonium (F4N1, a 3-hydroxychromone dye)

reorientation could easily come about because of the altered electronic distribution of the probe caused by the electrochromic mechanism. The observed fluorescence change of any particular probe could, thus, be the result of a combination of more than one overlapping voltage-response mechanism. A reorientation mechanism would be expected to have a much slower response time than a purely electro-chromic mechanism. Intramembrane reorientation would most likely occur on the nanosecond timescale.

Merocyanine dyes, for example, merocyanine 540 (see Fig. 2), also respond via a reorientation mechanism, either with or without an electrochromic component to the response. These dyes have a dipolar fluorophore, which would be expected to reorient itself along the electric field lines within the membrane on application of a membrane potential or on a change in intramembrane electric field strength. Upon reorientation, it has been suggested that merocyanine 540 also undergoes a change in its state of aggregation within the membrane. If a probe dimerizes, then the interaction between two adjacent fluorophore's electronic systems would also be expected to result in a change in the probe's fluorescence excitation or emission spectrum.

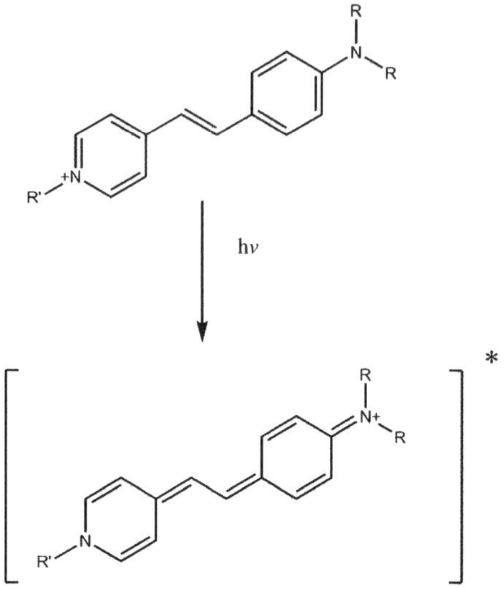

Fig. 3 Excited state charge
transfer in styrylpyridinium
dyes

Recently, a new class of electrochromic dyes based on the 3-hyrdroxyflavone
(3HF) or 3-hydroxychromone fluorophore was introduced by Demchenko, Klymchenko, and their colleagues [11–15] (see Fig. 2). Similar to the styrylpyridinium
dyes, the 3-hydroxyflavone dyes display a large charge shift on excitation. However, the important difference is that once the 3HF dyes have been excited, they
possess an excited state tautomeric equilibrium between the initial normal excited
state (N^*) and a tautomeric state (T^*), produced via an excited state intramolecular
proton transfer (ESIPT) reaction (see Fig. 4). Both the N^* and the T^* states of these
dyes are fluorescent, which results in two fluorescence emission bands (see Fig. 5).
The ESIPT reaction simultaneously causes the movement of a negative charge
between two oxygen atoms of the dye (see Fig. 4). As a result of this, the T^* state
has a smaller dipole moment than the N^* state. The energy of the T^* state is thus
only marginally affected by any electric field within the membrane whereas the
energy of the N^* state can be substantially changed. Any change in the intramembrane electric field strength hence causes a shift in the excited state equilibrium
between the N^* and T^* states, which causes a change in the relative intensities of the
two fluorescence emission peaks. The 3HF dyes, therefore, respond to changes in
the membrane potential or the intramembrane electric field strength with changes in
shape of their fluorescence emission spectrum. This is in contrast to the styrylpyridinium and annellated hemicyanine dyes, for which one expects only a wavelength
shift of fluorescence excitation and emission spectra for a pure electrochromic
mechanism. By coupling electrochromism to an excited state equilibrium, the 3HF
dyes, therefore, potentially could produce greater voltage-sensitive responses than the
other classes of fast dyes.

Because the fluorescence intensity changes observed in response to a change in
transmembrane potential are much smaller when using fast dyes in comparison to

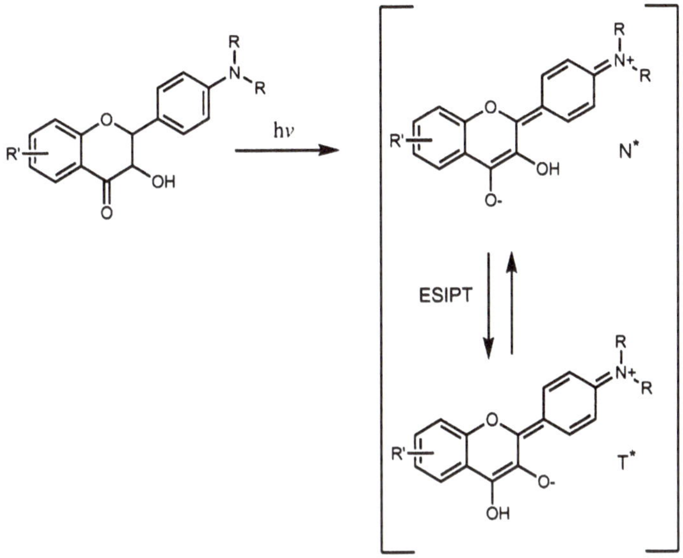

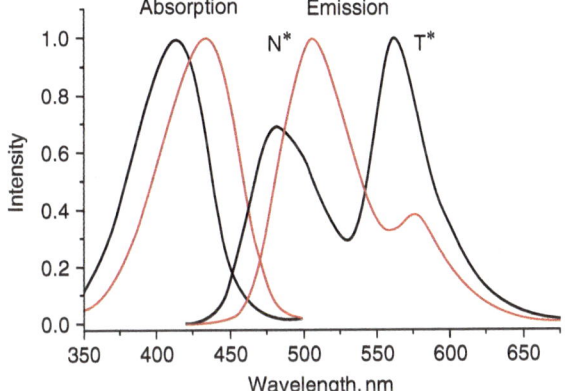

Fig. 4 Excited state intramolecular proton-transfer (ESIPT) mechanism of 3-hydroxychromone dyes

Fig. 5 Absorption and fluorescence emission spectra of the 3-hydroxychromone dye F4N1 in the absence (*black*) and presence (*red*) of a local electric field, which promotes the excitation charge transfer leading from the ground state to the N^* state. In the presence of the local electric field, the energy of the N^* state is reduced, causing a red shift of the N^* emission peak and an increase in its intensity relative to the T^* emission peak. The change in relative intensities of the N^* and T^* peaks reflects a shift in the excited state tautomeric equilibrium toward the N^* state

slow dyes, the choice of excitation and emission wavelengths is particularly important for work with fast dyes, so that the maximum possible response can be achieved. Electrochromism causes a wavelength shift in the excitation spectrum of the probe.

Therefore, if one excites a probe whose electric field sensing mechanism is based on electrochromism at the wavelength maximum of absorbance, the percentage change in absorbance will not be very great and hence neither will the fluorescence change. If one is particularly unlucky, one could even choose a wavelength near the absorbance maximum where, due to the shift in the excitation spectrum, there is a crossover in the excitation spectra with and without an electric field and absolutely no change in absorbance or fluorescence would be observed. In order to maximize the fluorescence response of electrochromic dyes, they should be excited on the long wavelength edge (i.e., the red edge) of their excitation spectra. This naturally decreases the fluorescence intensity detected, but it maximizes the change in fluorescence detected due to a change in electric field, which is more important.

3   Dual-Wavelength Ratiometry

A commonly used procedure in quantifying the fluorescence response of electric field sensitive dyes is dual wavelength ratiometry (recently published review [32] covers this issue). If the fluorescence intensity of a dye is measured at a single excitation wavelength and a single emission wavelength, the measured intensity will depend on the concentration of the dye. Therefore, any variation in the amount of dye added to a cell or membrane preparation would, under these conditions, alter the measured signal. However, if the fluorescence intensity is measured using either two different excitation wavelengths or two different emission wavelengths, then the ratio of the fluorescence intensities will be insensitive to small variations in the dye concentration. This is the major advantage of dual-wavelength ratiometry.

Two different ratiometric procedures exist: excitation ratiometry and emission ratiometry. The former requires excitation with two different wavelengths and the fluorescence emission to be recorded at a single fluorescence emission wavelength. Such a procedure can easily be followed if measurements are being performed on a suspension of cells or membranes in a fluorescence cuvette using a steady-state fluorescence spectrometer. However, if one wishes to image the membrane potential or the intramembrane electric field strength of cells or membranes using a fluorescence microscope, then excitation ratiometry has a number of disadvantages. First, if one were using a single light source, in order to determine the fluorescence intensity ratio, one would have to rapidly alternate between two different laser lines or, if using lamp excitation, one would have to rapidly exchange excitation filters. The need for rapidly alternating laser lines or filters could be avoided by using simultaneous irradiation with two different lasers or light sources. However, even in this case, the measured fluorescence ratio will depend on the relative intensities of the two light sources used. Therefore, the intensities of the two exciting light sources would need to be at least comparable (even if they are not the same). If this is achieved, a problem that can never be overcome in excitation ratiometry using a fluorescence microscope is that, because the two exciting light beams have different wavelengths, the focal plane of each beam will be in a slightly different position, and this will limit the spatial resolution possible. From the above discussion, it can be seen that excitation ratiometry is problematic when using a fluorescence microscope. Emission ratiometry, on the other hand, is technically far simpler and, therefore, preferable.

Emission ratiometry involves exciting with a single wavelength and measuring the fluorescence intensity ratio of two different emission wavelengths. In this method, no changing of laser lines or filters on the excitation side is necessary. Therefore, the measurements can be performed using a single laser or light source. It is only necessary to be able to record the fluorescence intensity at two separate emission wavelengths. This could be done using a single photodetector (e.g., a photomultiplier) if one alternated between two emission filters. Alternatively, and probably preferably, one could detect the intensity at both emission wavelengths simultaneously using two detectors.

Although emission ratiometry is for technical reasons preferable to excitation ratiometry, not all dyes are equally suited to both methods. The electrochromic mechanism of the styrylpyridinium and annellated hemicyanine dyes, which causes a shift in the fluorescence excitation spectrum, makes them more suited to excitation ratiometry. Emission ratiometry has been used for dyes of this class to quantify the transmembrane potential [16], but it has recently been shown that the method is not generally applicable to the quantification of intramembrane electric field strength because of complicating effects of orientational polarizability on the dyes' excited state relaxation prior to emission [17]. In contrast, the 3HF dyes, which show two fluorescence emission bands that change in intensity when the local intermembrane electric field strength changes, are potentially ideally suited to emission ratiometry. .

4 Photostability and Photoxicity

An important factor in the choice of a dye for its experimental application is its photochemical stability. In the case of the 3HF dyes, the excited state equilibrium between the N^* and the T^* states is an integral part of their voltage-sensing mechanism. However, one needs to avoid any photochemical reactions that might cause fluorescence changes that are unrelated to their voltage sensitivity. The styrylpyridinium dyes in particular have been found to undergo photochemical reactions leading, in some cases, to increase and, in other cases, decrease in their fluorescence, which occur even at a constant intramembrane electric field strength [18, 19]. This is most likely related to their *trans* double bonds, which are prone to photoisomerization. For this reason, Fromherz and coworkers [10] have recently produced a series of derivatives of the styrylpyridinium dyes with rigid polycyclic structures. These are the annellated hemicyanine or ANNINE dyes (see Fig. 2). For application in fluorescence microscopy with high intensity illumination, a high degree of photochemical stability would certainly be desirable. Otherwise, for kinetic experiments, it is necessary to carry out careful control experiments to know what the timescales of any photochemical reaction and voltage response are, so that they can be distinguished and if necessary to suppress the photochemical reaction by the use of neutral density filters and low light intensity excitation.

If one wishes to measure the membrane potential or intramembrane field strength on living cells, another important consideration is dye phototoxicity. Three mechanisms have been identified whereby electric field sensitive dyes, particularly cyanine, and merocyanine dyes could cause cell death [20]. One is via thermalization, i.e., the release of heat subsequent to light absorption by the dye. Instead of releasing the absorbed light energy via fluorescence, if a dye undergoes an exothermic reaction in the excited state (e.g., photoisomerization) or even radiationless conversion back to the ground state, this can cause localized heating, particularly if the dye is concentrated in the lipid membrane, which all electric field sensitive dyes are. A second mechanism of dye phototoxicity is via the photolytic production of toxins, i.e., the photolytic breakdown products of a dye may be toxic. The third mechanism, and possibly the most important mechanism, is via photosensitization. This involves the transfer of the excitation energy of a dye to normal triplet oxygen, thus producing the highly reactive singlet molecular oxygen, which can lead to cell membrane damage via lipid peroxidation. Regardless of the mechanism, photoxicity is a property that must be avoided in the application of any dye for the measurement of electric fields in living cells. It is, however, a highly desirable property if one wishes to use a dye for the photodynamic treatment of cancer. In fact, one of the earliest developed electric field sensitive dyes, merocyanine 540, although it has now lost favor as an electric field sensitive dye because of its phototoxicity, is now being widely used in the photodynamic treatment of a variety of cancers [21, 22]

5 Surface and Dipole Potential Measurements

Up to now in this chapter, we have concentrated on the measurement via electric field sensitive dyes of the transmembrane electrical potential, which by itself should produce a linear drop in the electrical potential across a membrane. However, at least through the lipid matrix of a cell membrane, the electrical potential, ψ, at any point does not change linearly across the membrane. Instead, it follows a complex profile (see Fig. 6). This is due to contributions other than the transmembrane electrical potential to ψ. The other contributions come from the surface potential and the dipole potential. Both of these can also be quantified via electric field sensitive dyes.

The transmembrane potential, $\Delta\psi$, the surface potential, ψ_s, and the dipole potential, ψ_d, can be defined as follows:

1. $\Delta\psi$ is the overall difference in potential between the two bulk phases separated by the membrane, which is due to differences in anion and cation concentrations of the two bulk phases;
2. ψ_s is the electrical potential due to lipids with charged headgroups or surface charges of membrane proteins, which controls the concentration of anions and cations at the membrane-solution interface; and

Fig. 6 The electrical
potential, ψ, profile across a
lipid bilayer. The
transmembrane potential, $\Delta\psi$,
is due to the difference in
anion and cation
concentrations between the
two bulk aqueous phases. The
surface potential, ψ_s, arises
from charged residues at the
membrane-solution interface.
The dipole potential, ψ_d,
results from the alignment of
dipolar residues of the lipids
and associated water
molecules within the
membrane

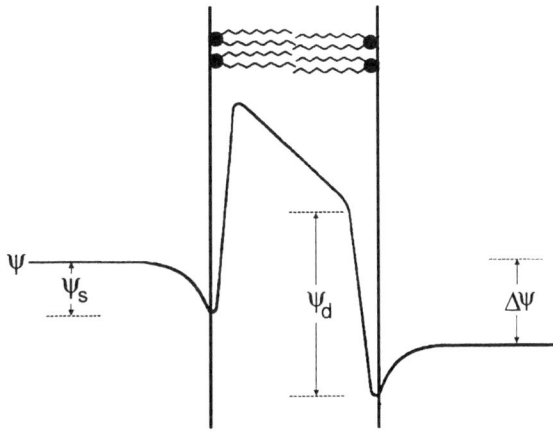

3. ψ_d is an electrical potential due to the alignment of dipolar residues of the lipids
   and/or water dipoles in the region between the aqueous phases and the hydro-
   carbon-like interior of the membrane.

Slow dyes that respond via a redistribution across the entire membrane (sometimes called Nernstain dyes) do so because of a change in the transmembrane
electrical potential. As such, they can only be used as probes of the transmembrane
potential and not as probes of the surface potential or the dipole potential. Dyes
whose electric field sensing mechanism involves a movement between the aqueous
medium and its adjacent membrane interface on one side of the membrane can, in
principle, respond to changes in both the transmembrane electrical potential and the
surface potential. Fast dyes that remain totally in the membrane phase (e.g.,
styrylpyridinium, annellated hemicyanine, and 3-hydroxyflavone dyes) respond to
their local electric field strength, whatever its origin. Therefore, these dyes can, in
principle, be used as probes of the transmembrane electrical potential, the surface
potential, or the dipole potential.

The femtosecond-picosecond response time of the fast dyes makes them suitable
for applications in which one needs to rapidly measure voltage changes, for example,
voltage transients. However, as mentioned earlier, they typically show only a 2–10%
change in fluorescence intensity for a 100 mV change in the transmembrane potential.
Therefore, for the monitoring of slow (i.e., second-minute) changes in the transmembrane potential, slow dyes with fluorescence responses of around 100% per 100 mV
change are preferable. However, the relatively low sensitivity of the fast dyes to the
transmembrane potential does not necessarily mean that they are also relatively
insensitive to the surface potential or the dipole potential. The fast dyes respond to
the gradient of the electrical potential within the membrane at the position where they
are located. Because the transmembrane potential drops across the entire membrane,
the gradient of the electrical potential it produces is not particularly large. In contrast,
the surface potential may and the dipole potential definitely does drop over a much
smaller distance, i.e., across the lipid headgroup region of the membrane in the case

of the dipole potential. As a consequence, the dipole potential (which for fully saturated phosphatidylcholine membranes has a value of around 300 mV, positive in the membrane interior) produces a much greater electric field strength within the headgroup region of the membrane than the transmembrane potential. The electric field strength produced by the dipole potential in the lipid headgroup region is in the range 10^8–10^9 V m^{-1} [23], whereas a total transmembrane potential of 100 mV across a membrane of thickness 4 nm would result in a field strength across the whole membrane of 2.5×10^7 V m^{-1}. This simple calculation demonstrates that a fast dye situated in the membrane at the appropriate position (i.e., in the headgroup region in the case of the dipole potential) would be expected to be much more sensitive to variations in the dipole potential than in the transmembrane potential. This has experimentally been proven to be the case.

The fast styrylpyridinium dye di-8-ANEPPS has been found to be a very sensitive probe of the dipole potential. It has been used to study the effects of varying lipid chain length and saturation as well as the structure of the lipid headgroup on the magnitude of the dipole potential [24, 25]. It has also been used to detect the binding of cholesterol and cholesterol derivatives [26] as well as various inorganic anions and cations to lipid membranes [27]. 3-Hydroxyflavone dyes have also been recently used for imaging the dipole potential of the plasma membrane of cells [14, 15]. These dyes have the advantage over di-8-ANEPPS for use in fluorescence microscopy, in that they allow emission ratiometric detection. Another styrylpyridinium dye, RH421, has been very useful in detecting electric field strength changes within the membrane adjacent to the ion pump the Na$^+$,K$^+$-ATPase and thus in resolving the mechanism of this important enzyme [28–30]. In some cases the relative fluorescence changes of RH421 observed due to activity of the Na$^+$,K$^+$-ATPase have exceeded 100%. Thus, data with very high signal-to-noise ratios can be achieved with fast dyes in particular applications.

However, although RH421 has been a very powerful tool in resolving the mechanism of the Na$^+$,K$^+$-ATPase, this does not necessarily mean that it can be used with success on all other ion pumps. For a fast electrochromic probe to respond to electric field strength changes, its chromophore must be located at a position in the membrane where the greatest changes in electric field strength are occurring. It so happens that RH421 and Na$^+$, K$^+$-ATPase are a good match. However, another membrane protein may cause electric field strength changes in another region of the membrane and hence a dye that has its chromophore at a position either deeper or shallower within the membrane may give better results. This can only be discovered by experimentation with a range of dyes.

6 Conclusion

The ideal electric field sensitive dye would respond to changes in intramembrane field strength or transmembrane electrical potential within femtoseconds; the magnitude of its absorbance or fluorescence response would be enormous; it would undergo no photochemical reactions unrelated to its electric field sensitive response and be completely resistant to photobleaching; and it would be universally applicable to any membrane preparation or cell type. Currently, no such dye exists.

Electric field sensitive dyes were first developed in the 1970's. Since then, many laboratories have invested much time and effort in developing new classes of dye and optimizing the structures of existing ones. This has continued up to the present day. The annellated hemicyanine dyes and the 3-hydroxyflavone dyes have been the most recent additions to the classes of electric field sensitive dyes. Many improvements have been made and researchers now have a wide range of dyes from which to choose. This wide range of choices would certainly be bewildering for anyone deciding to use electric field sensitive dyes in their research for the first time. One super dye would make things a whole lot easier. At this stage, it is virtually impossible to predict which particular dye will work best for a particular application. This can only come from trial and error. However, an understanding of the mechanisms of different dye classes and their strengths and weaknesses does at least help in making an educated guess as to which type of dye might be best and hence which one might try first. Hopefully, this chapter has contributed to a refinement of this educated guesswork.

References

1. Ross WN, Salzberg BM, Cohen LB et al (1977) Changes in absorption, fluorescence, dichroism and birefringence in stained giant axons: optical measurement of membrane potential. J Membr Biol 33:141–183
2. Zochowski M, Wachowiak M, Falk CX et al (2000) Imaging membrane potential with voltage-sensitive dyes. Biol Bull 198:1–21
3. Waggoner A (1976) Optical probes of membrane potential. J Membr Biol 27:317–334
4. Waggoner AS (1979) Dye indicators of membrane potential. Ann Rev Biophys Bioeng 8:47–68
5. Waggoner AS (1985) Dye probes of cell, organelle, and vesicle membrane potentials. In: Martonosi AN (ed) The enzymes of biological membranes, 2nd edn. Plenum, New York
6. Grinvald A, Frostig RD, Lieke E et al (1988) Optical imaging of neuronal activity. Physiol Rev 68:1285–1366
7. Loew LM (1988) How to choose a potentiometric membrane probe. In: Loew LM (ed) Spectroscopic membrane probes, vol 2. CRC Press, Boca Raton, FL
8. Loew LM (1994) Characterization of potentiometric membrane dyes. In: Blank M, Vodyanoy I (eds) Biomembrane electrochemistry. Washington DC, American Chemical Society
9. Fromherz P, Dambacher KH, Ephardt H et al (1991) Fluorescent dyes as probes of voltage transients in neuron membranes: progress report. Ber Bunsenges Phys Chem 95:1333–1345
10. Kuhn B, Fromherz P (2003) Anellated hemicyanine dyes in a neuron membrane: molecular Stark effect and optical voltage recording. J Phys Chem B 107:7093–7913
11. Klymchenko AS, Stoeckel H, Takeda K et al (2006) Fluorescent probe based on intramolecular proton transfer for fast ratiometric measurement of cellular transmembrane potential. J Phys Chem B 110:13624–13632
12. Demchenko AP, Mély Y, Duportail G et al (2009) Monitoring biophysical properties of lipid membranes by environment-sensitive fluorescent probes. Biophys J 96:3461–3470
13. Klymchenko AS, Demchenko AP (2002) Electrochromic modulation of excited-state intramolecular proton transfer: the new principle in design of fluorescence sensors. J Am Chem Soc 124:12372–12379

14. Klymchenko AS, Duportail G, Mély Y et al (2003) Ultrasensitive two-colour fluorescence probes for dipole potential in phospholipid membranes. PNAS 100:11219–11224
15. Shynkar VV, Klymchenko AS, Duportial G et al (2005) Two-colour fluorescent probes for imaging the dipole potential of cell plasma membranes. Biochim Biophys Acta Biomem 1712:128–136
16. Bullen A, Saggau P (1999) High-speed, random-access fluorescence microscopy: II. Fast quantitative measurement with voltage-sensitive dyes. Biophys J 76:2272–2287
17. Vitha MF, Clarke RJ (2007) Comparison of excitation and emission ratiometric fluorescence methods for quantifying the membrane dipole potential. Biochim Biophys Acta Biomem 1768:107–114
18. Amoroso S, Agon VV, Starke-Peterkovic T et al (2006) Photochemical behaviour and Na^+, K^+-ATPase sensitivity of voltage-sensitive styrylpyridinium fluorescent membrane probes. Photochem Photobiol 82:495–502
19. Pham THN, Clarke RJ (2008) Solvent dependence of the photochemistry of the styrylpyridinium dye RH421. J Phys Chem B 112:6513–6520
20. Davila J, Harriman A, Gulliya KS (1991) Photochemistry of merocyanine 540: The mechanism of chemotherapeutic activity with cyanine dyes. Photochem Photobiol 53:1–11
21. Pervaiz S, Hirpara JL, Clément M-V (1998) Caspase proteases mediate apoptosis induced by anticancer preactivated MC540 in human tumor lines. Cancer Lett 128:11–22
22. Nowak-Sliwinska P, Karocki A, Elas M, Pawlak A, Stochel G, Urbanska K (2006) Verteporfin, photofrin II, and merocyanine 540 as PDT photosensitizers against melanoma cells. Biochem Biophys Res Commun 349:549–555
23. Clarke RJ (2001) The dipole potential of phospholipid membranes and methods for its detection. Adv Colloid Interfac Sci 89–90:263–281
24. Clarke RJ (1997) Effect of lipid structure on the dipole potential of phosphatidylcholine bilayers. Biochim Biophys Acta 1327:269–278
25. Starke-Peterkovic T, Clarke RJ (2009) Effect of headgroup on the dipole potential of phospholipid vesicles. Eur Biophys J 39:103–110
26. Starke-Peterkovic T, Turner N, Vitha MF et al (2006) Cholesterol effect on the dipole potential of lipid membranes. Biophys J 90:4060–4070
27. Clarke RJ, Lüpfert C (1999) Influence of anions and cations on the dipole potential of phosphatidylcholine vesicles: a basis for the Hofmeister effect. Biophys J 76:2614–2624
28. Kane DJ, Fendler K, Grell E et al (1997) Stopped-flow kinetic investigations of conformational changes of pig kidney Na^+,K^+-ATPase. Biochemistry 36:13406–13420
29. Clarke RJ, Apell H-J, Kong BY (2007) Allosteric effect of ATP on Na^+,K^+-ATPase conformational kinetics. Biochemistry 46:7034–7044
30. Clarke RJ, Kane DJ (2007) Two gears of pumping by the sodium pump. Biophys J 93:4187–4196
31. Callis PR (2010) Electrochromism and solvatochromism in fluorescence response of organic dyes: a nanoscopic view. In: Demchenko AP (ed) Advanced Fluorescence Reporters in Chemistry and Biology I. Springer Ser Fluoresc 8:309–330
32. Demchenko AP (2010) The concept of lambda-ratiometry in fluorescence sensing and imaging. J Fluoresc DOI 10.1007/s10895-010-0644-y

Part IV
Fluorophores of Visible Fluorescent Proteins

Photophysics and Spectroscopy of Fluorophores in the Green Fluorescent Protein Family

Fabienne Merola, Bernard Levy, Isabelle Demachy, and Helene Pasquier

Abstract Proteins homologous to the green fluorescent protein (GFPs) form a large family of unconventional, genetically encoded fluorophores with widely diverse colors and applications, which have profoundly renewed the fields of biological imaging and drug screening. Their detailed spectroscopy stems from a complex interplay between the electronic properties of a relatively simple, yet flexible and multiprotonable chromophore formed after specific biosynthesis, and the spatial and dynamic organization of its protein carrier. Early experimental and theoretical studies of GFP from the *Aequorea victoria* jellyfish and of model synthetic compounds have revealed that chromophore twisting, *cis-trans* isomerization, proton transfer, and electron transfer are major excited state reactions that determine its photophysics and photochemistry. It has been found later that quite similar mechanisms are at work in several distant members of the GFP family, suggesting a unified picture that may guide the future development of new GFP-based biosensors.

Keywords *Cis–trans* isomerization · Electron transfer · ESPT · GFP · Photoconversion

Contents

F. Merola (✉)
Laboratoire de Chimie Physique, UMR 8000, Université Paris-Sud 11 and CNRS, Orsay, F-91405, France
e-mail: fabienne.merola@u-psud.fr

A.P. Demchenko (ed.), *Advanced Fluorescence Reporters in Chemistry and Biology I: Fundamentals and Molecular Design*, Springer Ser Fluoresc (2010) 8: 347–384, DOI 10.1007/978-3-642-04702-2_11, © Springer-Verlag Berlin Heidelberg 2010

1   Introduction to the Green Fluorescent Protein Family

1.1   Discovery, Functions, and Applications

The green fluorescent protein (GFP) has been originally found as an accessory color converting protein present in the tiny glowing photocytes of the *Aequorea victoria* jellyfish, where it works as an energy transfer acceptor of aequorin chemiluminescence [1, 2]. Together with its homolog from *Renilla* sea pansy, this protein was the first to be purified to homogeneity [3–5] and to be extensively characterized on the biochemical and photophysical levels [6]. After the cloning of its gene [7], the discovery that AvGFP green fluorescence builds up autocatalytically and in a host-independent manner in nearly any organism [8–10] inspired a wealth of groundbreaking techniques for biological imaging and molecular screening [11–14]. Meanwhile, GFP-like proteins were found in a large number of other, eventually nonluminescent, marine animals belonging to cnidarians and bilaterians, warranting their phylogenetic study as a GFP superfamily [15]. The biological functions associated to these many homologous forms are still being debated, and may be as diverse as signaling [2, 6], photoprotection [16, 17], antioxidant [18], or camouflage [19]. A recent report on the ability of various GFPs to behave as light-driven electron donors to coenzymes and other biological electron acceptors has opened exciting speculations on possible unexplored forms of light energy transduction in the animal world [20].

GFPs are the only known sources of visible fluorescence that are fully genetically encoded [10, 21], which makes these protein-based fluorophores amenable to the prolific tool box of genetic engineering and transgenesis. Nature has optimized many of them as bright light emitters operating in appropriate spectral windows. As such, their primary use in biotechnology has been to optically track with high molecular, spatial, and temporal specificity subcellular components such as well-identified proteins, specialized tiny organelles, or single chromosomic loci. Besides this specific tracking ability, the environmental sensitivity of GFP fluorescence has been exploited and tuned in various ways, for the development of specific chemical biosensors, able to report on intracellular pH, ion concentrations, cellular redox state, molecular association, enzymatic activities, or metabolite turnover. Finally, the specific photoreactions of some GFP variants has allowed spatially controlled, time-resolved photochemical manipulations, opening the way to detailed mechanistic and structural investigations of the living cell.

1.2 Diversity and Relationships Among the GFP Family

The GFP family has considerably expanded since its original discovery: in year 2007, over a thousand GFP variants and about 100 original gene sequences were identified or constructed according to the inventory of McNamara and Boswell [22]. The family now includes members emitting in all spectral ranges from the near UV to the near IR, as well as nonfluorescent proteins. Therefore, the bare GFP acronym (to which the shorter form FP is sometimes preferred) is not anymore uniquely associated neither to green color nor even to fluorescence, and will only be used here to refer to the whole family of structurally homologous chromophoric proteins. From a chemical and photophysical viewpoint, we will divide this family into proteins carrying the same fluorescent chromophore as AvGFP (green type GFPs, although not necessarily emitting green fluorescence), those carrying extended chromophores usually emitting further to the red (red type GFPs), and nonfluorescent chromoproteins (CPs), regardless of their chromophore structure.

All GFPs derive their chromophore from the initial intramolecular condensation and oxidation of a core tripeptide of general sequence X65-Y66-G67 (residue numbering corresponding to the AvGFP sequence), where G must be a glycine residue, Y any aromatic residue, although only tyrosine is found in nature, and residue X displays a wide variety of choices [10, 11, 21, 23, 24]. Chromophore formation and fluorescence requires the proper folding of a 25–30 kDa protein consisting of 220–240 amino acid residues, within which the tripeptide is forced to a specific reactive conformation, and the resulting chromopeptide is then rigidly embedded and protected from the bulk solvent [6, 11]. The first X-ray crystallographic structures of the protein [25, 26] revealed the typical GFP fold, built upon a closed 11-strand β-sheet, forming a so-called β-can, crossed over by a single α-helix carrying the chromophore moiety (Fig. 1). It was also soon recognized that the nearly complete amino-acid sequence of the protein was necessary and sufficient for fluorescence build-up [11, 27].

The original hydrozoan AvGFP, carrying the Ser65-Tyr66-Gly67 chromopeptide, has given birth, through site-directed mutagenesis, to a first generation of variants with different colors and photophysical responses, among which are

Fig. 1 Schematic view of the *Aequorea victoria* green fluorescent protein structure. Schematic representation showing the main polypeptide chain forming the 11 strand β-sheet barrel and central α-helix (*grey*), the chromophore buried in the barrel's interior (*green*), and the two highly conserved residues Arg 96 (*purple, on the left*) and Glu 222 (*purple, on the right*). Image built upon coordinates from the 1EMB entry of the Protein Data Bank [84]

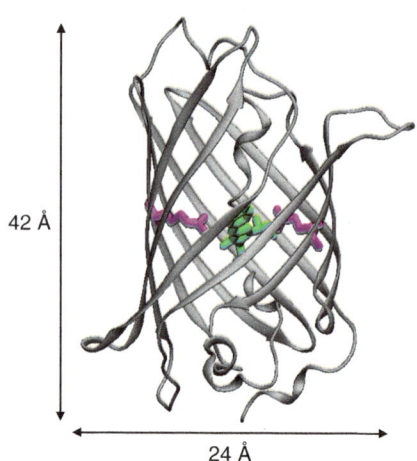

42 Å

24 Å

found the popular "enhanced" forms EGFP (enhanced green fluorescent protein, AvGFP-F64L/S65T, see Sect 3.2) [28], ECFP (enhanced cyan fluorescent protein, AvGFP-F64L/S65T/Y66W/N146I/M153T/V163A, carrying the Y66W mutation shifting emission to cyan), EYFP (enhanced yellow fluorescent protein, AvGFP-S65G/V68L/Q69K/S72A/T203Y with the T203Y mutation shifting emission to yellow) [11], and the photoactivatable variant PA-GFP (photoactivatable green fluorescent protein, AvGFP-T203H) [29], which are still of common use today in live cell imaging.

The anthozoan GFPs discovered later [30–32] are usually tight tetramer assemblies of subunits displaying the same β-can fold as AvGFP, but with rather distant peptide sequence homology. Some variants such as DsRed (*Discosoma* coral red fluorescent protein, with chromopeptide Gln65-Tyr66-Gly67), were found to emit bright red fluorescence. This was shown to result from an additional oxidation reaction within the polypeptide backbone, thereby extending the initial green chromophore conjugation [33–35]. Again, DsRed is the progenitor of many derived variants such as the mFruit series, obtained through sophisticated directed evolution techniques, which contributed to expand and cover the entire orange to red spectrum up to the near IR [36, 37]. Chromoproteins (CPs) are GFP homologs discovered in various branches of the cnidarians, which are naturally nonfluorescent, although they usually carry strongly absorbing chromophores and can sometimes be converted to fluorescent forms [38]. CPs play a crucial role in the pigmentation of corals and other marine animals, and, after genetic engineering, have been invaluable sources of photoconvertible fluorescent GFPs as well as far-red emitters for biological imaging [39, 40].

The unicity of the GFP family is better appreciated when knowing that all red GFPs mature from a green precursor carrying the same chromophore as AvGFP, to which they can eventually revert back [33, 41], while initially green GFPs can evolve in different ways toward red emission [20, 42–44]. Similarly, many chromoproteins can be turned fluorescent at alkaline pHs [45], upon photoactivation [46], or

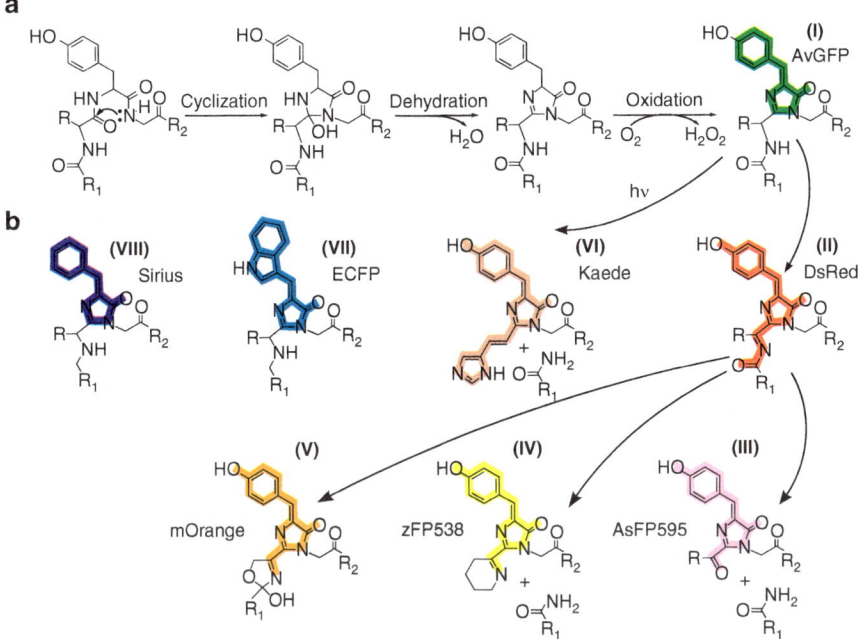

Fig. 2 Formation and chemical structures of different GFP chromophores. (**a**) Formation of the green type chromophore (I) as found in AvGFP. (**b**) Chromophore chemical structures arising from posttranslational processing or mutagenesis of the green form: (II) *red-type* chromophore as found in DsRed, (III) chromophore of photoactivatable AsFP595 from *Anemonia sulcata* (Met65-Tyr66-Gly67) displaying a keto-substituted chromophore formed upon cleavage of the acylimine bond [219], (IV) chromophore of yellow zFP538 (Lys65-Tyr66-Gly67) from button polyp *Zoanthus* carrying a cyclic imine proposed to arise from the lysine 65 amino group attack of its own α-carbon [220], (V) chromophore of the mOrange (Thr65-Tyr66-Gly67) mutant of DsRed [221] carrying an oxazole ring arising from threonine 65 attack at the preceding carbonyl carbon [37], (VI) chromophore of Kaede (His65-Tyr66-Gly67) obtained upon 400 nm light-induced conversion of a green type chromophore, (VII) chromophore structure after the Y66W mutation as in ECFP and (VIII) after the Y66F mutation as in Sirius. R is the amino acid side chain of residue at position 65, R_1 and R_2 stand for the peptidic main chains toward the N-terminus and C-terminus of the protein sequence, respectively

by single point mutations [40, 47, 48]. Overall, up to five distinct covalent structures of GFP-like fluorophores have been observed in naturally occurring proteins [49], while many more have been obtained from extensive mutagenesis (Fig. 2).

1.3   Focus and Plan

GFPs have received considerable attention recently [50–52], and a whole series of detailed reviews have appeared (see [53–55] and the 2009 themed issue in Vol 38 of

Chem Soc Rev). As our topic here is an introduction to the general photophysics of GFPs, we will first consider in Sect. 2 the electronic structure and properties of the green type chromophoric group, which is found throughout the entire GFP family. Sect. 3 will focus on the photophysics of the prototypic green type AvGFP protein, as the best characterized GFP to date from both the spectroscopic and structural point of views. The comparative study of the wild type form and some of its close variants sheds light on several crucial photophysical mechanisms, which have turned out to be central in the properties and applications of many green and red GFPs. In Sect. 4, we will review the photoreactions of several more or less distant GFPs collectively referred to as optical highlighters, which display remarkable and most useful photoswitching properties. Finally, we will discuss in Sect. 5 some important routes toward blue-shifted and red-shifted variants, in the process of expanding the GFP color palette.

2 The GFP Chromophore Structure and Properties

2.1 Formation and Posttranslational Chemistry of the HBI Chromophore

Only four residues are strictly conserved among all natural GFPs, regardless of their final color: Tyr66, Gly67, Arg96, and Glu222. The formation of the green type chromophore from the active tripeptide in freshly folded fluorescent proteins is robust and does not require cofactors or assisting enzymes. Except for Gly67 and Arg96, the process can accommodate numerous mutations inside and around the chromophore sequence [56]. However, it is driven by highly stereospecific chemistry (Fig. 2a), which proceeds along three major steps [10, 56, 57] similar to those involved in the active site tripeptide cyclization of histidine ammonia lyase and related enzymes [58]:

1. *Intrachain cyclization* is triggered by the bend imposed by the β-barrel structure on the central α-helix carrying the active tripeptide. Local protein conformation brings the Gly67 amide nitrogen in close proximity and proper orientation to the carbonyl carbon of Ser65 and prepares the initial nucleophilic attack leading to formation of an imidazolone ring [59]. During this initiation step, the electrostatic influence of the positively charged side chain of Arg96 was shown to play a crucial role [60].
2. *Dehydration* converting the imidazolone ring to imidazolinone seems to be sensitive to the aromatic nature of residue 66 [61, 62]. This step is thought to lead to the formation of an enolate intermediate, which can be trapped by reverse anaerobic chemical reduction of the mature chromophore using dithionite and other reducing agents [63].

3. *Oxidation* of the C_α–C_β bond of Tyr66 completes the conjugated system and leads to chromophore fluorescence. During this step, the carboxylate of Glu222 is thought to function as a general base facilitating proton abstraction from the α-carbon of Tyr66 [64].

The speed of chromophore formation and fluorescence build-up is slow (20–120 min) and is rate-limited by the oxidation step, which is accompanied by the stoichiometric production of hydrogen peroxide [10, 57, 65]. Alternative schemes have also been proposed, where chromophore oxidation may precede and possibly favor dehydration [57, 61, 66]. The conjugated moiety resulting from these autocatalytic steps is named p-hydroxybenzilidene-imidazolinone (HBI), and corresponds to the green type chromophore shared by many mature hydrozoan, anthozoan, and arthropod GFPs. The side chain of the amino-acid residue in position 65 is not part of the final conjugated system of HBI, explaining the large diversity of substitutions allowed at this position.

After formation of the green type chromophore, red type GFPs undergo further maturation to more red-shifted absorption and emission. In the case of DsRed homologs, this involves, as a first common step, the additional dehydrogenation of the C_α–N bond of residue 65, leading to the formation of an acylimine bridge extension –C = N–C=O [33]. The terminal carbonyl oxygen of the acylimine bridge stays out of the plane of the chromophore, but theoretical calculations confirm its participation in the conjugation [34]. The precursor for acylimine formation is thought to be the neutral form of the green type chromophore [67]. The process is very slow and prone to errors and dead-end trapping, leading in many cases to heterogenous final products mixing green and red emissions. The reactivity of the acylimine bridge in red GFPs is a rich source of further additional auto-catalytic chemistry, which frequently results in polypeptide chain fragmentation (Fig. 2b).

2.2 Multiple Configurations of the HBI Chromophore

Model analogs of the green type chromophore HBI have been chemically synthe-tized in different forms carrying blocking groups in place of the protein polypeptide chain [21, 24, 68, 69]. However, the covalent structure of HBI does not uniquely define its optical properties, because the molecule undergoes several protonation and conformational equilibria that directly affect its electronic structure.

First of all, the mesomerism of HBI is rendered complex by the presence of several protonable groups: actually, HBI might exist, depending on pH, under cationic, neutral, zwitterionic, anionic, and possibly enolic forms (Fig. 3a). The experimental pK_a's of model analogs of HBI in aqueous solutions have been studied. Titration curves follow two macroscopic transitions at pH 1.8 and pH 8.2, each corresponding to a single proton release [69]. Comparison of theoretical

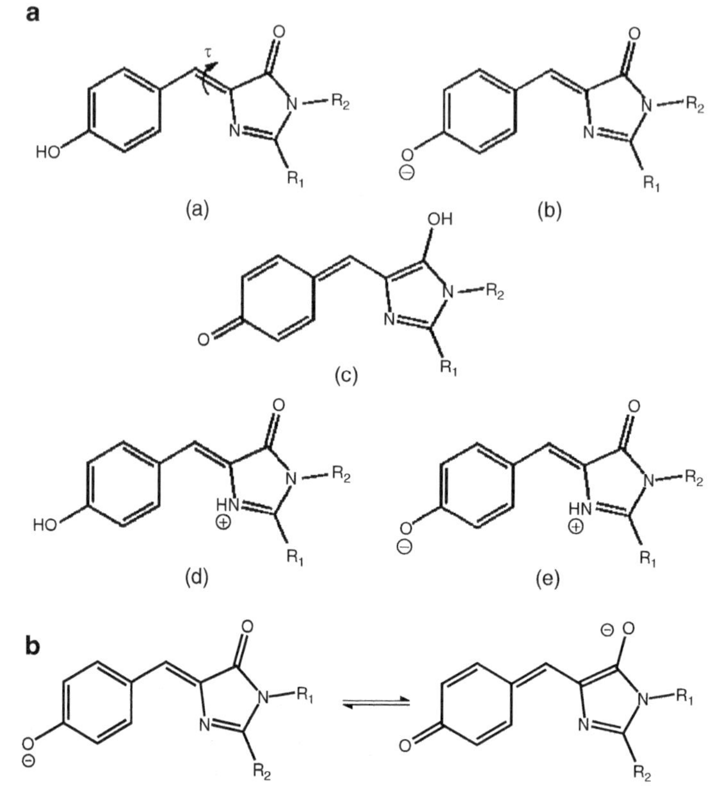

Fig. 3 Protonation states, isomerism and mesomerism of the HBI chromophore (*p*-hydroxybenzi-lidene-imidazolinone). The chromophore is shown in its most stable *Z* ("*cis*") conformation, conventionally associated to a 0° value of the dihedral angle τ, while the *E* ("*trans*") conformation corresponds to $\tau = 180°$. For model compound HBDI (4'-hydroxy-benzylidene-2,3-dimethyl-imidazolinone), $R_1 = R_2 = CH_3$, for chromophore in GFP, R_1, and R_2 stand for the peptidic main chains toward N-terminus and C-terminus, respectively. (**a**) Possible protonation states of HBI: (a) neutral, (b) anionic, (c) enolic, (d) cationic, and (e) zwitterionic. (**b**) Two resonance structures of the anionic form of HBI

computations with thoroughly assigned vibrational Raman and IR spectra indicate that these two transitions take place between the cationic and neutral form and the neutral and anionic form, respectively [69–71]. The absorption spectrum of the anionic form is significantly red-shifted as compared to the neutral form, with the spectrum of the cation staying in between (Fig. 4a). The solvent interactions of the phenolic hydroxyl group of HBI were shown to play a key role in these spectral shifts [72]. There is no experimental evidence for the HBI zwitterionic form, which is predicted from quantum chemical calculations to display an absorption spectrum at longer wavelengths than the anionic form [73]. It is interesting to note that the different ionic forms of HBI display different bond orders within the

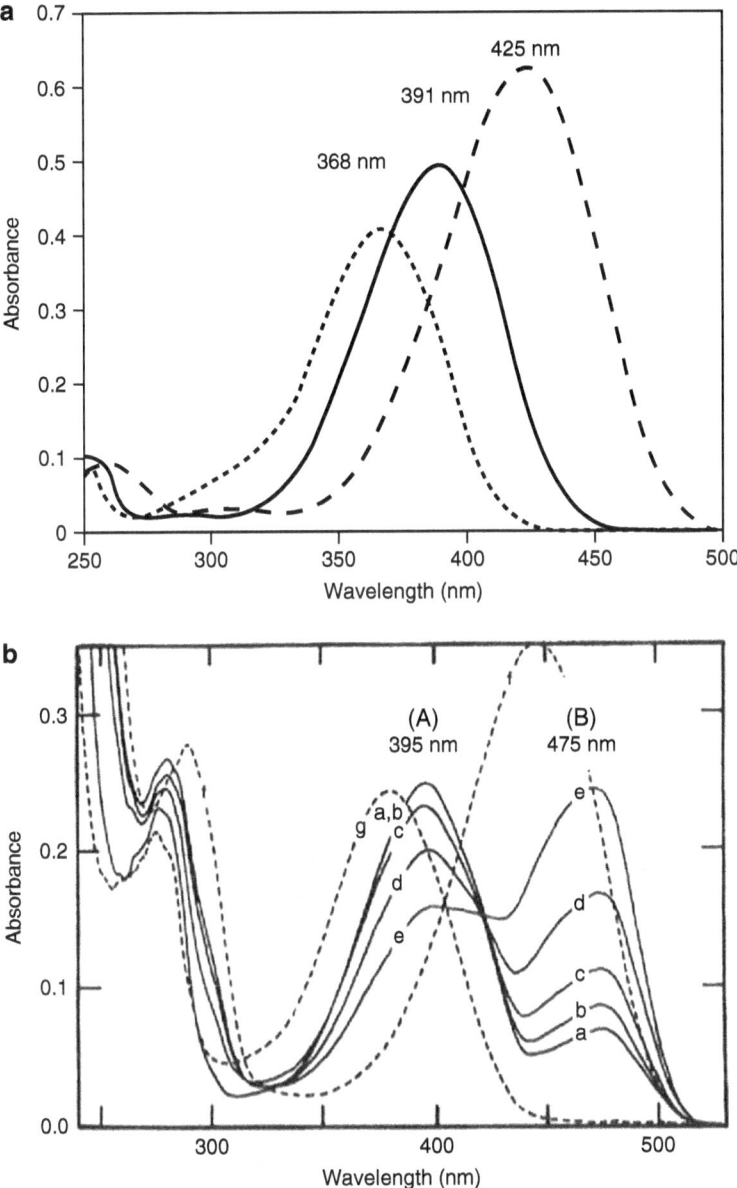

Fig. 4 Absorption spectra of the *green type* GFP chromophore as a function of pH. (**a**) Absorption spectra of model compound HBDI (4-hydroxybenzylidene-1,2-dimethyl-imidazolinone) in aqueous solution : cationic (–; 1 M HCl), neutral (....; acetate buffer, pH 5.5), and anionic (- -; 1 M NaOH). Reproduced with permission from [71]. (**b**) Absorption spectra of AvGFP as a function of pH : pH 5.46 (a), pH 8.08 (b), pH 10.22 (c), pH 11.07 (d), pH 11.55 (e), pH 13.0 (f), pH 1.0 (g). For curves (a–e) the buffer contained 0.01 M each sodium citrate, sodium phosphate and glycine. Sample f was in 0.1 M NaOH, and sample g was in 0.1 M HCl. Reproduced with permission from [6].

exocyclic ethylene bridge, with the anionic form being associated to a quinoid structure leading to a higher degree of conjugation (Fig. 3b).

Another possible source of modification of the HBI optical properties arises from *cis–trans* (or, more properly, *Z–E*) isomerization around its exocyclic ethylene bridge (dihedral angle τ as depicted in Fig. 3a) [74, 75]. The absorption spectrum of *trans* HBI in different solvents is red-shifted by 5–10 nm compared to that of the *cis* conformation [76]. While the *trans* conformation is thermodynamically unfavorable and contributes only a minor population at room temperature, *cis–trans* isomerization seems to take place regardless of the chromophore ionization state, and involves a relatively low energy barrier of about 50 kJ/mol [75], a value that appears significantly lower than initially predicted from quantum mechanics [77, 78].

It has been suggested that the elusive zwitterionic state [75], or a novel nucleophilic addition/elimination mechanism at the central carbon of the exocyclic bridge [79], or solvent-solute H-bonding interactions [76, 80] might play a role in modulating *cis–trans* interconversion. *Cis–trans* isomerization gives rise also to a remarkable intrinsic photochromism of HBI, as it can be easily and reversibly induced upon light absorption [74–76, 79, 80].

2.3 Electronic Transitions of HBI in the AvGFP Protein

The absorption spectrum of folded AvGFP in the visible region displays two well-separated peaks at 395 nm and 475 nm, whose relative intensity depends on pH (Fig. 4b). The chromophore states associated to these two peaks have been termed, respectively, A and B [81]. Although this has been initially debated [78, 82, 83], accumulated spectroscopic and theoretical evidences indicate that these two absorption bands, respectively, arise from the neutral (A) and anionic (B) forms of the HBI chromophore [10, 11, 69–71, 84–87].

If this assignment is correct, the neutral and anionic chromophore absorption bands are shifted, respectively, by 30 nm and 50 nm to the red, as compared to aqueous HBI, showing that the protein environment has a strong influence on the chromophore spectroscopy [88]. The molar absorption coefficients associated to the neutral A and anionic B chromophore in AvGFP are estimated to 25,000 and 50,000 M^{-1} cm^{-1}, respectively [89, 90], in qualitative agreement with the respective absorption strengths of neutral and anionic HBI. However, these absorption coefficients might be themselves strongly modulated by the protein matrix, as, for example, values approaching 100,000 M^{-1} cm^{-1} have been reported for the same anionic chromophore in the green forms of Kaede and Dronpa [91, 92]. The high intensity of the two bands indicates in both cases transitions of the π–π^* type. Polarization spectroscopy confirms that the associated transition moments approximately line with the connecting exocyclic bridge, as expected from its central role in chromophore conjugation [93, 94]. Stark spectroscopy indicates that excitation into the anionic B* state leads to a large increase in chromophore electric dipole (+7 D) [81], whereas excitation into the neutral A* state involves only a moderate change (+2 D) [95]. These experimental results are at odds with theoretical computations, which indicate that the first singlet excited state of the anion should have a rather moderate charge transfer character [96, 97].

2.4 HBI Excited State Twisting and Radiationless Decay

Most synthetic HBI derivatives in aqueous or organic solvents, as well as unfolded AvGFP and chromophore-containing peptide digests of AvGFP are nearly non-fluorescent at room temperature, while their fluorescence can be recovered at temperatures close to the solvent glass transition, suggesting a very efficient mechanism of radiationless decay [24, 68, 88]. The chromophore excited state decay in fluid solvents, as measured by time-resolved fluorescence techniques, is extremely fast and nonexponential, with a dominant component in the femtosecond range [98–101], while ultrafast polarization spectroscopy shows that the chromophore ground state is repopulated on a similar subpicosecond timescale [102, 103]. These results imply an efficient, noncollisional, internal conversion route between the S_1 and S_0 electronic states.

The most likely deactivation process in HBI is ultrafast excited state torsion around the exocyclic ethylene bridge, possibly leading to a conical intersection [78, 104]. Indeed, ethylene bridges are known to have a strongly decreased bond order in the excited state, associated to a preferred staggered, perpendicular orientation of their substituents. At some point along the rotation coordinate, the ground and excited state potential surfaces may approach or cross each other in a conical intersection, leading to rapid deactivation to the ground state. Indeed, some HBI analogs conformationally blocked by internal H-bonding or metal complexation were found to display significant fluorescence at room temperature [105, 106]. However, large excursions of the two chromophore aromatic moieties away from coplanarity do not appear compatible with the observed weak viscosity and temperature dependence of HBI fluorescence [107], although the viscosity dependence increases strongly close to the solvent glass transition [100]. Concerted multibond torsions and/or carbon pyramidalizations, preserving some rough planarity of the chromophore, such as the hula-twist mechanism, are often put forward [107, 108]. Current most sophisticated theoretical approaches of HBI excited state dynamics [109–112] have identified several possible reaction coordinates, involving either or both bonds in the exocyclic ethylene bridge, and have pointed to the critical role of solute–solvent interactions.

The protein matrix of AvGFP efficiently forbids significant torsional motions of the chromophore, leading to near-maximum and highly homogeneous green fluorescence emission (see Sect. 3.1). Failure to do so results in weakly or non-fluorescent GFPs [113–115], while it was shown recently that the differences in dihedral freedom around the exocyclic bridge correlate well with the respective quantum yield of a series of fluorescent protein variants [116].

3   Photophysics and Spectroscopy of AvGFP

3.1   Absorption and Emission Properties

The biophysical chemistry of AvGFP has been investigated in depth in early works (reviewed in [6, 11]). The AvGFP protein is a monomer in dilute solutions, but forms dimers at high concentrations (with a dissociation constant K_d of about 100 μM). The protein displays a remarkable stability toward thermal or chemical denaturation, is extremely resistant to enzymatic proteolysis, and the buried chromophore is highly protected against the action of freely diffusing fluorescence quenchers. Shimomura and others soon reported on the exquisite sensitivity of the two-band absorption spectrum of AvGFP to many environmental factors, such as temperature, pH, ionic strength, cryoprotectants, protein concentration, and illumination [3, 56, 117, 118]. The perturbations always consist of gross amplitude exchanges between the two A and B absorbance peaks at 395 nm and 475 nm, respectively, revealing in many cases a well-defined isosbestic point near 425 nm (Fig. 4b). These effects are interpreted in terms of conformation-induced shifts of the chromophore balance between its neutral and anionic forms [6, 11]. As a result of this general environmental sensitivity, quite dispersed values are reported in the literature for the apparent molar absorption coefficients of AvGFP, which vary according to the effective proportions of neutral and anionic chromophore.

Quite surprisingly, at room temperature, excitation into either peak A or B gives rise to a very similar green fluorescence emission peaking near 500 nm (Fig. 5), with, however, the fluorescence excited in peak B being slightly blue-shifted by a few nanometers, and being associated to a slightly higher anisotropy as compared to excitation in peak A [81]. In addition, excitation into the A band produces a faint blue emission centered at 450 nm (Fig. 5). The green fluorescence of AvGFP at 505 nm is characterized by a high quantum yield ($\Phi_F = 0.8$ [3, 119]) and a single exponential decay time (more than 95% of $\tau_F = 3.3$ ns [118, 120, 121]) which do not depend on the excitation band ([118] and unpublished results from our lab). The fluorescence excitation spectrum for emission at 505 nm approximately parallels the protein absorption spectrum. The AvGFP fluorescence quantum yield and lifetime display little variations with the protein environment, except for an expected temperature and refractive index dependence [122, 123], the later being well described by standard molecular photophysics [124]. The AvGFP protein can also reversibly lose its visible optical properties after treatment by a variety of reducing agents [9, 24, 63, 65, 125], i.e., the conjugated chromophore in its ground state behaves as an oxidant.

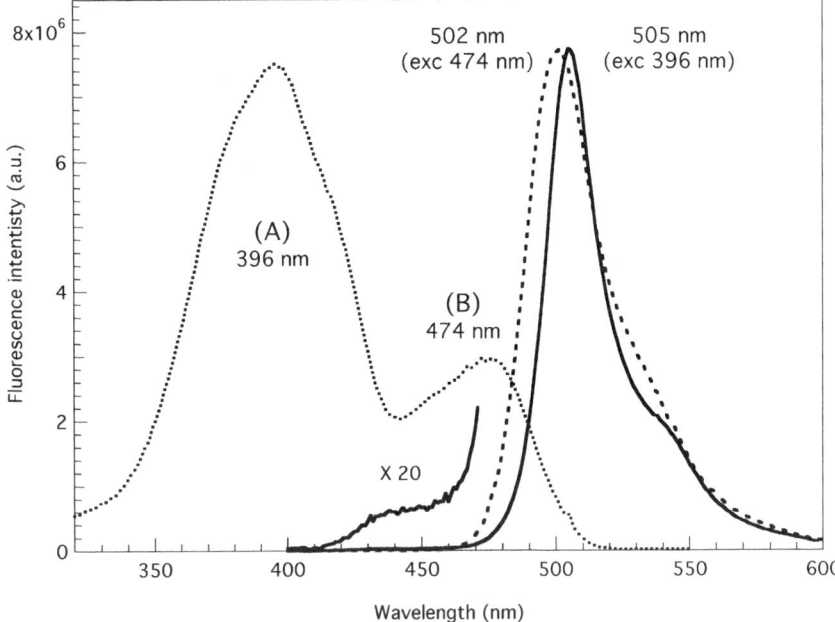

Fig. 5 Fluorescence emission and excitation spectra of AvGFP at 293K, 0.5 μM in 10 mM Tris buffer pH 8: (....) excitation spectrum for emission at 505 nm, (- - - -) emission spectrum for an excitation at 474 nm (B band), (—) emission spectrum for an excitation at 396 nm (A band). Fluorescence spectra are corrected only for variations in excitation intensity

3.2   Ground State Protonation Equilibria of the AvGFP Chromophore

In the AvGFP protein, the two A and B absorption bands, presumably arising from the neutral and anionic forms of the green chromophore, are maintained in nearly unperturbed proportions over a large range of pHs from 6 to 9 (Fig. 4b), with two macroscopic pKs at 5 and 11.5 [6]. The coexistence of two chromophore protonation states over such a broad range of pHs was another early puzzling observation. Based on estimates of their relative absorption strengths, the relative proportion of neutral versus anionic forms is evaluated to 6:1 at pH 7 [81, 89]. According to the X-ray crystallographic structures of AvGFP and its close variants [25, 26, 84, 126], the chromophore is involved in multiple hydrogen bonding with nearby residues and/or crystallographic water molecules. In particular, in the wild-type protein, a continuous wire of hydrogen bonds connects the phenol hydroxyl of Tyr66 to the carboxylate of Glu222, by the intermediate of the Ser205 hydroxyl and a crucial water molecule placed next to the chromophore (Fig. 6).

Therefore, it has been suggested that electrostatic repulsion by ionized Glu222 is a major stabilization factor favoring the neutral protonated form of the

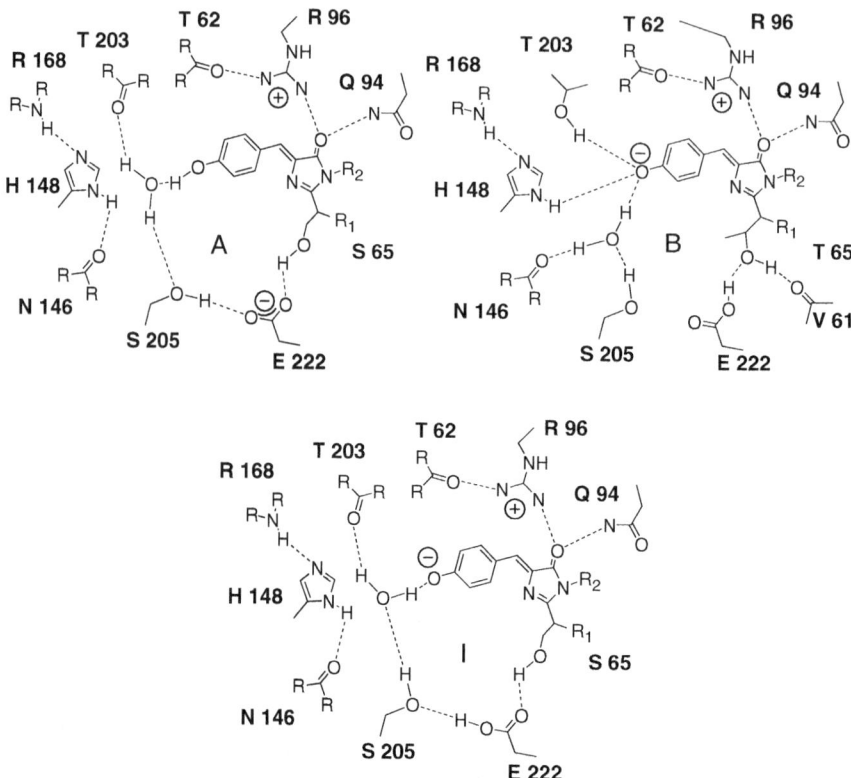

Fig. 6 Schematic diagram showing the chromophore hydrogen bonding networks in the different A, B, and I states of AvGFP. The crystallographic structure of AvGFP showing the hydrogen bonding of the neutral chromophore is taken as state A [84]. The crystallographic structure of AvGFP-S65T carrying an anionic chromophore at neutral pH is taken as representative of state B [25]. Inverting the H-bonding donor–acceptor directions within the structure of state A leads to the hypothetical structure of state I

chromophore [84]. In addition, the H-bonding status of the Glu222 carboxyle to Ser65 allows an easy switch between its neutral and ionized forms. Owing to this same proton wire, going from neutral to anionic chromophore can easily be achieved with conservation of total charge and number of protons, provided a concomitant change in Glu222 protonation. Electrostatic modeling [86] or more simple two site thermodynamic coupling models [127] clearly show that, under such a strong anticooperative coupling regime, a constant ratio of two protonation states of each partner can be maintained over a large range of pHs. Electrostatic modeling nicely accounts for the detailed pH responses of absorbance in AvGFP and various mutants, and allows the identification of crucial intermediates in the coupling [86]. This explains well how neutral and anionic chromophore states can indeed coexist in a delicate yet robust balance in wild-type AvGFP, and how this balance may be strongly shifted by conformational perturbations or mutations.

The early S65T mutation, originally introduced in the so-called "enhanced" form EGFP, leads to strong preferential absorption at 490 nm and to an inverted 20:1 predominance of the anionic form at neutral pHs [28]. The crystallographic structure of the S65T mutant indicates that the anionic form of the chromophore is now stabilized by new hydrogen bonds with the side chains of Thr203 and His148 as proton donors [25, 85]. Meanwhile, Glu222 is maintained neutral through H-bonding to Thr65, while the H-bonding proton wire is clearly interrupted between Ser205 and Glu222 (Fig. 6). All other mutations of Ser65 (G, A, C, V) lead to a similar destabilization of the chromophore neutral form at neutral pHs [28, 90, 128]. Similarly, mutations suppressing the major acidic partner of the proton wire at position 222 favor the anionic form [90, 129]. By contrast, mutants at position 203 [10, 29, 129, 130] or the so called GFPuv (AvGFP-F99S/M153T/V163A) [131, 132], show a predominant neutral chromophore.

Therefore, the protein environment is able to dramatically modify the ground state protonation behavior of the HBI chromophore, and mutations of AvGFP often have the correlative drawback of a strongly increased pH sensitivity of absorption and fluorescence properties close to neutral pHs [119]. Some variants were soon used to monitor intracellular pH [133], while several mutations were introduced around the chromophore to modulate the precise transition pK_a between 6 (S65T) and 7.8 (S65T/H148D) [85]. In some remarkable, extreme cases, the coupling of chromophore phenol ionization to that of nearby acidic residues and/or chromophore cis–trans isomerization (see Sect. 4.3), and possible related conformational rearrangements, lead to an inverted protonation scheme, where the neutral chromophore form is favored at high pHs: this was reported for ratiometric pHluorins [134], AsFP499 [89], and mKeima [135]. In addition, because the acid-base properties of the GFP chromophore are tightly connected to protein conformation, it is possible to couple many unrelated molecular recognition events to its absorption changes, giving rise to a variety of ratiometric intracellular chemical sensors, which can be tuned to the desired application range [134, 136–138].

3.3 Excited State Proton Transfer (ESPT) from the Neutral Chromophore

Excited state proton transfer (ESPT) from aromatic alcohols such as phenol and naphtol is a well-described mechanism arising from the increased acidity of these compounds in their excited state [139]. The demonstration of highly efficient ESPT upon excitation of the neutral phenol chromophore of AvGFP was provided by fluorescence up-conversion studies [81, 118], which brought insight into the early time evolution of its fluorescence emission. Upon excitation into the neutral A band at 400 nm, the weak blue emission at 450 nm and the strong green emission at 505 nm, respectively, decay and rise following very similar multiexponential responses, with typical time constants of 2 ps and 12 ps. The direct kinetic connectivity between the two processes, together with the observed strong isotopic effect on their rate constants, clearly point to an ultrafast ESPT reaction leading from the excited neutral form (presumably emitting at 450 nm) to the excited anionic form of the chromophore (responsible for emission at 505 nm).

Ultrafast ESPT from the neutral form readily explains why excitation into the A and B bands of AvGFP leads to a similar green anionic fluorescence emission [84]. Simplistic thermodynamic analysis, by way of the Förster cycle, indicates that the excited state protonation pK_a^* of the chromophore is lowered by about 9 units as compared to its ground state. However, because the green anionic emission is slightly different when it arises from excitation into band A or band B (Fig. 5) and because these differences are even more pronounced at low temperatures [81, 118], fluorescence after excitation of the neutral A state must occur from an intermediate anionic form I* not exactly equivalent to B*. State I* is usually viewed as an excited anionic chromophore surrounded by an unrelaxed, neutral-like protein conformation. The kinetic and thermodynamic system formed by the respective ground and excited states of A, B, and I is sometimes called the "three state model" (Fig. 7).

In AvGFP, the continuous proton wire connecting the chromophore phenol to a prepared Glu222 acceptor is likely an accelerating factor of ESPT. This role is suggested by time-resolved vibrational spectroscopy, showing that a glutamate, most likely Glu222, becomes protonated upon neutral chromophore excitation, a reaction that takes place on similar picosecond timescales [140, 141]. The complete kinetics along the GFP chromophore photocycle after ESPT has been studied by experimental [118, 132] and theoretical approaches [142, 143]. Pump-probe spectroscopy has brought insight into the time evolution of the I state: after ESPT,

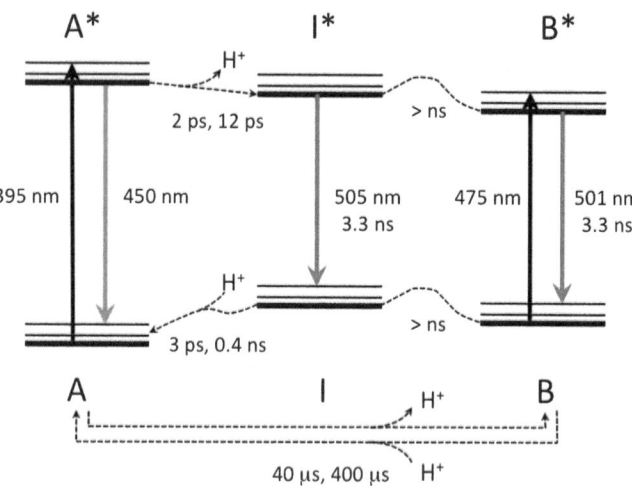

Fig. 7 Kinetic and thermodynamic relationships between the neutral A state, the anionic B state, and the intermediate I state of AvGFP in their ground and excited states at room temperature and neutral pH, according to ([81, 118, 132, 149, 150])

I* decays to the ground state by fluorescence emission with a lifetime of 3.3 ns (see above), then I returns to A in less than a nanosecond, a process that is slightly thermally activated, strongly sensitive to deuteration, and apparently involves several intermediates [118, 132, 144]. Because no production of the B state is observed during I* deexcitation and return to A, states I and B must exchange on timescales at least slower than the nanosecond.

It is thought that the I state mostly differs from the A state by the donor–acceptor polarity of its chromophore H-bonding network [81, 126], while the transformation of I into B would imply further conformational rearrangements, including the reorientation of T203 and H148 side chains as suggested by the structure of AvGFP-S65T and other mutants stabilized in the anionic state (Fig. 6). The 0–0 transitions energies of each protein state A, B, and I, and their thermodynamic relationships have been established by low temperature spectroscopy and spectral hole burning [144–146]. The mutant AvGFP-T203V/E222Q, presumably stabilized in the I state, has provided some insight into its possible room temperature spectral properties [147]. However, it remains unclear whether the I ground state is anything but a short-lived kinetic intermediate in the AvGFP photocycle, or whether it actually populates at room temperature, contributing to ground state heterogeneity in the native protein. In all cases, the original idea that it is both populated and in no-or-very-slow exchange (hours–days) with the equilibrium B state [81, 148] appears thermodynamically inconsistent, as the A and B states have been shown to be under submillisecond equilibrium [133, 149–152]. The experimental distinction between I, B, and possibly, in some cases, some amounts of the irreversibly photoconverted anionic form of AvGFP, which is easily formed at both cryogenic and room temperatures (see Sect. 4.1), may turn out to be extremely difficult, as these three forms are spectroscopically and structurally very similar.

The proton shuttling pathways in GFPs may be more extended and complex than currently acknowledged. The AvGFP chromophore proton wire might continue down to Glu5 toward the aqueous phase [153], and involve back and forth proton transfers [154]. Proteins in which this proton wire has been disrupted, such as the S65T mutants, still exhibit some ESPT although with slower time constants in the range of a fraction of nanosecond, and several examples are available of variants with "rewired" ESPT [155, 156]. However, at the high protein concentrations used in ultrafast spectroscopy studies, intramolecular ESPT may easily be mixed up with intermolecular FRET within GFP heterodimers mixing neutral and anionic chromophores, as was shown in the case of EYFP [157].

Even if no proton acceptor favoring rapid ESPT is present, the direct blue emission from the neutral chromophore at 450 nm remains usually rather weak. This may be due to the lower ability of the protein structure in maintaining a planar conformation of the neutral chromophore, when its phenol moiety is not hydrogen bonded, as well as to the higher flexibility of the protein matrix at the acid pHs usually required for stabilization of the neutral chromophore. However, there are several examples of a significant direct emission from the neutral chromophore in AvGFP variants, as shown by mutant S65G/T203V/E222Q [90], in the T203V/ S205V mutants [155, 158], in dual emission pH sensors deGFPs [159, 160], or in thiocyanate-bound EYFP [157]. Interestingly, green type anthozoan GFPs, such as asFP499, have also developed a proton wire, involving a rather different hydrogen bonding network, for the stabilization and control of chromophore protonation and efficient ESPT from the neutral form [89]. However, after maturation from the neutral green chromophore, the red type chromophore in DsRed appears to be mostly anionic at neutral pHs [34]. ESPT with a time constant of 300 ps was reported from the red-emitting chromophore of photoconverted Kaede [161], while an ultrafast ESPT with a time constant of 4 ps was reported for mKeima, a blue-absorbing, red-emitting variant engineered from a nonfluorescent chromo-protein [162].

4   Photoreactions in the GFP Family

4.1   Photoconversion of AvGFP

During its early use as an intracellular reporter, a peculiar enhancement of AvGFP fluorescence was observed after prior exposure to ultraviolet light [9]. The phe-nomenon was later explained by an increased absorption at 488 nm (Fig. 8a), likely as the result of a light-induced stabilization of the chromophore anion [56, 163]. This photoconversion of AvGFP is most easily obtained by UV light, but is also observed upon strong excitation into the A band at 400 nm, or by irradiation into the B band at 488 nm [87, 119, 163]. Despite early reports on some partial reversibility [81, 118, 144, 148], the process, as triggered by UV light at 254 nm, appears essentially irreversible [163], and is specifically associated to a decarboxylation of Glu222, as shown by mass spectrometry and X-ray crystallographic analysis of the photoproduct [164, 165].

The rate of AvGFP photoconversion strongly increases with shorter excitation wavelengths. Glutamate decarboxylation can proceed to 100% completion upon 254 nm illumination, while it is only partial with 400 nm or 480 nm light, probably because, on slower timescales, chromophore destruction leading to photobleaching takes place simultaneously [87, 166]. The photoconversion quantum yield is 0.03 upon 254 nm illumination, and 0.002 upon 400 nm illumination [167]. Photocon-version upon 254 nm illumination is thought to involve a higher excited state of the chromophore [87]. The process follows first order kinetics and depends linearly on light intensity at all illuminations wavelengths, showing that it is in all cases a one-photon reaction [87, 166]. A mechanism has been proposed by van Thor and co-workers that involves as a primary step electron transfer from Glu222 to the electron-deficient imidazolinone part of the excited chromophore (Fig. 8b), leading to the formation of a $^{\bullet}CH2$ radical intermediate, as expected in the so called Kolbe mechanism [166]. If this mechanism is correct, the initial electron transfer step must compete with the picosecond ESPT described above, which implies that it must take place on the nanosecond time range. The crystallographic structure of

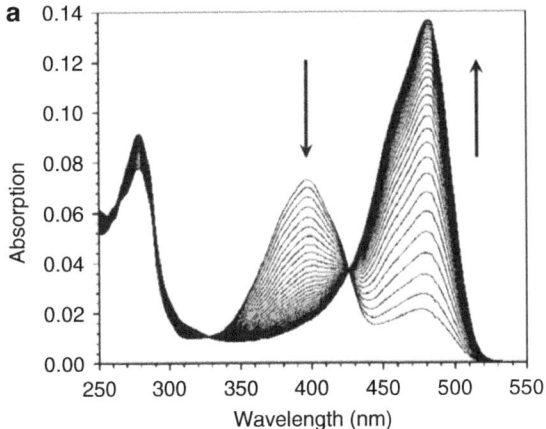

Fig. 8 Irreversible photoconversion of AvGFP. (**a**) Modification of the absorption spectra of AvGFP under UV light ($\lambda = 254$ nm, 100 s irradiation, 12.9 mW) at 293 K, pH 8.0, showing the increase in anionic B band (maximum at 483 nm). (**b**) Proposed Kolbe mechanism for Glu222 decarboxylation through transient formation of a •CH2 radical intermediate. Reproduced with permission from [166]

decarboxylated, photoconverted AvGFP reveals local structural rearrangements rather similar to those observed in the S65T mutant, with an anionic chromophore stabilized through H-bonds with the Thr203 and His148 side chains [166].

The highly efficient and selective photochemistry of AvGFP leading to Glu222 decarboxylation has been the first example among the diverse photoconversions later described in the GFP family. This property found early applications for intracellular time-tagging [168], and has led to the development of the photo-activatable variant PA-GFP [29], in which the absorption contrast between the native and photoconverted forms is maximized, opening the way to many elegant, dynamic, and high resolution cell imaging applications [14, 169].

4.2 Irreversible Photochemistry in GFPs

The remarkable resistance of GFPs to photobleaching, i.e., to a complete destruction of their chromophore conjugated system upon illumination, gives room to a wide variety of other highly valuable photoreactions, many of which are still awaiting description at the atomic level [41]. First of all, in a way similar to the photoconversion of AvGFP, several GFPs are known to undergo different types of specific, irreversible photochemistry. Glutamate decarboxylation seems to take place in AvGFP variants such as EYFP [170], as well as in DsRed, where Glu215 (structurally homologous to Glu222 in the green forms) was shown to be the modified residue [171]. Most remarkably, Glu215 has also been identified as the major protonable group controlling the pH-dependent spectral responses in some mFruit variants of DsRed [37]. A mechanism similar to that of AvGFP photoconversion is also suspected in the case of the irreversible cyan to green conversion of PS-CFP, a photoswitchable protein engineered from a jellyfish chromoprotein [172].

In other cases, photochemistry in green GFPs can lead to red-emitting forms, which reflect a direct modification and extension of the chromophore structure. Under anaerobic conditions, strong illumination of AvGFP into its anionic B absorption band leads to a bright red-emitting species with excitation peaking at 525 nm and maximum emission at 600 nm [42]. In the presence of various inorganic or biological oxidants, EGFP and several other green GFPs undergo a different type of "oxidative redding," which is also triggered by 488 nm laser light, but appears independent of the presence of molecular oxygen [20]. Yet another type of photochemistry originating from the green type chromophore is illustrated by a subfamily of naturally photoconvertible proteins, carrying the tripeptide His65-Tyr66-Gly67, to which belong Kaede [91], EosFP [173], and Dendra [174]. In this family, irradiation into the 400 nm band of the neutral chromophore leads to cleavage of the polypeptide backbone, and subsequent incorporation of the imidazole ring of His65 into a new orange-red-emitting three-cycle chromophore [43] (Fig. 2).

4.3 Reversible Photochromism

The recent years have seen the development of several reversibly switchable fluorescent proteins (RSFPs) in the GFP family, which are able to shift back and forth between spectrally different forms in a light-dependent manner. These reversible photoswitching properties have given way to new optical labeling and tracking methods for cell imaging, based on sequences of activation/inactivation of the fluorescence signals [46]. Indeed, model compounds of the HBI chromophore in solution can reversibly photoconvert between their *cis* and *trans* isomers, while in the dark, spontaneous equilibration between the two states is very slow (hours—days) and strongly dependent on solvent [76].

First of all, most natively bright green or red GFPs crystallize with a nearly coplanar *cis* chromophore conformation, while all nonfluorescent chromoproteins seem to bear a *trans* and noncoplanar chromophore [38, 49]. For AvGFP, there is, up to now, no indication of a chromophore conformation different from the *cis* isomer. However, various other situations have been reported in the GFP family, such as a *trans* coplanar fluorescent form in eqFP611 [175], mixed *cis–trans* configurations such as in HcRed [114], amFP486 [176], mKate [177], and mKeima [135], or distorted yet fluorescent chromophores observed in the mFruit series [37], in the blue-emitting mTFP series [178], and in a mutant of the chromoprotein Rtms5 [48].

The engineered photochromic proteins KFP1 (for "kindling fluorescent protein") and Dronpa are the most prominent examples of RSFPs. KFP1 is derived from the tetrameric purple chromoprotein asCP, also named asFP595 (chromopeptide Met65-Tyr66-Gly67), discovered in the tentacles of *Anemonia sulcata*. AsFP595 displays, after biosynthesis and red chromophore maturation (see Fig. 2), a dim red fluorescence at 595 nm ($\Phi < 0.001$), which can be strongly enhanced upon green light illumination. This red fluorescence then undergoes a slow spontaneous return to basal levels, or else can be instantaneously quenched by a flash of blue light. In addition, under more intense green light exposure, asFP595 photoactivation can become irreversible [32]. The X-ray crystallographic structures of an asFP595 mutant determined before and after in-situ photoswitching in the protein crystal clearly showed that the red fluorescence enhancement is to be ascribed to a transition from a *trans* (off) to a *cis* (on) conformation of the chromophore [179]. The next important member of the RSFP family is Dronpa (Cys65-Tyr66-Gly67), a monomerized variant of a *Pectiniidae* coral green fluorescent protein. Initially, Dronpa carries a bright ($\varepsilon_M = 95,000$ M^{-1} cm^{-1}, $\Phi_F = 0.85$), clearly anionic green type chromophore (excitation peak at 503 nm, fluorescence emission peak at 518 nm). However, prolonged or intense irradiation into its anionic excitation band converts Dronpa to a nonfluorescent neutral chromophore form, with peak absorption at 390 nm. Green fluorescence can then be easily reactivated by mild irradiation at 400 nm. The photoconversion quantum yields for the inactivation/activation of Dronpa are $\Phi_{off} = 3.2\ 10^{-4}$ and $\Phi_{on} = 0.37$, respectively. Several tens of (on/off) cycles can be performed with moderate fatigue of Dronpa fluorescence [92]. Again, the comparison of crystallographic structures of the bright [180] and dark [181] states of Dronpa shows that chromophore *cis* (on) to *trans* (off) isomerization is predominantly involved in the photoconversion, although some disorder and flexibility is seen in the conformation of the off state [115].

The reversible photophysical changes observed upon photoconversion of Dronpa are reminiscent of those reported earlier for E^2GFP (AvGFP-F64L/S65T/T203Y) [182]. Reversible light-induced absorbance or fluorescence changes have also been reported for other AvGFP variants, such as EYFP, Citrine, and ECFP [183], and more recently for EBFP (AvGFP-F64L/Y66F) and EYQ1 (AvGFP-F64L/T203Y/E222Q) [184]. Interestingly, the single mutation E222Q is apparently able to confer an efficient photochromism to different AvGFP variants [185]. The reversible bleaching of mTFP0.7, a teal variant derived from the cyan protein cFP484 from *Clavularia* coral, was also shown to arise from *cis* to *trans* isomerization [186]. Light or pH-induced *cis–trans* isomerization has been reported as well for several red GFPs [135, 187, 188], and thus might actually turn out to be a general property of the whole GFP family.

However, the sole *cis–trans* isomerization mechanism of the HBI chromophore, as described in Sect. 2.2, can clearly not account (1) for the large spectral shifts observed upon photoswitching and (2) for the nearly systematic association of the *trans* conformer with low fluorescence levels. For reasons that remain to be understood, it seems that isomerization toward the *trans* conformer in GFPs is frequently accompanied by switching to a protonated chromophore state, which implies large blue shifts in both absorption and emission. A coupling between chromophore isomerization and protonation has been unambiguously demonstrated by Raman spectroscopy studies of EYQ1 [184], and it has recently been proposed that protonation of the excited chromophore, possibly favored in the triplet state, might be the driving force in the *cis* to *trans* photoconversion process [115]. Some recent ab initio computations give ground to the idea that coupling between conformation and protonation might be an intrinsic property of the HBI group [112]. Quite interestingly, *cis–trans* isomerization can also be triggered in some GFPs by changes in pH, as shown in the chromoprotein Rtms5-H146S [45], in Cerulean [189], asFP595 [190], mKate [177], and mKeima [135]. The *trans* conformer is usually associated to the acid side of the transition, to the noticeable exception of mKeima [135]. In the latter case, it is striking to note that a protonated form of the chromophore remains associated to the *trans* conformation, resulting in a reverse protonation scheme of the chromophore as function of pH [135]. The reason why the *trans* chromophore conformer is usually not fluorescent remains another open question, as the brightness of eqFP611 shows that this is not necessarily the case [175]. The currently proposed explanations are similar to those accounting for the weak fluorescence of the neutral chromophore, and involve increased flexibility and deviation from chromophore planarity, due to loss of hydrogen bonding or π-stacking interactions and other unfavorable van der Waals contacts [115, 181, 186, 191]. Here again, mKeima stands as an unusual case, as excitation of the neutral *trans* form of the chromophore results in substantial red fluorescence following very fast ESPT [162].

4.4 Generality and Complexity of GFP Photoreactions

Depending on the photodynamic conditions, GFPs that undergo reversible photoswitching also display irreversible photoconversions, as exemplified by AsFP595. On the other hand, some AvGFP variants not originally identified as optical highlighters, such as ECFP and EYFP, are actually reported to undergo both reversible and irreversible phototransformations [183, 192, 193]. In this view, the contradictory reports about possible reversible components along AvGFP photoconversion experiments (see Sect. 4.1) would certainly deserve some reinvestigation. A remarkable example of the generality of the photoprocesses shared by members of the GFP family is found in IrisFP, a mutant of EosFP, which, in addition to the irreversible green to red photochemical maturation of its parent (see Sect. 4.2), displays reversible *cis–trans*/acid–base photoswitching in both its green and red forms [188]. The common mechanisms underlying these phenomena pertain to the peculiar isomerism and protonation properties of the various GFP chromophores, as well as to the unique conformational, hydrogen bonding and electron steering properties of their protein carriers, driving specific photochemistry and stabilizing new environments. The recent identification of distinct vibrational spectroscopic signatures for the *cis–trans* isomers and the protonation states of several GFP chromophores, reported recently from Raman spectroscopy studies [184] will be extremely useful in the detailed mechanistic investigation of these photoreactions. On the other hand, a better understanding of this complex photo-dynamics will certainly shed new light on the multiple interconverting states revealed earlier by single molecule spectroscopy studies of GFPs [150, 192].

5   Color Control in the GFP Family

5.1   *General Strategies*

The issue of spectral tuning in GFPs has been the subject of large scale efforts, such as those involved in the development of mFruits from their monomerized DsRed parent [37]. Random or semirandom mutagenesis followed by high throughput screening has been, up to now, the most efficient method of finding new GFP colors. Recent dedicated reviews provide a thorough insight into the variety of colors and underlying mechanisms observed in the GFP family [12, 49, 54, 194]. The GFP color palette now extends far away from the typical green AvGFP fluorescence, going from Sirius, emitting at 424 nm [195], to mPlum, emitting at 650 nm [36], recently challenged by RFP660 peaking at 660 nm [196]. Major spectral shifts toward the blue side have been obtained by replacement of the central residue Tyr66 of the chromopeptide by Phe, His, or Trp [10, 126]. Posttranslational chemistry extending the chromophore conjugation has opened the way toward red and far-red emission [30]. Fine tuning has then been obtained by different changes introduced in the chromophore environment.

The solvatochromism of the HBI chromophore absorption was studied in a wide variety of solvents [197]. While the absorption spectra of the neutral form showed little solvent dependence, the anionic form displays large and complex variations, which were traced to a combination of the acid–base and polar characteristics of the solvent. No solvent could reproduce, however, the strongly red-shifted spectra observed for the AvGFP chromophore as compared to aqueous HBI, which was ascribed to the polarizing effect of the positive charge of Arg96, and possibly also to a small breakdown in chromophore planarity. The detailed study of AvGFP time-resolved emission spectra (TRES) failed also to detect any time-dependent Stokes shifts, down to times as short as 2 ps, for both the neutral and the anionic fluore-scence emissions [130]. The absence of any dynamic Stokes shift in AvGFP, showing the absence of dielectric relaxation of the protein around the excited chromophore, suggests an exceptional local rigidity. A strong Stokes shift on the timescale of several tens of picoseconds has been observed in the case of mPlum, in contrast with some of its red type GFP parents, which was interpreted by a specific rearrangement of a few strongly interacting side chains [198, 199]. As a general rule, a global softening of the chromophore pocket that would allow significant nonspecific solvation of the chromophore excited state is likely to be incompatible with the high fluorescence quantum yields required for imaging applications.

Other, more specific ways of tuning the color of green type chromophore, by modulating its specific local interactions, have been more successful. Inserting an aromatic tyrosine residue in position of π-stacking with the chromophoric moiety in AvGFP has resulted in the yellow emitting variant EYFP [200], a solution that was also selected during GFP evolution [15]. A crystallographic study of the EYFP variant Citrine under hydrostatic pressure shows that subangstrom changes in the distance of the π-stacking interaction are sufficient to alleviate the red-shifting effect [201]. In AvGFP-S65T/H148D, a 15 nm red shift of the neutral chromophore absorption peak is believed to arise from the very short, low-barrier hydrogen bond formed between the chromophore hydroxyle and Asp148 [202]. Anthozoan GFPs carrying the green type chromophore, but emitting in the cyan region may provide interesting solutions for blue-shifting the chromophore spectra [178, 203]. In the cyan form amFP486 from *Anemonia majano* (chromopeptide Lys65-Tyr66-Gly67), the blue shift was ascribed to a stacking of the positively charged His199 on the chromophore phenolate, which also favors chromophore planarity and controls the fluorescence efficiency [176]. Similar influences were ascribed to the structurally homologous His197 and to His163 in the teal color of variants derived from cFP484 from *Clavularia* coral [204].

5.2   Blue and Cyan Variants

The cyan emitting form obtained by the mutation Y66W [10] deserves special attention, as it has led to the ECFP form (absorption at 437 nm, emission at 474 nm), whose variants are still intensively used as excitation energy donors in FRET imaging experiments for monitoring molecular associations in the living cell [205]. The chromophore of ECFP does not bear the deprotonable phenol that is crucial to the photophysics of most AvGFP variants, and displays a markedly different spectroscopy. It has been quickly recognized that, despite its prominent interest in biological applications, the properties of this variant are suboptimal. Indeed, while the brightness of the protein is relatively low ($\varepsilon_M = 32,000\,\mathrm{M}^{-1}\,\mathrm{cm}^{-1}$, $\Phi_F = 0.37$), the fluorescence emission is both spectrally and kinetically hetero-geneous. The fluorescence comprises two major decay times at 3.6 ns and 1.3 ns [206], arising from multiple emissive states associated to different spectra [121]. The complex fluorescence properties of ECFP, which are a major drawback in quantitative FRET analysis, stand in sharp contrast with the remarkable homoge-neity of AvGFP fluorescence [121]. Until recently, efforts devoted to improving the

ECFP properties had met rather mitigated results [207–209]. However, a new bright ECFP variant, mTurquoise, with a monoexponential fluorescence decay of 3.7 ns and $\Phi_F = 0.84$, has been recently reported [210], showing that indole-based chromophores can indeed reach high performances. Alternative strategies in the development of better energy transfer donors are based on engineering the natural cyan-emitting anthozoan GFPs [178, 204, 211], stabilizing the neutral, blue-emitting form of the original green type chromophore as in mKalama [158] and mTagBFP [212], or improving blue variants obtained after the Y66H or Y66F substitutions [158].

5.3   Red and Far-Red Emitters

The further development of red and far-red emitters is motivated by the need of efficient fluorophores for both multicolor and in vivo imaging in the depth of scattering tissues and entire organisms. The yellow variant EYFP has long remained the most red-shifted form that could be achieved by genetic engineering of proteins carrying the green type chromophore, and some recently developed red-emitting variants of AvGFP actually display a new modified chromophore [44]. The red fluorescent proteins discovered in anthozoans provide different routes toward spectral red-shifts, based on various chemical extensions of the chromophore. The major problem has been to decorrelate molecular association of these usually tetrameric assemblies from their color, as their quaternary structure also controls the efficiency of posttranslational maturation from the green to the red forms [54, 213]. Once this step has been secured, an astonishing variety of colors has been obtained, from orange to far-red [37, 199]. It has been recently shown that color hues in red GFPs are correlated to the magnitude of the quadratic Stark effects induced by the strong electric fields produced at the chromophore site by the protein matrix, a mechanism that might be generalized to the green type chromophore [214]. The photophysics of numerous red-emitting variants, although incorporating many common features with the green type chromophore, has only been partly unveiled today, and is likely to reveal surprising new mechanisms.

6   Conclusions and Prospects

The current issue in the development of new GFP-based biosensors is to achieve maximum chromophore brightness, while modifying the spectral ranges and controlling environmental sensitivity and photochemistry according to specific needs. Deciphering the molecular basis of green fluorescence in AvGFP has revealed a complex interplay between the intrinsic HBI chromophore and its protein carrier: due to the multiple protonation and conformational states of HBI, all chromophore fluorescence parameters, including transition energies, oscillator strengths, and competing nonradiative paths, can be strongly modulated by the protein matrix. Describing these chromophore–protein interactions up to an operative, predictive level, is an exciting challenge for both experimentalists and theoreticians.

Much knowledge has been gathered already, on the mechanisms that control these crucial properties. Owing to sophisticated spectroscopic methods, the ultrafast processes and fugitive intermediates leading to the preparation and fate of the GFP emissive state are being described in ever increasing detail. A wealth of crystallographic structures of GFP variants at different pHs or under different photoconverted forms are available for the assessment of mechanistic models. The GFP proteins are now amenable to detailed mass-spectrometry dissection, thereby allowing even larger scale structural studies of posttranslational modifications [215]. Behind the current picture of an extreme photophysical diversity, a few common mechanisms seem to be shared by many members of the GFP superfamily. The next years will certainly tell more about the generality of these findings, allowing a better control of the photoconversion and photochromic properties of GFPs. In this view, more studies will probably be needed to fully understand the GFP photodynamics in situations of high spatial or temporal excitation densities, as occurring under femtosecond pulsed excitation, confocal, or super-resolution microscopy [216–218]. Overall, GFP appears as an ideal test bench for the fundamental understanding of fluorescence emission and chemical reactivity within a protein architecture.

Acknowledgments The authors wish to thank Jacqueline Ridard, Marie Erard, Agathe Espagne, and Dominique Bourgeois for fruitful discussions and critical reading of the manuscript.

References

1. Shimomura O, Johnson FH, Saiga Y (1962) Extraction, purification and properties of aequorin, a bioluminescent protein from the luminous hydromedusan, *Aequorea*. J Cell Comp Physiol 59:223–239
2. Shimomura O (2005) The discovery of aequorin and green fluorescent protein. J Microsc-Oxf 217:3–15
3. Morise H, Shimomura O, Johnson FH, Winant J (1974) Intermolecular energy-transfer in bioluminescent system of *Aequorea*. Biochemistry 13:2656–2662
4. Prendergast FG, Mann KG (1978) Chemical and physical properties of aequorin and the green fluorescent protein isolated from *Aequorea forskalea*. Biochemistry 17:3448–3453
5. Ward WW, Cormier MJ (1979) An energy transfer protein in coelenterate bioluminescence. Characterization of the Renilla green-fluorescent protein. J Biol Chem 254:781–788
6. Ward WW (2006) Biochemical and physical properties of green fluorescent protein. Meth Biochem Anal 47:39–65
7. Prasher DC, Eckenrode VK, Ward WW, Prendergast FG, Cormier MJ (1992) Primary structure of the *Aequorea victoria* green-fluorescent protein. Gene 111:229–233
8. Inouye S, Tsuji FI (1994) *Aequorea* green fluorescent protein. Expression of the gene and fluorescence characteristics of the recombinant protein. FEBS Lett 341:277–280
9. Chalfie M, Tu Y, Euskirchen G, Ward WW, Prasher DC (1994) Green fluorescent protein as a marker for gene expression. Science 263:802–805

10. Heim R, Prasher DC, Tsien RY (1994) Wavelength mutations and posttranslational autoxidation of green fluorescent protein. Proc Natl Acad Sci USA 91:12501–12504
11. Tsien RY (1998) The green fluorescent protein. Annu Rev Biochem 67:509–544
12. Shaner NC, Patterson GH, Davidson MW (2007) Advances in fluorescent protein technology. J Cell Sci 120:4247–4260
13. Wolff M, Kredel S, Wiedenmann J, Nienhaus GU, Heilker R (2008) Cell-based assays in practice: cell markers from autofluorescent proteins of the GFP-family. Comb Chem High Throughput Screen 11:602–609
14. Betzig E, Patterson GH, Sougrat R, Lindwasser OW, Olenych S, Bonifacino JS, Davidson MW, Lippincott-Schwartz J, Hess HF (2006) Imaging intracellular fluorescent proteins at nanometer resolution. Science 313:1642–1645
15. Shagin DA, Barsova EV, Yanushevich YG, Fradkov AF, Lukyanov KA, Labas YA, Semenova TN, Ugalde JA, Meyers A, Nunez JM, Widder EA, Lukyanov SA, Matz MV (2004) GFP-like proteins as ubiquitous metazoan superfamily: Evolution of functional features and structural complexity. Mol Biol Evol 21:841–850
16. Dove SG, Hoegh-Guldberg O, Ranganathan S (2001) Major colour patterns of reef-building corals are due to a family of GFP-like proteins. Coral Reefs 19:197–204
17. Salih A, Larkum A, Cox G, Kuhl M, Hoegh-Guldberg O (2000) Fluorescent pigments in corals are photoprotective. Nature 408:850–853
18. Palmer CV, Modi CK, Mydlarz LD (2009) Coral fluorescent proteins as antioxidants. PLoS ONE 4:e7298. doi:7210.1371/journal.pone.0007298
19. Matz MV, Marshall NJ, Vorobyev M (2006) Symposium-in-print: green fluorescent protein and homologs. Photochem Photobiol 82:345–350
20. Bogdanov AM, Mishin AS, Yampolsky IV, Belousov VV, Chudakov DM, Subach FV, Verkhusha VV, Lukyanov S, Lukyanov KA (2009) Green fluorescent proteins are light-induced electron donors. Nat Chem Biol 5:459–461
21. Cody CW, Prasher DC, Westler WM, Prendergast FG, Ward WW (1993) Chemical structure of the hexapeptide chromophore of the *Aequorea* green-fluorescent protein. Biochemistry 32:1212–1218
22. McNamara G, Boswell C (2007) A thousand proteins of light: 15 years of advances in fluorescent proteins. In: Méndez-Vilas A, Díaz J (eds) Modern research and educational topics in microscopy. Formatex, Badajoz, Spain, pp 287–296
23. Shimomura O (1979) Structure of the chromophore of *Aequorea* green fluorescent protein. FEBS Lett 104:220–222
24. Niwa H, Inouye S, Hirano T, Matsuno T, Kojima S, Kubota M, Ohashi M, Tsuji FI (1996) Chemical nature of the light emitter of the *Aequorea* green fluorescent protein. Proc Natl Acad Sci USA 93:13617–13622
25. Ormo M, Cubitt AB, Kallio K, Gross LA, Tsien RY, Remington SJ (1996) Crystal structure of the *Aequorea victoria* green fluorescent protein. Science 273:1392–1395
26. Yang F, Moss LG, Phillips GN Jr (1996) The molecular structure of green fluorescent protein. Nat Biotechnol 14:1246–1251
27. Kahn TW, Beachy RN, Falk MM (1997) Cell-free expression of a GFP fusion protein allows quantitation in vitro and in vivo. Curr Biol 7:R207–R208
28. Heim R, Cubitt AB, Tsien RY (1995) Improved green fluorescence. Nature 373:663–664
29. Patterson GH, Lippincott-Schwartz J (2002) A photoactivatable GFP for selective photolabeling of proteins and cells. Science 297:1873–1877
30. Matz MV, Fradkov AF, Labas YA, Savitsky AP, Zaraisky AG, Markelov ML, Lukyanov SA (1999) Fluorescent proteins from nonbioluminescent Anthozoa species. Nat Biotechnol 17:969–973
31. Baird GS, Zacharias DA, Tsien RY (2000) Biochemistry, mutagenesis, and oligomerization of DsRed, a red fluorescent protein from coral. Proc Natl Acad Sci USA 97:11984–11989
32. Lukyanov KA, Fradkov AF, Gurskaya NG, Matz MV, Labas YA, Savitsky AP, Markelov ML, Zaraisky AG, Zhao XN, Fang Y, Tan WY, Lukyanov SA (2000) Natural animal

coloration can be determined by a nonfluorescent green fluorescent protein homolog. J Biol Chem 275:25879–25882

33. Gross LA, Baird GS, Hoffman RC, Baldridge KK, Tsien RY (2000) The structure of the chromophore within DsRed, a red fluorescent protein from coral. Proc Natl Acad Sci USA 97:11990–11995

34. Yarbrough D, Wachter RM, Kallio K, Matz MV, Remington SJ (2001) Refined crystal structure of DsRed, a red fluorescent protein from coral, at 2.0-A resolution. Proc Natl Acad Sci USA 98:462–467

35. Wall MA, Socolich M, Ranganathan R (2000) The structural basis for red fluorescence in the tetrameric GFP homolog DsRed. Nat Struct Biol 7:1133–1138

36. Wang L, Jackson WC, Steinbach PA, Tsien RY (2004) Evolution of new nonantibody proteins via iterative somatic hypermutation. Proc Natl Acad Sci USA 101:16745–16749

37. Shu X, Shaner NC, Yarbrough CA, Tsien RY, Remington SJ (2006) Novel chromophores and buried charges control color in mFruits. Biochemistry 45:9639–9647

38. Shkrob MA, Mishin AS, Chudakov DM, Labas YA, Lukyanov KA (2008) Chromoproteins of the green fluorescent protein family: Properties and applications. Russ J Bioorg Chem 34:517–525

39. Gurskaya NG, Fradkov AF, Terskikh A, Matz MV, Labas YA, Martynov VI, Yanushevich YG, Lukyanov KA, Lukyanov SA (2001) GFP-like chromoproteins as a source of far-red fluorescent proteins. FEBS Lett 507:16–20

40. Gurskaya NG, Fradkov AF, Pounkova NI, Staroverov DB, Bulina ME, Yanushevich YG, Labas YA, Lukyanov S, Lukyanov KA (2003) A colourless green fluorescent protein homologue from the non-fluorescent hydromedusa *Aequorea coerulescens* and its fluorescent mutants. Biochem J 373:403–408

41. Kremers GJ, Hazelwood KL, Murphy CS, Davidson MW, Piston DW (2009) Photoconversion in orange and red fluorescent proteins. Nat Meth 6:355–360

42. Elowitz MB, Surette MG, Wolf PE, Stock J, Leibler S (1997) Photoactivation turns green fluorescent protein red. Curr Biol 7:809–812

43. Nienhaus K, Nienhaus GU, Wiedenmann J, Nar H (2005) Structural basis for photo-induced protein cleavage and green-to-red conversion of fluorescent protein EosFP. Proc Natl Acad Sci USA 102:9156–9159

44. Mishin AS, Subach FV, Yampolsky IV, King W, Lukyanov KA, Verkhusha VV (2008) The first mutant of the *Aequorea victoria* green fluorescent protein that forms a red chromophore. Biochemistry 47:4666–4673

45. Battad JM, Wilmann PG, Olsen S, Byres E, Smith SC, Dove SG, Turcic KN, Devenish RJ, Rossjohn J, Prescott M (2007) A structural basis for the pH-dependent increase in fluorescence efficiency of chromoproteins. J Mol Biol 368:998–1010

46. Lukyanov KA, Chudakov DM, Lukyanov S, Verkhusha VV (2005) Innovation: photoactivatable fluorescent proteins. Nat Rev Mol Cell Biol 6:885–891

47. Bulina ME, Chudakov DM, Mudrik NN, Lukyanov KA (2002) Interconversion of Anthozoa GFP-like fluorescent and non-fluorescent proteins by mutagenesis. BMC Biochem 3:7. doi:10.1186/1471-2091-1183-1187

48. Prescott M, Ling M, Beddoe T, Oakley AJ, Dove S, Hoegh-Guldberg O, Devenish RJ, Rossjohn J (2003) The 2.2 A crystal structure of a pocilloporin pigment reveals a nonplanar chromophore conformation. Structure 11:275–284

49. Pakhomov AA, Martynov VI (2008) GFP family: structural insights into spectral tuning. Chem Biol 15:755–764

50. Chalfie M (2009) GFP: lighting up life (Nobel lecture). Angew Chem Int Ed Engl 48:5603–5611

51. Shimomura O (2009) Discovery of green fluorescent protein (GFP) (Nobel lecture). Angew Chem Int Ed Engl 48:5590–5602

52. Tsien RY (2009) Constructing and exploiting the fluorescent protein paintbox (Nobel lecture). Angew Chem Int Ed Engl 48:5612–5626

53. Bizzarri R, Serresi M, Luin S, Beltram F (2009) Green fluorescent protein based pH indicators for in vivo use: a review. Anal Bioanal Chem 393:1107–1122
54. Nienhaus GU, Wiedenmann J (2009) Structure, dynamics and optical properties of fluorescent proteins: perspectives for marker development. Chemphyschem 10:1369–1379
55. Tonge PJ, Meech SR (2009) Excited state dynamics in the green fluorescent protein. J Photochem Photobiol A-Chem 205:1–11
56. Cubitt AB, Heim R, Adams SR, Boyd AE, Gross LA, Tsien RY (1995) Understanding, improving and using green fluorescent proteins. Trends Biochem Sci 20:448–455
57. Wachter RM (2007) Chromogenic cross-link formation in green fluorescent protein. Acc Chem Res 40:120–127
58. Barondeau DP, Kassmann CJ, Tainer JA, Getzoff ED (2007) The case of the missing ring: radical cleavage of a carbon-carbon bond and implications for GFP chromophore biosynthesis. J Am Chem Soc 129:3118–3126
59. Barondeau DP, Putnam CD, Kassmann CJ, Tainer JA, Getzoff ED (2003) Mechanism and energetics of green fluorescent protein chromophore synthesis revealed by trapped intermediate structures. Proc Natl Acad Sci USA 100:12111–12116
60. Wood TI, Barondeau DP, Hitomi C, Kassmann CJ, Tainer JA, Getzoff ED (2005) Defining the role of arginine 96 in green fluorescent protein fluorophore biosynthesis. Biochemistry 44:16211–16220
61. Rosenow MA, Huffman HA, Phail ME, Wachter RM (2004) The crystal structure of the Y66L variant of green fluorescent protein supports a cyclization-oxidation-dehydration mechanism for chromophore maturation. Biochemistry 43:4464–4472
62. Barondeau DP, Kassmann CJ, Tainer JA, Getzoff ED (2006) Understanding GFP posttranslational chemistry: structures of designed variants that achieve backbone fragmentation, hydrolysis, and decarboxylation. J Am Chem Soc 128:4685–4693
63. Barondeau DP, Tainer JA, Getzoff ED (2006) Structural evidence for an enolate intermediate in GFP fluorophore biosynthesis. J Am Chem Soc 128:3166–3168
64. Sniegowski JA, Lappe JW, Patel HN, Huffman HA, Wachter RM (2005) Base catalysis of chromophore formation in Arg96 and Glu222 variants of green fluorescent protein. J Biol Chem 280:26248–26255
65. Reid BG, Flynn GC (1997) Chromophore formation in green fluorescent protein. Biochemistry 36:6786–6791
66. Zhang LP, Patel HN, Lappe JW, Wachter RM (2006) Reaction progress of chromophore biogenesis in green fluorescent protein. J Am Chem Soc 128:4766–4772
67. Verkhusha VV, Chudakov DM, Gurskaya NG, Lukyanov S, Lukyanov KA (2004) Common pathway for the red chromophore formation in fluorescent proteins and chromoproteins. Chem Biol 11:845–854
68. Kojima S, Ohkawa H, Hirano T, Maki S, Niwa H, Ohashi M, Inouye S, Tsuji FI (1998) Fluorescent properties of model chromophores of tyrosine-66 substituted mutants of *Aequorea* green fluorescent protein (GFP). Tetrahedron Lett 39:5239–5242
69. Bell AF, He X, Wachter RM, Tonge PJ (2000) Probing the ground state structure of the green fluorescent protein chromophore using Raman spectroscopy. Biochemistry 39:4423–4431
70. Schellenberg P, Johnson E, Esposito AP, Reid PJ, Parson WW (2001) Resonance Raman scattering by the green fluorescent protein and an analogue of its chromophore. J Phys Chem B 105:5316–5322
71. He X, Bell AF, Tonge PJ (2002) Isotopic labeling and normal-mode analysis of a model green fluorescent protein chromophore. J Phys Chem B 106:6056–6066
72. Webber NM, Meech SR (2007) Electronic spectroscopy and solvatochromism in the chromophore of GFP and the Y66F mutant. Photochem Photobiol Sci 6:976–981
73. Patnaik SS, Trohalaki S, Naik RR, Stone MO, Pachter R (2007) Computational study of the absorption spectra of green fluorescent protein mutants. Biopolymers 85:253–263
74. He X, Bell AF, Tonge PJ (2001) Photoconversion studies of green fluorescent protein and its model compounds. Biochemistry 40:8624

75. He X, Bell AF, Tonge PJ (2003) Ground state isomerization of a model green fluorescent protein chromophore. FEBS Lett 549:35–38
76. Voliani V, Bizzarri R, Nifosi R, Abbruzzetti S, Grandi E, Viappiani C, Beltram F (2008) Cis-trans photoisomerization of fluorescent-protein chromophores. J Phys Chem B 112: 10714–10722
77. Voityuk AA, Michel-Beyerle ME, Rosch N (1998) Structure and rotation barriers for ground and excited states of the isolated chromophore of the green fluorescent protein. Chem Phys Lett 296:269–276
78. Weber W, Helms V, McCammon JA, Langhoff PW (1999) Shedding light on the dark and weakly fluorescent states of green fluorescent proteins. Proc Natl Acad Sci USA 96:6177–6182
79. Dong J, Abulwerdi F, Baldridge A, Kowalik J, Solntsev KM, Tolbert LM (2008) Isomerization in fluorescent protein chromophores involves addition/elimination. J Am Chem Soc 130:14096–14098
80. Yang JS, Huang GJ, Liu YH, Peng SM (2008) Photoisomerization of the green fluorescence protein chromophore and the meta- and para-amino analogues. Chem Commun:1344–1346
81. Chattoraj M, King BA, Bublitz GU, Boxer SG (1996) Ultra-fast excited state dynamics in green fluorescent protein: multiple states and proton transfer. Proc Natl Acad Sci USA 93:8362–8367
82. Voityuk AA, MichelBeyerle ME, Rosch N (1997) Protonation effects on the chromophore of green fluorescent protein. Quantum chemical study of the absorption spectrum. Chem Phys Lett 272:162–167
83. El Yazal J, Prendergast FG, Shaw DE, Pang Y-P (2000) Protonation states of the chromophore of denatured green fluorescent proteins predicted by ab initio calculations. J Am Chem Soc 122:11411–11415
84. Brejc K, Sixma TK, Kitts PA, Kain SR, Tsien RY, Ormo M, Remington SJ (1997) Structural basis for dual excitation and photoisomerization of the *Aequorea victoria* green fluorescent protein. Proc Natl Acad Sci USA 94:2306–2311
85. Elsliger MA, Wachter RM, Hanson GT, Kallio K, Remington SJ (1999) Structural and spectral response of green fluorescent protein variants to changes in pH. Biochemistry 38:5296–5301
86. Scharnagl C, Raupp-Kossmann R, Fischer SF (1999) Molecular basis for pH sensitivity and proton transfer in green fluorescent protein: protonation and conformational substates from electrostatic calculations. Biophys J 77:1839–1857
87. Bell AF, Stoner-Ma D, Wachter RM, Tonge PJ (2003) Light-driven decarboxylation of wild-type green fluorescent protein. J Am Chem Soc 125:6919–6926
88. Ward WW, Cody CW, Hart RC, Cormier MJ (1980) Spectrophotometric identity of the energy transfer chromophores in *Renilla* and *Aequorea* green-fluorescent proteins. Photochem Photobiol 31:611–615
89. Nienhaus K, Renzi F, Vallone B, Wiedenmann J, Nienhaus GU (2006) Chromophore-protein interactions in the anthozoan green fluorescent protein asFP499. Biophys J 91: 4210–4220
90. Jung G, Wiehler J, Zumbusch A (2005) The photophysics of green fluorescent protein: influence of the key amino acids at positions 65, 203, and 222. Biophys J 88:1932–1947
91. Ando R, Hama H, Yamamoto-Hino M, Mizuno H, Miyawaki A (2002) An optical marker based on the UV-induced green-to-red photoconversion of a fluorescent protein. Proc Natl Acad Sci USA 99:12651–12656
92. Ando R, Mizuno H, Miyawaki A (2004) Regulated fast nucleocytoplasmic shuttling observed by reversible protein highlighting. Science 306:1370–1373
93. Rosell FI, Boxer SG (2003) Polarized absorption spectra of green fluorescent protein single crystals: Transition dipole moment directions. Biochemistry 42:177–183
94. Visser NV, Borst JW, Hink MA, van Hoek A, Visser AJWG (2005) Direct observation of resonance tryptophan-to-chromophore energy transfer in visible fluorescent proteins. Biophys Chem 116:207–212

95. Bublitz G, King BA, Boxer SG (1998) Electronic structure of the chromophore in green fluorescent protein (GFP). J Am Chem Soc 120:9370–9371

96. Helms V, Winstead C, Langhoff PW (2000) Low-lying electronic excitations of the green fluorescent protein chromophore. Theochem-J Mol Struct 506:179–189

97. Vallverdu G, Demachy I, Ridard J, Levy B (2009) Using biased molecular dynamics and Brownian dynamics in the study of fluorescent proteins. Theochem-J Mol Struct 898:73–81

98. Mandal D, Tahara T, Webber NM, Meech SR (2002) Ultrafast fluorescence of the chromophore of the green fluorescent protein in alcohol solutions. Chem Phys Lett 358:495–501

99. Gepshtein R, Huppert D, Agmon N (2006) Deactivation mechanism of the green fluorescent chromophore. J Phys Chem B 110:4434–4442

100. Kummer AD, Kompa C, Niwa H, Hirano T, Kojima S, Michel-Beyerle ME (2002) Viscosity-dependent fluorescence decay of the GFP chromophore in solution due to fast internal conversion. J Phys Chem B 106:7554–7559

101. Mandal D, Tahara T, Meech SR (2004) Excited-state dynamics in the green fluorescent protein chromophore. J Phys Chem B 108:1102–1108

102. Litvinenko KL, Webber NM, Meech SR (2001) An ultrafast polarisation spectroscopy study of internal conversion and orientational relaxation of the chromophore of the green fluorescent protein. Chem Phys Lett 346:47–53

103. Webber NM, Litvinenko KL, Meech SR (2001) Radiationless relaxation in a synthetic analogue of the green fluorescent protein chromophore. J Phys Chem B 105:8036–8039

104. Usman A, Mohammed OF, Nibbering ET, Dong J, Solntsev KM, Tolbert LM (2005) Excited-state structure determination of the green fluorescent protein chromophore. J Am Chem Soc 127:11214–11215

105. Chen KY, Cheng YM, Lai CH, Hsu CC, Ho ML, Lee GH, Chou PT (2007) Ortho green fluorescence protein synthetic chromophore; excited-state intramolecular proton transfer via a seven-membered-ring hydrogen-bonding system. J Am Chem Soc 129:4534–4535

106. Wu L, Burgess K (2008) Syntheses of highly fluorescent GFP-chromophore analogues. J Am Chem Soc 130:4089–4096

107. Litvinenko KL, Webber NM, Meech SR (2003) Internal conversion in the chromophore of the green fluorescent protein: Temperature dependence and isoviscosity analysis. J Phys Chem A 107:2616–2623

108. Maddalo SL, Zimmer M (2006) The role of the protein matrix in green fluorescent protein fluorescence. Photochem Photobiol 82:367–372

109. Martin ME, Negri F, Olivucci M (2004) Origin, nature, and fate of the fluorescent state of the green fluorescent protein chromophore at the CASPT2//CASSCF resolution. J Am Chem Soc 126:5452–5464

110. Toniolo A, Olsen S, Manohar L, Martinez TJ (2004) Conical intersection dynamics in solution: the chromophore of green fluorescent protein. Faraday Discuss 127:149–163

111. Altoe P, Bernardi F, Garavelli M, Orlandi G, Negri F (2005) Solvent effects on the vibrational activity and photodynamics of the green fluorescent protein chromophore: a quantum-chemical study. J Am Chem Soc 127:3952–3963

112. Olsen S, Lamothe K, MartiÃÃAnez TJ (2010) Protonic gating of excited-state twisting and charge localization in GFP chromophores: a mechanistic hypothesis for reversible photoswitching. J Am Chem Soc 132:1192–1193

113. Kummer AD, Kompa C, Lossau H, Pollinger-Dammer F, Michel-Beyerle ME, Silva CM, Bylina EJ, Coleman WJ, Yang MM, Youvan DC (1998) Dramatic reduction in fluorescence quantum yield in mutants of green fluorescent protein due to fast internal conversion. Chem Phys 237:183–193

114. Wilmann PG, Petersen J, Pettikiriarachchi A, Buckle AM, Smith SC, Olsen S, Perugini MA, Devenish RJ, Prescott M, Rossjohn J (2005) The 2.1 angstrom crystal structure of the far-red fluorescent protein HcRed: Inherent conformational flexibility of the chromophore. J Mol Biol 349:223–237

115. Mizuno H, Mal TK, Walchli M, Kikuchi A, Fukano T, Ando R, Jeyakanthan J, Taka J, Shiro Y, Ikura M, Miyawaki A (2008) Light-dependent regulation of structural flexibility in a photochromic fluorescent protein. Proc Natl Acad Sci USA 105:9227–9232

116. Megley CM, Dickson LA, Maddalo SL, Chandler GJ, Zimmer M (2009) Photophysics and dihedral freedom of the chromophore in yellow, blue, and green fluorescent protein. J Phys Chem B 113:302–308

117. Ward WW, Prentice HJ, Roth AF, Cody CW, Reeves SC (1982) Spectral perturbations of the aequoria green-fluorescent protein. Photochem Photobiol 35:803–808

118. Lossau H, Kummer A, Heinecke R, Pollinger-Dammer F, Kompa C, Bieser G, Jonsson T, Silva CM, Yang MM, Youvan DC, Michel-Beyerle ME (1996) Time-resolved spectroscopy of wild-type and mutant green fluorescent proteins reveals excited state deprotonation consistent with fluorophore-protein interactions. Chem Phys 213:1–16

119. Patterson GH, Knobel SM, Sharif WD, Kain SR, Piston DW (1997) Use of the green fluorescent protein and its mutants in quantitative fluorescence microscopy. Biophys J 73:2782–2790

120. Perozzo MA, Ward KB, Thompson RB, Ward WW (1988) X-ray diffraction and time-resolved fluorescence analyses of *Aequorea* green fluorescent protein crystals. J Biol Chem 263:7713–7716

121. Villoing A, Ridhoir M, Cinquin B, Erard M, Alvarez L, Vallverdu G, Pernot P, Grailhe R, Merola F, Pasquier H (2008) Complex fluorescence of the cyan fluorescent protein: comparisons with the H148D variant and consequences for quantitative cell imaging. Biochemistry 47:12483–12492

122. Suhling K, Siegel J, Phillips D, French PMW, Leveque-Fort S, Webb SED, Davis DM (2002) Imaging the environment of green fluorescent protein. Biophys J 83:3589–3595

123. Borst JW, Hink MA, van Hoek A, Visser AJWG (2005) Effects of refractive index and viscosity on fluorescence and anisotropy decays of enhanced cyan and yellow fluorescent proteins. J Fluoresc 15:153–160

124. Strickler SJ, Berg RA (1962) Relationship between absorption intensity and fluorescence lifetime of molecules. J Chem Phys 37:814–820

125. Inouye S, Tsuji FI (1994) Evidence for redox forms of the *Aequorea* green fluorescent protein. FEBS Lett 351:211–214

126. Palm GJ, Zdanov A, Gaitanaris GA, Stauber R, Pavlakis GN, Wlodawer A (1997) The structural basis for spectral variations in green fluorescent protein. Nat Struct Biol 4:361–365

127. Bizzarri R, Nifosi R, Abbruzzetti S, Rocchia W, Guidi S, Arosio D, Garau G, Campanini B, Grandi E, Ricci F, Viappiani C, Beltram F (2007) Green fluorescent protein ground states: the influence of a second protonation site near the chromophore. Biochemistry 46:5494–5504

128. Delagrave S, Hawtin RE, Silva CM, Yang MM, Youvan DC (1995) Red-shifted excitation mutants of the green fluorescent protein. Biotechnology (NY) 13:151–154

129. Ehrig T, O'Kane DJ, Prendergast FG (1995) Green-fluorescent protein mutants with altered fluorescence excitation spectra. FEBS Lett 367:163–166

130. Jaye AA, Stoner-Ma D, Matousek P, Towrie M, Tonge PJ, Meech SR (2005) Time resolved emission spectra of green fluorescent protein. Photochem Photobiol 82:373–379

131. Crameri A, Whitehorn EA, Tate E, Stemmer WP (1996) Improved green fluorescent protein by molecular evolution using DNA shuffling. Nat Biotechnol 14:315–319

132. Kennis JT, Larsen DS, van Stokkum IH, Vengris M, van Thor JJ, van Grondelle R (2004) Uncovering the hidden ground state of green fluorescent protein. Proc Natl Acad Sci USA 101:17988–17993

133. Kneen M, Farinas J, Li Y, Verkman AS (1998) Green fluorescent protein as a noninvasive intracellular pH indicator. Biophys J 74:1591–1599

134. Miesenbock G, De Angelis DA, Rothman JE (1998) Visualizing secretion and synaptic transmission with pH-sensitive green fluorescent proteins. Nature 394:192–195

135. Violot S, Carpentier P, Blanchoin L, Bourgeois D (2009) Reverse pH-dependence of chromophore protonation explains the large stokes shift of the red fluorescent protein mKeima. J Am Chem Soc 131:10356–10357

136. De Angelis DA, Miesenbock G, Zemelman BV, Rothman JE (1998) PRIM: proximity imaging of green fluorescent protein-tagged polypeptides. Proc Natl Acad Sci USA 95:12312–12316
137. Hanson GT, Aggeler R, Oglesbee D, Cannon M, Capaldi RA, Tsien RY, Remington SJ (2004) Investigating mitochondrial redox potential with redox-sensitive green fluorescent protein indicators. J Biol Chem 279:13044–13053
138. Nausch LWM, Lecloux J, Bonev AD, Nelson MT, Dostmann WR (2008) Differential patterning of cGMP in vascular smooth muscle cells revealed by single GFP-linked biosensors. Proc Natl Acad Sci USA 105:365–370
139. Arnaut LG, Formosinho SJ (1993) Excited-state proton-transfer reactions.1. Fundamentals and intermolecular reactions. J Photochem Photobiol A-Chem 75:1–20
140. Stoner-Ma D, Jaye AA, Matousek P, Towrie M, Meech SR, Tonge PJ (2005) Observation of excited-state proton transfer in green fluorescent protein using ultrafast vibrational spectroscopy. J Am Chem Soc 127:2864–2865
141. van Thor JJ, Zanetti G, Ronayne KL, Towrie M (2005) Structural events in the photocycle of green fluorescent protein. J Phys Chem B 109:16099–16108
142. Lill MA, Helms V (2002) Proton shuttle in green fluorescent protein studied by dynamic simulations. Proc Natl Acad Sci USA 99:2778–2781
143. Vendrell O, Gelabert R, Moreno M, Lluch JM (2008) Operation of the proton wire in green fluorescent protein. A quantum dynamics simulation. J Phys Chem B 112:5500–5511
144. Creemers TMH, Lock AJ, Subramaniam V, Jovin TM, Volker S (1999) Three photoconvertible forms of green fluorescent protein identified by spectral hole-burning (vol 6, pg 557, 1999). Nat Struct Biol 6:706
145. Seebacher C, Deeg FW, Brauchle C, Wiehler J, Steipe B (1999) Stable low-temperature photoproducts and hole burning of green fluorescent protein (GFP). J Phys Chem B 103:7728–7732
146. Bonsma S, Purchase R, Jezowski S, Gallus J, Konz F, Volker S (2005) Green and red fluorescent proteins: Photo- and thermally induced dynamics probed by site-selective spectroscopy and hole burning. Chemphyschem 6:838–849
147. Wiehler J, Jung G, Seebacher C, Zumbusch A, Steipe B (2003) Mutagenic stabilization of the photocycle intermediate of green fluorescent protein (GFP). Chembiochem 4:1164–1171
148. Striker G, Subramaniam V, Seidel CAM, Volkmer A (1999) Photochromicity and fluorescence lifetimes of green fluorescent protein. J Phys Chem B 103:8612–8617
149. Saxena AM, Udgaonkar JB, Krishnamoorthy G (2005) Protein dynamics control proton transfer from bulk solvent to protein interior: a case study with a green fluorescent protein. Protein Sci 14:1787–1799
150. Haupts U, Maiti S, Schwille P, Webb WW (1998) Dynamics of fluorescence fluctuations in green fluorescent protein observed by fluorescence correlation spectroscopy. Proc Natl Acad Sci USA 95:13573–13578
151. Schwille P, Kummer S, Heikal AA, Moerner WE, Webb WW (2000) Fluorescence correlation spectroscopy reveals fast optical excitation-driven intramolecular dynamics of yellow fluorescent proteins. Proc Natl Acad Sci USA 97:151–156
152. Mallik R, Udgaonkar JB, Krishnamoorthy G (2003) Kinetics of proton transfer in a green fluorescent protein: a laser-induced pH jump study. Proc Indian Acad Sci-Chem Sci 115:307–317
153. Agmon N (2005) Proton pathways in green fluorescence protein. Biophys J 88:2452–2461
154. Leiderman P, Huppert D, Agmon N (2006) Transition in the temperature-dependence of GFP fluorescence: From proton wires to proton exit. Biophys J 90:1009–1018
155. Shu X, Leiderman P, Gepshtein R, Smith NR, Kallio K, Huppert D, Remington SJ (2007) An alternative excited-state proton transfer pathway in green fluorescent protein variant S205V. Protein Sci 16:2703–2710
156. Stoner-Ma D, Jaye AA, Ronayne KL, Nappa J, Meech SR, Tonge PJ (2008) An alternate proton acceptor for excited-state proton transfer in green fluorescent protein: rewiring GFP. J Am Chem Soc 130:1227–1235

157. Shi X, Basran J, Seward HE, Childs W, Bagshaw CR, Boxer SG (2007) Anomalous negative fluorescence anisotropy in yellow fluorescent protein (YFP 10C): quantitative analysis of FRET in YFP dimers. Biochemistry 46:14403–14417

158. Ai HW, Shaner NC, Cheng Z, Tsien RY, Campbell RE (2007) Exploration of new chromophore structures leads to the identification of improved blue fluorescent proteins. Biochemistry 46:5904–5910

159. Hanson GT, McAnaney TB, Park ES, Rendell ME, Yarbrough DK, Chu S, Xi L, Boxer SG, Montrose MH, Remington SJ (2002) Green fluorescent protein variants as ratiometric dual emission pH sensors. 1. Structural characterization and preliminary application. Biochemistry 41:15477–15488

160. McAnaney TB, Park ES, Hanson GT, Remington SJ, Boxer SG (2002) Green fluorescent protein variants as ratiometric dual emission pH sensors. 2. Excited-state dynamics. Biochemistry 41:15489–15494

161. Hosoi H, Mizuno H, Miyawaki A, Tahara T (2006) Competition between energy and proton transfer in ultrafast excited-state dynamics of an oligomeric fluorescent protein red Kaede. J Phys Chem B 110:22853–22860

162. Henderson JN, Osborn MF, Koon N, Gepshtein R, Huppert D, Remington SJ (2009) Excited state proton transfer in the red fluorescent protein mKeima. J Am Chem Soc 131:13212–13213

163. van Thor JJ, Pierik AJ, Nugteren-Roodzant I, Xie A, Hellingwerf KJ (1998) Characterization of the photoconversion of green fluorescent protein with FTIR spectroscopy. Biochemistry 37:16915–16921

164. van Thor JJ, Georgiev GY, Towrie M, Sage JT (2005) Ultrafast and low barrier motions in the photoreactions of the green fluorescent protein. J Biol Chem 280:33652–33659

165. Henderson JN, Gepshtein R, Heenan JR, Kallio K, Huppert D, Remington SJ (2009) Structure and mechanism of the photoactivatable green fluorescent protein. J Am Chem Soc 131:4176–4177

166. van Thor JJ, Gensch T, Hellingwerf KJ, Johnson LN (2002) Phototransformation of green fluorescent protein with UV and visible light leads to decarboxylation of glutamate 222. Nat Struct Biol 9:37–41

167. van Thor JJ (2009) Photoreactions and dynamics of the green fluorescent protein. Chem Soc Rev 38:2935–2950

168. Yokoe H, Meyer T (1996) Spatial dynamics of GFP-tagged proteins investigated by local fluorescence enhancement. Nat Biotechnol 14:1252–1256

169. Lippincott-Schwartz J, Patterson GH (2008) Fluorescent proteins for photoactivation experiments. In: Sullivan KF (ed) Fluorescent proteins, 2nd edn. San Diego, Elsevier Academic Press Inc, pp 45–63

170. McAnaney TB, Zeng W, Doe CFE, Bhanji N, Wakelin S, Pearson DS, Abbyad P, Shi XH, Boxer SG, Bagshaw CR (2005) Protonation, photobleaching, and photoactivation of yellow fluorescent protein (YFP 10C): a unifying mechanism. Biochemistry 44:5510–5524

171. Habuchi S, Cotlet M, Gensch T, Bednarz T, Haber-Pohlmeier S, Rozenski J, Dirix G, Michiels J, Vanderleyden J, Heberle J, De Schryver FC, Hofkens J (2005) Evidence for the isomerization and decarboxylation in the photoconversion of the red fluorescent protein DsRed. J Am Chem Soc 127:8977–8984

172. Chudakov DM, Verkhusha VV, Staroverov DB, Souslova EA, Lukyanov S, Lukyanov KA (2004) Photoswitchable cyan fluorescent protein for protein tracking. Nat Biotechnol 22:1435–1439

173. Wiedenmann J, Ivanchenko S, Oswald F, Schmitt F, Rocker C, Salih A, Spindler KD, Nienhaus GU (2004) EosFP, a fluorescent marker protein with UV-inducible green-to-red fluorescence conversion. Proc Natl Acad Sci USA 101:15905–15910

174. Gurskaya NG, Verkhusha VV, Shcheglov AS, Staroverov DB, Chepurnykh TV, Fradkov AF, Lukyanov S, Lukyanov KA (2006) Engineering of a monomeric green-to-red photoactivatable fluorescent protein induced by blue light. Nat Biotechnol 24:461–465

175. Petersen J, Wilmann PG, Beddoe T, Oakley AJ, Devenish RJ, Prescott M, Rossjohn J (2003) The 2.0-A crystal structure of eqFP611, a far red fluorescent protein from the sea anemone *Entacmaea quadricolor*. J Biol Chem 278:44626–44631
176. Henderson JN, Remington SJ (2005) Crystal structures and mutational analysis of amFP486, a cyan fluorescent protein from *Anemonia majano*. Proc Natl Acad Sci USA 102:12712–12717
177. Pletnev S, Shcherbo D, Chudakov DM, Pletneva N, Merzlyak EM, Wlodawer A, Dauter Z, Pletnev V (2008) A crystallographic study of bright far-red fluorescent protein mKate reveals pH-induced cis-trans Isomerization of the chromophore. J Biol Chem 283:28980–28987
178. Ai HW, Henderson JN, Remington SJ, Campbell RE (2006) Directed evolution of a monomeric, bright and photostable version of *Clavularia* cyan fluorescent protein: structural characterization and applications in fluorescence imaging. Biochem J 400:531–540
179. Andresen M, Wahl MC, Stiel AC, Grater F, Schafer LV, Trowitzsch S, Weber G, Eggeling C, Grubmuller H, Hell SW, Jakobs S (2005) Structure and mechanism of the reversible photoswitch of a fluorescent protein. Proc Natl Acad Sci USA 102:13070–13074
180. Wilmann PG, Turcic K, Battad JM, Wilce MC, Devenish RJ, Prescott M, Rossjohn J (2006) The 1.7 A crystal structure of Dronpa: a photoswitchable green fluorescent protein. J Mol Biol 364:213–224
181. Andresen M, Stiel AC, Trowitzsch S, Weber G, Eggeling C, Wahl MC, Hell SW, Jakobs S (2007) Structural basis for reversible photoswitching in Dronpa. Proc Natl Acad Sci USA 104:13005–13009
182. Nifosi R, Ferrari A, Arcangeli C, Tozzini V, Pellegrini V, Beltram F (2003) Photoreversible dark state in a tristable green fluorescent protein variant. J Phys Chem B 107:1679–1684
183. Sinnecker D, Voigt P, Hellwig N, Schaefer M (2005) Reversible photobleaching of enhanced green fluorescent proteins. Biochemistry 44:7085–7094
184. Luin S, Voliani V, Lanza G, Bizzarri R, Nifosi R, Amat P, Tozzini V, Serresi M, Beltram F (2009) Raman study of chromophore states in photochromic fluorescent proteins. J Am Chem Soc 131:96–103
185. Bizzarri R, Serresi M, Cardarelli F, Abbruzzetti S, Campanini B, Viappiani C, Beltram F (2009) Single amino acid replacement makes *Aequorea victoria* fluorescent proteins reversibly photoswitchable. J Am Chem Soc 132:85–95
186. Henderson JN, Ai HW, Campbell RE, Remington SJ (2007) Structural basis for reversible photobleaching of a green fluorescent protein homologue. Proc Natl Acad Sci USA 104:6672–6677
187. Loos DC, Habuchi S, Flors C, Hotta J, Wiedenmann J, Nienhaus GU, Hofkens J (2006) Photoconversion in the red fluorescent protein from the sea anemone *Entacmaea quadricolor*: is cis-trans isomerization involved? J Am Chem Soc 128:6270–6271
188. Adam V, Lelimousin M, Boehme S, Desfonds G, Nienhaus K, Field MJ, Wiedenmann J, McSweeney S, Nienhaus GU, Bourgeois D (2008) Structural characterization of IrisFP, an optical highlighter undergoing multiple photo-induced transformations. Proc Natl Acad Sci USA 105:18343–18348
189. Malo GD, Pouwels LJ, Wang M, Weichsel A, Montfort WR, Rizzo MA, Piston DW, Wachter RM (2007) X-ray structure of Cerulean GFP: a tryptophan-based chromophore useful for fluorescence lifetime imaging. Biochemistry 46:9865–9873
190. Chudakov DM, Feofanov AV, Mudrik NN, Lukyanov S, Lukyanov KA (2003) Chromophore environment provides clue to "kindling fluorescent protein" riddle. J Biol Chem 278:7215–7219
191. Nifosi R, Tozzini V (2006) Cis-trans photoisomerization of the chromophore in the green fluorescent protein variant E(2)GFP: a molecular dynamics study. Chem Phys 323:358–368
192. Dickson RM, Cubitt AB, Tsien RY, Moerner WE (1997) On/off blinking and switching behaviour of single molecules of green fluorescent protein. Nature 388:355–358
193. Kirber MT, Chen K, Keaney JF Jr (2007) YFP photoconversion revisited: confirmation of the CFP-like species. Nat Meth 4:767–768

194. Day RN, Davidson MW (2009) The fluorescent protein palette: tools for cellular imaging. Chem Soc Rev 38:2887–2921
195. Tomosugi W, Matsuda T, Tani T, Nemoto T, Kotera I, Saito K, Horikawa K, Nagai T (2009) An ultramarine fluorescent protein with increased photostability and pH insensitivity. Nat Meth 6:351–353
196. Morozova K, Verkhusha VV, Perskyi Y (2009) Directed molecular evolution to develop new fluorescent proteins for biotechnological applications, Proceedings of the VIIth Parnas conference on biochemistry and molecular biology. Ukr Biokhim Zh 81(4):304
197. Dong J, Solntsev KM, Tolbert LM (2006) Solvatochromism of the green fluorescence protein chromophore and its derivatives. J Am Chem Soc 128:12038–12039
198. Abbyad P, Childs W, Shi X, Boxer SG (2007) Dynamic Stokes shift in green fluorescent protein variants. Proc Natl Acad Sci USA 104:20189–20194
199. Shu X, Wang L, Colip L, Kallio K, Remington SJ (2009) Unique interactions between the chromophore and glutamate 16 lead to far-red emission in a red fluorescent protein. Protein Sci 18:460–466
200. Wachter RM, Elsliger MA, Kallio K, Hanson GT, Remington SJ (1998) Structural basis of spectral shifts in the yellow-emission variants of green fluorescent protein. Structure 6:1267–1277
201. Barstow B, Ando N, Kim CU, Gruner SM (2008) Alteration of citrine structure by hydrostatic pressure explains the accompanying spectral shift. Proc Natl Acad Sci USA 105:13362–13366
202. Shu X, Kallio K, Shi X, Abbyad P, Kanchanawong P, Childs W, Boxer SG, Remington SJ (2007) Ultrafast excited-state dynamics in the green fluorescent protein variant S65T/H148D. 1. Mutagenesis and structural studies. Biochemistry 46:12005–12013
203. Malo GD, Wang M, Wu D, Stelling AL, Tonge PJ, Wachter RM (2008) Crystal structure and Raman studies of dsFP483, a cyan fluorescent protein from *Discosoma striata*. J Mol Biol 378:871–886
204. Ai HW, Olenych SG, Wong P, Davidson MW, Campbell RE (2008) Hue-shifted monomeric variants of *Clavularia* cyan fluorescent protein: identification of the molecular determinants of color and applications in fluorescence imaging. BMC Biol 6:13. doi:10.1186/1741-2007-1186-1113
205. Takanishi CL, Bykova EA, Cheng W, Zheng J (2006) GFP-based FRET analysis in live cells. Brain Res 1091:132–139
206. Tramier M, Gautier I, Piolot T, Ravalet S, Kemnitz K, Coppey J, Durieux C, Mignotte V, Coppey-Moisan M (2002) Picosecond-hetero-FRET microscopy to probe protein-protein interactions in live cells. Biophys J 83:3570–3577
207. Rizzo MA, Springer GH, Granada B, Piston DW (2004) An improved cyan fluorescent protein variant useful for FRET. Nat Biotechnol 22:445–449
208. Nguyen AW, Daugherty PS (2005) Evolutionary optimization of fluorescent proteins for intracellular FRET. Nat Biotechnol 23:355–360
209. Kremers GJ, Goedhart J, van Munster EB, Gadella TWJ (2006) Cyan and yellow super fluorescent proteins with improved brightness, protein folding, and FRET Forster radius. Biochemistry 45:6570–6580
210. Goedhart J, van Weeren L, Hink MA, Vischer NO, Jalink K, Gadella TWJ (2010) Bright cyan fluorescent protein variants identified by fluorescence lifetime screening. Nat Meth 7:137–139
211. Day RN, Booker CF, Periasamy A (2008) Characterization of an improved donor fluorescent protein for Forster resonance energy transfer microscopy. J Biomed Opt 13:031203. doi:10.1117/1111.2939094
212. Subach OM, Gundorov IS, Yoshimura M, Subach FV, Zhang JH, Gruenwald D, Souslova EA, Chudakov DM, Verkhusha VV (2008) Conversion of red fluorescent protein into a bright blue probe. Chem Biol 59:1116–1124

213. Campbell RE, Tour O, Palmer AE, Steinbach PA, Baird GS, Zacharias DA, Tsien RY (2002) A monomeric red fluorescent protein. Proc Natl Acad Sci U S A 99:7877–7882
214. Drobizhev M, Tillo S, Makarov NS, Hughes TE, Rebane A (2009) Color hues in red fluorescent proteins are due to internal quadratic stark effect. J Phys Chem B 113:12860–12864
215. Alvarez LA, Merola F, Erard M, Rusconi F (2009) Mass spectrometry-based structural dissection of fluorescent proteins. Biochemistry 48:3810–3812
216. Mondal PP, Diaspro A (2007) Reduction of higher-order photobleaching in two-photon excitation microscopy. Phys Rev E 75:6. doi:10.1103/PhysRevE.1175.061904
217. Hofmann M, Eggeling C, Jakobs S, Hell SW (2005) Breaking the diffraction barrier in fluorescence microscopy at low light intensities by using reversibly photoswitchable proteins. Proc Natl Acad Sci U S A 102:17565–17569
218. Patterson GH, Piston DW (2000) Photobleaching in two-photon excitation microscopy. Biophys J 78:2159–2162
219. Tretyakova YA, Pakhomov AA, Martynov VI (2007) Chromophore structure of the kindling fluorescent protein asFP595 from *Anemonia sulcata*. J Am Chem Soc 129:7748–7749
220. Remington SJ, Wachter RM, Yarbrough DK, Branchaud B, Anderson DC, Kallio K, Lukyanov KA (2005) zFP538, a yellow-fluorescent protein from *Zoanthus*, contains a novel three-ring chromophore. Biochemistry 44:202–212
221. Shaner NC, Campbell RE, Steinbach PA, Giepmans BN, Palmer AE, Tsien RY (2004) Improved monomeric red, orange and yellow fluorescent proteins derived from *Discosoma sp.* red fluorescent protein. Nat Biotechnol 22:1567–1572

Index

Printed by Printforce, the Netherlands